# TREES OF
# ANDHRA PRADESH, INDIA

# TREES OF ANDHRA PRADESH, INDIA

*T. Pullaiah*
and
*S. Sandhya Rani*

Department of Botany
Sri Krishnadevaraya University
Anantapur 515 003, A.P., India

*Regency Publications*
New Delhi

*ISBN 81-86030-92-1*

*Published by Regency Publications, 20/36-G, Old Market, West Patel Nagar, New Delhi 110 008 and Printed at Radiant Printers, New Delhi 110 008, Phones: 571 2539, 574 0038; Telefax: 578 3571.*

Website: *www.regencypublications.com*
Email: *regency @ satyam.net.in* and *info @ regencypublications.com*

# PREFACE

Trees are the dominant vegetation of a forest ecosystem and they play a crucial role in maintaining the ecological balance and also serve as a renewable resource for the sustainable growth and development of human kind. They have played an important role in the economy of the state and are an asset to the nation.

Natural resources survey like floristic studies, especially trees survey, play an important role in the economic development of developing country like India. The alarming rate at which this natural tree wealth is vanishing has become the compelling factors for a detailed botanical survey of all the tree species, particularly in Andhra Pradesh.

Indian trees (1906) by Brandis, Flora Sylvatica for southern India (1869–74) by Beddome are the only available reference for the tree floristic data of Andhra Pradesh state even today. Till date there is no comprehensive work on the tree Flora of the State. The present work was under taken therefore to provide a complete inventory of tree species occurring in the state of Andhra Pradesh.

In general the floristic data of trees will be useful for foresters, researchers, planners, Industrialists, naturalists, and policy makers for judicious utilization of the resources. It will also be useful in developing strategies for monitoring and conservation of precious tree genetic resources of Andhra Pradesh.

The authors are thankful to personnel of Girijan Co operative Corporation and forest department for their assistance in field work. The help and co operation received from Botanists at Herbaria, CAL, MH, BLAT, BSI and different Universities of Andhra Pradesh is thankfully acknowledged. They also thank Dr. K. Sri Rama Murthy for his help throughout this work.

# CONTENTS

# INTRODUCTION

*"Trees mean water, water means bread and bread is life"*
— K.M. Munshi

Multiple microevolutionary forces are known to contribute to the adaptive radiation and diversification of all the life forms to the desired habitats. The significance of bio-diversity is to exercise relative success in the biological environment with the elementary forces of continuous interactions among the organisms, the plant and animal species diversified so much to interact with the organic and inorganic spheres of the earth. The management of biodiversity and genetic resources is within the carrying capacity of the supporting ecosystem for the survival of man kind.

Of the various forms of life on earth, none are of great importance than are the members of Plant Kingdom. The very existence of human life on earth depends on plants, since they produce carbohydrates and oxygen primarily and sustain the whole Animal Kingdom. Plant resources, including forestry and agriculture, support the life of whole man-kind. Keeping this in view, it is no exaggeration to say that the plant kingdom play a dominant role in the life of man. The Science of Ayurveda clearly shows that the medicines prepared from trees are more effective than artificial medicines.

Trees are the main constituents of forests, which are considered as the most valuable natural resource and are of immense importance both to man and the nation. All the woody plants that reach a height of atleast 3 metres having a single stem with a definite crown shape (Lawrence 1951) come under trees.

Trees serve different purposes for different people. For our ancestors they were a source of shelter and fuel and sometimes an object of worship. Tree wealth plays a vital role in the life and economy of the Nation.

Trees are dependent on light to perform fundamental process of photosynthesis by which they synthesize food in their green parts, and as a consequence there is an intense competition to reach the light and avoid shade cast by other vegetation. Most plants therefore grow upwards and trees are the ones that have exploited this feature to the full by growing well above and dominating all other forms of vegetation. Light relations are dependent on the tree canopy. The canopy produces microclimate favourable for optimum growth of many plants. Trees have been able to dominate all other plants by virtue of its height and woody nature. The cell walls in the stem are thickened with lignin, a stiffening material, which provides them with great strength and durability, both of which are important characteristics. The pressure on the crown of a large tree in a gale, at the top of a long trunk which acts as a lever, is very considerable. The very long life span of trees, which may cross even hundred years, gives them time to grow to these great heights. This ability is gained not only by upward growth but also by the addition of a new layer of woody tissue under the bark each year, so that the tree is constantly expanding in diameter. The vital ring of living cells that produces this new layer annually is just under the bark, and if this is severely damaged the tree will suffer or die.

Trees comprising the forest cover, serve as substrate for millions of microbes, herbs, shrubs, climbers and diverse fauna. Trees protect life supporting system i.e., soil, and water ensuring cleaner air and water and soil building process, moisture conservation and reduce air and noise pollution. Trees are the sources of biomass, energy and also help in the proper distribution of rainfall.

Trees have played an important role in the economy of the state and are an asset to the nation. Besides meeting human needs like food (for example a variety of fruits, leaves), clothing, fuel, shelter and other basic products, they aid in preventing soil erosion. Trees are the source of raw materials for various industries and provide much needed wood, fuel, timber, paper, rayon, soft wood, tannin, essential oils, perfumes, gums, resins, fibres, wax, coir and drugs. Trees serve as host for many animal products like lac, honey, silk tasar etc. and also act as wind breakers and shelter belts, generate raw material for feeding the small scale industries.

Trees serve as natural land marks and memorials. Because they have more than the expected life span of man, they carry their associations through generations. There are trees with special religious, aesthetic or sentimental associations and have served as a source of inspiration to many Saints, Artists and Poets. (There are age-old trees that are known for their longevity.)

Trees are a rich resource base and have to be developed as part of the long term strategy.

Natural resources survey like floristic studies, play an important role in the economic development of developing country like India. The alarming rate at which this natural wealth is vanishing has become the compelling factor for a detailed botanical survey of all the plants, particularly in India. Earlier works do provide valuable floristic information but they are incomplete in many aspects as new information is being added continuously. This demands an immediate action to provide an up-to-date floristic information. Keeping this in view, several authors like Van Steenis (1968), Santapau (1952, 1962), Janaki Ammal (1954), Hutchinson (1964), Stace (1980), Jain (1978, 1982) have emphasized the importance of preparing Regional Floras. They aid in understanding the factors leading to the establishment of a stable ecosystem and also throw light on processes like isolation, evolution, speciation and endemism (Saldanha, 1980).

Heywood (1980) also emphasized, the indispensability of Plant taxonomy in providing a detailed account on the occurrence, distribution and identification of useful plants. In addition to the above fact, an adequate knowledge of the Floras and their taxonomy is an essential prerequisite for any measure of conservation.

Eventhough, the tropics are known for their rich diversified Flora, there is not any reliable source which provides full fledged information on a particular region of the tropics and also tropics as a whole. Realizing the indispensability of floristic studies, taxonomists have undertaken the great task of preparing regional Floras. As a result many regional Floras have been worked out but all the same much remains to be done. Flora Sylvatica for Andhra Pradesh is one such work. Eventhough district Floras and State Flora do provide information on trees there is no single work which is solely dedicated to the trees of Andhra Pradesh. The present work is taken up to fill this deficiency.

# TOPOGRAPHY AND GENERAL FEATURES

The State of Andhra was formed with Kurnool, as its capital on 1st October, 1953, but Andhra Pradesh was formed on November, 1956 comprising the then Andhra State and Telangana (Telugu speaking region) of the former State of Hyderabad. Hyderabad is the capital of Andhra Pradesh. Geometrically the State lies between the latitudes 12°37' and 19°54' North latitudes and 76°46' and 84°46' East longitudes. Geographically it occupies the middle portion of the eastern half of the Indian Peninsula with an area of 2,75,068 Sq. Kms. accounting for 8.4% of the entire country (Map 1 & 2). The State has land-boundaries with Orissa, Madhya Pradesh on the North; Maharashtra and Karnataka on the west; Tamil Nadu on the south and Sea boundary with the Bay of Bengal on the East. Among all States of India, only Andhra Pradesh has the longest coastline of 968 kilometers.

There are 23 districts and 1104 mandals in Andhra Pradesh in which Anantapur district is the biggest in area and Hyderabad district is the smallest.

Basing on physical, social and economic conditions, Andhra Pradesh can be divided into three regions. They are:

1. Coastal Andhra,
2. Rayalaseema, and
3. Telangana.

1. Coastal Andhra Pradesh comprises of districts viz., Srikakulam, Vizianagaram, Visakhapatnam, East Godavari, West Godavari, Krishna, Guntur, Prakasam and Nellore.

The area of Coastal Andhra region is 92,900 square kilometers. This region has the fertile deltas of the rivers, Godavari, Krishna and Penna. A major portion of food and commercial crops that are grown in Andhra Pradesh are from this region. Hence, the Coastal Andhra region is known as the 'Granary' of South India. As this region is agriculturally well-developed, it is equally well-developed in trade and transport sectors. Hence the highest concentration of population is found in this region. There are many industries here like Steel Plant, Oil Refinery, Ship-building, Bharat Heavy Plates and Vessels, Hindustan Zinc, Coromandel Fertilizers etc.

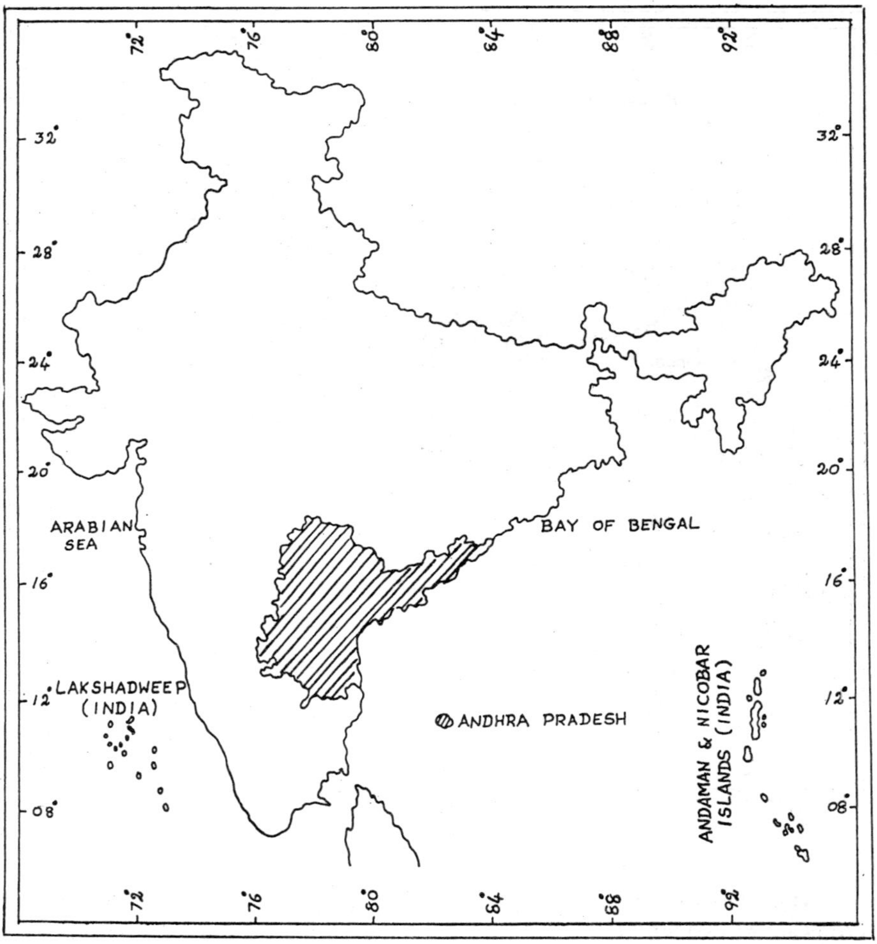


**Fig. 1.** Location Map of Andhra Pradesh in India.

2. Rayalaseema comprises of 4 districts, viz. Chittoor, Cuddapah, Kurnool and Anantapur.

Rayalaseema region extends over an area of 67,400 Sq. Kms. The density of population is very sparse, because of it, it is well known for droughts and famines since ancient times. Economically and industrially this region is very backward compared to the other two regions, and also it comprises of rocky area and infertile soils with very little and unsuitable rainfall.

3. Telangana comprises of 10 districts viz. Mahabubnagar, Hyderabad, Ranga Reddy, Medak, Nizamabad, Adilabad, Karimnagar, Warangal, Khammam and Nalgonda districts. This region extends to an area of 114,800

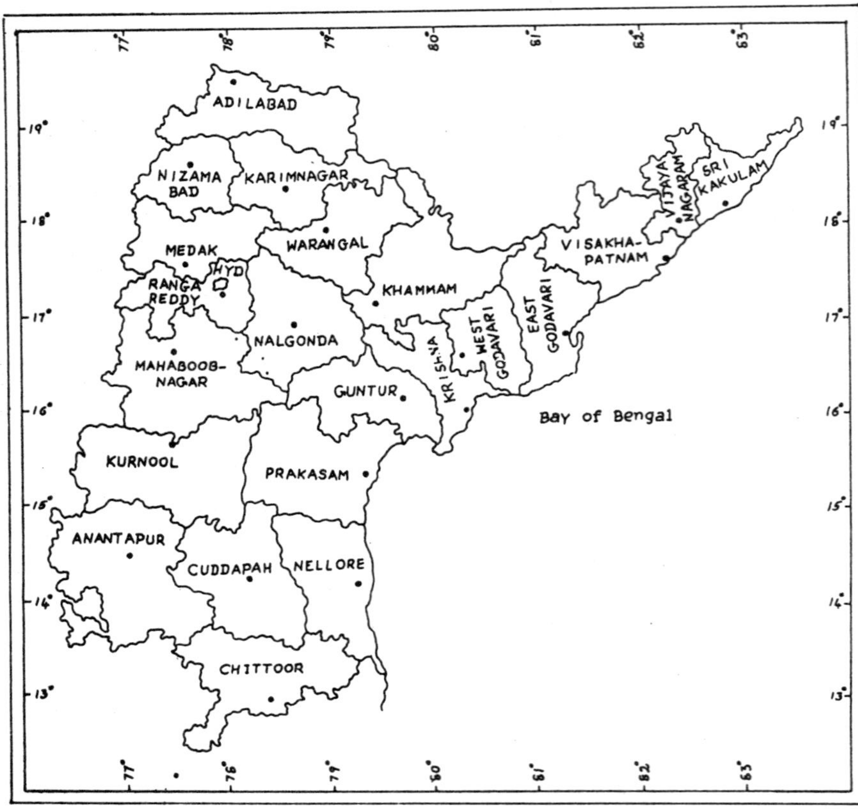


**Fig. 2.** Map of Andhra Pradesh showing Districts.

Sq. Kms. Most of the area is covered with dry and barren land. So upland crops are grown with the help of available tank water. Industrially this region is well-developed.

## TOPOGRAPHICAL DIVISIONS

Topographically 35% of the area in the State lies below 150 metres altitude forming the coastal plains. About 15% lies between 150 and 300 metres and nearly 37% between 300 and 600 metres both forming the plateau surface comprising of peneplained surfaces. About 13% of the area with an elevation above 600 metres is treated as hilly tract, in which 10% lies between 600 and 900 metres and the remaining at more than 900 metres above the sea level.

## GEOGRAPHICAL DIVISIONS

Geographically, the State can be divided into 3 natural regions.

### 1. The Coastal Plains

These plains lie between the Eastern Ghats and Bay of Bengal, stretching from Srikakulam district in North to Nellore district in south. The northern portion of the Coastal plain is narrow with an average width ranging between 30 and 40 Kms. The middle part is wider than the northern portion, with an average between 70 to 75 Kms and in some places 100 Kms even. It includes shallow freshwater lake of Kolleru (259 Sq. Kms) which is natural depression between Godavari and the Krishna deltas. The width of the coastal plain in the southern part is between 50 and 60 Kms. The Dolphin's nose, Visakhapatnam (357 m), Kondapalli (573 m), Simhachalam Cliff (244 m), Pulicat lake and Sriharikota island fall in this region.

### 2. The Eastern Ghats

Lying in between the coastal plains and the western plateau, the Eastern Ghats begin from Khondomal-Simlipal hills of Orissa, pass through Andhra Pradesh parallel to the East coast and in a South-west direction and merge with Western Ghats near Nilgiris of Tamil Nadu (Legris and Meher-Homji, 1982). They consist of different types of hills and present themselves as a broken chain of hills unlike Western Ghats. In between the Godavari and the Krishna rivers, there is an almost clear break in the chain of Eastern Ghats. The hills are closely formed and they rise high in the north Coastal Andhra, mostly above 750 m with a few peaks exceeding 1500 m elevation. The highest peak in these Ghats is Sambari Konda with an elevation of 1527 m near Gudem village in Visakhapatnam district (Subba Rao *et al.*, 1982).

The Eastern Ghat's river valleys are one of the most fertile regions of the country with several century-old civilizations and agricultural practices (Reddy, 1982). A number of small rivers and streams are born on these hills which, flowing eastward, provide water sources to the Coastal plains in the north Araku valley, celebrated for its natural splendour and scenic beauty. The northern hills of the Eastern Ghats are popularly known as Agency area which has been the abode of various hill tribes from very ancient times.

The hills of Eastern Ghats extending from the south of the Krishna river up to the Chittoor district, are variously called Velikondas, Palakondas, Nallamalais Erramalais, Lankamalais and Seshachalam. In the Kurnool district the chain of hills are crescent shaped, half-encircling the Nandyal valley. As the Kunderu river flows in this valley, the valley is also called Kunderu valley. To the east of Kunderu valley lie the Nallamalais and to the west lie the Erramalais. The Nallamalais are 112 Kms long and their height is 800 m. The region between the Nallamalais and Velikondas is a scanty rain area. There are no thick forests in this region. The Horsely hills (1268 m) near Madanapalli in the Chittoor district are a well known health resort. The Seshachalam hills are known as Tirupathi hills.

The Eastern Ghats in Andhra Pradesh pass through the districts of Vizayanagaram, Srikakulam, Visakhapatnam, East and West Godavari, Khammam, Krishna, Guntur, Prakasam, Kurnool, Mahabubnagar, Cuddapah, Anantapur, Nellore and Chittoor districts.

### 3. The Western Plateau

It is an ancient plateau made of rocks of Archaean origin. This plateau region consists of whole Telangana region, Anantapur and Kurnool districts except the Nallamalais of Rayalaseema. The plateau extends largely between 150–160 metres altitude except in Ranga Reddy, Mahabubnagar and in Western portion of Telangana region, where the plateau elevated between 600–900 metres.

Generally there is scanty rainfall in the southern part. Towards the northern part, a belt of Gondwanas, the coal deposits have been preserved along the lower Godavari trough and most of the area is covered by dry deciduous forests.

## B. CLIMATE

Andhra Pradesh is situated in tropical zone from the point of view of its latitudinal location. The State has tropical monsoon type of climate, which is generally hot and humid. The climate of the State is classified as tropical rainy climate represented by all regions of the State except the South-Western portion where the climate is of Hot-steppe (IMD, 1973, p.1). In Hot steppe areas the mean daily temperature is 18°C or less. After summer these places receive rainfall from monsoon. Maximum temperature in the summer season varying between 37° and 44°C and minimum temperature in the winter season ranging between 14°C and 19°C. In tropical rainy areas during the mean daily temperature is above 20°C, with an annual rainfall of less than 250 cm.

### Rainfall

There is a wide variation and uncertainty in the distribution of rainfall in the State (Tables 1 and 2). The State receives 1.7% of rainfall from the South-west monsoon during early June to the end of September and 25% of the rainfall from the North-east monsoon during October to December. The onset

**Table 1.** Rainfall Season-wise

(Rainfall in Millimetres)

| Year | South-west monsoon (June–Sept.) | North-west monsoon (Oct.–Dec.) | Winter period (Jan.–Feb.) | Hot weather period (Mar.–May) | Total |
|---|---|---|---|---|---|
| 1982–83 | 553 | 205 | 8 | 53 | 819 |
| 1983–84 | 986 | 227 | 28 | 37 | 1198 |
| 1984–85 | 510 | 158 | 23 | 43 | 734 |
| 1985–86 | 494 | 237 | 37 | 51 | 819 |
| 1986–87 | 597 | 189 | 14 | 68 | 868 |
| 1987–88 | 475 | 390 | 10 | 77 | 954 |
| 1988–89 | 994 | 091 | — | 58 | 1144 |
| 1989–90 | 896 | 088 | 37 | 322 | 1343 |
| 1990–91 | 647 | 283 | 11 | 41 | 982 |
| 1991–92 | 696 | 243 | 7 | 35 | 981 |
| 1992–93 | 555 | 221 | — | 61 | 837 |
| 1993–94 | 511 | 242 | 15 | 49 | 817 |

Source: *Bureau of Economics and Statistics*, Andhra Pradesh.

of the South-west monsoon is the chief rainy season of the State except the Southern-coastal belt. The State receives a remarkable amount of rainfall in the North-east monsoon period except the region of Telangana. The annual rainfall in the coastal region is between 70 to 150 cms, the northern and the central areas of the coastal region receives heavy rainfall from North-east monsoon while the southern part receives only 60% of it. The remaining parts of the State receive 60% of the rainfall from the South-west monsoon.

In Rayalaseema, the Western part is the driest, and ranks as the second driest part in the country with less than 60% of rainfall annually. From South-west monsoon, Rayalaseema receives 40 to 70% of rainfall annually, while from the North-east monsoon, the South-western parts of Rayalaseema receives 50 cms of rainfall and North-western parts receiving lesser.

The Northern part of Telangana, receives more rainfall than its South-western part, ranging between 75–100 cms during the South-west monsoon period.

Depressions are formed over the Bay of Bengal during the North-east monsoon seasons and south Coastal districts receive more rain and incur heavy losses whenever the cyclones lash the coast.

Anantapur district is the zone of the lowest rainfall with about 519 mm annually, the second lowest in the country (Ref. The Hindu, dated 23.01.1995) while the West Godavari district is the zone of highest rainfall in the State (Tables 1 and 2). On an average, there are 60 rainy days in a year in the State.

**Table 2.** Average Annual Rainfall — District-wise

(Rainfall in Millimetres)

| Sl. No. | District | 85–86 | 86–87 | 87–88 | 88–89 | 89–90 | 90–91 | 91–92 | 92–93 | 93–94 |
|---|---|---|---|---|---|---|---|---|---|---|
| 1. | Srikakulam | 1098 | 1485 | 1081 | 990 | 1575 | 1166 | 1155 | 1193 | 941 |
| 2. | Vizayanagaram | 1101 | 1315 | 1209 | 1063 | 1599 | 978 | 1093 | 1067 | 845 |
| 3. | Visakhapatnam | 1078 | 1240 | 1243 | 1114 | 1857 | 955 | 1094 | 1362 | 856 |
| 4. | East Godavari | 1016 | 1017 | 1166 | 1092 | 1988 | 988 | 1283 | 1084 | 937 |
| 5. | West Godavari | 1090 | 993 | 1113 | 1298 | 2185 | 968 | 1549 | 1052 | 808 |
| 6. | Krishna | 927 | 842 | 900 | 1264 | 1579 | 921 | 1121 | 872 | 704 |
| 7. | Guntur | 856 | 699 | 851 | 1082 | 1507 | 859 | 1112 | 864 | 728 |
| 8. | Prakasam | 764 | 633 | 913 | 709 | 902 | 942 | 959 | 654 | 890 |
| 9. | Nellore | 1103 | 807 | 1193 | 899 | 1007 | 1003 | 1515 | 854 | 1168 |
| **Coastal Andhra Avg.** | | **1006** | **1003** | **1074** | **1057** | **1577** | **976** | **1209** | **1000** | **876** |
| 10. | Kurnool | 557 | 385 | 791 | 914 | 743 | 609 | 652 | 774 | 673 |
| 11. | Anantapur | 297 | 334 | 718 | 757 | 785 | 482 | 574 | 500 | 593 |
| 12. | Cuddapah | 567 | 432 | 754 | 851 | 790 | 851 | 816 | 534 | 707 |
| 13. | Chittoor | 874 | 603 | 939 | 904 | 1016 | 904 | 1343 | 702 | 1108 |
| **Rayalaseema Avg.** | | **574** | **439** | **801** | **857** | **814** | **712** | **848** | **579** | **772** |
| 14. | Ranga Reddy | 575 | 553 | 975 | 1013 | 833 | 977 | 804 | 820 | 717 |
| 15. | Hyderabad | 552 | 684 | 975 | 859 | 1263 | 731 | 740 | 714 | 772 |
| 16. | Nizamabad | 857 | 816 | 1075 | 2144 | 1786 | 1736 | 857 | 780 | 904 |
| 17. | Medak | 701 | 628 | 877 | 1442 | 1446 | 1051 | 687 | 787 | 734 |
| 18. | Mahabub Nagar | 639 | 434 | 756 | 843 | 839 | 674 | 814 | 577 | 617 |
| 19. | Nalgonda | 711 | 547 | 726 | 876 | 1034 | 688 | 870 | 505 | 591 |
| 20. | Warangal | 841 | 908 | 1060 | 1387 | 1831 | 1010 | 792 | 799 | 869 |
| 21. | Khammam | 1071 | 1024 | 929 | 1397 | 1831 | 1217 | 1156 | 1014 | 893 |
| 22. | Karimnagar | 753 | 1096 | 860 | 1408 | 1393 | 1286 | 732 | 924 | 789 |
| 23. | Adilabad | 944 | 1224 | 867 | 2005 | 1527 | 1622 | 841 | 1056 | 941 |
| **Telangana Avg.** | | **764** | **792** | **910** | **1338** | **1344** | **1096** | **831** | **798** | **782** |
| **A.P. Average** | | **829** | **812** | **954** | **1144** | **1343** | **982** | **981** | **837** | **817** |

Source: *Bureau of Economics and Statistics*, Andhra Pradesh.

## Temperature

There is not much significant change in temperature of Andhra Pradesh from one place to another except in certain tiny patches of high altitudes. During rainy season the maximum temperature of the State varies between 29°–32°C with a minimum temperature of 20–25°C. In winter, the maximum temperature is about 29°C, sometimes it goes below 22°C. Summer will be fairly hot throughout the State. May being the hottest month with the maximum temperatures varies between 37–45°C, sometimes in some places it reaches above 46°C especially in Ramagundam of Khammam district.

## C. GEOLOGY

There are a wide variety of geological formations in the State. They vary from the oldest Dharwar Schists to the recent alluvial deposits. These rocks consist of minerals and they are well distributed throughout the State.

Dharwar Schists are known for their mineral ores. They possess Gold in Ramagiri of Anantapur district, Mica in Nellore, Hematite mineral in Karimnagar and Adilabad districts and Copper in Anantapur and Guntur districts.

In Rayalaseema and Telangana regions, the existence of Peninsular rock formations dominate containing granite, granodiorites and banded formations.

The charnokites and khondalites are found to spread in an extensive belt from Srikakulam to West Godavari. The charnokites bands are also found in Guntur, Krishna and Prakasam districts. They are rich in minerals like graphite and Manganese among charnokite rocks as well as Bauxite, Chromite and Iron among Khondalite rocks.

The Archaean of gneisses dominates in Telangana and Rayalaseema regions. The Puranas are seen in Cuddapah, Kurnool, Mahabubnagar, Nalgonda, Guntur, Ranga Reddy and Adilabad districts. They include limestones, sand stones, slates and shales.

The Gondwanas are fresh water sedimentary rocks deposited on the banks of Godavari and Pranahita rivers in eastern Telangana. Of these the lower Gondwanas contain coal seems among sandstones and shales. The coal seems at Kothagudem, Yellandu, Bellampalli and Godavarikhani are being exploited now. The Deccan formations, mostly containing greenish basaltic rocks, are found in the western and north-western parts of the State. The low hills comprising Rajahmundry sandstones and conglomerates represent the tertiary formations. In some places in western Telangana, Nellore and Visakhapatnam districts the laterites occur as over-caps of Deccan trap formations. Alluvium occurs extenisively in the deltas of Krishna and Godavari rivers, as also along valley of Vamsadhara and Nagavali. It consists of sand, gravel, silt and clay with silt forming a major part in the deltas.

## D. SOILS

A bulk of human population still continues to live where good soil help it to grow food; whereever deep and fertile soil is found and climate is favourable for natural vegetation. The climatic conditions i.e. temperature, wind, humidity, vegetation cover and chemical action play a vital role in soil formation. Soils are of 6 types in Andhra Pradesh namely Red, Black, Alluvial, Laterite, Coastal sandy, and Skeletal soils.

Red soils are common throughout the Andhra Pradesh, the humus from the forest litter imparts an acidic content through decomposition. The

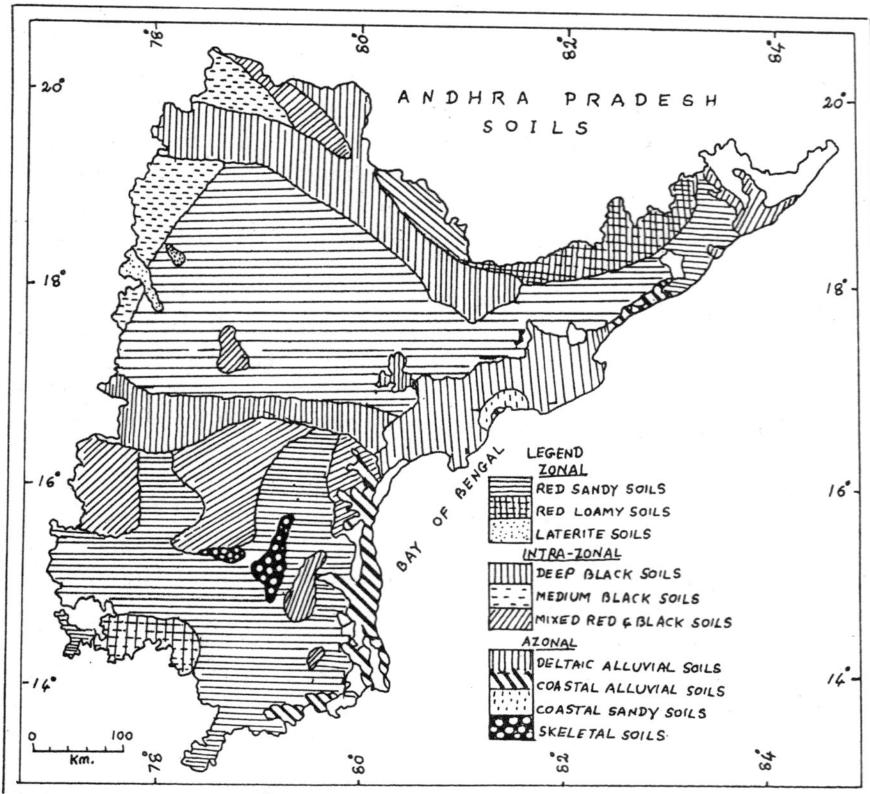


**Fig. 3.** Soils in Andhra Pradesh.

formation of red soil, is through rain which helps in the precipitation of silicates and sesquisoides and this is accelerated by the presence of lime. They occur widely in Rayalaseema and Telangana but appear rarely in the districts of Coastal Andhra. Their content of nitrogen and plant nutrients is poor, but are low to medium in the contents of their phosphates and medium to high in potash. Their moisture holding capacity is also poor.

Black soils are also called as black cotton soils. The black soils, rich in calcium and potash are found in western and north-western portions of the State. These are the second important soil group, which are fertile with retentive moisture but poor in nitrogen content. The black soils are of 3 types, namely the deep black, the light black and the mixed red and black soils.

The deep black soils are found in a broad belt along the Pranahita, Godavari and Krishna rivers. The light black soils are found in the North-western portion of the State (Adilabad, Medak, Nizamabad and Ranga Reddy districts) are developed from the Deccan trap rocks. The mixed red and black soils occur as broad patches between the Krishna and other perennial rivers. In the elevated plateaus, they are thin with a light colour and poor fertility but in low lying areas they are thick with deeper colour and rich fertility.

Alluvial soils are of two types, the Deltaic alluvial soils and Coastal alluvial soils. The Deltaic alluvial soils are highly fertile as they periodically add up the silt brought-down by the rivers. These occur extensively in the deltas of Godavari and Krishna, along the coast in Visakhapatnam and Srikakulam districts and along the valleys of Vamsadhara, Nagavali rivers and also along the Pennar in Nellore. The Coastal alluvial soils are lesser fertile, and not rich in plant nutrients and organic matters and made up of sandy loam. It stretches as a narrow belt and along the coast for a short break in Visakhapatnam district where the spurs of the Eastern Ghats meet the sea. Casuarina and Cashewnut plantations are raised extensively on this belt.

Laterite is a superficial alteration of rocks under certain meteorological conditions and occupies large areas of Srikakulam, Visakhapatnam, East Godavari, Medak, Hyderabad and Nellore districts. Laterite, which is formed by excess of precipitation and over evaporation, is rich in ferric oxides and poor in bases. They are generally poor in organic matter and plant nutrients like Nitrogen, Phosphate and Potash. The soils are well drained and spread expeditiously.

Coastal sand soils are found along the coast, in a narrow belt except in some places of Visakhapatnam district. These are less fertile than deltaic alluvial soil, not rich in plant nutritents and organic matters and made up of sandy loam. Casuarina and Cashew groves are extensively grown in these sandy soils.

Skeletal soils are infertile and are generally associated with the highly deteriorated hill-slopes and soils near the foot hills. They are found in the slopes of Nallamalais in Badvel, Giddalur of Cuddapah and Prakasam districts respectively and are also represented as a tiny patch in Jammalamadugu of Cuddapah district.

## E. RIVERS

A river is a sizable stream of fresh water flowing through a natural channel in the land. Most rivers flow on the surface, but in some areas they go underground for great distances.

Rivers are among the most natural forces in shaping the earth surface. In draining the land of surplus water, rivers wear down mountains, plateus and other high land forms. In a never ending process, eroded material is carried by rivers. Some is deposited to form flood plains in the valley, some form deltas at the river mouths, and some is deposited in the sea.

## River Systems

A river system is made up of a main river and all its tributaries or small streams. Andhra Pradesh is fortunate in having 75% of its territory covered by rivers, hence referred to as a river-state. Rivers constitute an immortal natural resource as they form the basis of irrigation, transportation etc., thus forming the lifeline of civilizations from time immemorial. Of late, the rivers have become the dumping centres for human and sewage wastes, industrial wastes etc., thus degrading the quality of the rivers. Indian rivers are classified into (1) the Himalayan group, (2) the Deccan group, (3) the rivers of inland drainage basin. The rivers, flowing through Andhra Pradesh fall into Deccan group of rivers, and all these rivers drain into the Bay of Bengal.

There are 7 major and 27 minor rivers. They are:

## The Godavari

It is the largest river system of Peninsular India and the second largest next to the Ganga system. It takes its origin at Triambak near Nasik in the Western Ghats in Maharashtra State. It flows over a distance of about 772 Kms in the State out of 1465 Kms of its total length before reaching the Bay of Bengal. The river receives about 80% of its water from its tributaries namely Indravathi, the Manes, the Manira, the Mule, the Penganga, the Pranhita, the Pravara, the Puma, the Sabari, the Wardha, and the Warigana. The river is fed by the two monsoons. It is also known as the "Ganga of the South" or "Dakshin Ganga"; and it is a perennial river of the south. But only 23 per cent of its catchment area spreads over Andhra Pradesh.

## The Krishna

It rises from a spring in the Western Ghats, near Mahabaleswar, Satara district, Maharashtra. It runs through the States of Maharashtra, Karnataka, and enters Andhra Pradesh in Makthal mandal of Mahabubnagar district. It is about 620 Kms in the State out of 1400 Kms of its total length. Andhra Pradesh shares about 29% of the drainage basin of this river. It flows through a plateau of 338 Kms in Kurnool and Nalgonda districts and nearly 193 Kms of it from Siddeswaram to Nandikonda lying in a narrow, continuous gorge. Finally it enters the plains at Pulichintala in Krishna district after flowing through Guntur district. It consists of 19 tributaries of which the Tunga-Bhadra, the Bheema and the Musi are the main tributaries.

### The Tunga-Bhadra

The Tunga-Bhadra is one of the important tributaries of the Krishna and it is the union of the Tunga and Bhadra river. It rises from the Varaha hills in Chikmagalur district of Karnataka State and mingle each other near Shimoga town in Karnataka. It enters Andhra Pradesh near Kosigi in Adoni mandal of Kurnool district and joins Krishna river near Moravakonda 48 Kms from Kurnool. It is the river which separates Kurnool district from the Telangana region.

### The Nagavali

It originates in the Eastern Ghats in Orissa State and enters Andhra Pradesh near Parvathipuram mandal in Srikakulam district. The major part of the river flows through Andhra Pradesh covering a length of 115 Kms. After running through Vizayanagaram and Srikakulam districts, it joins the sea near Mofaz-Bandar in Srikakulam district. Its chief tributary, the Swarnamukhi, also originates from Orissa.

### The Pennar

Also called the Pinakini, its source is Nandi hills in Kolar district of Karnataka State, and enters Andhra Pradesh near Chowluru in Anantapur district. It consists of six tributaries namely the Cheyyeru, the Chitravati, the Jayamangali, the Kumderu, the Papaghni and the Sagileru. After flowing through Anantapur and Cuddapah districts, it enters Nellore district through a fine gorge in the Veligonda at Somasila. It runs in a wide sandy bed for about 112 Kms in eastern direction through the mandals of Atmakur and Nellore and gets into the Bay of Bengal in the form of several streamlets at a distance of 29 Kms from Nellore. Its total length in Andhra Pradesh is almost 568 Kms. The river's total catchment area is 55,213 Sq. Kms.

### The Vamsadhara

As its banks are hemmed with bamboos, it is named Vamsadhara. It rises in the Eastern Ghats of Orissa State, enters the State near Pathapatnam Mandal of Srikakulam district. It runs between Narasampet and Srikakulam mandal and ultimately falls into the Bay of Bengal near Kalingapatnam. The total length of the river is about 185 Kms of which 90 Kms lie in Srikakulam district. The catchment area of the river is 4400 Sq. Kms.

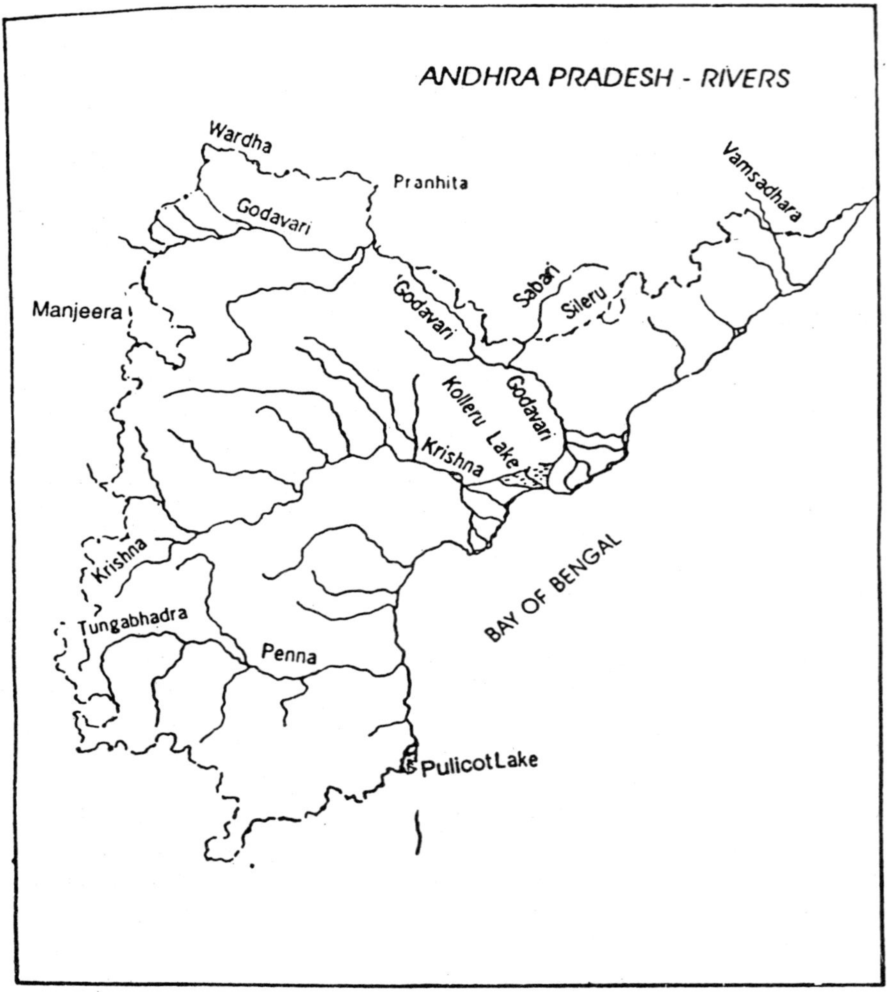


**Fig. 4.** Rivers in Andhra Pradesh.

## The Manjeera

Being the chief tributary to Godavari, this river rises in Pathoda taluk of Bir district in Maharashtra State. This river, after flowing through the districts of Osmanabad and Bidar enters Andhra Pradesh at Medak district. It joins the Godavari near Kandakurthi in Bodhan mandal of Nizamabad district. The total catchment area is about 30,821 Sq. Kms.

### The Gundlakamma

From Nallamalis, its source, in Kurnool district, it runs through Prakasam district and falls into the sea near Devarampadu village in Prakasam district.

### The Swarnamukhi

Its origin is in the Chandragiri hills of Chittoor district. It flows through Nellore district before merging into the sea near Siddavaram in Nellore district.

### Waterfalls

The important waterfalls in Andhra Pradesh are Kuntala and Potchera in Adilabad; Kailasakona, Kambakkam, Papanasanam, Talakona and Gunjana in Chittoor; Ethipotala in Guntur and Penchala Kona in Nellore district.

# FORESTS AND VEGETATION

## FORESTS

A forest is a community or an ecosystem of biotic and non-biotic components, consisting predominantly of trees or other woody vegetation growing more or less closely together. In addition to trees other biotic components are shrubs, grasses and other plants, big and small animals. The non-biotic components are soil and local climate prevailing within and in the vicinity of forests.

One million years ago when first man (Homo sapiens) came to exist on the planet, forests were plentiful and animal life bountiful and they required to be cleared for man's safety and existence, and the construction of dwellings and cultivate benefits of man. As civilisation progressed and man chose a settled life, clearence of forests for dwellings and cultivation and hunting animals for food has become indispensable.

With the advent of British rule, forests became a part of the Revenue Department, and there was no separate forest department till middle of last centuary. In 1855 for the first time an outline of permanent programme of forest administration in India was drawn up. The outline laid down the ruling principles of management of State forests namely State property and that in consequence, all trees living or dead were made public properties to which individual had no right or claim.

At present Forests are recognised as renewable natural resource and development of Forestry as a factor in agricultural progress, as a source of raw materials for industrial and other uses has been taken up.

In a forest ecosystem vegetation, climate and soil are in intimate dynamic relationship, however a complex one. Growth and distribution of different tree species is the resultant effect of interaction of climate and soil.

Forests provide five principal services; livelihood and way of life for forest dwellers; a diversity of plant and animal species; protection and nutrition of soils, and watershed services; regulation of climate; and wood. If the land is converted to an alternate use, most of these services are lost.

Forests contribute substantially to the economic development of our country by providing goods and services. The goods are not only woody

**Table 3.** Distribution of Geographical and Forest Areas, District-wise

| Sl. No. | District | Area (Sq. Km.) | Forest block (No.) | Forest area | Percentage to geographical area | Percentage forest area to total A.P. forest area |
|---|---|---|---|---|---|---|
| **Coastal Andhra** | | | | | | |
| 1. | East Godavari | 11161 | 161 | 3232.44 | 29.9 | 5.07 |
| 2. | Guntur | 11391 | 118 | 1619.41 | 14.2 | 2.54 |
| 3. | Krishna | 8727 | 60 | 664.28 | 7.6 | 1.04 |
| 4. | Nellore | 13076 | 268 | 2519.37 | 19.2 | 3.95 |
| 5. | Prakasam | 17626 | 130 | 4424.99 | 25.1 | 6.94 |
| 6. | Srikakulam | 5837 | 79 | 684.41 | 11.9 | 1.08 |
| 7. | Visakhapatnam | 10807 | 188 | 4411.66 | 39.3 | 6.92 |
| 8. | Vizayanagaram | 6539 | 103 | 1193.03 | 18.83 | 1.87 |
| 9. | West Godavari | 7742 | 50 | 811.66 | 10.4 | 1.27 |
| **Total** | | **92906** | **1157** | **19563.25** | **21.06** | **30.66** |
| **Rayalaseema** | | | | | | |
| 10. | Anantapur | 19130 | 103 | 1969.78 | 10.3 | 3.09 |
| 11. | Chittoor | 15152 | 184 | 4520.18 | 29.9 | 7.09 |
| 12. | Cuddapah | 15359 | 154 | 5002.95 | 32.7 | 7.84 |
| 13. | Kurnool | 17658 | 103 | 3515.49 | 19.8 | 5.51 |
| **Total** | | **67299** | **544** | **15008.4** | **22.3** | **23.53** |
| **Telangana** | | | | | | |
| 14. | Adilabad | 16128 | 235 | 7231.89 | 44.8 | 11.32 |
| 15. | Hyderabad | 217 | — | — | — | — |
| 16. | Karimnagar | 11823 | 201 | 2554.86 | 21.6 | 4.00 |
| 17. | Khammam | 16029 | 356 | 8436.94 | 52.6 | 13.52 |
| 18. | Mahabub Nagar | 18432 | 86 | 3032.51 | 16.5 | 4.75 |
| 19. | Medak | 9699 | 233 | 905.94 | 9.4 | 1.42 |
| 20. | Nalgonda | 14240 | 26 | 836.93 | 5.9 | 1.31 |
| 21. | Nizamabad | 7956 | 189 | 1812.15 | 22.6 | 2.84 |
| 22. | Ranga Reddi | 7493 | 127 | 730.75 | 9.7 | 1.15 |
| 23. | Warangal | 12846 | 112 | 3713.14 | 28.8 | 5.82 |
| **Total** | | **114863** | **1565** | **29242.08** | **25.46** | **45.82** |
| **Total Andhra Pradesh** | | **275068** | **3266** | **63813.73** | **2320.00** | **100.00** |

products obtained from the trunks of trees, but also products derived from other parts of trees and from the components of the forest community. Almost every religion in the world gives importance to forests and life in forests. Forests also have aesthetic and touristic values.

**Table 4.** Classification of Forest Area

| Sl. No. | Region | Forest area (Sq. Kms) | Percentage (%) |
|---|---|---|---|
| 1. | Legal | | |
| | Reserved | 50478 | 79.10 |
| | Protected | 12365 | 19.38 |
| | Unclassified | 970 | 1.52 |
| | **Total** | **63813** | **100.00** |
| 2. | Composition (Non Coniferous broad leaved) | | |
| | Sal | 47 | 0.07 |
| | Teak | 9145 | 14.33 |
| | Others | 54622 | 85.60 |
| | **Total** | **63814** | **100.00** |

**Table 5.** Actual Forest Cover by Density Classes of Andhra Pradesh

| Sl. No. | Forest | Area in Sq. Kms. |
|---|---|---|
| 1. | Dense Forest (Crown Density above 40%) | 25008 |
| 2. | Open Forest (Crown Density 10% to 40%) | 21870 |
| 3. | Mangroves | 378 |
| **Total** | | **47256** |

**Table 6.** Forest Cover Density, Districtwise

| Sl. No. | | No. | Area in Sq. Kms. |
|---|---|---|---|
| 1. | No. of districts having forest cover more than 33% | 3 | 17366 |
| 2. | No. of districts having forest cover between 19% and 33% | 5 | 14510 |
| 3. | No. of districts having forest cover 0.1 to 19% | 15 | 15414 |
| **Total** | | **23** | **47290** |

All above tables are taken from brouchre (Andhra Pradesh Forest) issued by Principal Chief Conservator of Forests, Hyderabad.

All these factors have made man realise that losing forests is losing an important feature of life. They are the abode of Flora and Fauna and possess the rare floral and faunal species. Hence they have to be protected and preserved.

Andhra Pradesh covers 8% of the total forest area of the country. About 65000 Sq. Kms of it is covered with forests, out of 2,75,000 Sq. Kms of its total area. It occupies fourth place in the country's state-forest area. Of the three regions of the State namely Coastal Andhra, Telangana and Rayalaseema is shown as 2, 2.9, 1.6 million hectares respectively. The distribution of forests in different districts is given in Table 2.

## BIOTIC INTERFERENCE

Man and his domestic animals have a profound influence on the vegetation. A number of wild animals also cause damage to vegetation at various stages of growth.

During the course of extensive and uninterrupted plant exploration of Andhra Pradesh State for five years, the author has noticed an alarming biotic interference on the forest vegetation in the State, the rich vegetation and the rare and valuable species have been disappearing owing to biotic interference. A large part of the forest cover has been denuded and deforestation has been going on in many parts of Andhra Pradesh.

The evergreen areas which were present once at Rampa hills in East Godavari, Araku Valley, Galikonda in Visakhapatnam; Gundlabrahmeswaram, Krishna Nandi and Upper Ahobilam in Kurnool district; Guvvalacheruvu, Balapalli in Cuddapah district; Kambakkam, Talakona, Tirumala hills in Chittoor district and, Penchalakona in Nellore district, are being converted into deciduous patches.

Deciduous forests are being converted into scrub jungles as in districts of Telangana, Anantapur, Krishna etc. In these areas, the appearence of drought indicators (thorny, spiny bushes) accounts for the influence of various biotic factors on the forest depicts the way it is being subjected to retrogression.

The agency area of Northern Circars, Nallamalais and some places of Telangana are usually inhabited by tribal population depending mostly on forest products for their living. The Girijans known by different sect names i.e., Chenchus, Koyas, Konda Reddis, Konda Doras etc., settled in the valleys of the hilly regions are mostly habituated with a type of cultivation called "Podu" in local language otherwise called shifting cultivation. It is a traditional form of cultivation on hill slopes, causing destruction of forest wealth and accelarating land degradation. It involves cutting of trees, bushes etc. on the hill slopes up to stump level in winter. Due to reduced forest area available to the tribal population, jhumming, which most of them still practise, has become unproductive and uneconomic and is resulting in severe destruction of forests in the hills.

Forest fire is an important factor which occurs either accidentally or by negligence of man which causes large scale damage to young seedlings, climbers, growing trees and forest soil. Forest fires are mainly caused by, apart from other factors, beedi leaf contractors to facilitate easy collection of produce and induce a new flush of leaf by running a ground fire. The tribals and other sportsmen in order to collect honey or trap animals like Lizard and other rodents like Hares, Rabbits and burrowing animals set the forest fires.

Illicit cutting and smuggling of timber also cause considerable damage to the forest besides heavy revenue loss to the State Government, as good number of sapling are also destroyed during such illicit cutting. The red sanders (*Pterocarpus santalinus*) and Sandalwood (*Santalum album*) in Seshachalam hills are now hardly seen owing to over exploitation and smuggling. Teak (*Tectona grandis*) population is also decreasing enormously due to illicit cutting. About 90% of trees are felled to meet the requirement of fuel.

Grazing and browsing by domestic animals are a constant and inevitable source of damage. Overgrazing of the wild land has resulted in the invasion by exotic weeds.

'Forest-species' richness is decreasing also either by intermediate tree harvesting or trampling. Plantations of coffee, casuarina, cashew nut etc. also affect the forest wealth directly or indirectly.

Majority of the hydro-electric projects will definitely rootout the natural habitats of plants. A large part of the forest area has been submerged by the Srisailam and Nagarjunasagar dams , besides the exploitaion by the people who got settled in the areas around the project. The Nagarjuna Fertilizers at Kakinada has a remarkable impact on the mangrove vegetation of Koringa, the richest mangrove vegetation in the country. The rich vegetation has an enormous damage owing to the establishment of Paper mills.

Majority of Paper mills, those of Government as well as private sector, have unrighteously cleared the bamboo forests of Rampa hills in East Godavari and Nallamalais in Kurnool districts. The Avanti Leathers Limited, at the foot of Kambakkam hills (Chittoor), the Panyam Cements Limited, at Cement Nagar ghat hills (Kurnool), the AP Rayon at Eturnagarm (Warangal), the Ibrahimpatnam Thermal project at Kondapalli forests (Krishna) with their chemical effluents have a devastating effect on the flora.

The mining operation for phosphate at Khasipatnam near Araku valley, for mica near Nellore, for graphite near Doramamidi (West Godavari district) and manganese mining in Visakhapatnam have had telling effect on the natural forest cover. Kondakarla lake and surrounding marshy vegetation have been affected by the Vizag Steel Plant and the zinc smelters and other subsidaries attached to it.

For the construction of the reservoirs and digging of canals under Telugu Ganga Project (Plan being implemented jointly by Tamil Nadu and Andhra Pradesh Governments) the vast area about 4805 hectares in Nallamalais, 1269 hectares in Erramalais in Kurnool district and 367 hectares in Chittoor, Cuddapah and Nellore districts had to be destroyed.

The ever increasing population has led to demand for more land for agricultural purposes as well as habitation. This has naturally resulted in encroachment and consequent destruction of forest areas. The total area of forests in Andhra Pradesh was 63.49 lakh hectares in 75–76 period, 61.43 lakh hectares in 82–83 period, whereas it is only 58.12 lakh hectares in 93–94 period. This goes to prove the gradual depletion of forest cover in the State.

## CONSERVATION

The preservation of biodiversity, is not only important for preservation by way of a gene bank but also for the protection of existing forests from further degradation and deterioration caused by all factors.

The foremost attempt in the conservation strategies should be avoidance of indescriminate felling and forest firing, collection and preservation methods of the seed output and germination of rare and endangered species. It is therefore considered very important to conserve existing bio-diversity in terms of increasing the population density.

Effective and substainable plantation programmes and construction of check dams, in order to increase the moisture levels of soil and of air and avoidance of harmful agrochemicals of the soil may help pollinators to score good seed material.

It is the best of times that the State Government has put lots of emphasis on clothing the land mass with vegetation cover and also should take every possible step to check population growth. The wasteland should be utilised for raising commerical crops. The Forest Department should be strengthened. The people should be in forestry activities. There is need for strong political will to save the forests of Andhra Pradesh from disaster.

### Wild Life Sanctuaries and Parks

Wild life is an integral part of the forests, their existence is also necessary for ecological or environmental balance. These are dedicated to the preservation of wild life and to represent ecological units in Andhra Pradesh.

## VEGETATION

The collective and continuous growth of plants in space is called vegetation. Thus vegetation is actually the totality of plant growth including large or some populations of each species intermixed in a region. It is an important natural resource and constitutes the basic productivity potential in which a world of rapidly growing population is deeply interested. According to Hills (1960) the total production possible from any specific locality is limited by

**Table 7.** Wild Life Sanctuaries and National Parks in Andhra Pradesh

| Name | Area in Sq. Kms. | Location (District) |
|---|---|---|
| **WILD LIFE SANCTUARIES** | | |
| Kawal | 893 | Adilabad; 50 Kms from the Mancherial Railway Station. |
| Eturnagaram | 803 | Warangal; 80 Kms from Khazipet. |
| Pakhal | 860 | Warangal; 60 Kms from Warangal. |
| Pocharam | 130 | Medak and Nizambad; 100 Kms from Nizamabad |
| Kinnerasani | 635.41 | Khammam Paloncha division. |
| Papikonda | 591 | Khammam; East & West Godavari borders along the river Godavari. |
| Nagarjunasagar | 3568 | Guntur, Kurnool, Srisailam Mahabubnagar and Nalgonda; 100 Kms south to Hyderabad. |
| Pranahita | 136 | Adilabad; 75 Kms from Mancherial in Adilabad district. |
| Kolleru (Bird) | 673 | West Godavari; 60 Kms from Vijayawada. |
| Nelapattu (Bird) | 0.83 | Nellore situated in Sullurpet. |
| Pulicat (Bird) | 600 | Nellore |
| Lanjamadugu | 38.66 | Adilabad and Karimnagar; 50 Kms from Mancherial in the Adilabad district. |
| Manjeera | 20 | Medak; 20 Kms from the (Crocodile) Sanga Reddy dam to Manjeera barrage near Sanga Reddy. |
| Coringa, Kakinada | 235.70 | East Godavari |
| Krishna (Crocodile) | 194.84 | Krishna & Guntur |
| Gundla Brahmeswaram (Wild life) | 1194.00 | Kurnool & Prakasam; 30 Kms from Nandyal. |
| Kaundinya (Wild life) | 374.70 | Chittoor |
| Rollapadu (Great Indian Bustard) | 6.14 | Kurnool |
| Sri Lanka Malleswara (Bustard) | 464.42 | Cuddapah |
| Sri Venkateswara (Wild life) | 525.70 | Cuddapah & Chittoor |
| Krishna (Crocodiles) | 194.84 | Krishna & Guntur |
| **NATIONAL PARKS** | | |
| Sri Venkateswara National Park, Tirupati | 1353.00 | Chittoor |
| Kasu Brahmananda Reddy National Park | 142.50 | Hyderabad |
| Mrugavani National Park, Chilkur | 360.00 | Ranga Reddy |
| Mahaveer Harina Vanasthali National Park | 1459.00 | Hyderabad |

# PLATES

1. *Dillenia bracteata* Wight

2. *Poyalthia cereasoides* (Roxb.) Beddome

3. *Cochlospermum religiosum* (L.) Alston

4. *Homalium nepalense* Benth.

5. *Shorea robusta* Gaertn. f.

6. *Shorea roxburghii* G. Don

7. *Shorea tumbuggaia* Roxb.

**PLATE I**

8. *Sterculia urens* Roxb.

9. *Hildegardia populifolia* (Roxb.) Schott & Endl. (Full view)

10. *Hildegardia populifolia* (Roxb.) Schott & Endl. (Fruits)

11. *Elaeocarpus tectorius* (Lour.) Poir.

12. *Sloanea sterculiacea* (Benth.) Rehder

13. *Erythroxylum monogynum* Roxb.

**PLATE II**

15. *Atalanta monophylla* (L.) Correa

16. *Naringi alata* (Wall. ex. Wight & Arn.) Ellis

14. *Aegle marmelos* (L.) Correa

17. *Protium serratum* (Colebr.) Engler

18. *Soymida febrifuga* (Roxb.) A. Juss. (Flowering)

19. *Soymida febrifuga* (Roxb.) A. Juss. (Fruiting)

20. *Trichilia connaroides* (Wight & Arn.) Bentvelzen

**PLATE III**

21. *Walsura trifoliata* (A. Juss.) Harms

22. *Cholorxylon swietenia* DC.

23. *Lepisanthes tetraphylla* (Vahl) Radlk.

24. *Buchanania lanzan* Spreng.

25. *Nothopegia heyeana* (Hook.f.) Gamble

26. *Semecarpus anacardium* L.f.

27. *Butea monosperma* (Lam.) Taub.

**PLATE IV**

28. *Pterocarpus marsupium* Roxb.

29. *Cassia fistula* L.

32. *Anogeissus latifolia* (Roxb. ex. DC.) Guill & Perr.

31. *Albizia procera* (Roxb.) Benth.

30. *Parkinsonia aculeata* L.

33. *Terminalia alata* Heyne ex Roth (part of the trunk)

34. *Terminalia alata* Heyne ex Roth (Flowering)

**PLATE V**

35. Fruits of different species of Genus *Terminalia.* A. *Terminalia alata* Heyne ex Roth B. *Terminalia arjuna* (Roxb. ex. DC.) Wight & Arn. C. *Terminalia bellirica* (Gaertn.) Roxb. D. *Terminalia chebula* Retz. E. *Terminalia coriacea* (Roxb.) Wight & Arn. F. *Terminalia pallida* Brandis G. *Terminalia paniculata* Roth

36. *Careya arborea* Roxb.

37. *Memecylon lushingtonii* Gamble

38. *Alangium salvifolium* (L.f.) Wangerin

39. *Gardenia resinifera* Roth

40. *Hymenodictyon orixense* (Roxb.) Mabb.

**PLATE VI**

41. *Mitragyna parvifolia* (Roxb.) Korth.

42. *Morinda pubescens* J.E. Smith

43. *Madhuca indica* J. Gmelin

44. *Diospyros montana* Roxb.

45. *Diospyros ovalifolia* Wight

46. *Olea glandulifera* Wall. ex G. Don

47. *Holarrhena pubescense* (Buch. Ham.) Wall. ex. G. Don

48. *Wrightia tinctoria* R. Br.

**PLATE VII**

49. *Strychnos nux-vomica* L
(Trunk with flowering twig)

50. *Strychnos nux-vomica* L.
(Fruiting)

51. *Cordia dichotoma* Forst. f.

52. Fruits of different species of family Bignoniaceae. A. *Radermachera xylocarpa* (Roxb.) Schum. B. *Jacaranda mimosifolia* D. Don C. *Stereospermum suaveolens* (Roxb.) DC. D. *Oroxylum indicum* (L.) Vent. E. *Dolichandrone falcata* (Wall. ex. DC.) Seem. F. *D. arcuata* (Wight) Clarke G. *D. atrovirens* (Roth) Spargue

53. *Callicarpa tomentosa* (L.) Murr.

54. *Premna hamiltonii* J.L. Ellis

## PLATE VIII

55. *Gyrocarpus americanus* Jacq. (Fruiting)

56. *Gyrocarpus americanus* Jacq. (Flowering)

58. *Cleistanthus collinus* (Roxb.) Hook. f.

57. *Santalum album* L.

59. *Margaritaria indica* (Dalz.) Airy-shaw

60. *Mallotus philippensis* (Lam.) Muell-Arg.

61. *Sapium eugeniifolium* Buch.-Ham.

**PLATE IX**

62. *Sapium insigne* (Royle) Trimen
(Trunk with leaves)

63. *Phyllanthus emblica* L.

64. *Antidesma ghaesembilla* Gaertn.

66. *Ficus auriculata* Lour. (Full veiw)

65. *Holoptelia integrifolia* (Roxb.) Planchon.

**PLATE X**

67. *Ficus auriculata* Lour. (Figs.)

70. *Ficus hispida* L.

68. *Ficus benghalensis* L. (Thimmamma Marrimanu) (Full view). The World's largest tree, Gootibayalu Anantapur District, figured in Guinnese records

71. *Ficus semicordata* Buch.-Ham. ex Smith (Trunk with flowering twig)

69. *Ficus benghalensis* L. A close view.

**PLATE XI**

the combined potential energy of macroclimatic and land-form features of the area. The actual production will, however, depend upon the dynamic of the ecosystem that develops within the region.

Andhra Pradesh has been affected worst by human influences since a long time. Due to this climax formations are very much altered and or destroyed for agriculture and other similar purposes. The present vegetation has suffered indeed from affects of plants, animals, soil, climate and man.

Due to climatic, edaphic and altitudinal differences the forest exhibits considerable variation in floristic composition, physiognomy and life-forms. The types of forests recognised are

1. Tropical semi-evergreen forests,
2. Tropical moist deciduous forests,
3. Dry deciduous forests,
4. Northern mixed dry deciduous forests,
5. Dry savannah forests,
6. Tropical dry evergreen forests,
7. Tropical dry evergreen scrub.

However, the general nomenclature for such forest types are as given by Champion and Seth (1968).

**1. Tropical Semi-evergreen Forests** (Moist deciduous forests mixed with evergreen elements)

These are mostly confined to the lower parmanently moist valleys in the hills and also met with in small patches not far away from the sea. These are characterised by large number of dominant, mostly evergreen but some deciduous, heavy climbers and luxurient epiphytic growth. These forests are denser and the trees of heights ranging from 20–30 m and girth of 1–2 m and above are met with.

This type of forests are found in patches of Gudem, Dharakonda, along Sileru, Sapparla, Chintapalli, Galikonda, Minumuluru, Pedavalasa, Thanjavanam of Visakhapatnam district. Besides these Borra blocks of Vijayanagram district, Tekkali and Pathapatnam of Srikakulam, Nulakamaddi, Maredumilli areas of East Godavari, Bhadrachalam and in Papikonda and Bison hills of West Godavari district.

*Michelia champaca, Mangifera indica, Artocarpus lakucha, Dillenia pentagyna, Firmiana colorata, Bridelia tomentosa, Xylia xylocarpa, Schleichera oleosa, Schrebera swietenoides* etc. form the top canopy, whereas small trees like *Polyalthia cerasoides, Macaranga peltata, Pittosporum nepaulense, Phoebe wightii, Murraya koenigi, Homalium napaulense, Cinnamomum caudatum* etc. form the middle canopy.

## 2. Tropical Moist Deciduous Forests

These forests are extensive and are found in continuation with semi-evergreen type close to the moist valleys. The upper storey in these forests consists of predominantly deciduous species but lower storey has evergreen trees, the forests have deciduous look. These are found where the soil conditions are comparatively dry due to lower rainfall (1016 mm) and an altitude of above 1000 m and in most ravines. Huge lianas going over most of the trees, prevent the healthy growth of this forest. Most of the important timber trees are found in these forests.

Several areas in these forests have been completely denuded or turned into open grass lands due to extensive biotic factors and thus the whole area look like a disjointed mixture of deciduous forest and savannahs.

For the sake of convenience, the forest is divided into following 3 sub-groups.

### A. Northern tropical moist deciduous forest

This type occurs in Srikakulam district.

The top canopy is predominated by the deciduous species of *Haldinia cordifolia, Lannea coromandelica, Pterocarpus marsupium, Syzygium cumini, Lagerstroemia parviflora, Terminalia alata, Anogeissus latifolia, Xylia xylocarpa, Albizia odorotissima,* and *Madhuca longifolia*. Middle storey comprises of *Buchanania lanzan, Dillenia pentagyna, Diospyros melanoxylon, Mallotus philippensis, Careya arborea, Syzygium operculatum, Bridelia airy-shawii, Strychnos nux-vomica, Phyllanthus emblica, Schleichera oleosa, Gardenia latifolia, Diospyros sylvatica, Litsea chinensis, Garcinia xanthochymus* and *Suregada angustifolia.*

Under storey is dominated by *Helicteres isora, Pavetta tomentosa, Holarrhena pubescens, Cipadessa baccifera, Ardisia solanacea, Antidesma diandra, A. ghaesembilla, Ixora pavetta, Canthium dicoccum, Morinda pubescens, Ochna obtusata, Grewia rothii, Polyalthia suberosa, Kydia calycina* etc.

Many woody and herbaceous climbers like *Gnetum ula, Bauhinia vahlii, Entada pursaetha, Calycopteris floribunda, Combretum ovalifolium, Dioscorea tomentosa, Olax scandens, Butea superba, Celastrus paniculatus* etc. grow densely on trees.

### B. South Indian tropical moist deciduous forests

These forests are found in parts of Gudem, Rampa agency, parts of West Godavari, between Rollapenta and Bairluti, Gundlabrahmeswaram, in Nallamalais, Talakona in Chittoor district.

In Nallamalais *Tectona gradis* is the predominant element along with rare species of *Sterculia villosa, Terminalia pallida* and *Syzygium alternifolium.*

*Terminalia alata, Anogeissus latifolia, Dillenia pentagyna, Haldinia cordifolia, Mitragyna parvifolia, Dalbergia latifolia, Albizia odoratissima, Schleichera oleosa* etc. form the top canopy.

The lower storey is mostly formed by *Pterospermum xylocarpum, Kydia calycina, Semecarpus anacardium, Strychnos potatorum, Phyllanthus emblica, Cleistanthus collinus, Holarrhena pubescens, Polyalthia cerasoides, Grewia tiliifolia, Careya arborea* are present in lower canopy.

In ravines, *Mangifera indica, Suregada angustifolia, Cleistanthus patulus, Linociera intermedia* are commonly found. *Alphonsea sclerocarpa Diospyros malabarica, Chukrasia tabularis* are restricted to Seshachalam hill Gangas.

**C. Southern tropical moist deciduous rivarian forests**

Along the rivers and streams in the plains, where alluvial soil is deposited, there are many plants predominant exclusively to these areas. This riparian vegetation is maintained by the interaction of constant erosion and redeposition of the soil going on the banks of rivers. It generally forms a very narrow belt along the banks; sometimes it may extend to the higher elevations. Trees may be evergreen or deciduous depending upon the region. This type of forest is present along the banks of the river Godavari and other hill streams in a narrow belt. The most common trees in these forests are *Anogeissus acuminata, Barringtonia acutangula, Vitex altissima, Vitex leucoxylon, Ficus benjamina, Holoptelia integrifolia, Streblus asper, Ardisia solanacea, Debregeasia longifolia, Ficus hispida, Pongamia pinnata, Schleichera oleosa, Syzygium cumini, Trema orientalis, Mangifera indica, Carallia brachiata* etc.

### 3. Dry Deciduous Forests

This type of forest widely spread over the State occurs wherever the climatic and soil conditions are poor. Generally trees shed their leaves by about December and looks very open between February and May. The tree-layer cannot be distinguished into distinct storeys.

Dry deciduous Sal forests lie in the north-western part of the district of Srikakulam that too restricted to Donubayi. The altitude varies from 150–500 m with a rainfall of 1000 mm–1200 mm and the soils are deep ferruginous sandy loams to clayey loams with good sub-soil drainage. *Shorea robusta* is the interesting and predominant tree of this Sal forest, along with other predominant trees like *Terminalia alata, Anogeissus latifolia, Garuga pinnata, Diospyros montana, Dillenia pentagyna* and *Callicarpa arborea*.

Common large tree taxa of non Sal dry deciduous forests are *Albizia odoratissima, Anogeissus latifolia, Chloroxylon swietenia, Terminalia alata,*

*Haldinia cordiflia, Schleichera oleosa, Hardwickia binata, Lagerstroemia parviflora, Sterculia urens, Dalbergia lanceolaria* etc.

Middle sized trees are *Bauhinia racemosa, Bridelia airy-shawi, Buchanania lanzan, Boswellia serrata, Ehretia laevis, Cassia fistula, Careya arborea, Cleistanthus collinus, Diospyros chloroxylon, Phyllanthus emblica, Ixora arborea, Cochlospermum religiosum, Strychnos potatorum, Wrightia tinctoria, Soymida febrifuga, Acacia leucophloea, Diospyros melanoxylon, Pterospermum xylocarpum* and *Ziziphus xylopyrus.*

Small trees are *Ixora pavetta, Premna tomentosa, Pleurostylia opposita, Manilkara hexandra, Ochna obtusata, Drypetes sepiaria, Holarrhena pubescens* etc.

*Garuga pinnata* spread over in the northern circars and Telangana region, *Dillenia pentagyna* in northern agencies and Nallamalais. *Xylia xylocarpa* forms the dominant component in forests of northern circars.

In Rayalaseema, dry deciduous forests are found in all hilly regions very considerably in condition, composition, and density. There are four endemic trees of interest that appear in these forests. They are *Pterocarpus santalinus, Terminalia pallida, Syzygium alternifolium* and *Shorea tumbuggaia*. Of these *Pterocarpus santalinus* is of high commercial value. *Boswellia ovalifoliolata* is found only in Chittoor and Cuddapah districts.

*Tectona grandis* forms almost pure crop in Telangana region, besides this *Butea monosperma* is also commonest in deciduous forests of these areas. *Dolichandrone falcata, Ougeinia oojeinensis, Morinda pubescens, Soymida febrifuga, Chloroxylon swietenia* are commonly found.

*Dendrocalamus strictus* and *Bambusa arundinacea* are also common elements of deciduous forests. A good variety of climbers, twiners and straggling shrubs, both woody and herbaceous are seen in these forests. The common species among them are *Butea superba, Cansjera rheedi, Celastrus paniculatus, Aspidopterys cordata, Combretum ovalifolium, Paracalyx scariosus, Dioscoria* sps, *Ventilago madraspatana, V. denticulata, Pueraria tuberosa, Mucuna* sps., *Capparis zeylanica* etc.

**4. Northern Mixed Dry Deciduous Forests**

This type of vegetation is covering the northern segments of the hills of Chittoor district, the southern section of the Kurnool district, Nidigal and Cherukupalli of Nalgonda district. Upper canopy is uneven and not very dense, rather open, formed by a mixture of tall deciduous species. The forest has been subjected to bio-edaphic factors which have resulted in making them more open due to clear felling of several tree species as compared to the moist deciduous forest, which grows as restricted patches along the dry deciduous forest. However, in some areas selected species are predominant, forming more or less pure associations. Along with few semi-evergreen tree taxa deciduous species are also present. The leaf-fall, characteristic of the deciduous forest, takes place usually for a few months every year and the various species of trees occupy distinct storeys in the forest.

Common amongst top storey are *Anogeissus latifolia, Hardwickia binata, Albizia amara, A. odoratissima, Terminalia chebula, T. coriacea, T. paniculata, Shorea tumbuggaia, Syzygium alternifolium, Bauhinia racemosa, Dalbergia lanceolaria, Mangifera indica, Pterocarpus marsupium, Protium serratum, Hymenodictyon excelsum, Bombax ceiba* etc.

*Strychnos potatorum, Dalbergia paniculata, Vitex altissima, Dolichandrone atrovirens, Gardenia gummifera, Alangium salvifolium, Ziziphus xylopyrus,* with bamboo bushes form the middle layer.

Lower layer is dominated by *Helicteres isora, Chomelia asiatica, Grewia rothii, Holarrhena pubescens* etc.

### 5. Dry Savannah Forests

These forests, formed as a result of intense biotic interference, are scattered throughout the Eastern Ghats. The stunted trees are *Phyllanthus emblica, Phoenix humilis, Pterocarpus marsupium* and *Terminalia chebula.*

### 6. Tropical Dry Evergreen Forests

This type occurs in coastal and plain areas with a rainfall of 635 mm and below at an altitude of less than 244 metres and on impoverished soils with practically no organic matter and where the top soil is practically non-existent. This type of vegetation is encountered along the Coast of Ramananthpuram in Visakhapatnam district, Pool bagh, Velagada of Vijayanagaram district, Tekkali and Pathapatnam of Srikakulam district, southern parts in Cuddapah district, Sriharikot Island near Nellore district and Mamandur valley in Chittoor division.

Common elements of these forests are *Manilkara hexandra, Acacia chundra, A. leucophloea, Chloroxylon swietenia, Dalbergia paniculata, D. lanceolaria, Diospyros chloroxylon, Erythroxylum monogynum, Gyrocarpus americanus, Strychnos nux-vomica, S. potatorum, Wrightia tinctoria, Ziziphus xylopyrus, Alangium salvifolium, Anona squamosa, Ochna obtusata, Catunaregum spinosa, Atalantia monophylla, Cordia dichotoma, Sapindus emarginatus* etc.

### 7. Tropical Dry Evergreen Scrub Forest

Scrub is a biotic forest which is formed due to secondary succession. They have developed due to the biotic influence of man and domesticated animals on the natural vegetation. The scrub type of forest develops in areas of low rainfall, rocky areas and poor soils. The vegetation presents a very open appearance so that the trees and shrubs are widely spaced. Plants are mostly thorny and practically devoid of leaves in summer. Scrubs are distributed in vast areas of exposed rocky boulders in the drier parts of Andhra Pradesh like the districts of Anantapur, Kurnool, Cuddapah, Guntur, Nalgonda, Medak and peripheries of forests in other districts.

There is practically no tree in these scrubs, if present these are usually stunted. The common plant species are *Euphorbia antiquorum*, *Gymnosporia senegalensis, Acacia chundra, Albizia amara, Balanites aegyptica, Flacourtia indica, Ziziphus oenoplia, Z. mauritiana, Dichrostachys cinerea, Maytenus emarginatus, Rhus mysorensis, Naringi alata, Diospyros chloroxylon, Wrightia tinctoria, Cassia fistula* etc.

## COASTAL VEGETATION

Coastal Andhra Pradesh approximately occupies two-thirds of the Peninsular India along the east coast. The vast sandy strip of limited width intercepted by extensive deltaic regions built by the rivers Godavari, Krishna and their tributaries, the rocky promontories of limited size projecting into the sea at Waltair and Pudimadaka and the large saline and mud flat noticeable at Kakinada, Machilipatnam, in the Coringa eastuarine region and in the vicinity of Pulicat lake on way to Sriharikota island.

The coastal vegetation of Andhra Pradesh can be divided into two subgroups: (1) Strand vegetation, (2) Estuarine vegetation.

### 1. Strand Vegetation

This type is characteristic with open, mat forming pioneer plants followed by scattered herbs, shrubs and trees dispersed along back-shore region. This is further divided into sub-strata types.

#### Sand strand

This type of vegetation occurs along the sandy beaches and can be divided into 4 zones namely — open pioneer, closed herbaceous, bushy and inner woodland zones.

Open pioneer zone is the first zone in the Upper tidal region introducing the drift line. In this zone, the vegetation is rather thin with grasses and herbs.

Closed herbaceous zone is rather dense and with some mat forming herbaceous species.

Bushy zone is a blend of herbs with sub-shrubby plants resulting in a mixed vegetation.

Inner woodland zone is chiefly dominated by tree species *Borassus flabellifer, Prosopis cineraria, Cocos nucifera, Thespesia populnea, Pongamia pinnata,* and *Calophyllum inophyllum*. This zone gradually mingles with the wastelands or cultivated fields in the fringed regions of Coastal belt.

**Rock strand**

This is limited to small bits in the coastal line where the inland hillocks and their rocky promontories jut into the sea at Waltair and Poodimadaka. The vegetation and flora are mostly a mixture of coastal and inland plants, occurring in the following zones.

i. The wave cut rocky humps and slopes

This zone, subjected to regular sea water inundation is seen exposed only during low tide and supports a rich growth of marine algae.

ii. Rocky relief sloping inland having the mantle of sand in crevices, pot-holes and crannies.

In this type *Blepharis repens, Euphorbia thymifolia, Portulaca tuberosa, Crotalaria hebecarpa, Vernonia cinerea, Hybanthus enneaspermus* are common.

iii. Inland gravelly/rocky habitat

The vegetation resembles a screw type characteristic with spiny thickets and shrubby herbs. The dominant shrubby plants are *Carissa spinarum, Toddalia asiatica, Ziziphus oenoplia, Maytenus emarginata, Dichrostachys cinerea, Euphorbia tirucalli* etc.

## 2. Estuarine Vegetation

The mangrove vegetation which develops along muddy tidal banks, is primarily restricted to the Godavari and Krishna. Tidal creeks are formed from which water moves in and out of the drain as tidal flows, changing periodically in varying times in the morning and in the evening. The alluvial soil deposited is subjected to tidal inundation. The trees of mangrove vegetation are evergreen and moderate in height. Trees are specially adopted to survive on tidal mud which is permanently wet with salt water and submerged by every tide. Stilt roots, pneumatophores, vivipary and thick coriaceous leaves are the predominant and specialised features of the forest. These are found on the mud banks of delta streams, the tail of islands and over more extensive areas near the sea face.

The mangrove vegetation in the Godavari and the Krishna estuarine formations covers an area of about 200 Sq. Kms and shows a regression in the number of mangrove species diversity in comparision to the Mahanandi tidal forests in Orissa.

The species of mangroves can be divided into various zones. The pioneer species is *Avicennia alba* after which come *Rhizophora apiculata, R. mucronata, Bruguiera gymnorhiza, Sonneretia apetala, Ceriops decandra,* etc. *Aegiceras corniculata, Exoecaria agallocha, Hibiscus tiliaceus, Xylocarpus granatum* and *Lumnitzera racemosa* prefer shallower waters towards the landside.

Another interesting feature of Andhra coast is existence of natural vegetation in Sriharikota island and in certain sandy tracts of Irakkam island in the Pulicat lake. Not very far from the coast line in Sriharikota, in some undisturbed areas, the vegetation is composed of tree species of *Strychnos nux-vomica, Terminalia arjuna, Sapindus emarginatus, Albizia amara,* and *Tamarindus indica.*

Much of the mangrove forests have been over-exploited for fuel and firewood and large deltaic areas have been reclaimed for agriculture. Resultantly the mangrove forests have dwindled and are in a state of secondary stage.

## Sandy Sea-shores

*Casuarina equisetifolia* is extensively planted along the sea-coast and for some kilometers to the interior. *Prosopis cineraria* has been introduced in the area.

# PAST AND PRESENT WORK

## PAST WORK

Botanically, Andhra Pradesh was paid little attention towards the tree Flora and it remains under examined despite the floristic work that has been done so far. The authors have not concentrated on tree floristic diversity.

Earliar works on trees of India include Silviculture of Indian trees (Troup, 1921), Palms of British India and Ceylon (Blatter, 1926), Trees and Shrubs of Punjab (Ireland, 1934), British Tress (Boulten, 1949), Trees of Calcutta and its neighbourhood (Benthall, 1946), Trees of India (MacCann, 1947), Flowering Trees and Shrubs in India (Cowen, 1950), Manual of Indian Forest Botany (Bor 1953), Some Beautiful Indian Trees (Blatter and Millard, 1954).

William Roxburgh made a survey of the Coromandel coast from 1795 to 1820 and published a book "Plants of the Coromandel coast," containing 300 coloured plates of plants with vivid description and names in Latin, Telugu and English languages in 12 volumes which formed the basis to his subsequent work to his Flora Indica.

Robert Wight (1796–1872) was very much fascinated by the plants around him. He made extensive plant collection from Samalkot and Rajahumundry area of East Godavari district and they were later passed on to Sir Willam Hooker. Wight in his monumental work "Icones Plantarum Indiae Orientalis" (1838–1853) presented 2,101 plates of Illustrations in 6 volumes out of which 1780 species were reported in Gamble's "Flora of the Presidency of Madras". He described about 38 genera and more than 3000 species new to science.

Hugh Francis Clake Cleghorn (1820–1895) made extensive collections in Madras Presidensy. His collections enriched herbaria at Edinburgh and Calcutta. His herbarium specimens, along with those of Dr. Drew and Sir W.Elliot, bearing the year 1853 under the Madras Museum Herbarium labels appear to have formed the nucleus of present Madras Herbarium located at Coimbatore under the Botanical Survey of India , Southern Circle.

Col. Richard Henry Beddome (1830–1911) joined the Forest service of Madras in 1856. He extensively collected plants from different parts of Andhra Pradesh, namely Godavari (1953), Hyderabad (1855), Krishna forest (1880), Nallamalais , Palakonda hills, Simhachalam, Visakhapatnam (1873) and made significant contribution, such as the Flora Sylvatica for southern India (1869–1874) and Icones Plantarum Indiae Orientalis (1874). He was also responsible for the initiation of forest management in the state and devoted his life time for the collection and study of Indian plants. He made contributions of about 6000 specimens consisting mostly of forest trees and shrubs to the Madras Herbarium.

Hooker's classical work entitled "Flora of British India" was a seven volume work and the various parts appeared between 1872 and 1897. Hooker visited India in 1847–1851 and along with Thomson made extensive explorations of the Himalayas.

Elliot (1859) in his Flora Andhrica gave vernacular and botanical list of plants commonly met within the telugu speaking districts of the Northern circars. His account on the plants of Andhra of his time was mostly borrowed from Roxburgh's work.

Walker (1849) and Bradley (1849) listed agricultural , medicinal and other economically important plants of Dawalatabad and Warangal districts of Hyderabad state.

Barber (1860–1888) was appointed Director of the Botanical Survey of South India to prepare the Flora of the Presidency of Madras. He collected plants from Nellore, Prakasam, Krishna, East and West Godavari, Chittoor, Cuddapah, Kurnool and Guntur districts. He also wrote sections of the Flora and forest in the District Gazetteers of the North circars

Walker collected plants during 1898–1918 from Nellore district. In the beginning of 19th century, Bourne collected plants from Biccavole in Godavari, Kovur and Polavaram. Indian Trees (1906) published by Brandis contains descriptions of tree species and also includes shrubs.

Lushinghton, conservator of forests, Waltair published (1915) vernacular list of trees, shrubs and woody climbers of Madras Presidency. In this book he discussed the distribution of wild and cultivated plants of Madras Presidency with the local names in all the languages spoken in the Presidensy.

Patridge (1911) brought a book "Forest Flora of Hyderabad state", enumerating 450 species belonging to 69 families besides giving local names and their economic importance. Khan revised his work (1953) and added 177 wild and cultivated taxa and gave more information on the vegetation.

Being forest officer James Sykes Gamble and Cecil Ernest Claude Fischer collected plants of Madras Presidency. Gamble made extensive collections in East and West Godavari, Srikakulam, Visakhapatnam, Krishna, Anantapur, Chittoor, Kurnool and Cuddapah districts housed in K, CAL, MH and DD. After retirement from service as a Forest officer he went back to London where he wrote the "Flora of Presidency of Madras" (1915–1935). Besides his own collections he included also those of Wight, Bourne, Barber, Ramaswamy, Narayana swamy and Rangachari (1868–1934) in that book. The first seven parts were published before his death, then Fischer continued his work and completed the Flora (Ulmaceae to Graminae). The Flora of Presidency of Madras immortalised their names in the annals Botany of India and it was warmly received by botanical work as one of the best regional Flora of India and the system of descriptive keys which could be referred to as a sort of full description of species. Gamble also published Mannual of Indian Timbers (1881).

Uhl (1930), Ramaswamy (1920), Narayanaswamy *et al.* (1947), Venkateswarlu (1947) and Rolla S. Rao (1958, 1964) have enriched the knowledge of Flora of Andhra Pradesh.

Sayeduddin (1935, 1938, 1941 a,b) and Santapau are the other contributors of Flora of Telangana region of Andhra Pradesh. Venkateswarlu (1947) listed 26 taxa of estuarial Flora of Godavari.

Botanical Survey of India, after re-organisation in 1954, took up the project of rewriting the "Flora of India", to which interesting and rare plant specimens from different regions were added on to the Flora of Andhra Pradesh by Rolla S. Rao, Subba Rao, Balakrishnan, D.C.S. Raju, Ellis, Sastry, Narayanaswamy, Ramaswamy etc.

Wagh's (1960) doctoral thesis entitled "Studies on the Flora of Andhra State" was based on his own collections in some selected areas of coastal districts and Rayalaseema and earlier works reports like Gamble, Fischer, Beddome etc.

Balakrishnan and Henry (1961) described a new tree taxon *Boswellia ovalifoliolata* from Tirumala hills in Chittoor District. Sebastine *et al.* (1960) recorded 268 species of Angiosperms from Narsapur forest in Medak district. A book of Medicinal plants of Andhra Pradesh was prepared and published in regional language by Rolla S. Rao and Hemadri (1979). Subba Rao (1960) prepared a tiny dissertation on the forest types of Andhra Pradesh.

Rolla S. Rao (1958 and 1964) published two papers on observations on the vegetation of Rampa and Gudem Agency tracts, which include the collections made by Ramaswamy in 1914 and Narayanaswamy in 1920 and 1947 from Rampa Agency area of East Godavari district.

Thothathri (1964) studied plants of Nagarjunakonda and its surroundings. Ellis (1966) explored Balapalli range of Seshachalam reserve forest in Cuddapah district and reported 334 species belonging to 78 families. D.C.S. Raju, (1966a) reported 170 species belonging to 50 families from Papi hills in West Godavari district. He also studied the Flora of Polavaram Agency (DCS Raju 1966b). In "Excursion Flora of Simhachalam hills" (DCS Raju 1966c) he listed 230 taxa of vascular plants. In 1968 Raju DCS gave an account of Godavari delta of West Godavari district.

Common trees by Santapau (1966) gives more information of common road side and avenue tree species, which are also common in Andhra Pradesh. The Flora of Tirupati hills was prepared by Naidu and Prakash Rao (1967, 1969) and Naidu *et al.* (1971) of which 838 taxa were reported.

Venkateswarlu et al. (1972) published "Flora of Visakhapatnam" and enumerated 570 species belonging to 90 families occurring in Visakhapatnam city and its environs. Subba Rao (1977) published a paper on the Flora of Visakhapatnam district.

Rao and Sastry (1970) have contributed to the Flora of coastal Andhra Pradesh. They distinguished main types of coastal vegetation namely strand and esturine. In 1972 Rao and Shanware studied Coringa estuary of the Godavari.

Sudhakar (1979) made a detailed survey of upper East Godavari and enumerated 1074 species. Ramachandrachari (1980) studied the Flora of Achampet taluk in Mahabub Nagar district and reported 477 species. His collections included 29 new records to the forests of Telangana region.

Subba Rao and Kumari (1981) published a paper entitled "Some interesting plants common to Western Ghats and Eastern Ghats" of which, 33 tree species belonging to 26 genera and 17 families are from Andhra Pradesh and includes newly reported 5 tree taxa namely *Nothopegia racemosa* (Dalz.) Ramam. *Syzygium malabaricum* Bedd., *Ixora brachiata* Roxb. *Glochidion ellipticum* Wight and *Debregeasia longifolia* (Burm. f.) Wedd.

Narayana Rao *et al.* (1981) surveyed Tirumala hills and reported 1500 species under 176 families of which 1480 taxa are Angiosperms. They published a booklet, "Plant Wealth of Tirumala".

P.N. Rao *et al.* (1981) brought out a useful review of the floristic research work in Andhra Pradesh which summarized the contributions of older works of Roxburgh (1795–1819), Elliot (1859), Beddome (1869–74), Hooker f. (1872–1897), Gamble (1884–85), Gamble and Fischer (1915–1935), Bradley (1849), Walker (1849) and Patridge (1911) as well as the recent works on Flora of Andhra Pradesh.

V.S. Raju (1983) listed 10 additions to the Flora of Andhra Pradesh. Ellis made several intensive collections from Nallamalis and enumerated 743 species under 419 genera belonging to 109 families in 2 volumes of Flora of Nallamalis (Ellis 1987–1990).

Subbarayudu (1989) surveyed Flora of Palakonda hills of Cuddapah district and reported 339 species under 254 genera pertaining to 75 families. Banerjee *et al.* (1989) brought out Mangroves in India and in it he reported about 59 species falling under 41 genera pertaining to 17 families of which 26 species under 21 genera belonging to 17 families are distributed in Andhra Pradesh alone.

Venkateswara Reddy (1990) explored the Flora of Guvvalacheruvu RF in Cuddapah district and identified 393 taxa. Reddy *et al.* (1991) in their book Forest Flora of Andhra Pradesh listed 2141 wild and cultivated taxa spread over 211 families, occurring in forests of Andhra Pradesh.

During the period of 1979–95, intensive exploration were done by botanists in the state. This resulted in the publication of a number of district

Floras, such as Flora of Srikakulam district by Rolla S. Rao and Hara Sreeramulu (1986), Flora of West Godavari district by Rolla S. Rao *et al.*, (1986) reported 209 tree taxa, Flora of Anantapur district by Pullaiah and Yesoda (1989) reported 129 tree taxa, Flora of Adilabad district by Pullaiah *et al.* (1992) reported 147 tree taxa, Flora of Kurnool by Raju and Pullaiah (1995) reported 211 tree taxa, Flora of Nizamabad by Pullaiah and Ravi Prasad Rao (1995) reported 122 tree taxa, Krishna district by Lakshmi Narayana *et al.* (1997) reported 144 tree taxa and Flora of Medak district by Pullaiah *et al.* (1997) reported 101 tree taxa.

Some Floras of districts have been in the form of Doctoral theses; these are Flora of Nalgonda district by Raghava Swamy (1986) reported 103 tree taxa; Systematic studies on the Flora of Prakasam district by Krishna Mohan (1987) reported 160 tree taxa, Flora of Cuddapah district by A.M. Rao (1989), Flora of Mahabub Nagar district by Raghava Rao (1989), Flora of Guntur district by Ramakrishnaiah (1990) reported 145 tree taxa. Contributions to the Dicotyledonous Flora of Visakhapatnam by Subba Rao (1991), Flora of Chittoor district by Rangacharyulu (1991) reported 251 treea taxa and Flora of Nellore district by Srinivasa Rao (1991).

Pullaiah (1997), Pullaiah and Alimoulali (1997), Pullaiah and Chennaiah (1997), and Pullaiah and Surya Prakash Babu (1998) brought out Flora of Andhra Pradesh in 4 volumes and enumerated 2502 species belonging to 1014 genera and 172 families. The largest families are Leguminosae with 300 species and Poaceae with 279 species.

Botanical Survey of India published 3 volumes of Flora of India from Ranunculaceae to Erythroxylaceae in 1993. Volumes 11 and 12 were later brought out in 1996 covering the family Asteraceae.

Kumar and Manoranjan Bhanja (1992) of Andhra Pradesh forest department published (1993) forestry seed Manual of Andhra Pradesh. This book provides information about ecology, phenology, morphology of fruit and seed and germination details of 100 tree taxa.

Thammanna *et al.* (1994) reported 1110 taxa of wild and cultivated in Angiospermic Wealth of Tirumala. Raju and Pullaiah (1994) described a new species *Dimophocalyx kurnoolensis* from oak forest in Kurnool district.

Rama Rao and Henry (1996) enumerated 501 species of ethnobotanical interest and published in a book 'The Ethnobotany of Eastern Ghats in Andhra Pradesh'.

Suryanarayana and Sreenivasa Rao reported (1996) *Dolichandrone spathacea* (L.f) Schum. as a new record to Andhra Pradesh collected from Sriharikota Island.

Though a good number of taxonomists have worked on floristic studies of different areas in Andhra Pradesh, a comprehensive data on Flora Sylvatica for Andhra Pradesh is not available.

## PRESENT WORK

The Present research work is formulated on the basis of an analytical study of the authors' own collections from the different parts of districts of the State of Andhra Pradesh, the herbarium specimens and the literature obtainable from various Universities of Andhra Pradesh and also the institutions of Botanical Survey of India.

The main Herbaria consulted are:

1. Central National Herbarium, Howrah (CAL),
2. Blatter Herbarium, St. Xavier's College, Bombay (BLAT),
3. Madras Herbarium, BSI Southern Circle, TNAU Campus, Coimbatore (MH),
4. Forest Research Institute Herbarium, Dehradun (DD),
5. Herbarium of BSI Western Circle, Pune, (BSI),
6. Sri Krishnadevaraya University Herbarium, Department of Botany, Anantapur (SKU),
7. Sri Venkateswara University Herbarium, Department of Botany, Tirupati (SVU), and
8. Visvodaya College, Department of Botany, Venkatagiri, Nellore district (VV).

## FIELD STUDY

Intensive field works have been carried out in different places in different seasons, covering almost all the important vegetational areas and all districts of Andhra Pradesh. Field works were carried out in such a way that they cover almost all seasons of a year to procure different developmental stages of plants, viz., vegetative, flowering and fruiting, collections were made from a wide range of habitats i.e., valleys, hill slopes, hill-tops, plain forests, and non-forest areas of the State. Every plant was collected in quadruplicate and every attempt was made to study the habit, habitat, flower colour and aroma, associated plants, abundance, general distribution and phenology, local names, economic importance of the individual species. Information on medicinal uses of some plants were collected from local people. Field numbers were given for every specimen collected and all the above information was recorded carefully in the field itself. Photographs of important plants like endemic, rare or any economically useful and general vegetation types were also taken during field work.

### Herbarium Preparation

The collected specimens were poisoned, pressed, dried and stitched on herbarium sheets according to the methodology described by Santapau (1955) and Jain and Rao (1977). The specimens were collected in well-tied thick polythene bags. The collected specimens were first poisoned with saturated mercuric chloride containing ethyl alcohol. Later they were kept in between the blotting papers and subjected to proper pressing with the help of Iron plant-presser. The blotting papers were changed for every 24 hours and the specimens were spread and placed in between the fresh blotters in the presser. The procedure is continued until the specimens dry up completely. Finally, they were mounted on the standard herbarium sheets measuring about 28 × 42 cm with glue and were stitched. The labels with detailed information were affixed on the right hand corner at the base of the sheets.

Final identification of the plants was made after carefully dissecting the floral parts using stereo-microscope and compound microscope. In the case of dried specimens, plant parts were boiled for sometime and worked out. For details, Regional floras like Gamble (1957 repr. ed.), Matthew (1983), monographs and other relevant literature were referred constantly for final identification of plants. Various parts of the dissected floral parts were affixed on a small adhesive paper board and placed it in a small pocket which is affixed on the left hand bottom side of the concerned herbarium sheet. Illustrations were drawn after critical study. The identifications of the specimens was confirmed by comparing the specimens with the authentic specimens available at Botanical Institutions.

All the mounted herbarium sheets were classified Genera-wise and were kept in species-folders after their identification which were housed in the Herbarium at Department of Botany, Sri Krishnadevaraya University, Anantapur for future use and reference.

## PLAN OF SYSTEMATIC ENUMERATION

The present Flora sylvatica for Andhra Pradesh deals with wild and naturalised tree taxa of *Angiosperms* encountered in the State.

Under systematic treatment, the families are arranged according to Bentham and Hooker's system of classification (1862–1883) with certain exceptions to accommodate recent changes in the arrangement of families.

At the beginning a comprehensive key to the families of tree flora of Andhra Pradesh is given and also keys have been provided for all the genera and species in each family. Distinct keys, based on vegetative, flowering and fruiting characters, have been prepared for all the species of concerned genera to facilitate easy identification of tree taxa at any season of the year in the field or in the laboratory. Keys are purely dichotomous indented type, based mostly on macroscopic characters. The genera under a family and the species of a genus, are arranged in alphabetical order.

In the treatment of individual species, the following details are listed.

## 1. Citation

The valid name of the plant is given first followed by its author name and place of publication. The valid name is followed by basionym if necessary and or synonyms mostly to connect with the correct name of the taxon for the names published in Flora of British India, Flora of the Presidency of Madras (Gamble), Indian trees by Brandis, Flora of India (1993, 1, 2, 3 Vols.) and Flora Sylvatica for Southern India (Beddome). References to standard illustrations, monographs and manuals are also cited wherever necessary. For Gamble the page number in parentheses represent the 1957 edition whereas the page number outside the parenthesis represent the original edition (1915–1935).

### Description

The terminology followed is based on Lawrence (1951) and Radford *et al.* (1974). The usual sequence followed is (a) deciduous or evergreen, height; crown nature; bark nature, wood characters; branchlets nature, (b) Leaves texture, shape, length × breadth, apex, margin, base; venation nature; petiole length, (c) Flower colour, position, type and length of inflorescence, (d) Fruit colour, shape, size, texture, nature or if any specific characters; seed number, size, shape or any other important characters. All the measurements are given in metric system.

Ecology, followed by vegetative propagation in few cases, phenology (leaf-fall, new foliage, flowering and fruiting), Vernacular names, English names (some plants) are also given.

### Uses

Most of the economic and medicinal uses of plants were taken from literature.

### Distributional Pattern in the State of Andhra Pradesh

Locality, district, the parentheses (abbreviated) in which the plant collected followed by collector's name (abbreviated) with field number have been recognised.

### India and World Distribution

Distributional pattern of species in different states of India and different countries of world is given.

## ABBREVIATIONS USED IN GENERAL TEXT AND ENUMERATION

| | |
|---|---|
| *auct. non* | : *auctorum non*; not of authors |
| *auct. pl.* | : *auctorum plurimorum;* of most authors |
| BLAT | : Blatter Herbarium, St. Xavier's College, Mumbai. |
| CAL | : Central National Herbarium, Howrah, Calcutta. |
| ca. | : Circa, about |
| cm | : Centimetre |
| DD | : Forest Research Institute, Herbarium, Dehradun. |
| E | : East |
| ed. | : Editor, edition |
| *emend.* | : *emendavit*; emended |
| *et al.*, | : *et alia* (and others of people) |
| etc. | : et cetera; and other things. |
| FBI | : Flora of British India<br>Vol. 1. 1872–1875;<br>Vol. 2. 1876–1879;<br>Vol. 3. 1880–1882;<br>Vol. 4. 1883–1885;<br>Vol. 5. 1886–1890;<br>Vol. 6. 1890–1894;<br>Vol. 7. 1897. |
| Fig. | : Figure |
| Fls. | : Flowering |
| Frs. | : Fruiting |
| Gamble | : Gamble, J.S. and C.E.C. Fischer,<br>Flora of the Presidency of Madras,<br>3 Vols. (1915–1936) in 10 parts.<br>Part I–VII by Gamble and<br>VIII–X by Fischer.<br>Vol. 1 includes<br>Families Rananculaceae–Caprifoliaceae;<br>Vol. 2 includes<br>Rubiaceae–Euphorbiaceae and<br>Vol. 3 includes<br>Ulmacea–Graminae. |
| m | : Metre(s) of height of the tree. |
| MH | : Madras Herbarium, BSI<br>Southern Circle, TNAU Campus<br>Coimbatore. |
| *ibid.* | : *ibedem*; the same, in the same place. |
| Ic | : Icone; illustration |

| | |
|---|---|
| ICBN | : International code of Botanical Nomenclature. |
| incl. | : *inclosus*; including |
| *loc. cit.* | : *loco citato*, at the place cited |
| nom. | : *nomen ambiguum*; doubtful name |
| *nom. illegit.* | : *nomen illegitimum*; illegitimate name |
| *nom. nud.* | : *nomen nudum*; name published without description and without reference to previous published description. |
| *nom. superfl.* | : a name superfluous when it was published; in most cases it is a name based on the same type as an other earlier specific name. |
| op. cit. | : *opere citato*; in the work cited. |
| *p.p.* | : *pro parte*, in part |
| RF | : Reserve Forest |
| *sensu lato* | : in the broad sense |
| sp. | : Species |
| spp. | : Species (plural) |
| ssp. | : Subspecies |
| suppl. | : Supplement |
| SVU | : Sri Venkateswara University, Tirupati. |
| syn. | : Synonymum; synonym |
| t. | : *tabula*; plate |
| var. | : *variaetas* (Variety) |
| vern. | : vernacular name, local name. |
| viz. | : videlicit; namely |
| vol.(s) | : Volume(s) |
| VV | : Visvodaya College Herbarium, Department of Botany, Visvodaya Govt. College. Venkatagiri, Nellore District, AP. |

## ABBREVIATIONS — PLANT EXPLORERS

| | |
|---|---|
| BR | : B. Raviprasad Rao |
| BSMD | : B.S.M. Dutt |
| CAB | : C.A. Barber |
| CBDM | : Colonel Beddome |
| CECF | : C.E.C. Fischer |
| CP | : C. Prabhakar |
| CS | : C. Subbarayudu |
| DAM | : D. Ali Moulali |
| DCSR | : D.C.S. Raju |
| DR | : D. Rangachari |
| EC | : E. Chennaiah |

EKK : E.K. Krishnan
ESR : E. Sreenivasa Rao
GNR : G. Narasimha Rao
GVS : G.V. Subba Rao
HS : H. Santapau
JLE : J.L. Ellis
JSG : J.S. Gamble
KCJB : K.C. Jacob
KH : K. Hanumanthappa
KMS : K.M. Sebastine
KRK : K. Ravikumar
KS : K. Subramanyam
KSM : K. Sri Rama Murthy
MCB : M. Chandra Bose
MRRP : M.R. Rajendra Prasad
MSNS : M.S. Narayanaswamy
NPB : N.P. Balakrishnan
NRR : N. Rama Rao
NY : N. Yesoda
PRS : P. Rajasekhara Reddy
PSPB : P. Surya Prakash Babu
PVK : P. Venkanna
RVK : R. Vijayakumar
R.V. Reddy : R. Venkateswara Reddy.
RVR : R.R. Venkata Raju
SKW : S.K. Wagh
**SSR : S. Sandhya Rani**
STR : S.T. Ramachandrachari
TAR : T.A. Rao
TDCK : T. Dharmachandra Kumar
TP : T. Pullaiah
TRS : T. Ravishankar
VNS : V. Narayanaswamy
VRK : V. Ramakrishnaiah
VSR : Vatsavaya S. Raju.

## ABBREVIATIONS DISTRICTS

ADB : Adilabad
ATP : Anantapur
CDP : Cuddapah
CTR : Chittoor
EG : East Godavari

| | |
|---|---|
| GTR | : Guntur |
| HYD | : Hyderabad |
| KMM | : Khammam |
| KNR | : Karimnagar |
| KNL | : Kurnool |
| KSN | : Krishna |
| MBNR | : Mahabubnagar |
| MDK | : Medak |
| NLG | : Nalgonda |
| NLR | : Nellore |
| NZB | : Nizamabad |
| PKSM | : Prakasam |
| RR | : Ranga Reddy |
| SKLM | : Srikakulam |
| VSKP | : Visakhapatnam |
| VZN | : Vizayanagaram |
| WG | : West Godavari |
| WGL | : Warangal |

# FLORISTIC ANALYSIS

Intensive floristic studies during the present work yielded a total number of 531 tree species (including cultivated and ornamental taxa) belonging to 274 genera spread over 76 families. The details of representative number of taxa, genera and families under different groups are given below.

| GROUP | Families | | Genera | | Species | |
|---|---|---|---|---|---|---|
| | Total | Percentage | Total | Percentage | Total | Percentage |
| Dicotyledons | | | | | | |
| Polypetalae | 50 | 65.78 | 157 | 57.50 | 291<br>(66) | 54.82<br>(70.22) |
| Gamopetalae | 13 | 17.12 | 67 | 24.52 | 132<br>(18) | 24.85<br>(19.14) |
| Monochlamydae | 11 | 14.47 | 43 | 15.63 | 101<br>(8) | 19.02<br>(8.51) |
| Monocotyledons | 2 | 2.63 | 7 | 2.45 | 7 | 1.31<br>(2.12) |
| TOTAL | **76** | **100.00** | **274** | **100.00** | **531**<br>**(102)** | **100.00**<br>**(100.00)** |

Number of species in parentheses of all the tables, represent the cultivated or ornamental tree taxa.

It is evident from the data that the ratio of species belonging to monocotyledons and dicotyledons is 73:1 of species.

Eleven families of tree Angiosperms as given on page 46 represent 14 or more species in the state of Andhra Pradesh and their order of dominance is given along with the number of genera.

As analysis of the well represented/dominant genera in the district as given below, indicates that as many as 13 genera have 6 or more species.

| S. No. | Familes | No. of genera | No. species |
|---|---|---|---|
| 1. | EUPHORBIACEAE | 21 (2) | 46 (2) |
| 2. | MIMOSACEAE | 11 (6) | 32 (8) |
| 3. | RUBIACEAE | 19 (1) | 32 (1) |
| 4. | MORACEAE | 4 (2) | 25 (3) |
| 5. | CAESALPINIACEAE | 10 (4) | 24 (12) |
| 6. | RUTACEAE | 12 (2) | 21 (6) |
| 7. | FABACEAE | 10 (5) | 20 (6) |
| 8. | BIGNONIACEAE | 12 (9) | 19 (10) |
| 9. | VERBENACEAE | 7 (1) | 17 (1) |
| 10. | MELIACEAE | 12 (1) | 14 (2) |
| 11. | MYRTACEAE | 5 (3) | 14 (6) |

## DOMINANT GENERA REPRESENTING 6 OR MORE SPECIES IN THE PRESENT FLORA

| S. No. | Genus | No. of species |
|---|---|---|
| 1. | *Ficus* | 20 (2) |
| 2. | *Acacia* | 15 (3) |
| 3. | *Diospyros* | 12 |
| 4. | *Grewia* | 11 |
| 5. | *Terminalia* | 9 (1) |
| 6. | *Cassia* | 9 (7) |
| 7. | *Cordia* | 8 (1) |
| 8. | *Wendlandia* | 7 |
| 9. | *Vitex* | 6 |
| 10. | *Bridelia* | 6 |
| 11. | *Dalbergia* | 6 (2) |
| 12. | *Bauhinia* | 6 (1) |
| 13. | *Albizia* | 6 |

Thirtyfive families are represented by single genus and 21 families are represented by single species.

Although the Flora of Andhra Pradesh (Pullaiah, 1997; Pullaiah and Alimoulali, 1997; Pullaiah and Chennaiah, 1997; Pullaiah and Surya Prakash Babu, 1998) has been published, few of wild tree taxa have not been recorded. These are:

*Amoora canarana* (Turz.) Hiern.
*Bischofia javanica* Bl.
*Crateva adansonii* DC. ssp. *odora* (Buch.-Ham.) Jacobs
*Diospyros assamilis* Bedd.
*D. cordifolia* Roxb.

*Dolichandrone spathacea* (L.f.) K. Schum.
*Isonandra villosa* Wt.
*Macaranga indica* Wt.
*Mallotus resinosus* (Blanco) Merr.
*Maytenus bailadillana* (Narayan & Mooney) Raju and Babu
*Nothopegia racemosa* Dalz.
*Olea dioica* Roxb.
*Premna serratifolia* L.
*Prunus jenkinsii* Hook. f.
*Salix tetrasperma* Roxb. (Buch.-Ham.) M. Jacobs.
*Wendlandia bicuspidata* Wt. & Arn.
*Wendlandia heynei* (Roem. & Schult.) Santapau and Merchant.

## ENDEMIC TREE TAXA

### a. Endemic Tree Taxa Restricted to Andhra Pradesh

*Boswellia ovalifoliolata* Balakr. & Henry.
Chittoor and Cuddapah districts.
*Croton scabiosus* Bedd.
Cuddapah district.
*Dimorphocalyx kurnoolensis* Venkata Raju & Pullaiah
Kurnool district.
*Premna hamiltonii* (Buch.-Ham.) Ellis
Northern circars, Nallamalais, Rampă hills of Godavari.
*Pterocarpus santalinus* L.f.
Cuddapah, Chittoor, Nellore districts.
*Syzygium alternifolium* (Wt.) Walp.
Chittoor, Cuddapah, and Gundlabrahmeswaram of Kurnool district.
*Terminalia pallida* Brandis
Chittoor, Cuddapah and Kurnool districts.

### b. Endemic Tree Taxa of Eastern Ghats Occurring in Andhra Pradesh

*Actinodaphne madraspatana* Bedd.
Cuddapah, Nellore, Chittoor districts of Andhra Pradesh; Tamil Nadu.
*Alphonsea madraspatana* Bedd.
Visakhapatnam and Cuddapah districts of Andhra Pradesh:
Tamil Nadu and Orissa.
*Albizia thompsonii* Brandis
Anantapur, Mahabubnagar and Cuddapah districts of
Andhra Pradesh; Orissa and Tamil Nadu.
*Cordia domestica* Roth

Cuddapah district of Andhra Pradesh, Tamil Nadu.
*Croton evolutior* (C.B. Clarke) Gamble
Kunool and Krishna districts of Andhra Pradesh and Tamil Nadu.
*Eriolaena lushingtonii* Dunn.
Kurnool district of Andhra Pradesh; Tamil Nadu.
*Hildegardia populifolia* (Roxb.) Schott & Endl.
Anantapur and Cuddapah districts of Andhra Pradesh;
Dharmapuri of Tamil Nadu.
*Maytenus bailadillana* (Narayan & Mooney) Raju & Babu
Visakhapatnam of Andhra Pradesh; Orissa and Madhya Pradesh.
*Wendlandia angustifolia* Wt. ex Hook. f.
Cuddapah district of Andhra Pradesh, Tamil Nadu.
*Wendlandia gamblei* Cowan
East Godavari and Visakhapatnam districts of Andhra Pradesh.

### c. Endemic Plants of Peninsular India Occurring in Andhra Pradesh

*Acacia campbelli* Arn.
Godavari to Nellore, Kurnool, Cuddapah in AndhraPradesh; Deccan and Carnatic.
*Ficus dalhousiae* Miq.
Cuddapah, Nellore and Chittoor districts of Andhra Pradesh; Western Ghats, Nilgiris.
*Isonandra villosa* Wt.
Veligonda hills (Nellore district) of Andhra Pradesh.
*Neolitsea foliosa* (Nees) Gamble var. *caesia* Gamble
Northern circars, Visakhapatnam, Godavari districts, Seshachalam hills of Andhra Pradesh; Arcot district of Tamil Nadu; hills of Karnataka, Coimbatore, Travancore of Western Ghats.
*Polyalthia cerasoides* (Roxb.) Beddome.
Peninsular India extending up to 1200 m. Kanara,
Travancore, Annamalais (Coimbatore).
*Shorea roxburghii* G. Don
Chittoor, Cuddapah districts rare in Anantapur district of
Andhra Pradesh; Karnataka, Kerala and Tamil Nadu.

### d. Threatened Tree taxa

*Acacia campbelli* Arn.
*Aegle marmelos* (L.) Cor.
*Albizia thompsoni* Brandis
*Alphonsea madraspatana* Bedd.
*A. sclerocarpa* Thw.

*Aphanamixis polystachya* (Wall.) Parker
*Antidesma bunius* Spreng.
*A. menasu* Miq.
*Artocarpus lakucha* Buch.-Ham.
*Bridelia pubescens* Kurz.
*Carallia brachiata* (Lour.) Merr.
*Chukrasia tabularis* Adr. Juss.
*Chysophyllum roxburghii* Don
*Dillenia aurea* Sm.
*Diospyros candolleana* Wt.
*D. cordifolia* Roxb.
*D. ebenum* J. Koenig
*Dolichandrone spathacea* (L.f.) K. Schum.
*Elaeocarpus lanceifolia* Roxb.
*E. lucidus* Roxb.
*Epiprinus mallotiformis* (Muell.-Arg.) Croiz.
*Eriolaena lushingtonii* Dunn
*Erythrina fusca* Lour.
*Euodia lunu-ankenda* (Gaertn.) Merr.
*Ficus dalhousiae* Miq.
*Flacourtia jangomas* (Lour.) Raeusch
*Carcinia xanthochymus* Hook. f. ex T. And.
*Glochidion ellipticum* Wt.
*Heterophragma quadriloculare* (Roxb.) K. Schum.
*Hildegardia populifolia* (Roxb.) Schott & Endl.
*Hymenodictyon obovatum* Wall.
*Isonandra villosa* Wt.
*Litsea laeta* (Nees) Hook. f.
*Madhuca longifolia* (J. Koenig.) Macbr.
*Mallotus resinosus* (Blanco.) Merr.
*Margaritaria indica* (Dalz.) Airy Shaw
*Memecylon lushingtonii* Gamble
*Moringa concanensis* Nimmu ex Gibs.
*Nothopegia racemosa* Dalz.
*Oroxylum indicum* (L.) Vent.
*Pamburus missionis* (Wt.) Swingle
*Persea macrantha* (Nees) Kosterm.
*Premna serratifolia* L.
*Santalum album* L.
*Saraca asoca* (Roxb.) de Wilde
*Schrebera swietenoides* Roxb.
*Scyphiphora hydrophyllacea* Gaertn.
*Shorea robusta* Gaertn.

*S. tumbuggaia* Roxb.
*Symplocos cochinchinensis* (Lour.) S. Moore
*Syzygium calophyllifolium* Walp.
*S. heyneana* (Duthie) Wall. ex Gamble
*S. samarangense* (Bl.) Merr. & Perr.
*Wendlandia angustifolia* Wt. ex Hook. f.

## OVER EXPLOITED TREE TAXA

*Anogeissus latifolia* (Roxb. ex DC.) Wall. ex Guill. & Perr.
*Cochlospermum religiosum* (L.) Alston
*Chloroxylon swietenia* DC.
*Chukrasia tabularis* Adr. Juss.
*Dalbergia latifolia* Roxb.
*Diospyros melanoxylon* Roxb.
*Dolichandrone atrovirens* (Roth) Sprague
*Givotia moluccana* (L.) Sreem.
*Gmelina arborea* Roxb.
*Haldinia cordifolia* (Roxb.) Ridsd.
*Pterocarpus marsupium* Roxb
*P. santalinus* L.
*Santalum album* L.
*Shorea robusta* Roxb. ex Gaertn.
*Soymida febrifuga* (Roxb.) Adr. Juss.
*Sterculia urens* Roxb.
*Tectona grandis* L.f.
*Terminalia paniculata* Roth
*T. coriacea* (Roxb.) Wt. & Arn.

## TIMBER AND MINOR FOREST PRODUCE YIELDING TREE TAXA

### Timber Yielding Trees

*Tectona grandis, Shorea robusta, S. tumbuggaia, S.roxburghii, Dalbergia latifolia, Anogeissus latifolia, Santalum album, Alangium salvifolium, Chloroxylon swietenia, Pterocarpus marsupium, Bombax ceiba, Acacia nilotica, Soymida febrifuga, Boswellia serrata, Erythrina suberosa, Garuga pinnata, Chukrasia tabularis, Albizia odoratissima, Terminalia paniculata* etc.

## Fibre and Floss Yielding Trees

*Hardwickia binata, Cordia dichotoma, Bauhinia* spp, *Sterculia villosa, Helicteres isora, Grewia tiliifolia, Butea monosperma, Trema orientalis, Laportea crenulata, Hibiscus tiliaceus, Caryota urens* (Leaves), *Borassus flabellifer, Bombax ceiba, Ceiba pentandra, Cochlospermum religiosum* etc.

## Fuel Wood Yielding Trees

*Casuarina equisetifolia, Tamarindus indica, Mangifera indica, Acacia auriculiformis, A. catechu, Haldinia cordifolia, Chloroxylon swietenia, Diospyros* spp., *Shorea robusta, Tectona grandis, Terminalia paniculata, Xylia xylocarpa, Kydia calycina, Acacia nilotica, Albizia lebbeck, Dalbergia sissoo, Syzygium cumini, Terminalia arjuna, T. coriacea, Ziziphus mauritiana, Tamarix aphylla, Ailanthus excelsa, Boswellia serrata, Bombax ceiba, Buchanania lanzan, Garuga pinnata, Kydia calyçina, Madhuca longifolia* etc.

## Fodder Yielding Trees

*Acacia catechu, A. nilotica, Albizia lebbeck, Anogeissus latifolia, Azadirachta indica, Bombax ceiba, Butea monosperma, Cochlospermum religiosum, Dalbergia sissoo, Ficus benghalensis, F. religiosa, Gmelina arborea, Haldinia cordifolia, Hibiscus tiliaceus, Kydia calycina, Lagerstroemia parviflora, Leucaena leucocephala, Melia azedarach, Miliusa tomentosa, Michelia champaca, Polyalthia longifolia, Prosopis chilensis, Shorea robusta, Terminalia arjuna, T. bellirica, T. coriacea, Trema orientalis, Ziziphus mauritiana* etc.

## Oil Yielding Trees

*Madhuca* spp., *Shorea robusta, Azadirachta indica, Pongamia pinnata, Jatropha curcas, Santalum album, Schleichera oleosa, Garcinia indica, Mimusops elengi, Anacardium occidentale, Sapindus emarginatus, Atalantia monophylla, Givotia moluccana, Acacia farnesiana, Boswellia serrata* etc.

## Edible Fruits and Leaves from Trees

Leaves in *Leea indica, Antidesma acidum, Erythroxylum monogynum Bauhinia malabarica, Moringa oleifera.* Flowers in *Madhuca indica*. Fruits and nuts in *Anacardium occidentale, Annona squamosa, Syzygium alternifolium, S. cumini, Mangifera indica, Artocarpus heterophyllus, Diospyros melanoxylon, Ziziphus mauritiana, Moringa oleifera, Tamarindus indica, Limonia acidissima, Ziziphus xylopyrus, Buchanania lanzan, B. axillaris, Ehretia laevis, Ficus* spp, *Grewia* spp., *Phoenix sylvestris, Spondias pinnata, Aegle marmelos, Alangium salvifolium, Dillenia indica, D. pentagyna, Anthocephalus chinensis, Antidesma acidum, Cordia dichotoma, Phyllanthus emblica, Borassus flabellifer, Xylia xylocarpa, Careya arborea* etc.

## Arrack and Toddy Yielding Trees

*Phoenix sylvestris, Borassus flabellifer, Caryota urens, Madhuca indica, Acacia leucophloea, A. nilotica* etc.

## Dye Yielding Trees

(Bark) *Erythoxylum monogynum, Mimusops elengi, Lannea coromandelica, Casu ırina equisetifolia, Terminalia coriacea*; (wood) *Pterocarpus santalinus, Acacia catechu, Artocarpus lakucha, A. heterophyllus, Adenanthera pavonia*; (Root) *Morinda pubescense, Punica granatum*; (flower) *Butea monosperma, Michelia champaca, Toona ciliata*, (leaf) *Terminalia catappa, Wrightia tinctoria, Commiphora berryi;* (fruits) *Mallotus philippensis, Thespesia populnea, Semecarpus anacardium, Nothopegia heyneana* etc.

## Tannins

(Bark) *Rhizophora mucronata, Phyllanthus emblica, Lannea coramandelica, Syzygium cumini, Lagerstroemia parviflora, Careya arborea, Xylia xylocarpa, Anacardium occidentale, Acacia nilotica, Shorea robusta, Terminalia arjuna, T. chebula, T. bellirica, T. pallida, Acacia nilotica, Oroxylum indicum, Diospyros malabarica, Caesalpinia coriaria, Anogeissus latifolia, Cassia fistula, Phyllanthus emblica* etc.

## Gums and Resin Yielding Trees

*Acacia nilotica, Macaranga peltata, Lannea coromandelica, Azardirachta indica, Acacia catechu, Anogeissus latifolia, Aegle marmelos, Anacardium occidentale, Limonia acidissima, Pterocarpus marsupium. Sterculia urens, Bauhinia semla, Cochlospermum religiosium, Garuga pinnata, Gardenia gummifera, Jatropha curcas, Shorea roxburghii, S. tumbuggaia, S. robusta, Boswellia serrata* etc.

### Avenue Trees

Some tree species are planted along roadsides for shade and such tree species commonly found in the state are *Albizia labbeck, Azadirachta indica, Bauhinia purpurea, Cassia javanica, C. roxburghii, C. siamea, Dalbergia sissoo, Delonix regia, D. elata, Erythrina variegata, Gliricidia sepium, Millingtonia hortensis, Peltophorum pterocarpum, Tamarindus indica, Samanea saman, Ficus benghalensis, Pongamia pinnata, Eucalyptus* spp. etc.

## INTERESTING WILD TREE TAXA AS AN ORNAMENTAL VALUE

*Anthocephalus chinensis, Bauhinia variegata, Butea monosperma, Callicarpa arborea, C. tomentosa, Cochlospermum religiosum, Careya arborea, Dillenia bracteata, D. indica, Diospyros malabarica, Euphorbia antiquorum, Ficus auriculata, Firmiana colorata, Gardenia latifolia, Gmelina arborea, Helicteres isora, Holarrhena pubescens, Mallotus philippensis, Memecylon umbellatum, Naringi crenulata, Ochna obtusata, Saraca asoca, Shorea roxburghii, Tamilnadia uliginosa, Wrightia tinctoria* etc.

# ARTIFICIAL KEY TO FAMILIES

1. Root system characterized by a tap root. Leaves usually reticulately veined. Flowers typically 4–5 merous rarely 3–merous. Embryo usually with 2–cotyledons ............... ***Dicotyledons***
  2. Perianth biseriate, usually as calyx and corolla:
    3. Inner whorl of perianth or petals free:
      4. Carpels free or cohering at the apex:
        5. Stipules present ............................................................... ***Magnoliaceae***
        5. Stipules absent:
          6. Leaves small–medium sized, to 20 cm long, margin entire, secondary nerves not parallel; petiole not sheathy. Flowers 3–merous. Fruit without calyx ....................................................................... ***Annonaceae***
          6. Leaves large, to 70 cm long, margin serrate or dentate, secondary nerves parallel and prominent; petiole sheathy. Flowers 5–merous. Fruit with accrescent calyx ........................................................ ***Dilleniaceae***
      4. Carpels united:
        7. Torus usually expanded in a disc:
          8. Placentation parietal. Fruit an elongated capsule. — Seeds winged ............................................................................... ***Moringaceae***
          8. Placentation and fruit not as above:
            9. Leaves covered with glands ........................................ ***Rutaceae***
            9. Leaves not covered with glands:
              10. Flowers with staminal tube:
                11. Stamens usually 8–10. Leaves usually oblique. Inflorescence terminal or axillary ............................................***Meliaceae***
                11. Stamens 5. Leaves equilateral. Inflorescence leaf–opposed. ....................................................................... ***Leeaceae***
              10. Flowers without staminal tube:
                12. Plants armed:
                  13. Stamens 5, enclosed by cucculate petals. Ovary 5-celled. Leaves unifoliate ...................................... ***Rhamnaceae***
                  13. Stamens 10, not enclosed by lanceolate petals. Ovary 2–4-celled. Leaves bifoliate ............................. ***Balanitaceae***
                12. Plants unarmed:
                  14. Stamens numerous. Ovary deeply lobed; styles connate.– Calyx persistent ............................................ ***Ochnaceae***
                  14. Stamens 1–10.Ovary not deeply lobed; style not connate:
                    15. Leaves simple:
                      16. Fruit samaroid. Leaves glaucous beneath ....... ............................................................ ***Aceraceae***
                      16. Fruit not samaroid. Leaves green beneath:
                        17. Carpels enclosed by the enlarged disc which becomes a false fruit.– Flowers 1–7 in each axil .......................................... ***Siphonodontaceae***

17. Carpels free from disc or not enclosed by disc and not form a false fruit:
    18. Leaves 5-angular or deeply lobed. Inflorescence leaf–opposed. – Small trees .................... ***Vitaceae***
    18. Leaves entire. Inflorescence axillary or terminal:
        19. Fruit berry. Flowers in axillary clusters or umbels .................... ***Hippocrateaceae***
        19. Fruit drupe. Flowers in cymes or panicles:
            20. All stamens fertile:
                21. Drupe on a fleshy hypocarp. Stamens 10. Panicle dioecious (flowers unisexual, sometimes bisexual, rarely polygamous) .................... ***Anacardiaceae*** *p.p.*
                21. Drupe not on a fleshy hypocarp. Stamens 5. Cymes bisexual ..... ***Celastraceae***
            20. All stamens not fertile:
                22. Sepals and petals persistent in drupe .................... ***Sabiaceae*** *p.p.*
                22. Sepals and petals not persistent:
                    23. Stamens inserted at the margin or base of the disc, petals regular, without claw and any other scales.................... ***Anacardiaceae*** *p.p.*
                    23. Stamens arise from the inside of the disc or behind it, petals irregular, clawed, inside often with basal scales, if petals absent panicles slender and pendulous. .................... ***Sapindaceae*** *p.p.*

15. Leaves compound:
    24. Seeds winged .................... ***Flindersiaceae***
    24. Seeds not winged:
        25. Fruit berry with acute angle. Flowers dark purple.–cultivated .......... ***Averrhoaceae***
        25. Fruit drupe. Flowers usually cream or whitish:
            26. Flowers bisexual, sepals and petals persistent .................... ***Sabiaceae*** *p.p.*
            26. Flowers polygamo–dioecious, sepals and petals deciduous:
                27. Bark greenish, smooth. Fruit with 2–5 pyrenes .................... ***Burseraceae***
                27. Bark greyish or dark, rough. Fruit without pyrenes:
                    28. Drupe lobed. Plants neither acrid nor resinous ................***Sapindaceae*** *p.p.*
                    28. Drupe not lobed. Plants acrid and resinous ................***Anacardiaceae*** *p.p.*

7. Torus not expanded into a disc:
    29. Ovary superior:
        30. Fruit a legume. Leaf base pulvinous:
            31. Flowers very small, rarely medium sized, characteristically are in spikes or heads; corolla valvate, actinomorphic; stamens definite to indefinite ........ ***Mimosaceae***
            31. Flowers small–large, in cymes, corymbs, umbells, and panicles; corolla imbricate, zygomorphic; stamens definite:
                32. Corolla caesalpinaceous, ascendingly imbricate in bud; stamens free. Flowers mostly yellow .................... ***Caesalpiniaceae***
                32. Corolla papilionaceous, descendingly imbricate in bud; stamens mono or diadelphous. Flowers red, purple or white .................... ***Fabaceae***
        30. Fruit not a legume. Leaf base not pulvinous:
            33. Mangroves. Leaves reduced to scales .................... ***Tamaricaceae***
            33. Non-mangroves. Leaves well-developed:
                34. Trees with sap or latex:
                    35. Fruit prickled, seeds covered with orange coloured dye. Anthers horse-shoe shaped .................... ***Bixaceae***
                    35. Fruit neither prickled nor with any dye; Anthers not horse-shoe shaped.
                        36. Leaves entire. Ovary with 1–3 carpels, ovule 1, placentation axile to basal – latex yellow. Stamens connate into central column or 4–5 bundles .................... ***Clusiaceae***
                        36. Leaves palmi–lobed. Ovary with 5 carpels, ovules numerous on parietal or axile to parietal placentation:
                            37. Latex red or orange. Leaves spreading. Stamens numerous, free from petals. Fruit 5-valved capsule. Seeds woolly........ .................... ***Cochlospermaceae***

37. Latex milky. Leaves terminally clustered as a crown. Stamens 10; inserted on petals. Fruit fleshy berry. Seeds not woolly ........................................................ ***Caricaceae***

34. Trees without sap or latex:

38. Ovules on parietal placentation, if ovules on axile placentae, stamens are free and numerous.

39. Petals cruciform, flowers 4-merous, andro or gynophore or two are distinctly present ............................................................................ ***Capparaceae***

39. Petals not cruciform, flowers 4–5–merous, without distinct gynophore or androphore:

40. Stamens 5, petals reflexed, 6 mm long. Seeds with resinous pulp ..... ............................................................................................***Pittosporaceae***

40. Stamens 6–numerous, petals 0, if present 2 mm long. Seeds without resinous pulp ............................................................................ ***Flacourtiaceae***

38. Ovules on axile or pendulous placentation:

41. Calyx persistent in fruit:

42. Fruit winged, (wings formed by enlarged calyx lobes) Connectives of anthers produced beyond anther lobe. Trees resinous ..................... ***Dipterocarpaceae***

42. Fruit not winged. Anther without appendages. Trees not resinous:

43. Petals clawed; stamens numerous, free, inserted at the base of calyx tube. Fruit capsular; seeds winged. Leaves opposite .......................***Lythraceae***

43. Petals ligulate or base with 2–5 lobes, stamens few, connate at the base around the ovary, not on a calyx tube. Fruit berry; seeds not winged. Leaves alternate ........................................................... ***Erythroxylaceae***

41. Calyx not persistent in fruit:

44. Filaments of anthers free, rarely slightly connate at the base in a ring or in 3–5 groups

45. Anthers opening by pores or valves; petals not clawed and glandular but ciliate or lacinate if entire fruit berry; stamens arising from inside the disc. Fruit drupe or 4–5-valved, echinate capsule ........... ***Elaeocarpaceae***

45. Anthers opening by slits; petals clawed with a gland inside, not cilitate; stamens borne on torus. Fruit capsular or separating into cocci ........ .................................................................................... ***Tiliaceae***

44. Filaments of anthers united into a column or cup:

46. Anthers 2-celled.–Fruit capsular or with 5 spreading or spirally twisted follicles. Stamens 10-numerous ........................................... ***Sterculiaceae***

46. Anthers 1-celled.–Stamens numerous:

47. Stamens polyadelphous; anthers smooth. Pericarp of fruit woolly. Leaves digitately compound ........................................ ***Bombacaceae***

47. Stamens monadelphous, anthers rough. Pericarp of fruit not woolly. Leaves simple ................................................................. ***Malvaceae***

29. Ovary semi-inferior or inferior:

48. Stamens numerous:

49. Anthers didymous. Carpels free or connate ................................................ ***Rosaceae***

49. Not as above:

50. Calyx lobes valvate:

51. Mangroves. Ovary free from calyx. Petals often absent. Carpel cells not superposed ........................................................................ ***Sonneratiaceae***

51. Terrestrials. Calyx adnate to ovary. Petals present and crimpled. Carpel cells superposed in two series ........................................................... ***Punicaceae***

50. Calyx lobes imbricate:

52. Secondary nerves close and form a distinct intramarginal vein. Stamens inflexed in bud. Flowers small to medium-sized ........................... ***Myrtaceae***

52. Secondary nerves not close and without intramarginal vein. Stamens erect in bud.Flowers large, if small inflorescence pendulous................***Lecythidaceae***

48. Stamens few up to 12:
53. Ovules 6–12 on axile or central placentae; petals 4, blue; connectives of anthers spurred at base ........ *Melastomataceae*
53. Ovules 1 or 2 up to 6, pendulous; petals 5–7, white to yellowish, never blue; connectives not spurred:
54. Petals absent; stamens 2–seriate. Flowers in distinctly simple or branched spikes or heads. Drupe winged or ribbed when dry ........ *Combretaceae*
54. Petals present; stamens not 2–seriate. Flowers in cymes or condensed – capitate or umbellate. Fruit not as above.
55. Leaves alternate. Ovary inferior. Fruit berry, crowned by calyx lobes ........ *Alangiaceae*
55. Leaves decussate. Ovary semi-inferior. Fruit drupaceous, not crowned by calyx lobes ........ *Rhizophoraceae*

3. Inner whorl of perianth or petals united:
56. Ovary inferior:
57. Intra or Interpetiolar stipules present. Flowers in heads, cymes, corymbs or panicles. Stamens as many as corolla lobes. Leaves opposite ........ *Rubiaceae*
57. Stipules absent. Flowers in spikes. Stamens numerous. Leaves alternate ... ........ *Symplocaceae*
56. Ovary superior:
58. Leaves opposite:
59. Flowers actinomorphic; all stamens equal:
60. Stamens 2 ........ *Oleaceae*
60. Stamens as many as corolla lobes:
61. Ovary sub-apocarpous with numerous ovules on marginal or parietal placentation. Corolla coronary. Fruit follicle, if drupe flowers yellow ........ *Apocynaceae*
61. Ovary syncarpous, few ovules on axile or basal placentation. Corolla without corona. Fruit berry or drupe:
62. Flowers 4–merous:
63. Flowers cream coloured in panicles. Drupe 1-ovuled. Branchlets glabrous ........ *Salvadoraceae*
63. Flowers purple in umbellate cymes. Drupe 4-ovuled. Branchlets fulvous tomentose........ *Verbenaceae (Callicarpa)*
62. Flowers 5–8-merous:
64. Fruit berry. Leaves 3–5-nerved from base. Cymes short, corolla lobes reflexed; ovary 2-celled ........ *Loganiaceae*
64. Fruit drupe. Leaves penninerved. Cymose panicles large, corolla lobes erect; ovary 4-celled ........ *Verbenaceae (Tectona)*
59. Flowers zygomorphic; stamens didynamous:
65. Leaves compound. Stamens 5, 1 reduced. Stigma spoon shaped. Fruit capsular; seeds winged ........ *Bignoniaceae*
65. Leaves simple. Stamens 4, without reduced stamen. Stigma entire to fid. Fruit drupaceous or baccate; seed not winged ........ *Verbenaceae*
58. Leaves alternate or subopposite:
66. Flowers unisexual.–Fruit with persistent calyx ........ *Ebenaceae*
66. Flowers bisexual:
67. Stamens opposite to the petals:
68. Ovary 1-celled, ovules on free central placentation. Stamens as many as corolla lobes. Plants without latex ........ *Myrsinaceae*
68. Ovary 4–10-celled, ovules on axile placentation. Stamens as many as or multiples of corolla lobes. Plants laticiferous ... ........ *Sapotaceae*
67. Stamens alternate to the petals:

69. Carpels many-ovuled. Corolla lobes erect. Fruit many–seeded berry ........................................................................................ ***Solanaceae***
69. Carpels 1-ovuled. Corolla lobes reflexed. Fruit 2–4-seeded drupaceous nutlets ........................................................................................ ***Cordiaceae***

2. Perianth 1-whorled, sometimes absent:
70. Stamens 6–20 or more, usually in 4–series; staminodes glandular ............. ***Lauraceae***
70. Stamens not more than 6, glandular.
71. Fruit samaroid. Flowers polygamous – Leaves lobed .................... ***Hernandiaceae***
71. Fruit not samaroid. Flowers not as above:
72. Flowers bisexual:
73. Leaves dissected, ovary superior ............................................ ***Proteaceae***
73. Leaves entire, ovary semi-inferior ....................................... ***Santalaceae***
72. Flowers unisexual:
74. Ovary 3-celled:
75. Leaves digitately 3-foliolate ......................................... ***Bischofiaceae***
75. Leaves simple or pinnate ............................................ ***Euphorbiaceae***
74. Ovary 1-celled:
76. Flowers zygomorphic ........................................................ ***Salicaceae***
76. Flowers actinomorphic:
77. Ovary 2-ovuled:
78. Leaves reduced to scales. Sepals absent in female flowers ... ........................................................................ ***Casuarinaceae***
78. Leaves well developed. Sepals present in both flowers. ............. ........................................................................ ***Stilaginaceae***
77. Ovary 1-ovuled:
79. Plants with latex ....................................................... ***Moraceae***
79. Plants without latex:
80. Filaments inflexed in bud with reversed anthers ...... ........................................................................ ***Urticaceae***
80. Filaments not inflexed ...................................... ***Ulmaceae***

1. Root system fibrous. Leaves usually with parallel veins. Flowers 3–merous or with perianth in multiple of threes. Embryo with one cotyledon ................................... ***Monocotyledons***
81. Leaves large, pinnately or palmately divided. Flowers in spadix. Stem caudex ........................................................................................ ***Arecaceae***
81. Leaves small, linear, entire. Flowers in spikelets. Stem culm .............................. ***Poaceae***

# SYSTEMATIC ENUMERATION

## DILLENIACEAE

## **DILLENIA** Linnaeus

### KEY 1 (Vegetative)

1. Trees evergreen. — branchlets tomentose:
   2. Leaves pale yellowish pubescent beneath, broadly elliptic-oblong, obovate, or ovate-oblong, brownish-green when dry ........ ***D. bracteata***
   2. Leaves strigose beneath, oblong-lanceolate, green when dry ........ ***D. indica***
1. Trees deciduous:
   3. Branchlets grey or tawny silky; leaves pubescent beneath, base unequal ........ ***D. aurea***
   3. Branchlets glabrous; leaves glabrous on both surfaces, base equal ........ ***D. pentagyna***

### KEY 2 (Flowering)

1. Flowers white appearing with the leaves:
   2. Flowers solitary, 10–15 cm across ........ ***D. indica***
   2. Flowers few in racemes, 2.5–6 cm across ........ ***D. bracteata***
1. Flowers yellow, on leafless branches:
   3. Flowers 3–7 in umbellate fascicles on woody shoots, 2–3 cm across ........ ***D. pentagyna***
   3. Flowers solitary, axillary, 10–12.5 cm across ........ ***D. aurea***

### KEY 3 (Fruiting)

1. Fruiting carpels 5, rarely 6:
   2. Fruit 3 cm across ........ ***D. bracteata***
   2. Fruit 1.5 cm across ........ ***D. pentagyna***
1. Fruiting carpels >7:
   3. Fruit greenish-yellow, 8– 10 cm dia, carpels 14–20 ........ ***D. indica***
   3. Fruit yellow, 3–5 cm in dia., carpels 8–12 ........ ***D. aurea***

**Dillenia aurea** Sm. in Exot. Bot. 2: 65. t. 92 & 93. 1806; FBI 1: 37; Subba Rao & Kumari in Bull. Bot. Surv. India 12: 208. 1970; Brandis 4; Flora of India 1: 153. 1993. *D. pulcherrima* Kurz. in J. Asiat. Soc. Beng. 40: 46. 1871.

Deciduous, spreading tree, up to 12 m tall; bark light coloured, smooth; wood reddish-brown; young branchlets grey or tawny pubescent or silky. Leaves coriaceous, obovate or broadly oblong, 30–50 × 10–18 cm, silky above, pubescent beneath in young, apex rounded or obtuse, margin spinulose-denticulate, base often unequal, channeled and sheathing to 6 cm long petiole; secondary nerves 20–30 pairs, close. Flowers golden yellow, 10–12.5 cm across, solitary. Fruit yellowish-orange, subglobose, to 3–5 cm across; seeds dark brown, numerous, obovoid, glossy, surrounded by viscous edible pulp.

Rare, in hilly forests of Visakhapatnam district. Leaf fall: Jan.-Feb. New foliage: May. Fls.: Mar.–Apr. Frs.: May–June.

Wood moderately hard and heavy; not useful, only used as fuel wood, and makes a good charcoal. Fruits edible and is greedily eaten by wild elephants which destroy the trees to obtain them.

Vangasara forest (VSKP), *SSR* 19655; Chintapalli (VSKP), *GVS* 30072 (MH & CAL).

INDIA : Uttar Pradesh, West Bengal, Bihar, Orissa, Madhya Pradesh and Andaman & Nicobar Islands.

WORLD : Myanmar and N.E. Thailand, Nepal to Bhutan.

**Dillenia bracteata** Wight Icon. 2: 4. t. 358. 1840; FBI 1: 37, Gamble 1: 7(5); Brandis 3; Flora of India 1: 154. 1993.

Evergreen tree, up to 12 m tall; bark greyish, rough; branchlets grey-silky tomentose. Leaves coriaceous, broadly elliptic-oblong, obovate, ovate-oblong, 13–35 × 5–16 cm, glabrous above, pubescent beneath, brownish-green when dry, apex obtuse or emarginate or bluntly acuminate, margin obscurely serrate or crenate, base acute or cuneate; secondary nerves 16–21 pairs; petiole to 1.6 cm long. Flowers white, 2.5–6 cm across in few-flowered racemes. Fruit sub-globose, carpels 5 rarely 6, 3 cm across; seed ovoid, arillate.

Rare, near streams of mixed deciduous forests of Chittoor district. Fls.: Apr.-May. Frs.: July-Aug.

Kambakkam (CTR), *SSR* 16152 & *DR* 2365 (SVU).

INDIA : Tamil Nadu and Kerala — Endemic.

**Dillenia indica** L. Sp. Pl. 535. 1753; FBI 1: 36; Gamble 1: 6(5); Brandis 3; Flora of India 1: 155. 1993. *D. speciosa* Thunb. in Trans. Linn. Soc. London 1: 200. 1791.

Tall; evergreen tree, up to 20 m tall; crown dense, round; bark smooth, peeling off in small, thin, hard scales; wood pale to reddish brown, somewhat twisted grained and rather coarse textured; branchlets shiny and silky. Leaves coriaceous, oblong-lanceolate, 25–50 × 5–12 cm, glabrous above, strigose beneath, dark green, apex acute to acuminate, margin crenate-serrate, each serrature with apicule, base attenuate and decurrent; secondary nerves 30–40 pairs; petiole hairy, 2.5–5 cm long, slightly winged. Flowers white, fragrant, solitary, 10–15 cm across, terminal, pendulous. Fruit greenish-yellow with large fleshy accrescent calyces (as an outer covering), globular, indehiscent, 8–10 cm in dia.; seeds brownish, reniform, compressed, margin hairy, 5 × 3.5 mm, embedded in a glutinous pulp.

Occasional, in valleys and near streams of moist and evergreen forests. It is a shade bearer and moderately fast growing tree. It thrives best in dampy regions and deep fertile soils. Vern: Kalinga, Pedda Kalinga, Revadi chettu, Uvva, Uppu ponna. Eng.: Elephant apple.

Bark and leaves are astringent. Raw accrescent calyces are eaten after cooking. Fruit juice with sugar is used as a cooling beverage. The wood is not much used, except for making planks, rafters, tool handles and gunstocks, also it makes good firewood and charcoal. Dry leaves are substitute for sand paper in polishing ivory and horn.

Tirumala hills (CTR), *SSR* 19682; Anantagiri (VSKP), *NPBK* 11009 (MH) and *GVS* 22679 (MH); Rampa hills (Godavari), *MSR* 1544 (CAL).

INDIA : Uttar Pradesh, West Bengal, Assam, Meghalaya, Orissa, Madhya Pradesh, Maharashtra, Karnataka, Tamil Nadu and Kerala.

WORLD : Sri Lanka, Bangladesh, Myanmar, Thailand, Cambodia, Vietnam to Malaysia.

**Dillenia pentagyna** Roxb. Pl. Cor. t. 20. 1795; FBI 1: 38; Gamble 1: 6(5); Brandis 4; Flora of India 1: 156. 1993; Beddome Fl. Sylv. S. India t. 104. 1870.

Deciduous tree, up to 20 m tall; trunk straight, cylindrical, crown rounded; bark grey or pale brown with shallow depressions; wood reddish-grey. Leaves coriaceous, oblong-lanceolate, 15–70 × 7–25 cm, smooth and shining, glabrous, apex obtuse, margin crenate, base cuneate; secondary nerves 30–50 pairs, often forking towards the margins; petiole to 6 cm . Flowers yellow, 2–3 cm across, fragrant, 3–7 in fascicles on woody shoots. Fruit yellowish-orange, sub-globose, 1.5 cm across, drooping, indehiscent; carpels 5, fleshy; seeds 1 or 2 in each carpel, ovoid, 3 × 2 mm, black, exarillate.

Occasional, in moist deciduous forests of Eastern Ghats at 500–1000 m. It is light demander, sensitive to frost and coppices well. Leaf fall: Feb.–May, New foliage: June; Fls.: Feb.–Apr. Frs.: May–June. Vern: Revadi chettu, Chinna Kalinga, Parudu, Uvva Kaya, Ratheggi.

Flower buds and fleshy sepals of fruits are edible. Wood durable but liable to wrap and split and difficult to work, suitable for match-splits, making, and it gives an excellent charcoal. Bark fibre is used for cordage. The large leaves serve as a substitute for roof thatch.

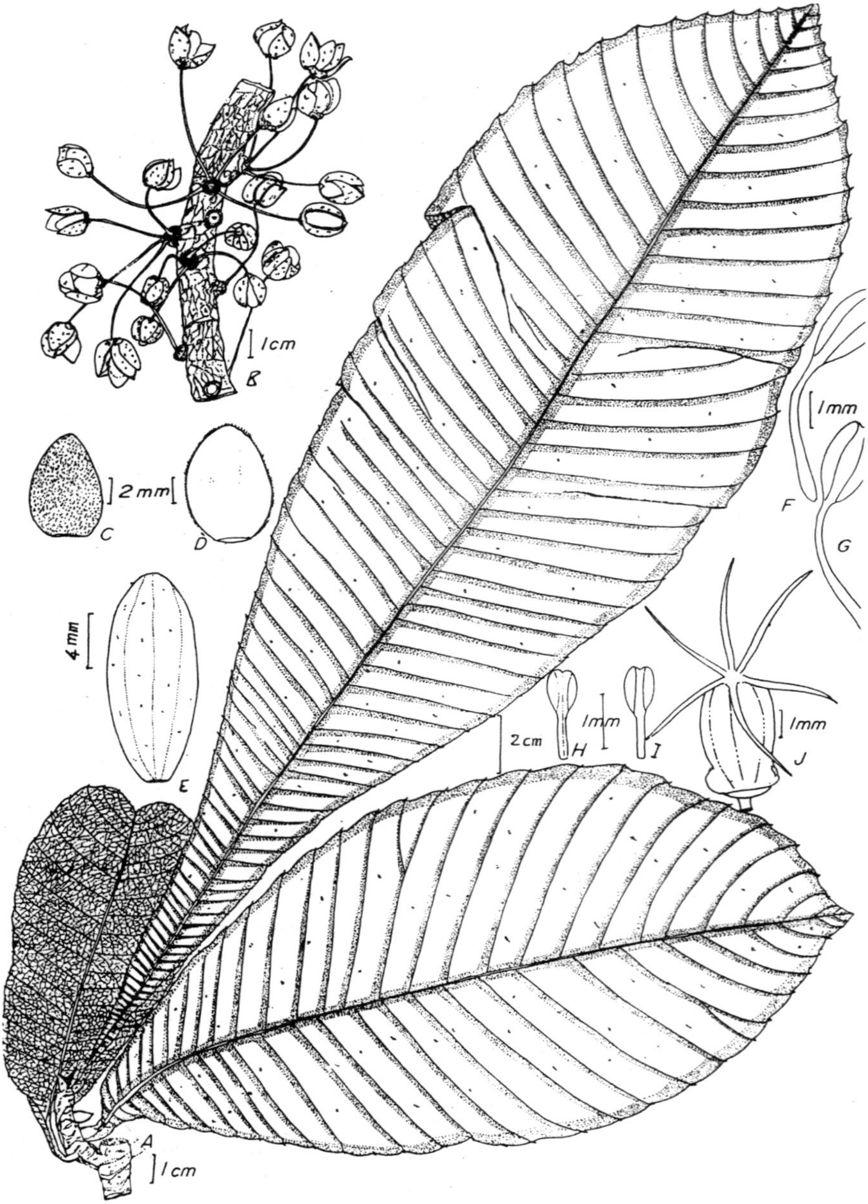


**Fig. 5.** *Dillenia pentagyna* Roxb.
A. Twig, B. Flowering branch, C. & D. Sepals, E. Petal, F & G. Curved stamens, H & I. Erect stamens J. Pistil.

Minumuluru (VSKP), *SSR* 15951; Talakona (CTR), *SSR* 18596; Godavari, *CBDM* 212 (MH); Maredumilli (EG), *GVS* 24262 (MH); Rakthachandanam thota (SKLM), *GVS* 62364; Papikonda, Polavaram Agency (WG), *DCSR* 494 (MH & CAL); Ramannapenta (KNL), *JLE* 23815 (MH).

INDIA : Tropical and sub-tropical regions up to 900 m in Eastern Himalayas, North-Eastern regions, Gangetic plains, Deccan and Western Ghats, Uttar Pradesh, Bihar, West Bengal, Sikkim, Assam, Meghalaya, Orissa, Madhya Pradesh, Maharashtra, Goa, Karnataka, Kerala and Andaman & Nicobar Islands.

WORLD : Myanmar, China, Vietnam, Thailand, Java, Celebes and lesser Sunda Islands.

## MAGNOLIACEAE
### **MICHELIA** Linnaeus

**Michelia champaca** L. Sp. Pl. 536. 1753; FBI 1: 42; Gamble 1: 9(6); Brandis 8; Flora of India 1: 175. 1993.

Densely foliaceous, evergreen tree, up to 20 m tall; bark grey, thick, smooth; wood light olive-brown or yellowish, even-grained; branches ascending, spreading, forms a close crown. Leaves thinly coriaceous, ovate-lanceolate, 10–20 × 4–8 cm, glabrescent and shining above, minutely puberulous beneath, apex long acuminate, margin wavy, base sub-acute; secondary nerves 10–18 pairs, petiole 1–8 cm long. Flowers pale yellow, becoming orange, fragrant, solitary, 5 cm across, axillary. Fruit dark brown, an aggregate of follicles, laxly arranged, woody, ovoid, white speckled; seeds numerous, brown, subglobose, 8–10 × 5–7 mm, angular, enclosed by scarlet red aril.

Planted in cool valleys, near and around temples and gardens for the sake of flowers. It occurs on deep, moist, well drained, good quality soil. It is sensitive to frost. Fls.: Peak during Mar.–May, less during rest of the year Frs.: July. Vern: Champakamu, Sampangi, Sampenga, Gandhaphali, Hemapushpamu, Kanjanamu, Hemangamu, Chempeyamu. Eng.: Golden Champa, Yellow Champa.

Flowers and fruits used in the treatment of dyspepsia, fever and gonorrhoea. Fruits and seeds are used for healing cracks in feet. Flowers yield champa oil. Wood is not very durable, used for making posts, boards, veneers, furniture, decorative fittings, carriage and ship buildings. It is also suitable for making pencils, aircraft construction, battery separators and tea chest plywoods.

INDIA : Often planted. Tropical and sub-tropical forests up to 600 m, and Himachal Pradesh, Uttar Pradesh, Assam, Nagaland, Manipur, Mizoram, Tripura, Meghalaya and Orissa.

WORLD : Bhutan, Bangladesh, Myanmar, S.W. China to Malaysia, Nepal and Thailand.

## ANNONACEAE

1. Ripe carpels united into a large fleshy syncarpium. Outer whorl of petals keeled at base ........ ***Annona***
1. Ripe carpels free or distinct and form stalked monocarps. Outer whorl of petals not keeled at base:
   2. Connectives produced above the anther lobes. Monocarps numerous, glabrous:
      3. Seeds 6–12 in 2 rows. Leaf base oblique ........ ***Cananga***
      3. Seeds 1–5. Leaf base acute or rounded ........ ***Polyalthia***
   2. Connectives not produced above the anther lobes. Monocarps few, tomentose or pubescent;
      4. Leaves glabrous. Flowers < 1 cm across; outer petals equal to inner ones ........ ***Alphonsea***
      4. Leaves hairy. Flowers > 1 cm across; outer petals shorter than inner ones ...... ***Miliusa***

## **ALPHONSEA** J.D. Hooker & T. Thomson

1. Petiole to 5 mm. Sepals orbicular, reflexed. Stamens 15. Monocarps (Ripe carpels) tuberculate, rusty tomentose, globose or subglobose; stalk 5–8 mm long ........ ***A. sclerocarpa***
1. Petiole 6–8 mm. Sepals ovate, erect. Stamens 12. Monocarps (Ripe carpels) smooth, yellow tomentose, oblong; stalk 1–2 mm long ........ ***A. madraspatana***

**Alphonsea madraspatana** Bedd. Ic. t. 92. 1868–1874; FBI 1:89; Gamble 1: 23(16); Brandis 18; Flora of India 1: 211. 1993.

Evergreen, small or medium sized tree, young branches glabrous. Leaves elliptic-oblong, orbicular, ovate, 2.4–11 × 2–5.3 cm, glabrous or slightly puberulous, glossy, apex obtuse, rarely subacute, margin entire, base rounded or cuneate; secondary nerves 5 pairs; petiole 6–8 mm, glabrous. Flowers bright yellow, 1–6, 2 mm across in supra axillary, leaf-opposed cymes. Monocarps 3–6, yellow-tomentose, smooth, ovoid-globose, 1–2 cm long; stalk 1–2 mm long; seeds few in two rows.

Rare, in forests of Visakhapatnam and Cuddapah districts, in ravines and along streams, up to 1000 m. Fls.: May., Frs.: Aug.

Nattavararam (VSKP), *CAB* 1842; Krishnapuram (VSKP), *CAB* 1923; Sudikonda (EG), *GVS* 24498 (MH).

INDIA : Orissa and Tamil Nadu—Endemic.

**Alphonsea sclerocarpa** Thw. Enum. Pl. Zeyl. 11. 1858; FBI 1: 89; Gamble 1: 23(16); Flora of India 1: 211. 1993.

Densely foliaceous, evergreen tree, up to 20 m tall; bark and branches rugose, glabrous. Laves chartaceous and light green when young, coriaceous and dark green later, thin, oblong, elliptic-lanceolate, 3–7 × 1.5–4.6 cm,

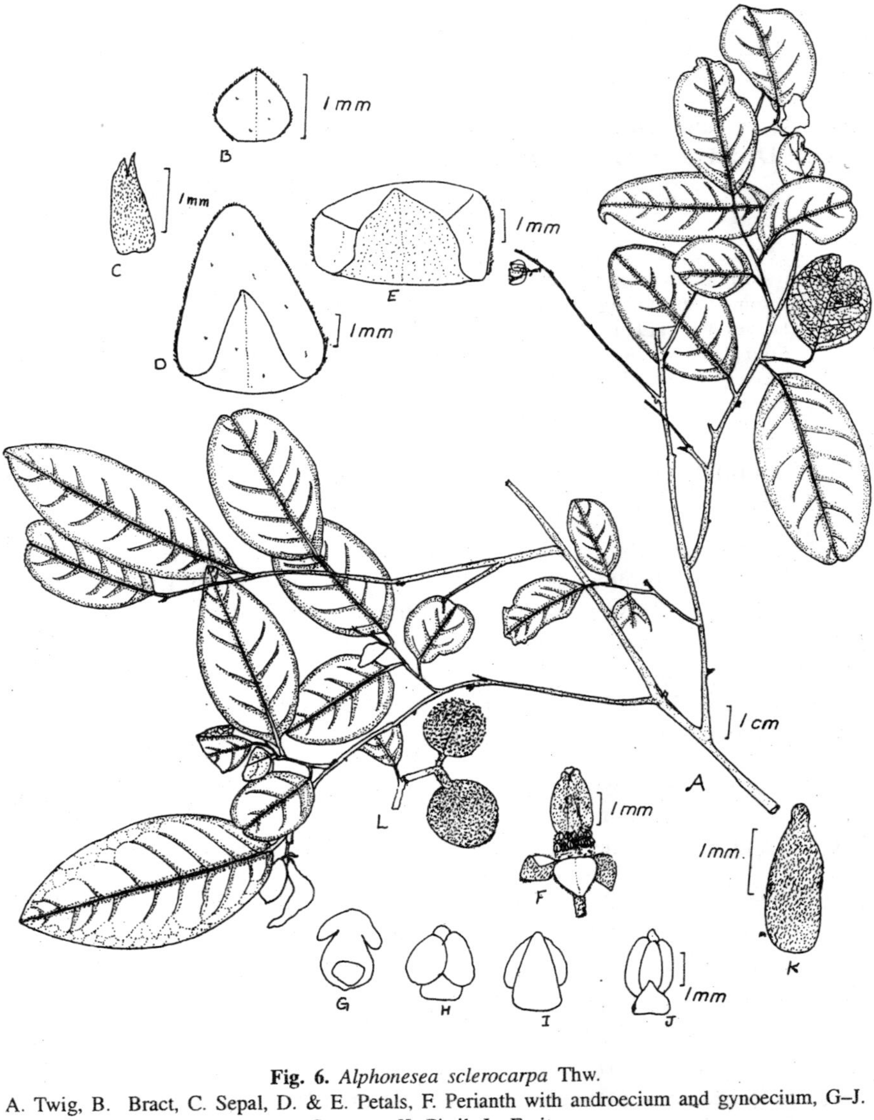


**Fig. 6.** *Alphonesea sclerocarpa* Thw.
A. Twig, B. Bract, C. Sepal, D. & E. Petals, F. Perianth with androecium and gynoecium, G–J. Stamens, K. Pistil, L. Fruits

glabrous, apex obtusely acute, margin entire, base obtuse or cuneate, petiole to 5 mm. Flowers yellowish, 9 mm across, in leaf-opposed or extra axillary cymes. Monocarps 3–5 in a cluster; stalk 5–9 mm long, globose or subglobose, 2 × 1 cm, warty, rusty tomentose; seeds more than 2.

Occasional, along streams in deep valleys of moist deciduous forests of East Godavari, Nellore, Chittoor and Cuddapah districts at 600–800 m. Fls.: Feb.–Apr. Frs.: July–Sept. Vern: Putusu Mamidi.

Talakona (CTR), *SSR* 13719, 13789, 15556; Udayagiri (NLR), *CECF* 4149 (CAL); Seshachalam hills (CDP), *CECF* 4634 (CAL); Rapur (NLR), *ASR* 6434 & 6469 (VV).

INDIA : Tamil Nadu and Kerala.
WORLD : Myanmar, Sri Lanka.

## ANNONA Linnaeus

1. Fruit softly spinescent. Inner petals nearly equal to outer ones ........................ ***A. muricata***
1. Fruit spinescent. Inner petals smaller than outer ones:
   2. Leaves 4–8 cm long, apex obtuse. Flowers greenish, solitary or few together in axils. Fruit glaucous, 5–10 cm in dia., tuberculate with rounded tips .................. ***A. squamosa***
   2. Leaves 10–18 cm long, apex acuminate. Flowers pale green, extra-axillary or in terminal cymes, or 2–3 together. Fruit yellowish-brown to reddish-green, 10–20 cm in dia., smooth .................................................................................................. ***A. reticulata***

**Annona muricata** L. Sp. Pl. 1: 536. 1753. Brandis 22; Flora of India 1: 206. 1993.

Evergreen small tree, branchlets glabrous. Leaves obovate-oblong, 11–14 × 4–5 cm, glabrous, apex shortly acuminate, margin entire, base acute, petiole 4–5 mm. Flowers yellowish, solitary, trimerous, axillary. Fruit 6–8 cm long, ovoid; carpels dull green, tuberculate, covered with long, curved with fleshy spinules; seeds many, reddish-brown.

Cultivated, Fls.: & Frs.: Apr.–Oct. Eng.: Prickly custard apple. Fruits used for sherbets and refreshing drinks.

INDIA : Assam, Maharashtra, Karnataka, Tamil Nadu and Andaman & Nicobar Islands.
WORLD : Central America and West Indies. Cultivated widely in tropics.

**Annona reticulata** L. Sp. Pl. 537. 1753; FBI 1: 78; Gamble 1: 20(14); Brandis 22; Flora of India 1: 207. 1993.

Small, evergreen tree, up to 7 m tall; branchlets glabrous. Leaves membranous, oblong-lanceolate, 10–18 × 2–5 cm, glabrous above, scattered hairy beneath, apex shortly acuminate, margin entire, base cuneate or rounded; secondary nerves 15–18 pairs; petiole to 1.8 cm. Flowers greenish, 2.5 cm long, 2–4, on lateral pedicels. Fruit yellow or yellowish, red when ripe, sub globose or ovoid, 10–16 cm in dia., nearly smooth, with impressed lines; seeds numerous, blackish, smooth, arillate.

Commonly cultivated in gardens. Fls.: and Frs.: nearly through the year. Vern: Ramaphalamu, Ramasithaphalamu. Eng.: Bullock's heart. Netted Custard apple.

Fruits sweet, edible, used in medicine as blood purifier and also to cure dysentry.

A.U. campus (VSKP), *SSR* 19666; Punyagiri (VSKP), *GVS* 23313 (CAL, MH); Araku (VSKP), *TP & EC* 7346; Tekrial (NZB), *TP & BR* 6420; Venkatagiri (NLR), *ASR* 6538 & 8664 (VV).

INDIA : Cultivated and often naturalised, up to 1200 metres. Delhi, Uttar Pradesh, Bihar, West Bengal, Assam, Orissa, Madhya Pradesh, Maharashtra, Karnataka, Kerala and Tamil Nadu.
WORLD : Native of Central America and West Indies.

**Annona squamosa** L. Sp. Pl. 537. 1753; FBI 1: 78; Gamble 20(14); Brandis 22; Flora of India 1: 207. 1993.

Deciduous tree, up to 8 m tall; bark grey, thin. Leaves thin-coriaceous, elliptic or oblong-lanceolate, pellucid dotted with a peculiar smell, 4–8 × 2–3 cm, glabrescent, glaucous beneath, apex acute, entire, base cuneate; petiole to 1.5 cm. Flowers greenish, solitary, on leaf-opposed peduncles or 2–4 in a cluster. Fruit a fleshy syncarp, cordate or ovoid, 5–10 cm dia., warty, greyish black when ripe, pulp sweet; seeds numerous, embedded in pulp, angular, brownish black, 1 × 0.5 cm, shiny, arillate.

Introduced, and now naturalised everywhere in the plains to 250 m, cultivated in gardens for its fruits. It grows on a variety of soils. It is recommended for aforestation of bare, eroded and denuded areas. It is light demander, drought resistant and good coppicer. Leaf fall: Feb. New foliage: Mar.–Apr. Fls.: Mar.–May. Frs.: Sept.–Dec. Vern.: Sithaphalamu, Ganda Gathram, Sitapandu. Eng.: Custard apple, Sugar apple, Sweet sop.

Fruit sweet, tasty, and edible. It increases muscular strength and enriches blood. Powdered seeds are said to possess insecticidal properties, especially used to kill lice on the head. Leaves used in toothache. The root is purgative and the bark is antidote to snake bite.

S.K University (ATP), *NY* 691; Jaladurgam RF (KNL), *RVR* 1881; Macherla RF (GTR), *VRK* 3967; Diguvametta (PKSM), *RVK* 15758; Mannanur (MBNR), *TDCK* 16811; Udayagiri (NLR), *ASR* 4402 (VV); Bhakarapet (CTR), *DR* 1842 (SVU).

INDIA : Delhi, Uttar Pradesh, Bihar, West Bengal, Assam, Orissa, Madhya Pradesh, Rajasthan, Gujarat, Maharashtra, Karnataka, Kerala, Tamil Nadu.
WORLD : Native of Central America, now widely cultivated in tropics.

## CANANGA (A.P. deCandolle) J.D. Hooker et T. Thomson

**Cananga odorata** (Lam.) Hook. f. & Thoms. Fl. Ind. 130. 1855; FBI 1: 56. Flora of India 1: 255. 1993. *Uvaria odorata* Lam. Encycl. 1: 595. 1789. *Canangium odoratum* Baillon Ann. IV. t. 67. 1893.

Evergreen tree, 10–25 m tall; trunk straight, branched at the top; young branches minutely pubescent. Leaves elliptic to oblong, 10–17 × 4–7 cm, glabrous, apex narrowly acuminate, margin entire, base obtuse or rounded, oblique; secondary nerves 7–10 pairs, 1 cm apart; petiole to 1 cm. Flowers greenish-yellow, very fragrant, 4 cm long, few to numerous in shortly pedunculate pendulous racemes. Monocarps 10–12, black, oblong-ovoid or globose, pulpy, 1–2.5 cm long; stalk 1–2 cm long; seeds 6–12 in 2 rows, transversely flat, pale brown with pitted surface.

Cultivated, grown as an ornamental. (Rao. Rolla *et al.*, 1986). The tree grows well on rocky or light soils. Fls.: July, Frs.: Oct.–Nov. Vern.: Apurvachandanamu. Eng.: Ilang-Ilang.

Flowers yield an essential oil, commonly called Cananga oil, which is extensively used in high class perfumery, face powder, and in the fixation of floral odours. Leaves also yield an oil which is used medicinally for headache, ophtalmia and gout. Decoction of bark is used against rheumatism, ulcers and fever. Wood is suitable for posts and house-hold implements and for packing cases.

INDIA : Cultivated throughout the country, sometimes naturalised.
WORLD : Native of Myanamar, Indochina, Thailand, Malaysia to Australia and New Zealand.

## **MILIUSA** Leschenault ex A. de Candolle

1. Leaves ovate-oblong, glabrescent above, tomentose beneath, base rounded or subacute. Sepals ca. 1 mm long; sepals and outer petals linear-lanceolate, inner petals not saccate at base. Monocarps 2-seeded ........ ***M. velutina***
1. Leaves oblong-elliptic, tomentose on both surfaces, base obliquely cordate. Sepals 2–4 mm long, sepals and outer petals ovate, inner petals saccate at base. Monocarps 3–4-seeded ........ ***M. tomentosa***

**Miliusa tomentosa** (Roxb.) Sinclair in Gard. Bull. Striat. Settlem. 14: 378. 1955; Flora of India 1: 221. 1993. *Uvaria tomentosa* Roxb. Pl. Cor. t. 35. 1795. *Saccopetalum tomentosum* (Roxb.) Hook. f. & Thoms. Fl. Ind. 152. 1855; FBI 1: 88; Gamble 1: 22(16); Brandis 11; Beddome Fl. Sylv. S. India t. 39. 1870.

Large, deciduous tree, up to 20 m tall; bark blackish-brown, deeply longitudinally fissured; wood yellowish, slightly pink or brown. Leaves thick-coriaceous, oblong-elliptic, 8–13 × 2–5.5 cm, glabrescent above, tomentose beneath, apex acute, margin entire, base rounded; petiole 2–5 mm. Flowers greenish, 1.5 cm across, solitary or in leaf opposed or terminal pairs. Monocarps dark purple, subglobose, 8–15 in a ring, 2–3 cm in dia; stalk, 1–1.5 cm long, velvety brown, seeds 3–4.

Common, in hill slopes and of plateues of deciduous forests, at 500–1000 m. Leaf fall: Feb.–Mar., New foliage: May or June. Fls.: May–June,

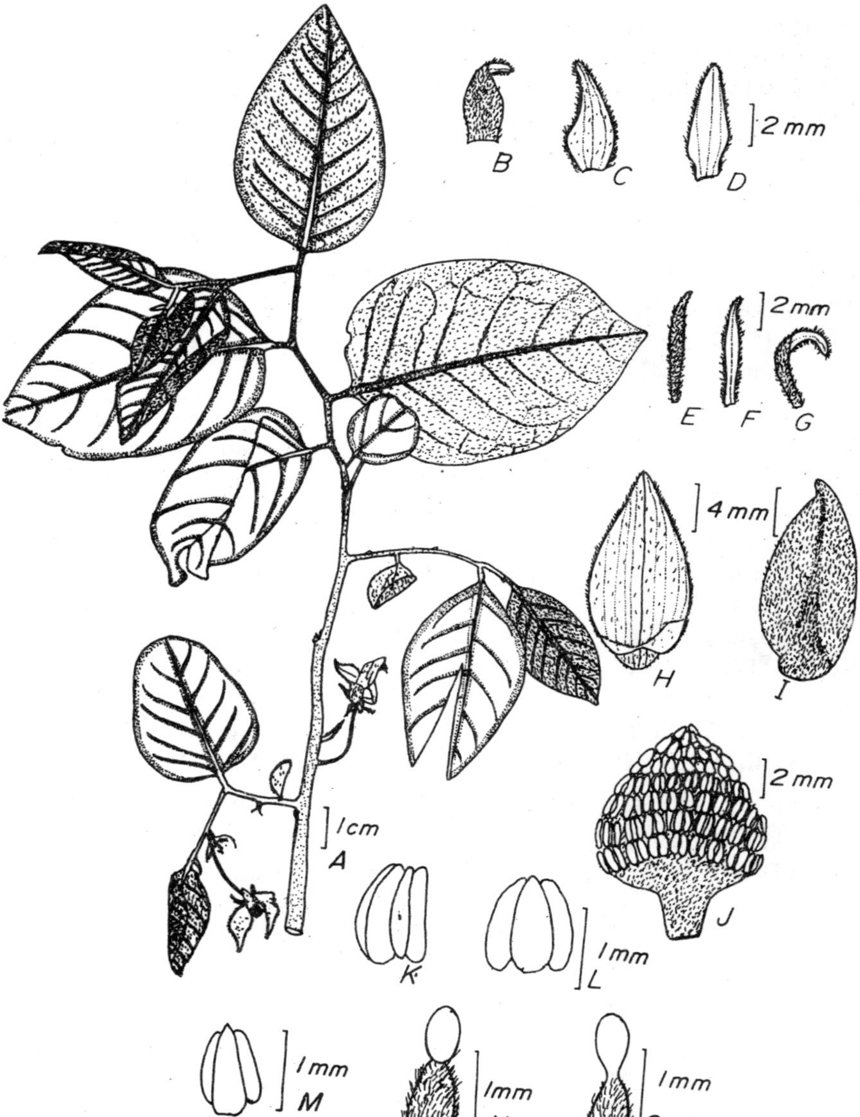


**Fig. 7**. *Milusa tomentosa* (Roxb.) Sinclari
A. Twig, B–D. Sepals, E–G. Outer petals H. & I. Inner petals, J. Stamens and Pistils, K–M. Stamens, N. & O. Pistil.

Frs.: June–July. Vern.: Budda Duduga, Barre Duduga, Pedda chiluka duduga, Nalla Chilka Dudi, Gadida lotta, Guthe chettu.

Wood is strong and durable, used for house poles.

Donubai (SKLM), *SSR* 15592; Rollapenta (KNL), *SSR* 15993; Anantagiri (RR), *SSR* 18509; Eturnagaram (WGL), *SSR* 18552; Ichampalli RF (ADB), *GO & DAM* 5149; Manchippa RF (NZB), *BR* 9214; Lambasingi (VSKP), *GVS* 47387 (MH); Cheedipalem (EG), *GVS* 27438 (Cal); Marri Pakala (EG), *GVS* 27413 (CAL); Tirumala (CTR), *DR* 2116 (SVU).

INDIA : Rajasthan, Bihar, Orissa, Maharashtra, Karnataka, and Tamil Nadu.

WORLD : Nepal and Sri Lanka.

**Miliusa velutina** Hook. f. & Thoms. Fl. Ind. 151. 1855 & FBI 1: 87; Gamble 21(15); Flora of India 1: 222. 1993. *Uvaria velutina* Dunal Monogr. Annon. 91. 1817.

Deciduous tree, up to 12 m tall; bark greyish, fairly smooth, often fluted, young branches yellowish tomentose. Leaves coriaceous, oblong-elliptic, 5–15 × 2.5–6.5 cm, smooth, velvety tomentose on both surfaces, apex obtuse, acute or acuminate, margin entire, base obtuse to subcordate; secondary nerves 9–11 pairs; petiole to 5 mm. Flowers pale yellow, few in 5–8 cm long, leaf-opposed cymes. Monocarps bluish-purple, 10–15, ellipsoid, 1–2.3 cm in dia; stalk to 4 cm long, tomentose; seeds 1–2.

Occasional, in moist deciduous forests of Visakhapatnam and Nellore districts. Fls.: Feb.–May. Frs.: June–October. Vern.: Pedda Chilka Duduga, Nalla duduga.

G. Madugula (VSKP), *SSR* 15982; Owk RF (KNL), *RVR* 3157; Chintapalli (VSKP), *GVS* 30089 (CAL); Urlakonda (NLR), *MSR* 1313 (CAL).

INDIA : Jammu & Kashmir, Uttar Pradesh, Bihar, West Bengal, Assam, Meghalaya, Orissa, and Madhya Pradesh.

WORLD : Myanmar, Cambodia, Laos, Vietnam, Thailand.

## **POLYALTHIA** Blume

### KEY 1 (Vegetative)

1. Crown conical, leaves glossy, undulate margined-cultivated .......... ***P. longifolia***
1. Crown other than conical, leaves not glossy, entire margined—wild:
  2. Leaves obtuse at apex, slightly narrowed at base .......... ***P. suberosa***
  2. Leaves acute-acuminate at apex, base acute-rounded:
    4. Leaves glabrous; secondary nerves 5–6 pairs .......... ***P. korinti***
    4. Leaves glabrescent, or puberulous to pubescent beneath; secondary nerves 8–12 pairs .......... ***P. cerasoides***

### KEY 2 (Flowering)

1. Flowers numerous in sub-umbellate clusters .......... ***P. longifolia***
1. Flowers solitary or 2–5:

2. Petals equal in 2-series; stamens 1.5 mm long .......... ***P. cerasoides***
2. Petals unequal in 2-series; stamens 1 mm long:
   3. Flowers extra-axillary; connectives of stamens flat at top .......... ***P. suberosa***
   3. Flowers axillary; connectives of stamens convex at top .......... ***P. korinti***

KEY 3 (Fruiting)

1. Monocarp 1–2 cm long, stalk 1.5–2 cm long:
   2. Monocarps ca. 1 cm long, 5 mm across, red, apiculate; seed smooth, ellipsoid .......... ***P. cerasoides***
   2. Monocarps ca. 2 cm long, 1 cm across, yellowish, black later, obtuse; seed slightly grooved, ovoid .......... ***P. longifolia***
1. Monocarp 5–7 mm long, stalk 0.5–1 cm long:
   3. Monocarps bright crimson .......... ***P. korinti***
   3. Monocarps purple .......... ***P. suberosa***

**Polyalthia cerasoides** (Roxb.) Bedd. Fl. Sylv. S. India t. 1: 1869; FBI 1: 63; Gamble 1: 17(12); Brandis 14; Flora of India 1: 271. 1993. *Uvaria cerasoides* Roxb. Pl. Cor. t. 33. 1795.

Semi-evergreen tree, up to 13 m tall; bark light grey, rough, lenticellate; wood deep brown; young branches densely tomentose. Leaves chartaceous, oblong-lanceolate, 3.5–22 × 2–6 cm, glabrous, dark green and glossy above, puberulous to pubescent beneath, apex acuminate, margin entire, base obtuse; petiole to 4 mm. Flowers greenish, 1–1.2 cm across, axillary or extra axillary. Monocarps 35–40 in a cluster, reddish, ellipsoid to broadly ovoid, apiculate, ca 1 cm long, glabrous; stalk 1–2 cm long; seed one, brown, globose, 6 mm.

Common, in hilly slopes, along ravines in dry deciduous forests. Leaf fall: Jan. New foliage: Apr. Fls.: Apr.–July, Frs.: May–Aug. Vern.: Gutti, Dudduga, Chilka dudduga, Chitti dudduga, Guva, Tella chilaka dudi.

Timber used for carpentary and shipping purposes. Ripe carpels are edible.

Bairluti (KNL), *TP & SSR* 15212; Donubai RF (SKLM), *SSR* 15587; Pakhal (WGL), *SSR* 18525; Eturnagaram (WGL), *SSR* 18550; Kalasamudram RF (ATP), *EC* 2926 & 2971; Rampa (EG)', *VNS* 614 (CAL); Guvvalacheruvu RF (CDP), *R.V. Reddy* 8522; Diguvametta (PKSM), *RVK* 15737; Mallalatheertham (MBNR), *TDCK* 15390; Palamaner (CTR), *DR* 2048 (SVU); Nellore (NLR), *MSR* 1265 (CAL).

INDIA : Bihar, West Bengal, Assam, Arunachal Pradesh, Orissa, Tamil Nadu and Kerala.

WORLD : Myanmar, Thailand, Cambodia and Vietnam.

**Polyalthia korinti** (Dunal) Thw. Enum. Pl. Zeyl. 398. 1864; FBI 1: 64; Gamble 1. 16(11); Brandis 15; Flora of India 1: 273. 1993. *Guatteria korinti* Dunal Monog. Anon. 133. 1817.

Small tree, 3–5 m tall; spreading; bark grey; young branches glabrous or subpilose. Leaves coriaceous, elliptic or ovate-elliptic, 3–10 × 1.8–5 cm, glabrous, glossy, apex acute to shortly acuminate, margin entire, base acute; secondary nerves 4–6 pairs, tertiary nerves are more prominent; petiole 4–8 mm. Flowers bright green, solitary, rarely 1–3, axillary, 1.4–1.8 cm dia. Monocarps numerous, bright crimson, 5 × 4 mm, ellipsoid or subglobose; stalk to 8 mm; seed 1, smooth, shiny grey.

Rare, in dry evergreen forests of Chittoor and Visakhapatnam districts. Fls.: Apr.–May, Frs.: July–Aug.

Kambakkam to Thandipal (CTR), *MCB* 4510 (MH); Tirumala (CTR), *DR* 2197 (SVU).

INDIA : Tamil Nadu and Kerala.
WORLD : Sri Lanka.

**Polyalthia longifolia** (Sonn.) Thw. Enum. Pl. Zeyl. 398. 1864; FBI 1: 62; Gamble 1: 16(11); Brandis 15; Flora of India 1: 274. 1993; Beddome Fl., Sylv. S. India t. 28. 1870. *Uvaria longifolia* Sonn. Voy. Aux. Indes 2: 288. Pl. 131. 1782.

Tall; evergreen tree, up to 20 m tall; trunk straight; crown conical; bark brown, slightly rough; young branches spreading, pendulous. Leaves shining, oblong-lanceolate to narrow-lanceolate, 14–20 × 2.6–3.6 cm, glabrous, apex long-acuminate, margin undulate, base cuneate; petiole to 5 mm. Flowers yellowish-green, numerous in subumbellate clusters. Monocarps yellowish, numerous, ovoid-ellipsoid, 3 × 1.5 cm, obtuse at both ends, glabrous, reddish to black later; stalk 1–1.5 cm long; seed 1, almost ovoid, smooth or slightly grooved, pale brown, glossy.

Common, planted in gardens, avenues and parks. Fls.: Mar.-May. Frs.: June-Aug. Vern.: Aswarthamu, Asokamu, Naramamidi, Devadaru. Eng.: Indian fir, Mast tree.

Bark is used as antipyretic agent.

S.K. University Campus (ATP), *NY* 581 & *TP* 2946.

INDIA : Cultivated throughout in gardens and also as an avenue tree, sometimes self regenerating.
WORLD : Said to be native of South India and Sri Lanka, introduced to many tropical countries.

**Polyalthia suberosa** (Roxb.) Thw. Enum. Pl. Zeyl. 398. 1864; FBI 1:65; Gamble 1: 16(12); Brandis 14; Flora of India 1: 298. 1993. *Uvaria suberosa* Roxb. Pl. Cor. t. 34. 1795.

Evergreen, small tree, 4–6 m tall; branching from base; bark corky, deeply fissured, reddish inside; young branches tomentose. Leaves chartaceous, oblong or lanceolate, 3.6–11 × 1.3–4.4 cm, glabrous and shining above, paler and slightly pubescent along nerves beneath, apex acute or obtuse, margin entire, base subacute; secondary nerves 10 pairs, irregular, not very prominent, petiole 2–3 mm. Flowers greenish-yellow, linear, 8–10 mm long, solitary, extra axillary. Monocarps many, purple, puberulous, subglobose, 8 × 4 mm; stalk slender, 0.5–1 cm long; seeds 1–2, globose, smooth.

Occasional, in dry deciduous forests near streams. Fls.: Apr.-May. Frs.: Throughout the year. Vern.: Chilakadudduga.

Salur (SKLM), *NPBK* 973 (CAL); Polavaram Agency (EG), *DCSR* 328 (CAL); S. Kota (VSKP), *GVS* 21818 (CAL); Maredumilli (EG), *GVS* 24292 (CAL); Ichapur (SKLM), *GVS* 62489 (CAL).

INDIA : West Bengal, Assam, Meghalaya, Orissa, Madhya Pradesh, Tamil Nadu and Kerala.

WORLD : Sri Lanka, Myanmar, Thailand, China, Malaysia and Philippines.

## CAPPARACEAE

1. Branches thorny (usually absent in *C. grandis*). Leaves simple ........................... ***Capparis***
1. Branches not thorny. Leaves 3-foliolate:
   2. Leaflets elliptic-oblong, apex acute or rarely retuse with mucronate tip. Flowers in racemes, pedicel 2–3 cm long; petals absent. Fruit oblong or ovoid, moniliform ........................................ ***Maerua***
   2. Leaflets ovate-lanceolate, apex long-acuminate. Flowers in corymbs; pedicel 3–7 cm long; petals present. Fruit oblong-ellipsoid or oblong-ovate, globose .................. ***Crateva***

## **CAPPARIS** Linnaeus

### KEY 1 (Vegetative and Fruiting)

1. Leaves early caducous, fleshy, linear spathulate, tip spinous; petiole 0; cataphylls confined to the base. Berries 1.5–2 cm across; stipe thin, to 2.5 cm long ............................... ***C. decidua***
1. Leaves not caducous, not fleshy, other than linear spathulate, tip without spine; petiolate. Berries 2.5–4 cm across; stipe woody, 3–4 cm long:
   2. Stipular thorns present. Leaves dimorphic, linear, linear-oblong or oblong-lanceolate, glabrous, midrib raised, 5–7-nerved from base; Berries beaked with 5–10 longitudinal ridges, warty; seeds 7 or 8, to 9 × 6 mm, embedded in white or creamy pulp ........................................ ***C. divaricata***
   2. Stipular thorns usually absent. Leaves not dimorphic, sub-orbicular or elliptic-ovate, velvety pubescent; secondary nerves 5–10 pairs from flat midrib. Berry not beaked and without ridges, smooth; seeds 2–6, 11 × 9 mm, embedded in pink pulp ............ ***C. grandis***

### KEY 2 (Flowering)

1. Flowers solitary—greenish yellow ........................................ ***C. divaricata***
1. Flowers in corymbs:
   2. Flowers greenish-yellow, not showy; sepals subequal, neither saccate nor hooded, sepaloid; stamens 45–50 ........................................ ***C. grandis***

2. Flowers coral pink, showy; sepals unequal, one is saccate and hooded, another one is petaloid; stamens 10–16 .......... *C. decidua*

**Capparis decidua** (Forsk.) Edgew. in J. Linn. Soc. Lond., Bot. 6: 184. 1832; Flora of India 2: 265. 1993. *Sodada decidua* Forsk, Fl. Aeg.-Arab. 81. 1775. *Capparis aphylla* Roth, Nov. Sp. Pl. Ind. Or. 238. 1821; FBI 1: 174; Gamble 1: 45 (32); Brandis 33.

Small showy tree, up to 7 m tall; trunk short, branches numerous, green, divaricating, forming a dense rounded crown; wood light yellow, close grained. Stipular thorns yellowish, straight, 3–6 mm long, cataphylls confined to base. Leaves on young shoots only, caducous, fleshy, 3–18 × 1–3 mm, linear-spathulate, glabrous, apex acute, spiny, margin entire, base acute; petiole short or 0. Flowers showy, reddish, 1.5–2.5 cm across, many in corymbs from the old branches, or from short lateral shoots. Fruit scarlet red, globose, beaked, 1.4–2 cm across, smooth with thin pericarp; seeds 1–4, reniform, 2.5 × 2–4 mm.

Occasional, in semi-arid scrub jungle of Telangana (Khan, 1953). Often planted as a sand binder to arrest spreading of deserts. Fasciation of twigs is common and vegetative propagation is by root suckers. Fls.: Feb.-May. Frs.: Sept.–Nov. Vern: Kariramu.

Bark is an anthelmintic. The plant yields a carminative, tonic, used as an aphrodisiac, and appetite. Flower buds and fruits are used as vegetables and pickles. Young shoots, leaves, roots and fruits in powdered form are used for treating boils, skin eruptions, rheumatism, etc. Wood is resistant to white ants and hence serves as a raw material making for rafters, beams, ploughs and country boards.

INDIA : Throughout the country except Jammu & Kashmir, N.E. hill states, Andaman & Nicobar Islands and other high humid region.

WORLD : Pakistan, extending to Middle East, West Asia and N.E. tropical Africa.

**Capparis divaricata** Lam. Encycl. 1: 606. 1785; FBI 1: 174; Brandis 33; Flora of India 2: 266. 1993. *Capparis stylosa* DC. Prodr. 1: 246. 1824; Gamble 1: 45(32).

Small tree 2–6 m tall; much branched, crown compact; bark deeply fissured; young branches glabrous or pubescent. Stipular thorns divaricate, to 4 mm. Leaves coriaceous, variable,linear-oblong, elliptic or lanceolate, 4–5.5 × 1.4–3 cm, glabrous, tomentose when young, apex acute or obtuse with mucro, margin entire, base rounded, 5–7-nerved from base; secondary nerves 1–2 pairs above basal ones, reticulations obscure; petiole 4 to 6 mm. Flowers greenish or reddish, solitary, axillary. Fruits reddish, globose, beaked, 3.5–4.5 × 2.5–4 cm, longitudinally ribbed, and tuberculate; stipe woody, 3-4 cm long; seeds 6–8, embedded in a creamy pulp, 8–9 × 4–6 mm.

Occasional, in scrubs or deciduous forests on black cotton or laterite soil, to 300 m. Fls.: Feb.–Apr. Frs.: July–Oct. Vern.: Badreni.

Loyapalle (GTR), *VRK* 3997; Kammavaripalli (KNL), *TP & GO* 3379; Erramalais (KNL), *TP & RVR* 1317; Tirupati (CTR), *DR* 2308 (SVU).

INDIA : Maharashtra, Karnataka, Tamil Nadu, Pondichery and Kerala.
WORLD : Sri Lanka, in dry localities.

**Capparis grandis** L.f. Suppl. Pl. 263. 1781; FBI 1: 176; Gamble 1: 46(33); Flora of India 2: 274. 1993.

Deciduous tree, up to 10 m tall; bark greyish, characteristically deeply furrowed, upper part unarmed, lower part armed, lenticellate; branchlets velvety-tomentose. Leaves thin-coriaceous, broadly ovate or obovate, 3–7 × 2–5 cm, dull green above, pale and grey silky-pubescent beneath, apex acute or obtuse with mucro, margin entire, base cuneate; secondary nerves 5–10 pairs, close, parallel; petiole to 1 cm. Flowers creamy, fragrant, 2.5 cm across, in axillary or terminal corymbs. Berry globose, 2.2–2.8 cm across, smooth, purple; seeds 2–6, embedded in pink pulp, to 1.2 × 1 cm.

In dry deciduous forests on rocky soils. Fls.: Throughout the year, peak period in Feb.–July and Nov.–Dec. Frs.: June–Jan. Vern: Regutti, Guti, Ragota, Nallauppi, Duduppi, Remide, Oridona.

Bark and leaf infusion used internally for swellings and eruptions.

Bogada (PKSM), *SSR* 16106; Gani RF (KNL), *RVR* 3117; Bhadrachalam (KMM), *JSG* 15822 (DD); Ferrangayya Kuntal (CDP), *EKK* 77835 (DD); Kondavedu Fort (GTR), *JSG* 17490 (MH); Cheedipalem (EG), *GVS* 27454 (MH); Mannanur (MBNR), *TDCK* 16806.

INDIA : Rajasthan, Gujarat, Maharashtra, Goa, Karnataka, Tamil Nadu, Pondicherry and Kerala.
WORLD : Sri Lanka, Myanmar, extends to Laos, Vietnam and Thailand.

## CRATEVA Linnaeus

1. Mostly leafless when flowering. Flowers many (up to 100) in corymb; rachis 2–3 cm long. Fruit orange-yellow; turning to reddish on maturity .................. ***C. adansonii*** sub sp. ***odora***
1. Mostly leaf-bearing when flowering. Flowers few (12–30) in corymbs; rachis 10 cm long. Fruit yellowish-grey to brown on maturity .......................................................... ***C. magna***

**Crateva adansonii** DC. sub sp. **odora** (Buch.-Ham.) Jacobs in Blumea 12:198. 1964. *C. odora* Buch.-Ham. in Trans. Linn. Soc. London 15: 188. 1827; Flora of India 2: 322. 1993. *C. religiosa* var. *roxburghii* (R.Br.). Hook. f. Thomson in Fl. Brit. India. 1:172. 1872. *C. religiosa* sensu Dunn in Gamble, Fl. Pres. Madras 1:47. 1915, *p.p.* non Forst. f. 1786.

Small tree; branchlets glabrescent. Leaves 3-foliolate; petiole 7–10 cm, glandular at tip; leaflets subcoriaceous, ovate to elliptic-lanceolate; central leaflets up to 12 × 7 cm with abruptly acuminate apex; lateral leaflets 6–8 × 3.5–5 cm with acute tip, base oblique and tapered, glabrous; secondary nerves 4–6 pairs; petiolule 6–10 mm; Flowers greenish-white, fragrant, 2.5–3.5 cm across in terminal corymbs. Berries orange-yellow or reddish brown, globose, 2.5–4 × 2 cm; seeds brown, broadly elliptic, 5–6 × 2 cm, smooth, embedded in yellow pulp.

Occasional, in deciduous forests of Godavary Valley and upper Godavary districts of Eastern Ghats (Ram Rao & Henry, 1996). Leaf fall: Dec. New foliage: Apr. Fls.: Feb.–Apr. Frs. May–Nov. Vern.: Mugalina, Muvva.

Wood used for agricultural implements, toys and planks. Fruits edible when cooked. Bark extract with pepper used to cure chest pain and menstrual disorders by local people.

INDIA : In open forests or scrub jungles, up to 750 m Throughout India except in the humid hill states of N.E. India and Kashmir.
WORLD : Myanmar and Sri Lanka.

**Crateva magna** (Lour.) DC. Prodr. 1: 243. 1924; Jacobs in Steenis Fl. Males (Ser. 1) 7: 822. 1976 (in addendum) et in Blumea 12. 206. 1964; Flora of India 2: 324. 1993. *Capparis magna* Lour. Fl. Cochinch. 1: 731. 1790. *Crateva nurvala* Buch.-Ham. in Trans. Linn. Soc. London 15: 121. 1827 "*narvala*" *C. religiosa* Forst. f. var. *nurvala* (Buch.-Ham.) Hook. f. & Thoms in FBI 1: 172. 1872. *C. religiosa* sensu Dunn in Gamble Fl. Madras. Pres. 1: 47 (34) 1918. *p.p.* non. Forst. f. 1786. Beddome Fl. Sylv. S. India t. 116. 1871.

Densely foliaceous, deciduous tree, up to 10 m tall; bark grey, fairly smooth with horizontal wrinkles; wood yellowish-white, close grained; branchlets greyish-brown when dry, lenticellate. Leaves 3-foliolate, petiole 5.5 cm; leaflets chartaceous, ovate-lanceolate, 5.5–8.6 × 2.5 cm, glabrous, pale beneath, apex abruptly acuminate, margin entire, slightly undulate, base attenuate; petiolule 5 mm. Flowers greenish-white in dense terminal corymbs. Berry oblong-ellipsoid or oblong-ovate, 4–6 × 3–5 cm, woody, yellowish-grey; seeds numerous embedded in pulp, dark brown, 8–12 × 5–9 mm, tubercled.

Common, in deciduous or semi-evergreen forests, along streams up to 1000 m, often planted in temples and gardens. Fls.: Apr.–May. Frs.: Apr.–Aug. Vern.: Tella ulimidi, Uski, Bilvaram, Urimidi, Voolemeri, Magalingamu.

Young berries are edible. Flowers are astringent. Leaves are bitter and used for treating skin ailments. Root bark extract administered to relieve gastric troubles by Konda-reddis and Valmikis. Bark promotes appetite, it is antipyretic and anthelmintic. Leaves used as a remedy for swelling of the feet. Leaf juice is given orally in rheumatism. Wood shiny, moderately hard, smooth, not durable and used for making chappels, combs and other small articles.

Tirupati (CTR), *SSR* 19683; Johrapuram (KNL), *AA* 2087; Manchippa (NZB), *BR* 9024; Sileru (VSKP), *GVS* 47415 (MH & CAL); Rampa hills (EG), *JSG* 16028 (CAL).

INDIA : Peninsula, Western India, Gangetic plains and Eastern India up to Tripura and Manipur.

WORLD : Bangladesh, Myanmar, China and Sri Lanka extending to Malaysia and Indonesia.

### **MEARUA** Forsskal

**Maerua apetala** (Roth) Jacobs in Blumea 12(2): 202. 1964; Flora of India 2: 329. 1993. *Capparis apetala* Roth, Nov. Pl. Sp. Ind. Or. 288. 1821. *Niebuhria linearis* DC. Prodr. 1: 229. 1829; FBI 1: 171; Brandis 31. *N. apetala* Dunn in Gamble 1: 41 (30) 1915.

Deciduous tree, up to 5 m tall; bark pale. Leaves thick, sub-coriaceous, 3-foliolate, petiole to 7.5 cm; leaflets linear-oblong and lanceolate, 3–6 × 1–3 cm, apex acute, tip pointed, margin entire, base rounded; secondary nerves 3–6 pairs, obscure; petiolule to 5 cm. Flowers fragrant, creamish tinged with purple, axillary or terminal corymbs. Berry oblong-ovoid, 1–2-ridged, 1.5–2 cm long; seeds are embedded in a scarlet pulp, muricate, 6 mm, white.

Scrub forests and dry regions, scattered and infrequent. Fls.: Feb.-Apr. Frs.: Apr.-May. Vern.: Pilli adugu, Pilli chettu.

Paste of root bark is applied for leucoderma and the extract given orally for the same by Chenchus and Lambadis. Tender leaves ground with spices and the paste made into pills are given orally for nervous disorders and foot pains by Chenchus.

Srisailam (KNL), *SSR* 15560; Chintala (PKSM), *SSR* 16116; Muchukota RF (ATP), *TP & NY* 220; Macherla RF (GTR), *VRK* 2924; Palakonda hills (CDP), *CS* 9368; Dornal (PKSM), *KTH* 9668 (CAL); Kulipad (CDP), *JSG* 11222 (CAL).

INDIA : Tamil Nadu—Endemic.

## BIXACEAE

### **BIXA** Linnaeus

**Bixa orellana** L. Sp. Pl. 512. 1753; FBI 1: 190; Gamble 1: 51(57); Brandis 39; Beddome Fl. Sylv. S. India t. 79. 1870; Flora of India 2: 381. 1993.

Evergreen tree, up to 5 m tall; young branches densely, dark silky, wood pale yellow, soft. Leaves alternate, coriaceous, cordate, 5.4–16.5 × 6–11.5 cm, glabrous on both surfaces, apex acuminate, margin entire, base cordate; 5-nerved; petiole to 9 cm. Flowers purplish, large in terminal panicles, showy, fragrant. Capsule reddish brown, ovoid, 2 valved, soft bristly, 3.5 × 1.5 cm; seeds trigonous, rounded and grooved on the back, covered with a red pulp (originally fleshy papillae on the testa).

Exotic, American species cultivated in some areas for its dye. Occasionally wild as an escape. Fls.: and Frs.: Aug.–Feb. Vern.: Jabaru Kaya, Jafra; Eng.: Annatto tree, Arnatto.

The red coating on the seeds yields the well known 'Annato' dye which is great demand especially for colouring food stuffs, butter, cheese, chocholates, cosmetics, oils, varnishes, soaps and also used for colouring the palms (hands) in Russia. Leaves used as febrifuge.

Maredumilli (EG), *SSR* 15234 & *GVS* 65522 (MH); Anjodigadda (VSKP), *TP & EC* 7432.

Native of tropical America, now widely spread in tropics and cultivated in many parts of India.

**Note** : It occurs in two forms, one with white flowers and green capsules and another with pink flowers and reddish brown capsules.

## COCHLOSPERMACEAE

### **COCHLOSPERMUM** Humboldt, Bonpland et Kunth *nom. cons.*

**Cochlospermum religiosum** (L.) Alston in Trimen Handb. Fl. Ceylon 6 (Suppl) 14. 1931; Flora of India 2: 383. 1993. *Bombax religiosum* L. Sp. Pl. 512. 1753. *Cochlospermum gossypium* DC. Prodr. 1: 527. 1824; FBI 1: 190; Gamble 1: 50 (36); Brandis 38.

Deciduous tree, up to 12 m tall; often with crooked branches; bark smooth, ash coloured; wood soft. Leaves near the ends of branches, thin coriaceous, 3–5-lobed, lobes ovate or lanceolate, 5–15 × 7–20 cm, glabrous above, densely brownish tomentose beneath, apex acute-acuminate, margin entire, base obtuse, 5–7-nerved from base, petiole to 20 cm. Flowers golden yellow, 15 cm across in terminal, sub-corymbose panicles. Capsule large, deep brown, obovoid, 14 × 8 cm, striate outside, 3–5-valved; leathery; seeds numerous, curved, cochleate, to 6 × 5 mm, covered with a white silky wool.

Common, in tropical dry deciduous forests in dry gravelly soils, stony hills, lower slopes at 300–900 m. Leaf fall: Jan.–Feb., New foliage appears from May. Fls.: Feb.–Apr. Frs.: Mar.–June. Vern.: Konda gogu, Konda Buruga, Adavi buruga, Adavi gogu, Kongu Akshpthamu. Eng.: Silk-cotton tree.

Used in local medicines as a substitute for the gum of *Sterculia urens*. Bark yields an edible gum called "Katara gum" which is used in Cough and Gonorrhoea, Syphilis, Asthma. Wood fibrous, quite valueless. Cotton from seeds little used for stuffing beds and pillows.

Vanipenta (KNL), *SSR* 15502; G. Madugula (VSKP), *SSR* 15983; Vengalamma cheruvu RF (ATP), *TP & NY* 2119; Palkonda range (CDP), *KMS* 7805 (CAL); Sadasiva Kona (CTR), *DR* 1798 (SVU); Pakhal RF (WGL), *ANH* 15948 (MH); Peddamantanala (PKSM), *RVK* 15777; Farahabad (MBNR), *TDCK* 13666; Venkatagiri (NLR), *ASR* 2618 (VV).

INDIA : Uttar Pradesh, Madhya Pradesh, Bihar, Orissa, Maharashtra and throughout South India.

WORLD : Sri Lanka, Vietnam, Cambodia, Thailand, Malaysia, Indonesia and Philippines.

**Note** : This species renders an elegant look in full bloom to the whole hill slopes when deciduous forests look monotonous in hot seasons.

## FLACOURTIACEAE

1. Flowers unisexual:
   2. Berry globose—Plants armed ........ ***Xylosma***
   2. Berry lobed ........ ***Flacourtia***
1. Flowers bisexual:
   3. Petals absent, axil fleshy coloured.-stamens 8, alternating with scale like staminodes ........ ***Casearia***
   3. Petals present, axil not coloured:
      4. Sepals and petals 7 each, pubescent, calyx tube adnate to the ovary, stamens 7, staminodes glandular, opposite to the petals. Plants unarmed ........ ***Homalium***
      4. Sepals and petals 5 each, glabrous, calyx tube free from the ovary, stamens numerous, staminodes 0. Plants armed ........ ***Scolopia***

## **CASEARIA** N.J. Jacquin

### KEY 1 (Leaf)

1. Leaves tomentose beneath ........ ***C. tomentosa***
1. Leaves hairy or glabrous beneath:
   2. Leaves broadly elliptic, margin crenate-serrate, base rounded or unequally obtuse ........ ***C. graveolens***
   2. Leaves narrowly elliptic, margin serrate, base acute ........ ***C. zeylanica***

### KEY 2 (Flowering)

1. Pedicel 4–5 mm long ........ ***C. tomentosa***
1. Pedicel 1–2 mm long:
   2. Flowers yellowish, not foetid ........ ***C. zeylanica***
   2. Flowers greenish, foetid ........ ***C. graveolens***

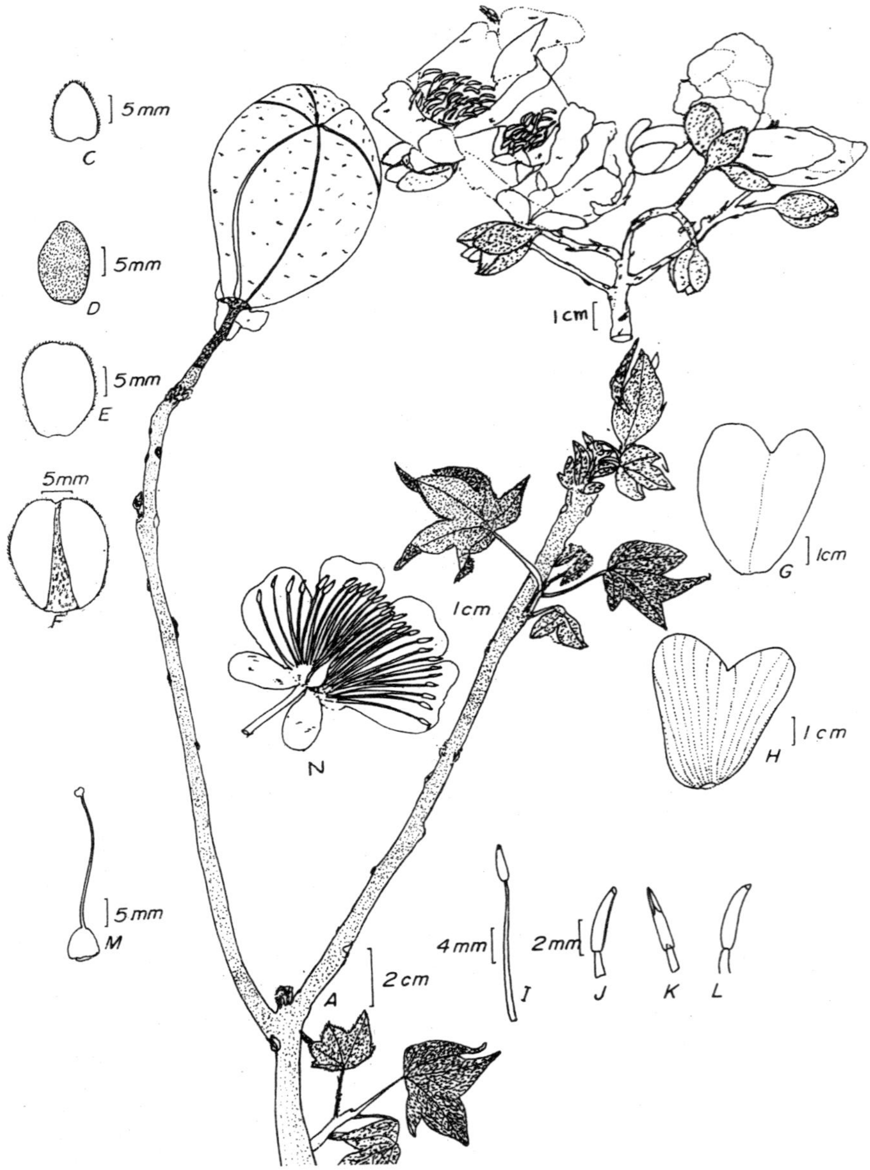


**Fig. 8.** *Cochlospermum religiousum* (L.) Alston
A. Twig, B. Inflorescence, C–F. Sepals, G. & H. Ptetals, I. Stamen J, K, & L. Anthers, M. Pistil, N. Flower.

KEY 3 (Fruiting)

1. Capsule orange ............................................................ *C. zeylanica*
1. Capsule yellow:
   2. Fruits in dense clusters ............................................................ *C. graveolens*
   2. Fruits few in clusters ............................................................ *C. tomentosa*

**Casearia graveolens** Dalz. in Hooker J. Bot. Kew. Gard. Misc. 4: 107. 1852; FBI 2: 593; Gamble 1: 520 (367); Flora of India 2: 394. 1993.

Small, deciduous tree, up to 6 m tall; bark dark grey; wood light yellow, even grained; branchlets glabrous. Leaves thick, broadly elliptic, 9.5–13 × 5.5–7 cm, glabrous, apex shortly acuminate, margin crenate-serrulate, base rounded or unequally obtuse; secondary nerves 8–12 pairs, widely upcurved, conspicuous beneath; petiole to 1 cm. Flowers pale green, foetid, villous, in clusters on leafless axils. Capsule yellow, ellipsoid or subglobose, 3-valved, 2–5 × 1–1.5 cm; seeds 10–12, ovoid, compressed with scarlet aril.

Wide spread at low elevations in both deciduous and mixed forests, in open hills, valleys and ravines. Fls.: Feb.-Mar. Frs.: Apr. onwards.

Wood moderately hard.

Galikonda (VSKP), *SSR* 15930; Singapuram (SKLM), *GVS* 62332 (CAL); Gummada-Donubai (SKLM), *GVS* 2420 (MH); Sriharikota (NLR), *JSG* 12832 (MH); Kutranda (EG), *GVS* 27256 (MH).

INDIA : Himachal Pradesh, Punjab, Uttar Pradesh, Bihar, West Bengal, Sikkim, Assam, Meghalaya, Orissa,Madhya Pradesh, Gujarat, Maharashtra, Karnataka, Tamil Nadu and Kerala.

WORLD : Pakistan, Nepal, Bhutan, Bangladesh, Myanmar, Indo-China and Thailand.

**Casearia tomentosa** Roxb. Fl. Ind. 2: 431. 1832; FBI 2: 593; Gamble 1: 521 (368); Brandis 343; Flora of India 2: 392. 1993. *C. elliptica* Willd. Sp. Pl. 2: 628. 1799, nom. superfl. (based on *Anavonga lanceolata* Lam).

Small deciduous tree, up to 8 m tall; bark thick, pale; wood yellowish-white, close-grained; branches spreading, branchlets tomentose; all parts bitter. Leaves thin-coriaceous, elliptic-oblong, or ovate-lanceolate, 8–17 × 3–4.8 cm, pubescent to velvety tomentose, apex obtuse to acute, margin coarsely serrate, base oblique, sub-truncate; secondary nerves 7–10 pairs; petiole to 8 mm. tomentose. Flowers greenish-white, 5–8 mm across, numerous in dense axillary glomerules. Capsule yellow, glossy, ellipsoid, 2.5–3 × 2–3 cm, 6-ribbed, 3-valved; seeds numerous, embedded in a red pulp, acrid, agglomerate.

Occasional, along streams in dry deciduous forests; up to an elevation of 900 m. Fls.: Feb.–Apr. Frs.: Apr.–Aug. Leaf fall: Feb.–Mar., red before falling. Vern.: Chilaka-duddi.

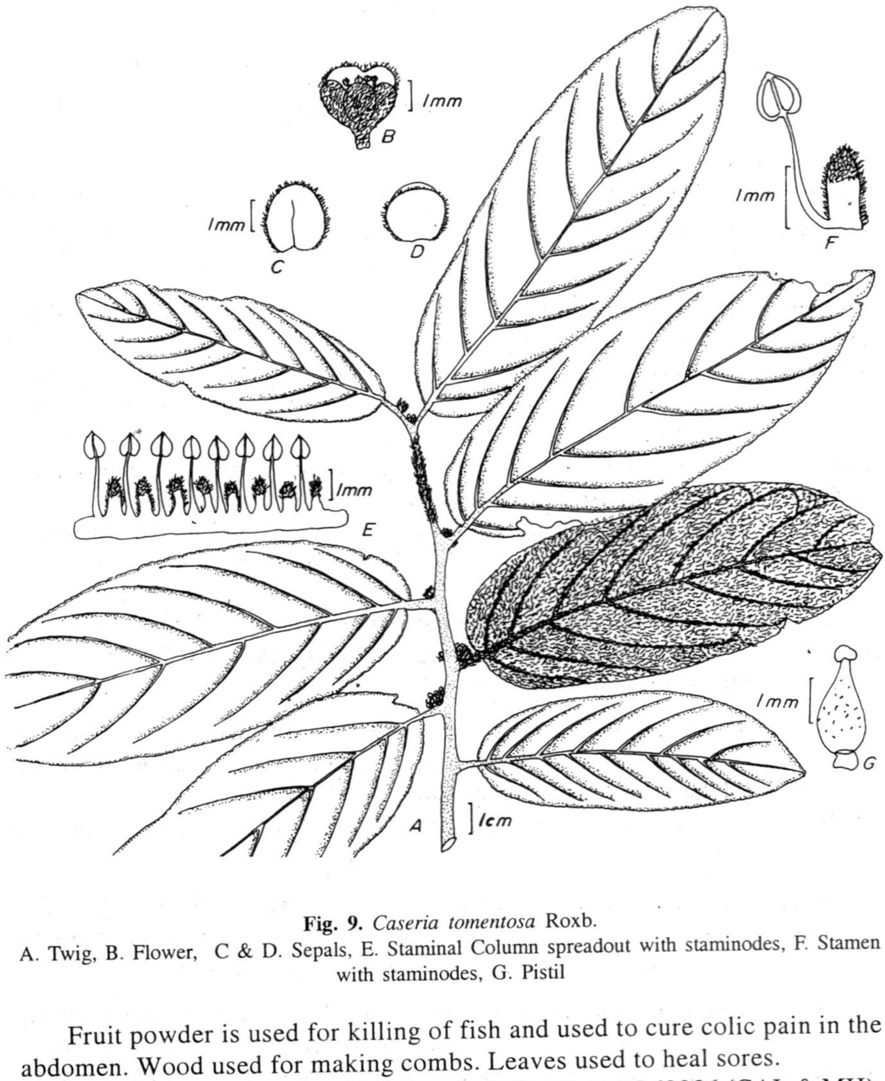


**Fig. 9.** *Caseria tomentosa* Roxb.
A. Twig, B. Flower, C & D. Sepals, E. Staminal Column spreadout with staminodes, F. Stamen with staminodes, G. Pistil

Fruit powder is used for killing of fish and used to cure colic pain in the abdomen. Wood used for making combs. Leaves used to heal sores.

Talakona (CTR), *SSR* 15554; Korada (SKLM), *GVS* 62386 (CAL & MH); Bairluti (KNL), *JSG* 18710 (CAL); Punyagiri (VSKP), *GVS* 23111 (MH); Addatheegala (EG), *MSR* 1703 (CAL); Naidupet (NLR), *ASR* 6210 (VV).

INDIA : Common throughout from sub-Himalayan tract to Peninsular India.

WORLD : Sri Lanka, Pakistan, Malaysia to Australia.

**Casearia zeylanica** (Gaertn.) Thw. Enum. Pl. Zeyl. 19. 1858; Alston in Trimen. Handb. Fl. Ceyl. 6 (Suppl.) 131. 1931; Flora of India 2: 399. 1993. *Vareca zeylanica* Gaertn. Fruct. 290. t. 60. 1788. *C. esculenta* Roxb. Fl. Ind. 2. 1832; FBI 2: 592; Gamble 1: 321 (368); Brandis 343.

Evergreen, small tree, up to 6 m tall; bark white; young shoots angular, appressed hairy. Leaves thick, subcoriaceous, narrowly elliptic-lanceolate, 9–15 × 4.5–5 cm, glabrous, shining, often becoming blackish green on drying, apex obtusely acuminate, margin serrate, base acute or cuneate; nerves irregularly branched and form a dense imperfect reticulum; petiole to 8 mm. Flowers yellowish, 3–4 mm across, glabrous in axillary fascicles. Capsule ellipsoid, 2–3 valved, 3 cm long, stalked, aril bright red.

Common, in dry evergreen forests of Nellore and Srikakulam districts. Fls.: Feb.–Mar. Frs.: May–Sept. Vern.: Konda jungare. Leaves are used as vegetable and in local medicine.

Seethampet forest (SKLM), *SSR* 15578; Sriharikota (NLR), *CECF* 4680, 4682 (CAL); *JSG* 12832 (CAL).

INDIA : East-Coast of Peninsular India and South-Westwards to Annamalai hills up to 800 m, Tamil Nadu and Kerala.

## FLACOURTIA L. Herit.

1. Leaves elliptic-oblong or obovate, apex acute or obtuse. Racemes pubescent; style little or not connate at base. Berry 5–10 mm across ..... ***F. ramontchi***
1. Leaves ovate-oblong or oblong-lanceolate, apex obtusely acuminate. Racemes glabrous; style connate into a distinct column. Berry 1.5–2.5 cm across ..... ***F. jangomas***

**Flacourtia jangomas** (Lour.) Raeusch Nom. Bot. ed. 3: 290. 1797; Flora of India 2: 403. 1993. *Stigmarota jangomas* Lour. Fl. Cochinch. 2: 634. 1790. *Flacourtia cataphracta* Roxb. ex Willd. Sp. Pl. 4. 830. 1806; FBI 1: 193; Gamble 1: 54(39).

Deciduous tree, 7–10 m tall; bark smooth; wood reddish, close grained; spines simple or branched. Leaves thinly-chartaceous, ovate-oblong, or oblong-lanceolate, 5–10 × 2–5 cm, glabrous, and shiny green above, minutely puberulent along midrib beneath, apex obtusely acuminate, margin crenate to crenate-serrate, base cuneate or rounded, 3–5-ribbed; secondary nerves 4–5 pairs; petiole 4–10 mm, puberulous. Flowers greenish-white, 5–6 mm across, few, deciduous in irregular glabrous racemes or corymbose 1–1.5 cm long fascicles. Berry red or purple then blackish, subglobose, 1.5–2.5 cm in dia; seeds 4–5.

Rare, in moist deciduous forests of Visakhapatnam. Fls.: June. Frs.: Oct.–Jan. Vern.: Kuragayi, Talispatram. Eng.: Puneela plum.

The leaves and bark are used to treat bleeding gums and toothache. Fruits are acidic, but edible and can be made into jelly. The wood is used for agricultural implements.

There are no specimens in MH, CAL, DD, or BLAT herbaria, but it is included on the Authority of Gamble (1918).

Native of North and East Bengal.

INDIA : Bengal, Assam, Bihar, Orissa, Uttar Pradesh, Maharashtra.
WORLD : Widely cultivated in South-East Asia and East Africa.

**Flacourtia ramontchi** L.' Herit. Strip. Nov. 3: 59. t. 30 & 30 B. 1786; FBI 1: 193; Gamble 1: 54 (39); Brandis 40; Flora of India 2: 402. 1993. *F. sapida* Roxb. Pl. Cor. t. 69. 1796 & Flora India 3: 835. 1832.

Deciduous tree, up to 12 m tall; wood red and close-grained. Leaves clustered at the ends of older branches, membranous to sub-coriaceous, variable, obovate, elliptic or oblong, 2.5–9 × 1.5–4 cm, more or less pubescent along midrib and nerves on both surfaces, apex acute or obtuse, margin sub-entire to crenate, base acute or cuneate; 3–5 pliveined; petiole to 1.6 cm, pubescent. Flowers yellowish-green, small in compact racemes or small paniculate racemes. Drupe dark red or purple, globose, 7 × 6 mm, stones 10, superposed in two tiers.

Occasional, in high hills of deciduous and semi-evergreen forests. Leaf fall: Dec.–Jan. New foliage: Mar.–Apr. Fls.: Oct.–Feb. Frs.: Mar.–May. Vern.: Kanru, Pulivelaga, Pulielka, Pulregu, Putika Kada, Nelli, Rartha Kandragu, Nakka neredo. Eng.: Governer's plum, Madagarscar plum.

Fruits are sweet and edible, increase appetite and digestion and also given in jaundice. Wood hard and durable, but liable to splint, used for agricultural implements and posts.

Tirumala hills (CTR), *SSR* 13771; Tirumala-Papavinasanam (CTR), *SSR* 15283; Rollapenta (KNL), *SSR* 15513; Donubai RF (SKLM), *SSR* 15589; Gangapur RF (MDK), *BR & CP* 11433; Kavu Kona (CDP), *R.V. Reddy* 8O71; Way to Kunthala (ADB), *GO & DAM* 5122; Gandivet RF (NZB), *BR* 9548.

INDIA : Common throughout.
WORLD : Madagascar, Malayan Archipelago.

**Note** : *Flacourtia ramontchi* L.' Herit. is distinct from the species *F. indica* (Burm. f.) Merr. by the characters of small trees with larger leaves, distinct paniculate inflorescences and found in dry deciduous and mixed forests. In *F. indica* plants are usually twiggy shrubs with smaller leaves, few flowers in axillary clusters and found in dry thickets and scrub forests.

## HOMALIUM N. Jacquin

1. Leaves glabrous; petiole 0.7–1.3 cm long, puberulous. Flowers greenish-white; racemes simple, pendulous, glabrous; sepals 4–6; staminal filaments 2–3 mm long ... ***H. ceylanicum***
1. Leaves glabrescent; petiole 1.5–2.7 cm long, densely tomentose. Flowers whitish cream in erect tomentose panicles; sepals 6–8; staminal filaments 1–1.6 mm long ..... ***H. napaulense***

**Homalium ceylanicum** (Gard.) Benth. in J. Linn. Soc. Bot. 4: 35. 1860 (as "*zeylanicum*") FBI 2: 596; Gamble 1: 522 (369); Brandis 345; Flora of India 2: 410. 1993; Beddome Fl. Sylv. S. India t. 210. 1872. *Blackwellia ceylanica* Gard. in Calc. J. Nat. Hist. 7: 452. 1846.

Evergreen tree, 10–20 m tall; bark grey, smooth, peeling off in irregular flakes; wood brownish-red; young branches glabrous. Leaves sub-coriaceous, ovate, ovate-elliptic or broadly elliptic, 5–12 × 3–6 cm, glabrous, glossy above, apex acuminate, margin repando-crenate, base obtuse or cuneate; secondary nerves 5 pairs, tertiary and quarternary veins forming fine reticulum; petiole to 7–13 mm. Flowers greenish-white, foetid, 3–4 mm across; racemes simple, pendulous, appressed, to 12 cm long.

Rare, in semi-evergreen forests up to 850 m in Chittoor district. Fls.: Mar.-May. Vern.: Manthralamukhi.

Wood is hard, heavy and durable, used for beams, poles, mats and building purposes.

Talakona (CTR), *SSR* 13718; Kambakkam (CTR), *DR* 1363 (SVU); Tirumala (CTR), *DR* 2063 (SVU).

INDIA : Karnataka, Tamil Nadu and Kerala.
WORLD : Sri Lanka.
**Note** : Rarely found in Andhra Pradesh State in the interiors of narrow gorges of the hills of Tirumala, Talakona and Kambakkam in Chittoor district. It grows very tall. Reasons for its rarity in this area may be due to lack of fruit set.

**Homalium nepaulense** (DC.) Benth. in J. Linn. Soc. Bot. 4: 34. 1866; FBI 2: 596., as "*nepalense*" Gamble 1: 552 (369); Brandis 345; Flora of India 2: 413. 1993. *Blackwellia napaulense* DC. Prodr. 2: 54. 1825.

Tree, up to 15 m tall; bark greyish white, smooth; wood greyish red, darker in the centre, even grained; branchlets puberulous. Leaves coriaceous, ovate-elliptic, 9–15 × 5–8 cm, glabrescent, greyish after drying, apex narrowly, shortly acuminate, margin coarsely serrate, base acute or rounded; secondary nerves 6–8 pairs, conspicuous; petiole to 2.7 cm. Flowers whitish-cream, foetid, sessile in 5–10 cm long, densely ferruginous, tomentose axillary panicles. Fruit not seen.

Common, in moist deciduous forests of Visakhapatnam and East Godavari districts up to 1500 m Fls.: Apr.–June.

Wood moderately hard, strong and durable, used for posts, rafters and building purposes.

Sunkarimetta (VSKP), *SSR* 15911; Anantatgiri (VSKP), *SSR* 19613;Rampa hills (EG), *JSG* 16010 (MH); Anantagiri (VSKP), *GVS* 19575 (MH); Inspection Bungalow, Anantagiri (VSKP), *GVS* 19375 (MH).

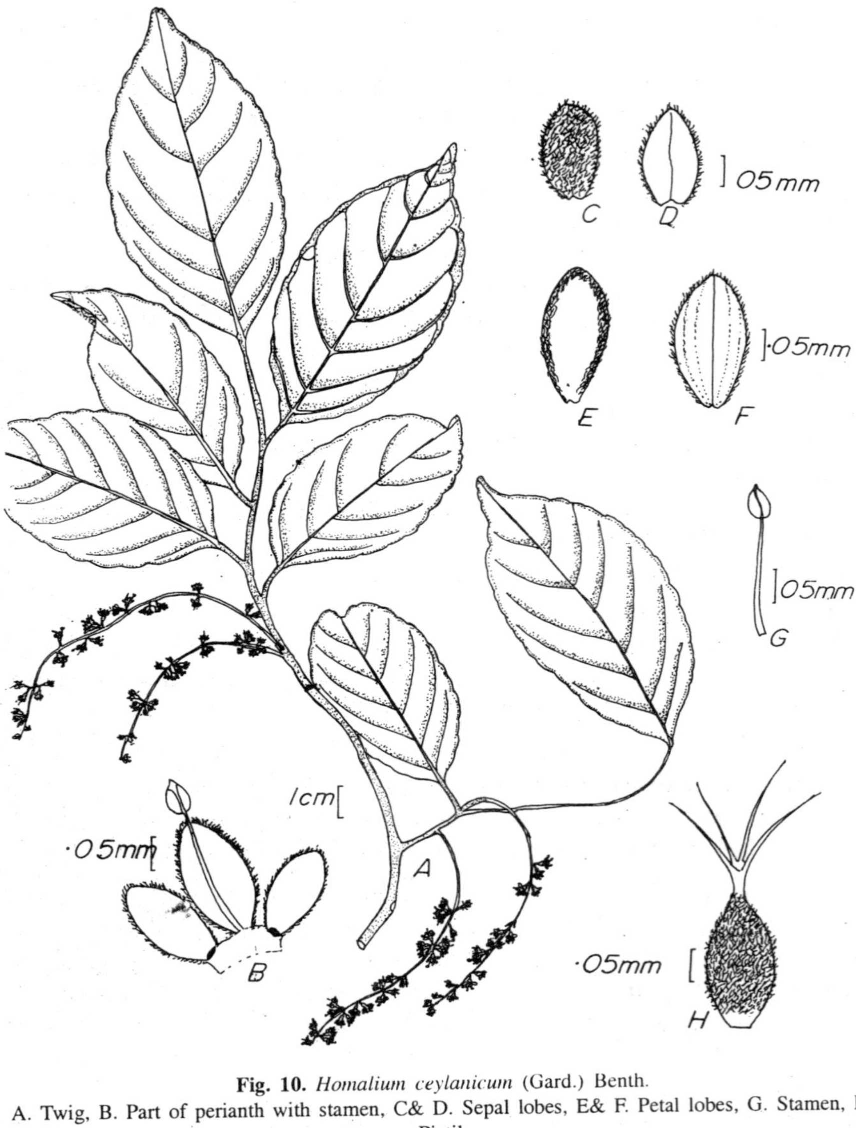


**Fig. 10.** *Homalium ceylanicum* (Gard.) Benth.
A. Twig, B. Part of perianth with stamen, C& D. Sepal lobes, E& F. Petal lobes, G. Stamen, H. Pistil.

INDIA : Tropical and sub-tropical mixed and evergreen forests, 700–2100 m Nagaland, Bihar (Singhbum), Orissa.

WORLD : Nepal.

**Fig. 11.** *Homalium nepaulense* Benth.
A. Twig, B. Flower front view, C & D. Sepal lobes, E& F. Petal, G. Stamen, H. Pistil.

## SCOLOPIA Schreber

**Scolopia crenata** (Wt. & Arn.) Clos in Ann. Sci. Nat. Bot. Ber. 4. 8. 250. 1857, excl. specimen Philippin.; Bedd. Fl. Sylv. S. India t. 78. 1871; FBI 1: 191; Gamble 1: 32 (38); Brandis 39; Flora of India 2: 430. 1993. *S. crenata* var. *brevifolia* N. Mukherjee in J. Bombay Nat. Hist. Soc. 69. 394. 1973. *Phoberos crenatus* Wt. & Arn. Prodr. Fl. Ind. Orient. 20. 1834.

Densely foliaceous, evergreen tree, 6–10 m tall; crown compact, conical; bark grey, rough and pustular by raised lenticels; wood reddish brown, close-grained; young branches glabrous, pale brownish; spines simple, straight, up to 2 cm long. Leaves coriaceous, variable, ovate, ovate-lanceolate, 5–12.5 × 3–5.5 cm, glabrous, green and glossy above, paler beneath, apex shortly, obtusely acuminate, margin glandular, serrate or crenate, base narrowed or rounded; secondary nerves 5–8 pairs, irregular; petiole 6–8 mm, minutely biglandular at junction of lamina. Flowers white, fragrant, 7–8 mm across in lax, puberulous, 3–5 cm long, terminal racemes. Berry shiny green, globose, 1.2–1.5 cm across, apiculate; seeds 2–6, ovoid.

Occasional, along the water courses on hills of moist deciduous forests at 500–850 m. Fls.: Mar.–May. Frs.: July–Oct. Vern.: Purusha kesari; Eng.. Potato of Mysore.

Wood very hard.

Tirumala (CTR), *SSR* 19601; Mutinayani Cheruvukota (CTR), *GVS* 45939 (MH); Tirumala (CTR), *DR* 1582 & 2136 (SVU).

INDIA : Karnataka, Tamil Nadu and Kerala—Endemic.

## XYLOSMA Forster *nom. cons.*

**Xylosma longifolium** Clos in Ann. Sci. Nat. Bot. Ser. 4. 8: 23; FBI 1: 194; Gamble 1: 53 (33); Brandis 40; Flora of India 2: 456. 1993.

Evergreen tree, up to 15 m tall; trunk often armed with sharp, stout, 2–3 cm long thorns; bark smooth with long black warts. Leaves coriaceous, elliptic-lanceolate, oblong-lanceolate, 4.5–11.5 (22) × 1.6–5 (6.5) cm, glabrous, shining, dark green above, blackish green on drying, apex shortly acuminate, margin serrate, base acute or cuneate; secondary nerves 10–12 pairs, arching, little raised beneath; petiole 5 mm. Flowers pale yellowish green, 5 mm across, in axillary pubescent racemes forming dense glomerules. Berry red, globose, 2–4 mm dia.; seeds 3–8, angled.

Occasional, in moist evergreen ravines in Visakhapatnam district at higher elevations. Fls.: Oct.–Jan. Frs.: Feb.–Apr.

Palakonda hills (VSKP), *JSG* 14013 (MH & DD); Minumuluru (VSKP), *GVS* 29592 (MH).

INDIA : Jammu & Kashmir, Himachal Pradesh, Punjab, Haryana, Uttar Pradesh, Bihar, West Bengal, Arunachal Pradesh, Assam, Meghalaya, Manipur, Tripura, Orissa, and Tamil Nadu.
WORLD : Pakistan, Nepal, Indo-China, Thailand and China.

## PITTOSPORACEAE
### PITTOSPORUM Banks ex Solander

**Pittosporum nepaulense** (DC.) Rehder & Wilson in Pl. Wilson 3: 336. 1916; Flora of India 2: 445. 1993. *Senacia napaulensis* DC. Prodr. 1: 347. 1824. *Pittosporum floribundum* Wt. & Arn. Prodr. Fl. Ind. Orient. 154. 1834; FBI 1: 199; Gamble 1: 55 (40); Brandis 42.

Evergreen tree, up to 10 m tall; bark light grey, smooth, lenticillate, slightly peeling; sap watery, drying sticky; wood whitish brown; branchlets sub-verticillate. Leaves loosely clustered at the ends of branches, thinly coriaceous, oblong to lanceolate or oblanceolate, 3.5–12 × 1.7–4 cm, glabrous, dark green and shining above, pale green beneath, apex acute, margin slightly undulate, base cuneate; secondary nerves 7–11 pairs; petiole to 1.5 cm. Flowers yellow, small in terminal or axillary panicles. Capsule orange-yellow, globose, pisciform, which at maturity turns jet black and retains its apiculation, 2-valved, striated inside, 4–5 mm in dia; seeds 4–6, orange with resinous pulp.

Rare, on high hills of moist deciduous forests. Fls.: May–Aug. Frs.: Oct. Vern.: Chettu Kasirda, Rakamuki. Eng.: Ginger tree.

The bark is bitter, but aromatic and has ginger-like smell when freshly cut and also possesses narcotic properties; it is used for skin diseases, leprosy, chest infections, inflammations, rheumatic pains, and fevers. Wood is moderately hard and suitable for making small toys and is also used as fuel.

Tirumala hills (CTR), *SSR* 13739 & 15264; Anantagiri (VSKP), *SSR* 19624; Galikonda (VSKP), *GVS* 21645 (MH); Chintapalli (VSKP), *JLE* 37116 (MH); Rampa (EG), *VNS* 486 (CAL); Horsely hills (CTR), *CECF* 4645 (MH).

INDIA : Throughout, except Rajasthan.
WORLD : Nepal, Bhutan, China (Tibet) and Madagascar.

## TAMARICACEAE
### TAMARIX Linnaeus

**Tamarix aphylla** (L.) Karsten. Deutsche. Fl. 641. 1882; Flora of India 3: 22. 1993. *Thuja aphylla* L. Cent. Pl. 1: 32. 1755, p. p. *T. articulata* Vahl, Symb. Bot. 2: 48. t. 32. 1791; FBI 1: 249; Brandis 46.

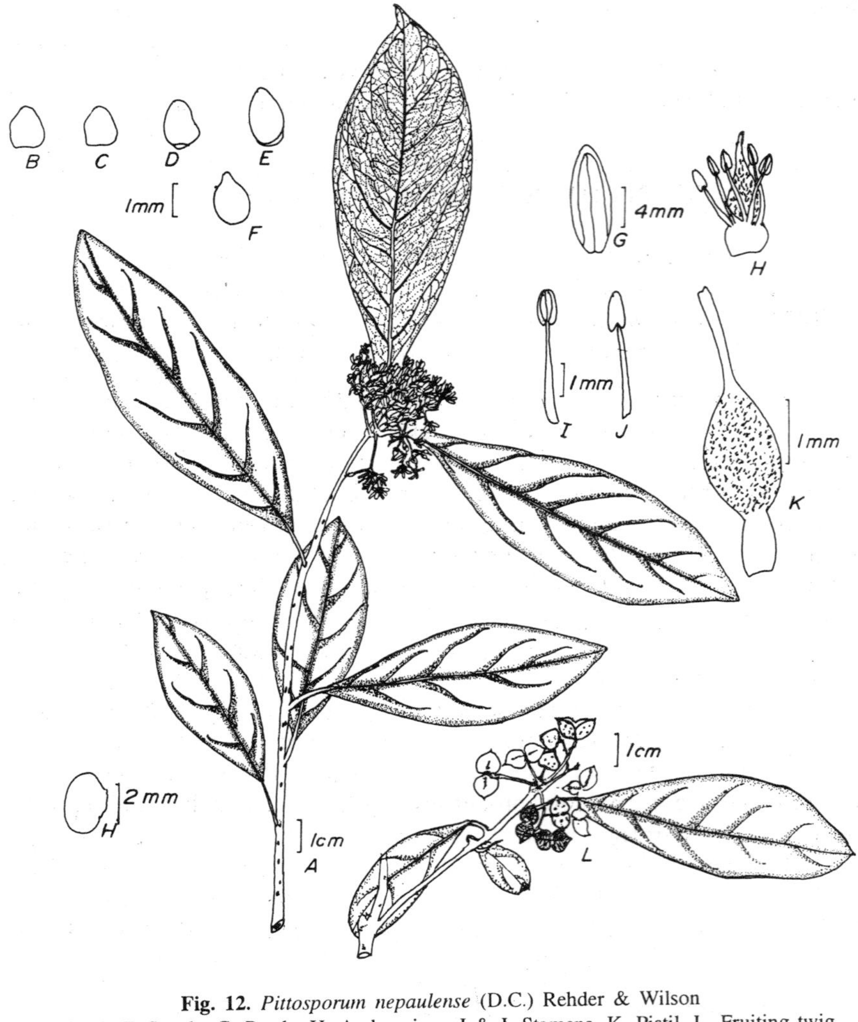


**Fig. 12.** *Pittosporum nepaulense* (D.C.) Rehder & Wilson
Twig, B–F. Sepals, G. Petals, H. Androecium, I & J. Stamens, K. Pistil, L. Fruiting twig.

Small to moderate sized tree, 6–7 m tall; trunk erect; wood white; branchlets jointed at the base of leaf sheath. Leaves vaginate, triangular, minute or very small, apex acuminate, margin entire, base broad. Flowers pinkish white, fragrant, 3 mm across in simple or compound racemes. Capsule 7 mm long, 3-gonous; seeds terete, 5 mm long, pilose, coma 2–3 mm long.

In sandy, saline habitats along streams and rivers. Introduced, and cultivated in plains, especially on the coasts as a wind break and sand binder. It is also recommended for planting in shifting sands. The extremities of branchlets and the leaves on older branchlets are shed during the cold season, new shoots and leaves come out about May. It coppices well. Fls.: May–Aug. Frs.: Oct.–Feb. Vern.: Erravsaru, Errashisisaru, Farash, Sarru, Lal-Thau.

The timber is strong, used in house construction, agricultural implements, furniture etc. Baskets are made out of twigs and the wood is valued as cheap fuel. The bark is employed in treatment of eczema and other skin diseases.

Krishna River bed-Vijayapuri South (GTR), *VRK* 6763.

INDIA : Punjab, Haryana, Delhi, Uttar Pradesh, Rajasthan, Gujarat and Tamil Nadu.

WORLD : Pakistan, Afghanistan, South-West Asia and East, North and West Africa.

## CLUSIACEAE

1. Lateral veins close, and differentiated into secondary and inter-secondary nerves. Ovary one-celled .......... ***Calophyllum***
1. Lateral veins not close, and differentiated into secondary and inter-secondary nerves. Ovary bi-multicelled:
   2. Calyx of 4–5 sepals .......... ***Garcinia***
   2. Calyx bursting into 2-valves .......... ***Mammea***

### **CALOPHYLLUM** Linnaeus

**Calophyllum inophyllum** L. Sp. Pl. 513. 1753; FBI 1: 273; Gamble 1: 76 (55); Brandis 54. Flora of India 3: 92. 1993. *C. brintagor* Roxb. Fl. Ind. 2: 606. 1832.

Densely foliaceous, evergreen tree, up to 10 m tall; latex milky; crown spreading; bark pale-grey, smooth, often fissured; wood reddish-brown; branchlets glabrous, compressed. Leaves thick-coriaceous, elliptic-oblong or obovate, 11–13.5 × 7–8 cm, glabrous, shining, dark green above, apex silightly emarginate, margin wavy, base rounded or obtuse; midrib very distinct and raised below, lateral nerves numerous, very close, parallel; petiole to 1.2 cm. Flowers white, fragrant, 2 cm dia., 10–15-flowered in axillary, 8–12 cm long racemes. Drupe yellowish, globose, 2.8 × 2.5 cm, apiculate, pericarp thick; seed pulp scanty; seed solitary, 1.2 cm across.

Cultivated as an ornamental tree. Fls. & Frs.: Throughout the year. Vern.: Ponna, Punnaga chettu, Namare. Eng.: The Dielo Oil Tree, The Alexandrian Laurel.

The timber is durable under water and is used for making knees for boats, railway sleepers, for construction and for domestic vessels and bowls etc. Seed oil known as Dielo Oil is used in skin diseases, rheumatism, burns and painting wood and also as a lubricant, a substitute for Castor oil. The leaves used to cure migraine and inflammed eyes. The decoction of flowers is given as a remedy for syphilis, eczema and insanity. Fruits eaten by bats and squirrels. Bark boiled in water is used for dyeing fish nets. Bark yields an aromatic yellowish-green resin and used as astringent and purgative.

A.U. Campus (VSKP), *TP & EC* 7323; Sriharikota (NLR), *ASR* 367 (VV).

INDIA : A littoral species, common on sandy coast and laterite soils just above high tide level in Orissa, Maharashtra, Karnataka, Tamil Nadu, Kerala, Lakshadeep and Andaman & Nicobar Islands.

WORLD : Mozambique, Madagascar, Mascerene Isles and also cultivated in many parts of the world.

## GARCINIA Linnaeus

1. Branchlets slender, latex white or yellow. Leaves ovate-oblong or elliptic, 5–20 cm long, apex obtuse; secondary nerves obliquely parallel. Flowers in pseudo spikes; pedicel 2.5–3.8 cm. Berry 4 cm in dia., oblong, yellowish; seeds 2.5 cm long ............................ ***G. spicata***

1 Branchlets thick, compressed, latex milky or pale green, turning yellow on exposure. Leaves linear-oblong or oblong-lanceolate, 26 to 38.5 cm long, apex acuminate, secondary nerves subparallel. Flowers in fascicles; pedicel 0.6–1.2 cm. Berry 6 cm in dia., sub-globose, dark yellow; seeds 3–5 cm long ........................................................................ ***G. xanthochymus***

**Garcinia spicata** (Wt. & Arn.) Hook. f. in J. Linn. Soc. Bot. 14: 486. 1785; Gamble 1: 74 (53); Brandis 49; Flora of India 3: 124. 1993. *Xanthochymus spicatus* Wt. & Arn. Prodr. Fl. Ind. Orient. 102. 1834. *Garcinia ovalifolia* Hook. f. var. *spinuta* (Wt. & Arn.) T. And. in Hook. f. Fl. Brit. India 1: 269. 1874.

Moderate sized, evergreen tree, trunk straight; branches spreading; bark brownish, thick, smooth; wood yellowish-white, medium coarse to fine textured; branchlets ridged or angular, slender; latex white or yellow, sticky, thick, scanty. Leaves thickly-coriaceous, elliptic-oblong or ovate-oblong, 5–20 × 3–9 cm, glabrous, green and glossy above, paler beneath, apex obtuse, margin entire or wavy, base obtuse; secondary nerves numerous, prominent, obliquely parallel; petiole 1.2 cm, stout. Flowers pale green or white in axillary, 10–15 cm long pseudo-spikes. Fruits yellowish green, broadly oblong, 4 cm in dia., smooth; seeds 1–3, oblong, to 2.5 cm long, laterally flattened.

Occasional, along the coast but abundant at Sriharikota Islands in Nellore district. Fls. & Frs.: Mar.–July. Vern.: Pidaetha.

The fruits are edible. Sticky pulp of the young fruits afford a chrome-yellow pigment. Bark contains colouring matter. "Fukuji" used as mordant dye stuff in Japan. Timber is hard and strong, useful for general construction purposes, where strength is the main criterion.

Sriharikota (NLR), *YYR* 2056 & *JSG* 12669 (DD).

INDIA : In Western Ghats, Eastern Coast and North-east India at low elevations. Assam, Meghalaya, Orissa, Maharashtra, Karnataka, Tamil Nadu and Kerala.
WORLD : Sri Lanka.

**Garcinia xanthochymus** Hook. f. ex T. And. in Hook. f. Fl. Brit. India 1: 269. 1874; Brandis 49; Flora of India 3: 129. 1993. *Xanthochymus pictorius* Roxb. Pl. Cor. t. 196. 1805 (Sphalm *X. tinctorius* DC. 1824) non *Garcinia pictoria* Roxb. 1824. *G. tinctoria* (DC.) W.F. Wight in U.S.D.A. Bur. Pl. Industr. Bull. 137: 50. 1909; Gamble 1: 74 (55).

Moderate-sized tree, 15–20 m tall; crown dense, pyramidal; branchlets drooping, 6–8 angular, often dilated just below nodes; bark blackish, exfoliating in small round flakes; wood yellowish-brown to dark greyish-brown; latex milky or pale green, turning yellow on exposure. Leaves thick-coriaceous, variable in shape and size, oblong-lanceolate or linear-oblong, 26.8–38.5 × 6.4–10.5 cm; glabrous, dark green and glossy above, paler beneath, apex acuminate, margin entire, reflexed, base obtuse, rounded or acute; secondary nerves to 25 pairs or numerous, sub-parallel and joined to form a intramarginal vein; petiole 1–2.5 cm, rugose, angular, thick, stout. Flowers white or cream coloured, polygamous, 1.5–2 cm in dia., 4–10 in axillary fascicles. Berry dark yellow, subglobose, 6 cm in dia., pointed, with persistent calyx; seeds 1–4, brown, oblong, 3.5 × 1.8 cm.

Rare, in East Godavari and Srikakulam districts, sometimes grown in gardens as an ornamental tree for its dense foliage. Fls. & Frs.: Throughout the year. Vern.: Ivarumidi, Iwara mamidi, Chitakamaraku, Thamalamu, Lollorimanu, Sikatimramu, Pidatha. Eng.: Mysore Gamboge.

Fruits acidic, pleasant and edible, they are used for making sherbeths, medicaments, preservatives and jams. The gum resin from the stem, bark and fruit make a good water colour (gamboge) used in dyeing. Timber is strong, useful for general construction purposes, where strength is the main criterion.

Vathangi (EG), *JSG* 12677 (MH).

INDIA : In evergreen, semi-evergreen, moist deciduous, and in the lower hill forest up to 1400 m. Sikkim, Assam, Tripura, Meghalaya, Orissa, Maharashtra, Goa, Karnataka, Tamil Nadu, Kerala and Andaman & Nicobar Islands.
WORLD : Bangladesh, Myanmar, China, Thailand, and Malay Peninsula.

## **MAMMEA** Linnaeus

**Mammea surgia** (Buch.-Ham. ex Roxb.) Kosterm. in Comm. For. Res. Inst. Indones. Bogor. 72: 33. f. 19. 1961; Flora of India 3: 132. 1993.

*Calophyllum surgia* Buch.-Ham. ex Roxb. Flora Ind. 2: 608. 1832. *Mammea longifolia* (Wt.) Planch. & Triana in Ann. Sci. Nat. Ser. 4, 15: 240. 1861. *Ochrocarpus longifolius* (Wt.) T. Anders. in Fl. Brit. India 1: 270. 1874; Gamble 1: 75 (54); Brandis 53.

Evergreen tree, 12–16 m tall; bark rough, exfoliating in irregular strips; wood red, smooth, sap milky; branchlets obscurely 4-angled. Leaves thickly coriaceous, oblong to oblong-lanceolate, 15–20 × 5–6 cm, glossy, dark green, glabrous, apex oblong-obtuse or shortly or bluntly pointed; margin entire, base obtuse; secondary nerves numerous, parallel, faint, very slender; petiole to 1 cm. Flowers white or pinkish, unisexual or often bisexual, numerous, 8–12 mm in dia., sweet scented in axillary fascicles. Berry obliquely ovoid, stipitate, 2.5–3 × 1–1.5 cm, tipped with hard point, pulp juicy; seeds 1–4, 2 × 1 cm.

Cultivated in Northern Circars and especially in Kakinada (Wagh, 1960). It is planted as an avenue tree and in household gardens. Fls. & Frs.: Feb.-May. Vern.: Suraponna, Sura punnaganu.

Wood hard, sometimes used for building purposes, and rarely as fuel. flower buds used for dyeing silk, and yield a perfume. Flowers used in Hindu worship and for decorating hair. Fruits edible. Seeds yield viscid gum.

INDIA : In evergreen and deciduous forests up to 600 m. Maharashtra, Karnataka, Tamil Nadu, Kerala. Cultivated in West Bengal, Assam, Orissa and Uttar Pradesh.

WORLD : Bangladesh.

## DIPTEROCARPACEAE

### **SHOREA** Roxburgh

### KEY 1 (Vegetative)

1. Leaf base cordate; secondary nerves thick, usually 1.5 cm apart; petiole to 2 cm long; leaves brownish-green when dry; bark reddish brown or grey .................................... ***S. robusta***
1. Leaf base rounded or subcordate; secondary nerves slender, 0.8–1 cm apart; petiole to 5 cm long; leaves yellowish-green when dry:
   2. Bark light-grey, fleshy, smooth or narrowly fissured; branchlets glabrous.Leaves elliptic-oblong or oblong, apex acute or retuse; secondary nerves 11–16 pairs, 1 cm apart .................................... ***S. roxburghii***
   2. Bark dark brown, thick, rough, longitudinally fissured; branchlets tomentose. Leaves ovate or ovate-elliptic, apex acuminate; secondary nerves 8–10 pairs, 0.7 cm apart .................................... ***S. tumbuggaia***

### KEY 2 (Flowering & Fruiting)

1. Panicles glabrous; stamens ca 15, anther connective appendages filiform. Fruit glabrous .................................... ***S. roxburghii***

1. Panicles pubescent or tomentose; stamens 20–50, anther connective appendages ciliate. Fruit pubescent or tomentose:
   2. Flowers yellow or cream coloured. Fruit belly 1.5 cm long, acute; wings 11-nerved, 2 smaller, 3-larger, 3.5–7 × 0.5–1.5 cm .......... ***S. robusta***
   2. Flowers white. Fruit belly 2 cm long, acuminate; wings ca. 9-nerved, unequal, 3.5–4.5 × 1–1.4 cm .......... ***S. tumbuggaia***

**Shorea robusta** Roxb. ex Gaertn. f. Fruct. 3: 48. t. 186. 1805; FBI 1: 306; Gamble 1: 83 (60); Brandis 70; Flora of India 3:237. 1993; Beddome Fl. Sylv. S. India t. 4. 1869. *Vatica robusta* Steud. nom. ed. 2: 745. 1841.

Very large, deciduous tree; crown spreading; bark reddish-brown or grey with deep longitudinal furrows; wood dark brown, coarse and cross grained; branchlets stellate pubescent. Leaves thick-coriaceous, broadly oblong, 8.5–14.5 × 7.4–9.4 cm, glabrous above and tomentose or glabrescent beneath, glossy, apex obtusely acuminate, margin thick, slightly undulate, base cordate; secondary nerves 10–12 pairs, 1.6–1.8 cm apart; petiole up to 2 cm. Flowers cream or pale yellow, small, dense in slender, terminal or lateral, drooping, lax panicles. Fruit reddish to pale yellowish-green, ovoid, 7.5 cm long; wings 5, 3-larger and 2-smaller, linear-oblong or spathulate; seed one, ovoid with fleshy unequal cotyledons.

Rare, in moist forests of Srikakulam district, up to 900 m. It grows well in well-drained, moist, deep sandy loam, with good sub-soil drainage. It is light demander, fire-resistant and good coppicer. Leaf fall: Mar. New foliage: Apr. Fls.: Feb.–Apr. Frs.: June–July. Vern.: Saluva, Guggilam, Gugal, Sal, Thamba. Eng.: The Sal tree, Indian dammer tree, Common sal.

The leaves are considered medium to poor fodder and used in bidi-making and also preparation of plates and cups. The wood being strong and durable, ranks next to teak and is considered one of the highly coveted constructional and domestic wood. It is also used for making electrical transmission poles, sleepers, floor, bridges, carriages and wagon building, boat and ship building, agricultural implements etc. It also makes a very good fuel wood. The tree yields an oleo-resin called as sal used in indigenous system of medicine, menorrhagia, enlargment of the spleen, dysentry, itching and burns as an incense and as a hardening agent in making shoe polishes, carbon paper, type writer ribbons. Tannin obtained from the bark, twigs and leaves is used for tanning leather, paper, cellulose etc. The fatty oil from the seed (sal butter) is used for cooking and lighting purposes. The tree is one of the primary hosts for tasar silk worm.

Donubai (SKLM), *SSR* 15582; Gummada (SKLM), *GVS* 62418 (MH); Palakonda hills (VSKP), *JSG* 3432 (MH).

INDIA : Terai region of tropical Himalayas from Himachal Pradesh to Assam and to Tripura, West Bengal, Bihar and Orissa, Eastern districts of Madhya Pradesh extending further south to the Eastern Ghats in Andhra Pradesh up to an altitude of 900 m.

WORLD : Nepal and Bhutan.

**Shorea roxburghii** G. Don, Gen. Syst. 1: 813. 1831; Flora of India 3: 239. 1993. *S. talura* Roxb. Fl. Ind. 2: 618. 1812; FBI 1: 304; Gamble 1: 83 (60); Brandis 70. *Vatica laccifera* Wt. & Arn. Prodr. Fl. Ind. Orient. 84. 1834. *Shorea laccifera* Heyne ex Beddome Fl. Sylv. S. India t. 6. 1869.

Deciduous, resiniferous tree, up to 25 m tall; bark grey, fairly smooth or narrowly fissured; wood compact, dark yellowish; branchlets glabrous. Leaves coriaceous, elliptic-oblong, 8.5–20 × 4–8.5 cm, glabrous above, glabrescent beneath, dark green, apex acute or retuse, margin undulate to entire, base rounded; secondary nerves 11–16 pairs, parallel, bifurcate near margins, 1 cm apart; petiole to 2 cm. Flowers whitish, fragrant in dense, 5–10 cm long, terminal or lateral drooping, lax, glabrous panicles. Fruit belly ovoid, 2 × 1 cm, glabrous, tapering above into a slender, remnant style; wings 5, unequal; 3-longer, 8–10 cm long, spathulate, obtuse, saccate at base with slender parallel veins, two shorter, 3.5 cm long; seed solitary.

Occasional, in ravines, valleys (sites of streams and lower slopes of hills) of moist deciduous forests; especially in Seshachalam hill ranges and Veligonda hills up to 1000 m. It is lower sometimes gregarious and occurs in patches of varying extent. The tree tends to become stunted on dry ridges and plateaus. Leaf fall: Dec.-Jan. New foliage: Feb. Fls.: Feb.–Mar. Frs.: Apr.–May. Vern.: Jalari, Jagigiri; Eng.: Lac tree of South India.

Flowers used to decorate hair by women. Wood hard, much used for house building, especially for beams, posts, door and window frames, and for agricultural implements. Oleoresin is used in the indigenous system of medicine as an external stimulent, heals wounds, burns, pains, itching and used to control dysentry. It makes an excellent fuel wood.

Talakona (CTR), *SSR* 15552; Tirumala hills (CTR), *SSR* 15970; Gundumala RF (ATP), *NY* 1237; Venkatagiri (NLR), *ASR* 6800 (VV).

INDIA : Tamil Nadu (North Arcot and Chingleput districts).

**Shorea tumbuggaia** Roxb. Fl. Ind. 2: 617. 1832; FBI 1: 306; Gamble 1: 83 (60); Beddome Fl. Sylv. S. India t. 5. 1870; Flora of India 3: 241. 1993. *Vatica tumbuggaia* (Roxb.) Wt. & Arn. Prodr. Fl. Ind. Orient. Fl. Ind. Orient. 84. 1834.

Deciduous tree, up to 20 m tall; bark dark brown, thick, rough, longitudinally fissured; wood brown, branchlets tomentose. Leaves thick-coriaceous, 6–18 × 3–10 cm, ovate, glabrous on both surfaces, apex acuminate, margin entire, base cordate or subcordate; secondary nerves ca. 9 pairs, 7 mm apart, petiole 1.5–4.5 cm. Flowers white, fragrant, in 15–20 cm long terminal panicles. Capsule yellowish-white, belly (seed part) 2 cm long, ovoid, pubescent; wings unequal, spathulate, to 5 cm long, nerved.

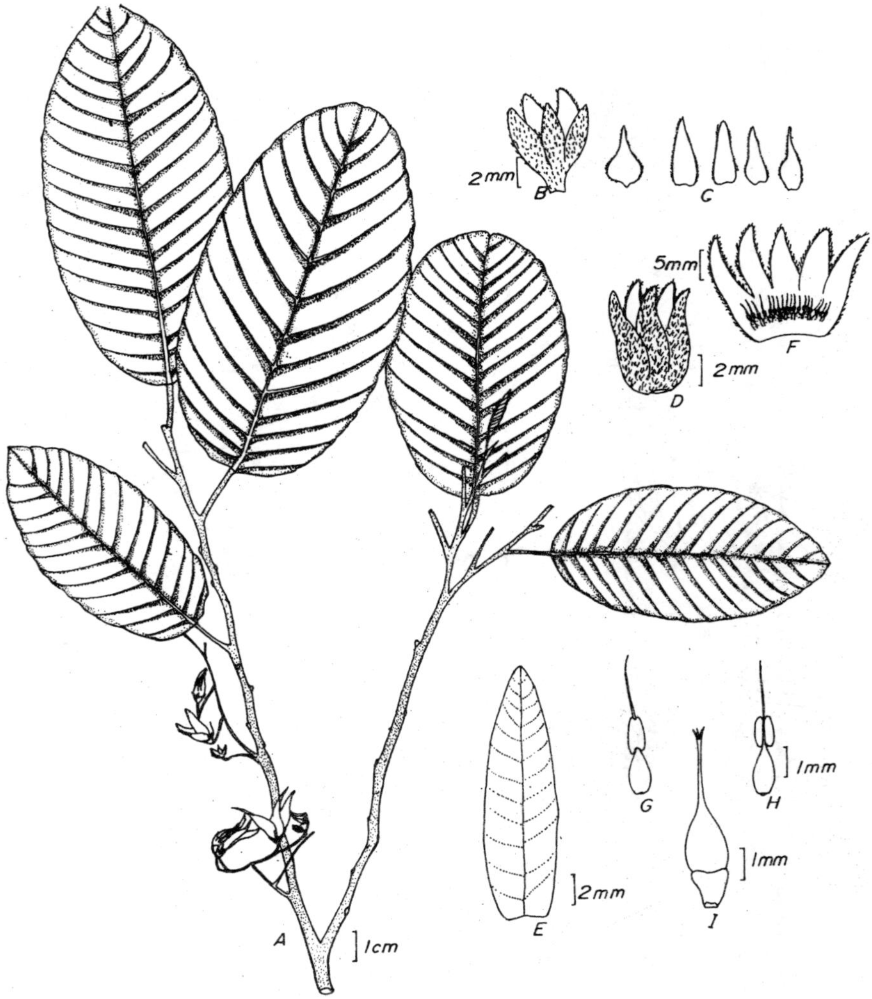


**Fig. 13.** *Shorea roxburghiii* G. Don
A. Twig, B. Calyx, C. Sepals, D. Corolla, E. Petal, F. Corolla with stamens, G & H. Stamens, I. Pistil.

Scarce, chiefly in hilly areas and to some extent in gregarious patches, mostly on the upper slopes and crests of hills and on pleateaus of Seshachalam and Veligonda hills up to 1000 m. Leaf fall: Dec.–Jan. New foliage: Apr. Fls.: Mar.–Apr. Frs.: June–Aug. Vern.: Thamba jalari, Thamba, Guggilam, Nalladammara, Nalla guggilam. Eng.: Green Dammer tree.

Wood used for constructional work, especially for beams, posts, door and window frames, and making agricultural implements and also for flag-poles in temples. Oleo-resin is used in the indigenous system of medicine as an external stimulant.

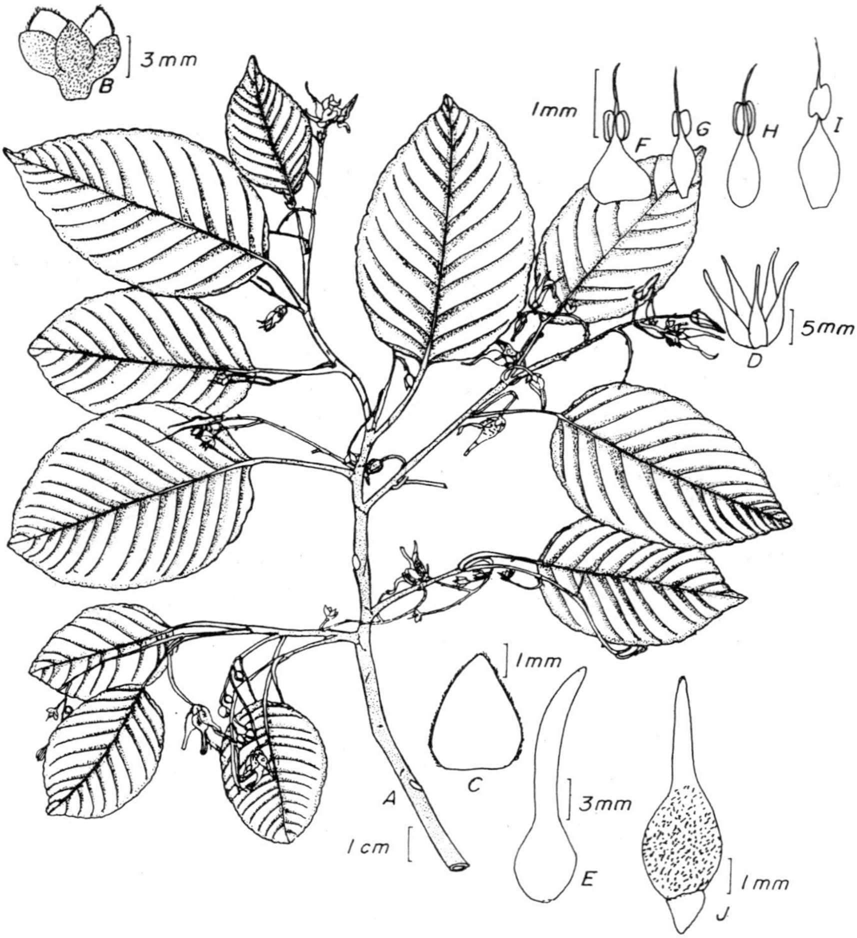


**Fig. 14.** *Shorea tumbuggaia* Roxb.
A. Twig, B. Calyx, C. Sepal, D. Corolla, E. Petal, F–I. Stamens, J. Pistil.

Tirumala (CTR), *SSR* 16155; Veligonda hills (NLR), *MSR* 1361 (CAL), Balapalli (CDP), *CECF* 4734 (CAL).

INDIA : Tamil Nadu — Endemic.

## MALVACEAE

1. Fruit winged. Flowers small in panicles. Leaves with small gland on the under surface of the mid vein ........ ***Kydia***
1. Fruit not winged. Flowers large, solitary or in racemes. Leaves without glands.
    2. Calyx woody, capsule indehiscent ........ ***Thespesia***
    2. Calyx not woody, capsule dehiscent ........ ***Hibiscus***

## HIBISCUS Linnaeus *nom. cons.*

1. Leaves hairy on both surfaces, margin dentate. Flowers white or purple, epicalyx foliaceous. Capsule globose, loculicidally 5-valved; seeds globose or subspherical, smooth .......... ***H. platanifolius***
1. Leaves glabrous above, tomentose beneath, margin crenate-entire. Flowers yellow, epicalyx cupular. Capsule ovoid, loculicidally 10-valved; seeds reniform, slightly warty .. ***H. tiliaceus***

**Hibiscus platanifolius** (Willd.) Sweet. Hort. Brit. ed. 1. 51. 1826–1827; Gamble 1: 98 (70); Flora of India 3: 338. 1993. *Pavonia platanifolia* Willd. in Ges. Naturf. Freude Berlin Mag. Nevesten Entdeck Gesammten Naturk. 4: 220. 1810. *Hibiscus collinus* Roxb. Fl. Ind. 3. 198. 1832; FBI 1: 338.

Tree, up to 8 m tall; bark greenish, rough; branchlets warty, stellate pubescent. Leaves entire to 3-lobed, thin coriaceous, cordate, 5–8 × 5–9 cm, apex acuminate, margin distantly dentate, base cordate, sparsely stellate-pilose above, densely pubescent on veins beneath, 5-nerved at base; petiole to 5.8 cm. Flowers solitary, axillary, whitish, fading to purple by noon. Capsule depressed-globose, 2–3 cm long, beaked, brown with woolly hairs, valves glossy above; seeds smooth, globose or subspherical, 4–5 mm long, brownish.

Occasional, in deciduous forests and evergreen forests up to 600 m above, hills on exposed slopes on either along border or along streams. Sometimes cultivated as an ornamental for its showy flowers. Fls.: Oct.–Nov. Frs.: Throughout the year. Vern.: Konda gogu.

Bark yields strong fibre used as cordage.

Talakona (CTR), *SSR* 13788; Penchalakona (NLR), *BSN* 5085 (VV); Balapalli (CDP), *JSG* 11104 (CAL).

INDIA : West Bengal, Orissa, Maharashtra, Karnataka, Tamil Nadu and Kerala.

WORLD : Sri Lanka and Nepal.

**Hibiscus tiliaceus** L. Sp. Pl. 695. 1753; FBI 1: 343; Gamble 1: 98 (70) ssp. *tiliaceus* Brandis 75; Flora of India 3: 322. 1993. *Paritium tiliaceum* (L.) St. Hill. Fl. Bras. Merid 1: 256. 1825; Rakshit & Kundu in Bull. Bot. Surv. India. 12: 158. 1972.

Evergreen, much branched, small tree, up to 10 m tall; branches close to the ground; bark rough, fibrous; wood whitish-grey. Leaves chartaceous to coriaceous, cordate to ovate, 4–12 cm in dia., glabrous above, whitish beneath with grey tomentum, apex acuminate, margin entire-crenate, base deeply cordate; 7-nerved, secondary nerves 3 pairs above basal ones; petiole to 2.5 cm. Flowers yellow,7.5–10 cm in dia., in axillary or terminal, few-flowered, pubescent racemes or in cymes. Capsule globose to ovoid, 10-valved, 1–2 cm across, beaked, tomentose; seeds 5–8, blackish-brown, reniform, slightly pilose.

Occasional, in borders of mangrove swamps and tidal streams. Fls. & Frs.: Throughout the year. Vern.: Etagogu. Eng.: Coast cotton tree, Yellow Mallow tree.

Bark yields dark fibre which is used locally for making ropes and cordage. The fibre is more resistant to water than jute and sun hemp. Bark, roots, leaves and flowers are reported to have medicinal properties. Leaf infusion is employed to clean ulcers and wounds. Bark powder is given as an emetic. Wood is light, soft and flexible, durable in sea water and useful for planking and light boats. Flowers are boiled in milk, and used as a remedy for earache.

Coringa river estuary (EG), *TAR* 7288 (CAL); Coringa (EG), *M.R. Mandal* 9340 & 9347 (CAL); Coringa (EG), *ARKS* 7369 & 7663 (CAL).

INDIA : Throughout Coastal areas, sometimes planted in interior areas.
WORLD : Throughout subtropics.

## **KYDIA** Roxburgh

**Kydia calycina** Roxb. Pl. Cor. t. 215. 1811; FBI 1: 348; Gamble 1: 93 (67); Brandis 78; Flora of India 3: 344. 1993.

Deciduous tree, up to 15 m tall; bark grey, peels off in irregular flakes; wood greyish-white, soft, no heartwood, faintly pink at the margins, medullary rays joined by white transverse bars of parenchyma resembling wood of Annonaceae; branchlets stellate-woolly. Leaves coriaceous, entire — 3-lobed, rounded, cordate, 4–12 × 3.5–15 cms, thinly scattered hairs above, grey tomentose beneath, apex rotund, obtuse, margin wavy to obscurely crenate-dentate, base truncate-cordate, palmately 7-nerved; petiole to 6.3 cm. Flowers 1.5 cm across, pinkish, polygamous in axillary panicles. Capsule globose, winged, ca. 5 mm across, brown, 3-loculed with 2 seeds in each locule; seeds reniform, 2.5 × 2 mm, brownish, glabrous, striate.

Occasional, in deciduous and semi evergreen forests, on hill slopes up to 1200 m. It is light demander and quick growing tree. It is fairly frost-hardy and drought resistant. It coppices well. Leaf fall: Feb. New foliage: Apr. to May. Fls.: July–Oct. Frs.: Nov.–Feb. Vern.: Konda-podari, Potri chettu; Botka, Konda patti, Eruku Tada, Pacha Botuku, Pedda Kunji, Potari.

Bark yields strong fibre, used locally for making coarse ropes and fishing nets. Leaves are chewed when there is a deficiency of saliva. Bark yieds a maroon dye used for colouring clothes. Flowers are made into garland and decorate the house; leaf paste is applied to releive body pains. Leaves are lopped for fodder. Wood used in building construction, making of match boxes, splints and light packing cases.

Upper Ahobilam (KNL), *TP & SSR* 15207, & *SSR* 15285; Rampa (EG), *VNS* 359 (CAL); Sunkarimetta (VSKP), *NPBK* 10915 (MH & CAL), way to Velgode (KNL), *JLE* 2221 (MH); Venkatagiri (NLR), *ASR* 632 & 8712 (VV).

INDIA : Throughout.
WORLD : Pakistan, Bhutan, Myanmar and China.

### THESPESIA Solander ex Correa

**Thespesia populnea** (L.) Soland ex. Cor. in Ann. Mus. Natl. Hist. Nat. 9: 290. t. 8. 1807; FBI 1: 345, *p.p.* Gamble 1: 101 (72); Brandis 75; Beddome Fl. Sylv. S. India t. 63. 1870; Flora of India 3: 352. 1993. *Hibiscus populneus* L. Sp. Pl. 694. 1753.

Evergreen tree, up to 12 m tall; wood dark coloured; branchlets with peltate-scales. Leaves cordate, 7–13 × 6–12.5 cm, with minute peltate scales, rarely glaucous beneath, apex acuminate, margin entire, base cordate, 7–9-nerved, lateral nerves 3 pairs above the basal ones, petiole to 10 cm. Flowers yellow, change to purplish pink when withering, large, axillary, solitary or 2 together. Capsule globose, to 2.5 cm dia., with minute peltate scales, 5-valved, surrounded at the base by the persistent calyx, black when ripe; seeds 2–4, ovoid, 1 cm long, silky, channelled along the back.

Plains, occasional in villages and roadsides especially Muslim tombs. Fls. and Frs.: All round the year. Vern.: Ganga Ravi, Gange Reni, Muni Ganga Reni. Eng.: Bhendi Tree, Umbrella tree, portia tree, Tulip trees. It is one of the fast growing tree. Bark extraction is given to control dysentry.

Peddagutta (NZB), *BR* 9202; Hindupur (ATP), *NY* 1071; Nizampatnam (GTR), *VRK* 6818.

INDIA : Planted throughout.
WORLD : Widely distributed in the tropies.

## BOMBACACEAE

1. Armed. Flowers red—solitary or paired ........ ***Bombax***
1. Unarmed. Flowers white yellowish:
    2. Trunk straight and long, base buttressed; bark green; branches tiered. Petiolule 0.5–1 cm long. Flowers clustered; stamens 5. Capsule ellipsoid to fusiform, 7–12.5 × 3–4 cm; seeds ca. 6.5 mm long. ........ ***Ceiba***
    2. Trunk irregular and short, base not buttressed; bark grey; branches not tiered. Leaflets sessile. Flowers solitary, stamens numerous. Capsule oblong-obovoid, 20-40 cm long; seeds ca. 4 cm long ........ ***Adansonia***

## ADANSONIA Linnaeus

**Adansonia digitata** L. Sp. Pl. 1190. 1753; FBI 1: 348; Gamble 1: 102 (73). Flora of India 3: 404. 1993.

Deciduous tree, up to 20 m tall; trunk short; branches spreading; bark smooth, greyish, often with purplish tinge or brown; wood grey; young branches pubescent. Leaves clustered towards the ends of the branches, digitate, 5–7-foliolate, petiole to 15 cm; leaflets obovate-oblong or lanceolate, 5–15 × 3–6 cm, glabrous above, downy beneath, apex caudate-acuminate, margin entire, base cuneate and decurrent; petiolule 0 or short. Flowers white, 15 cm dia., solitary, axillary, pendulous on 90 cm long peduncles. Capsule pale brown, oblong to oblong-ovoid, woody, 20–40 cm long, velvety tomentose, filled with tough fibres, longitudinally sulcate, apex rounded; seed brown, numerous, reniform, glossy, hard, immersed in fleshy pulp.

Introduced, planted in roadsides and compounds. Fls.: Apr.–June. Frs.: Sept.–Apr. Vern.: Brahma Malika. Eng.: Baobab, Monkey bread tree.

Leaves are eaten as vegetable and applied locally for a variety of inflammatory conditions, including insect bites. The bark is hard and woody, yields a flexible, moderately strong and durable fibre used for making ropes, socks and bags, horse girths, elephant saddler, baskets, mats etc. Capsules used as floats for fishing nets. The fruit pulp is also used as remedy in dysentry. The seeds are made into necklace. Wood is light and soft used in wooden trays, canoes, rafts, floats for fishing nets and for charcoal. It is a poor fuel and suitable for making ceiling and fibre-boards.

INDIA : Cultivated in Uttar Pradesh, West Bengal, Rajasthan, Gujarat, Maharashtra, Karnataka and Tamil Nadu.

WORLD : Native of tropical Africa; introduced in Madagascar and tropical Asia.

## BOMBAX Linnaeus

**Bombax ceiba** L. Sp. Pl. 511. 1753. *Bombax malabaricum* DC. Prodr. 1: 479. 1824; FBI 1: 349; Gamble 1: 99 (71); Flora of India 3: 398. 1993; Beddome Fl. Sylv. S. India t. 82. 1871. *Salmalia malabarica* (DC.) Schott & Endlicher Melet Bot. 35. 1832.

Deciduous tree, up to 30 m tall; trunk straight, usually buttressed, whorled; prickles straight, stout, hard, conical on trunk; branches spreading; bark greyish, smooth; wood soft, dark pink streaked with pale pink and white; young branches warty, puberulous. Leaves crowded at the ends of branches, digitately 5–7-foliolate, thin-coriaceous, elliptic-lanceolate, up to 16 × 7.5 cm, terminal odd one larger, basal ones smaller, apex cuspidate, margin entire, base acute, petiole 12.5–16.5 cm, petiolule to 1.5 cm. Flowers large, 12–15 cm across, solitary or paired, bright crimson, stamens polyadelphous. Capsule oblong to ovoid, 10–18 × 3–5 cm, dark reddish brown, loculicidally 5-valved, valves woody, rough; seeds many, obovoid, smooth, dark brown, enveloped in dense, white, silky floss, oily.

Usually occurring along the streams in mixed deciduous forests up to 1500 m, often cultivated as an avenue tree. It grows in any soil, either through cuttings or through seeds. It is strong light demander. Leaf fall: Dec.–Jan. New foliage: Mar.–Apr. Fls.: Feb.–Apr., very showy and noticed from great distances. Fruits ripen from Mar.–May. Vern.: Mundh, Buruga, Pinna Burugh, Salmalia. Eng.: Silk-cotton tree, Red silk cotton tree.

Gum (known as Mochras) exudes from the trunk, is used in medicine. The wood is soft and light but hard and durable under water, therefore, used in the construction of sea-going boats, and also used for making plywood, packing boxes, match boxes and splinters. The cotton surrounding the seeds is used for stuffing pillows and beds, but it is not so good as that of *Eriodendron*. Buds and flowers used as vegetables. Bark used in menorrhogae.

Pakhal (WGL), *SSR* 18524; Gowraram RF (NZB), *TP & BR* 6386; Kadam forests (ADB), *GO & DAM* 5193; Chelama (KNL), *Range Officer* 78052 (DD); Sudekulapadu (PKSM), *RVK* 15862; Talakona (CTR), *DR* 2421 (SVU).

INDIA : Throughout, often cultivated as an avenue tree.
WORLD : Throughout tropical Asia, extending eastward to New Guinea.

## CEIBA P. Miller

**Ceiba pentandra** (L.) Gaertn., Fruct. Sem. Pl. 2: 224. t. 1133. 1791; Flora of India 3: 400. 1993. *Bombax pentandrum* L. Sp. Pl. 511. 1753. *Eriodendron pentandrum* (L.) Kurz. in J. Asiat. Soc. Bengal 43. 113. 1874; Gamble 1: 100 (72). *E.anfractuosum* DC. Prodr. 1: 479. 1824; Beddome Fl. Sylv. S. India. Anal. Gen. t. 4 f. 2. 1871.

Deciduous tree, up to 30 m tall; trunk straight, tapering, prickly when young, base buttressed; bark smooth, green; young branches horizontal in whorls of 5–7 in all directions. Leaves 5–9-foliolate, petiole to 13 cm; leaflets thin-coriaceous, usually oblanceolate or elliptic, 6–24 × 4 cm, glabrous, apex acuminate, margin entire, base obtuse-cuneate, 12–15 nerved; petiolule 0.5–1 cm. Flowers yellowish, usually in clusters of 3–10, axillary or grouped toward ends of leafless branchlets. Capsule ellipsoid-fusiform, indehiscent, 7.5–12.5 × 3.5–4 cm, green when young, browning with maturity; seeds many, pendulous, subglobose, 0.65 cm, black lined with copious white silky hairs.

Introduced, planted throughout the state, mostly on the road sides, in the backyards, gardens etc. Recommended for plantation on deep and porous sandy loams or on alluvial flats. Leaf fall: Dec.–Jan. New foliage: Mar.–May. Fls.: Dec.–Feb. Fruits ripen in Mar.–Apr. Vern.: Kadami, Tella Buruga, Eng. White cotton tree, Kapok-floss tree.

Gum exudes from the trunk is an astrigent used as tonic in bowel complaints. Leaves are used to cure boils and leprosy. Floss is used for stuffing cushions, pillows and mattresses. Fibre is brittle, smooth and slippery, so this is unsuitable for textile industries. Seeds yield oil. The press cake and seed are used as fodder for cattle. It is used for match industries. The wood is light and soft, and is suitable for making panelling and toys.

Macherla RF (GTR), *VRK* 3937; Anantapur town (ATP), *TP* 1298..

## STERCULIACEAE

### KEY 1 (Flowering)

1. Flowers unisexual or polygamous; petals absent:
   2. Flowers yellow, yellowish or greenish:
      3. Ovaries free; style short, spreading or incurved—Mangroove plants ......... ***Heritiera***
      3. Ovaries united, style long, connate ......... ***Sterculia***
   2. Flowers scarlet:
      4. Flowers pendulous; calyx tubular, inflated towards apex ......... ***Firmiana***
      4. Flowers erect; calyx not tubular, lobes free nearly to the base, not inflated ......... ***Hildegardia***
1. Flowers bisexual; petals present:
   5. Staminal column bears fertile anthers throughout; staminodes absent ......... ***Eriolaena***
   5. Staminal column bears fertile anthers at apex only; staminodes present:
      6. Staminal column free from carpels; flowers yellow ......... ***Guazuma***
      6. Staminal column adnate to gynophore and forming a stalk; flowers other than yellow:
         7. Flowers regular ......... ***Pterospermum***
         7. Flowers irregular:
            8. Flowers red; calyx 2 cm long, 2-lipped; petals reflexed ......... ***Helicteres***
            8. Flowers pinkish; calyx 6–7 mm long, 5-lobed: petals folded ......... ***Kleinhovia***

### KEY 2 (Fruiting)

1. Fruit follicle:
   2. Follicles winged, thin, membranous, reticulately nerved; seeds 2 per follicle:
      3. Follicles oblong, opening long before maturity ......... ***Firmiana***
      3. Follicles oblique, not opening before maturity ......... ***Hildegardia***
   2. Follicles not winged, thick, not membranous and reticulately nerved; seeds more than 10:
      4. Follicles cylindrical, spirally twisted ......... ***Helicteres***
      4. Foliicles oblong or ovoid, spreading ......... ***Sterculia***
1. Fruit other than follicle:
   5. Fruit samara with an ellipsoid nut ......... ***Heritiera***
   5. Fruit capsule:
      6. Seeds winged:
         7. Capsule 5–10-valved ......... ***Eriolaena***
         7. Capsule 5-valved ......... ***Pterospermum***

6. Seeds not winged:
   8. Capsule 5-winged, inflated, membranous, smooth; seed one in each locule ... ........................................................................................ ***Kleinhovia***
   8. Capsule septicidally 5-valved, not inflated, woody, tubercled; seeds numerous ... ........................................................................................ ***Guazuma***

## ERIOLAENA A.P. deCandolle

1. Leaves glabrescent above, slightly rusty pubescent beneath; petiole 2–10 cm long; base 7–9 nerved; Buds ovoid, bracteoles very near to the sepals; staminal column with stamens only at apex. Capsule up to 10-locular, tubercled, pointed, 4 cm long ............ ***E. hookeriana***
1. Leaves very white pubescent beneath; petiole ca 2 cm long; base 5–7 nerved. Buds oblong; bracteoles not near the sepals; staminal column with series of fertile stamens. Capsule 5–8 locular, not tubercled, acuminate, < 3 cm long ................................................ ***E. lushingtonii***

**Eriolaena hookeriana** Wt. & Arn. Prodr Fl. Ind. 70. 1834; FBI 1:370. Gamble 1:110 (78); Brandis 87; Flora of India 3:415. 1993; Beddome Fl. Sylv. S. India Anal. Gen. t. 5.f.2. 1871.

Deciduous tree, up to 10 m tall; wood red, branchlets apically tomentose. Leaves coriaceous, broadly ovate with small tubercles on surface, 12 cm in dia., thinly hairy above, rusty tomentose beneath, apex acute, margin dentate, base cordate; 7–9-nerved, conspicuous beneath; petiole 2–10 cm. Flowers yellow, 8 mm across, numerous in axillary peduncled cymes. Capsule ovoid, valves 8–10, to 4 × 2 cm, pointed, woody, tubercled, downy; seeds numerous, wings papery, brown, to 1.4 cm.

Rare, in deciduous forests up to 900 m on slopes in full sun. Leaf fall: Feb.–Mar. New foliage: Apr.–June Fls.: Apr–June Frs.: Sept.–Jan. Vern.: Gurraputhada, Nara Botaka, Botta. Wood is strong, used for furniture.

Maredumilli (EG), *SSR* 15236; Horsely konda (CTR), *JSG* 15722 (MH); Aklaspur (KNR), *GVS* 25602; Anantagiri (VSKP), *NPBK* 794 (CAL); Yarlagadda (EG), *VNS* 363 (CAL); Near Microwave station (CTR), *DR* 1535 (SVU).

INDIA : Uttar Pradesh, Bihar, Sikkim, Orissa, Rajasthan, Gujarat, Maharastra, Tamil Nadu.

**Eriolaena lushingtonii** Dunn in Kew. Bull. 1915: 88. 1915 & in Gamble 109 (78); Flora of India 3: 417. 1993.

Tree, up to 7 m tall; bark greyish red; young branches pubescent. Leaves membranous, cordate-orbicular, 4 × 3.6 cm, stellate hairy above, white pubescent beneath, apex acute or shortly acuminate, margin shortly, irregularly dentate, base cordate, 5–7-nerved, petiole to 1.7 cm, densely stellate pubescent. Flowers yellow, axillary on 2–3-flowered cymes. Capsules ovoid, ca 1–1.5 × 0.5–0.7 cm, woody, pubescent, valves 5–8; seeds 5–8 in each locule, winged on one side top.

Rare, in open slopes of moist decidous forests between 350 and 900 m. Fls.: and Frs.: June–Aug.

Pothurajupenta (KNL), *SSR* 15987; Chalama (KNL), *JLE* 16715; Rollapenta (KNL), *JLE* 42284 (MH).

INDIA : In open slopes of moist deciduous forests of Karnataka, Tamil Nadu and Kerala.

## FIRMIANA Marsili

**Firmiana colorata** (Roxb.) R. Br. in Bennet & R.Br. Pl. Jav. Rar. 235. 1844; Gamble 1: 107 (75); Flora of India 3: 420. 1993. *Sterculia colorata* Roxb. Pl. Cor. t. 25. 1795; FBI 1: 359; Brandis 84.

Deciduous tree, up to 12 m tall; trunk erect, butressed; lenticelled; stem often fluted; bark ashy grey, roughish; wood grey, with streaks of cream and yellow; youngest shoots clothed with grey stellate pubescence. Leaves crowded apically on branchlets, coriaceous, shallowly 3-lobed, 12–30 × 6–20 cm; dark green above, pale green beneath, apex acute to acuminate, margin entire, base cordate; main lateral nerves 3–5, palmately veined, primary veins slightly raised on the lower surface; petiole to 13 cm. Flowers polygamous, 8 mm across, scarlet, very showy in terminal, many-flowered coral like panicles, with red tomentose. Follicles 5, stipitate, coppery when young, oblong, lanceolate, obtuse, membranes, 7–10 × 3–4.5 cm, follicle bearing 2 seeds near the base; seed ovoid, yellow, compressed, wrinkled or smooth.

Occasional, in the slopes of deciduous and moist deciduous forest at 600–1200 m. Leaf fall: Dec.–Feb.; New foliage: June; Fls.: Mar.–Apr. Frs.: Apr.–June Vern.: Kharka, Maraka, Karu boppaja. Eng.: The coloured sterculia.

Wood very soft and light reported to be suitable for paper pulp. Bark yields an inferior fibre used for ropes.

Simhachalam (VSKP), *SSR* 15561; Kalasamudarm (ATP), *EC* 2930; Way to Ramannapenta (KNL), *JLE* 23821; Karai Road (CDP), *CBDM* 4840 (MH); Palkonda hills (VSKP), *JSG* 14023 (DD).

INDIA : Throughout including Andaman Islands.
WORLD : Sri Lanka, Nepal, Bhutan, Bangladesh, Myanmar, Thailand, Indo-China, Malay Peninsula and Indonesia.

**Note** : The flowers are structurally bisexual but functionally unisexual in male, anthers are smaller than those in the female flowers but in the later the anthers do not open.

## GUAZUMA P. Miller

**Guazuma ulmifolia** Lam. Encycl. 2: 52. 1759; Robyns in Ann. Missouri Bot. Gard. 51: 102. 1964. Flora of India 3: 424. 1993. *Theobroma guazuma* L. Sp. Pl. 782. 1753. *Guazuma tomentosa* Kunth, Nov. Sp. 5. 320. 1823; FBI 1: 375; Gamble 1: 111 (79); Beddome Fl. Sylv. S. India t. 107. 1871.

Evergreen tree, up to 12 m tall; bark fissured in older parts; branchlets tomentose to velvety. Leaves elliptic-lanceolate, 8–16 × 3–6.5 cm, scabrous above, pubescent beneath; apex acuminate, margin serrulate, base obliquely cordate, 5-nerved; petiole 2–3 cm. Flowers yellow, numerous, 8 mm across, in 1–3 cm long, axillary or terminal panicles. Capsule subglobose, to 2 cm dia., covered with obtuse black tubercles resembling small ripe mulberry in shape, size and odour, septicidally 5-valved; seeds numerous, globose, 1 mm in dia.

Introduced, often run wild, usually in the vicinity of towns and villages. Fls.: Apr.–July Frs.: Feb.–June, not dehiscent. It remains on the trees and falls at the next flowering period. Vern.: Bhadraksha, Rudraksha. Fruits are edible.

Kurnool town (KNL), *AA* 2205; Sriharikota (NLR), *YYR* 2164; Coringa forest rest house (EG), *GVS* 687005 (MH); Andhra University campus (VSKP), *TP & EC* 7320; Hyderabad, *KMS* 5998 (MH); Salur (SKLM), *NPBK* 998 (CAL); Kambakkam (CTR), *DR* 2349 (SVU).

INDIA : Almost throughout in the plains (often cultivated).
WORLD : Native in tropical America and Java.

## HELICTERES Linnaeus

**Helicteres isora** L. Sp. Pl. 963. 1753; FBI 1: 365; Gamble 1: 107 (77); Brandis 88; Flora of India 3: 426. 1993. Beddome Fl. Sylv. S. India Anal. Gen. t. 5. f. 1. 1871.

Deciduous tree, up to 7 m tall; bark grey; wood white; branchlets apically stellate tomentose. Leaves thin-coriaceous, orbicular-obovate, 2.7–15 × 2–11.5 cm, scabrid above, pubescent beneath, apex shortly acuminate, margin serrate, base oblique; petiole to 1.5 cm. Flowers bright red, 2.5 cm across, 2–6 in axillary clusters. Follicles cylindrical, spirally twisted, beaked, 4–6 × 1 cm, brown, stellate tomentose; seeds numerous, 2 mm long, angular, wrinkled, tomentose.

Common, along water courses in dry deciduous forests up to 1500 m as an under growth and forests clearings. It coppices well. Leaf fall: Mar. New foliage: Apr. Fls.: Apr.–Dec. Frs.: Oct.–Jan. Vern.: Pedda Shamala,

Gubathada, Chemali Nara, Thada, Potra, Nuli Tada, Adavi chamanti. Eng.: East Indian Screw Tree.

Bark yields greyish green, soft, silky, more durable fibre. Leaves and tender branches are lopped for fodder. Dry fruits and bark are used in their poultic. The soft wood used as fuel. The charcoal from this wood is said to be used in making gun powder.

Talakona (CTR), *SSR* 13784, 15551; Lower Ahobilam (KNL), *TP & RVR* 2587 & 2766; Ethipothala (GTR), *VRK* 3567; Chennupalli RF (CDP), *EKK* 79516 (DD); Pakhal RF (WGL), *KMS* 11602 (MH); Polavaram (EG), *CAB* 4982 (MH); Vempalli RF (ADB), *GO* 4441; Gundaram RF (NZB), *BR* 9682; Pedamantanala (PKSM), *RVK* 15780; Udayagiri (NLR), *BSN* 6140 (VV).

INDIA : Throughout.
WORLD : Pakistan, Sri Lanka, Nepal, Bhutan, Bangladesh, S. China, S.E. Asia and Australia.

## HERITIERA Linnaeus

**Heritiera littoralis** Dryand ex Ait. Hort. Kew. ed. 1. 3: 546. 1789; FBI 1: 363; Gamble 1: 104 (74); Brandis 85; Flora of India 3: 429. 1993.

Medium sized, evergreen tree; bark pale brown, vertically and superficially fissured; wood yellowish-red, straight or interlocked grained and fine textured; branchlets lepidote. Leaves broadly elliptic or ovate-elliptic, 7–24 × 10–22 cm, glabrous and glossy above, hairy beneath, apex acute or slightly acuminate, margin entire, base rounded or slightly cordate, often oblique; secondary nerves 10 or more pairs. Flowers bright yellow, tinged red within, small, densely pubescent in axillary, lax panicles. Samara light brown, ellipsoid, 3–8 × 2–5 cm, woody, glabrous and shining, keeled and narrow beaked; seeds ellipsoid, 2–3 cm broad.

Occasional, along coast. Fls.: June–Aug. Frs.:Sept.–Feb. Vern.; Adavibadamu. The twigs are used as tooth brushes and the decoction of the seeds is given for diarrhoea and dysentry; seeds are edible. Wood is durable, chiefly employed for building boats, ships etc. It is also used for bridges, piling, beams, poles, furniture etc. It makes a good fuel.

It is included here on the authority of Gamble and Wagh.

INDIA : Mangrove forests of West Bengal, Orissa, extending inland up to Meghalaya; Tamil Nadu, Kerala, Karnataka and Andaman and Nicobar Islands.
WORLD : Myanmar, Coast of East Africa, Thailand, Indo-China, Hong Kong, Taiwan, Malaysia, tropical Australia, Pacific Islands up to Hawais and New Caledonia.

## HILDEGARDIA H.W. Schott et Endlicher

**Hildegardia populifolia** (Roxb.) Schott & Endl., Melet. Bot. 33. 1832. Flora of India 3: 432. 1993. *Sterculia populifolia* Roxb. Fl. Ind. 3. 148. 1832; FBI 1: 361 *(populnifolia)*, Gamble 1: 107 (76).

Deciduous tree, up to 20 m tall; bark smooth, wood whitish; branchlets angular, tubercled. Leaves thin-membranous, cordate, 9.5–11.5 × 8–11 cm, glabrous, apex long-acuminate, margin entire, base deeply cordate, 7-nerved, lateral nerves 4-pairs above basal nerves, raised; petiole 5–7 cm, whitish. Flowers scarlet, polygamous in axillary and terminal, up to 15 cm long, spreading panicles. Fruit rusty brown, 2-follicled; follicle obliquely lanceolate, membranous 10.2 cm long, base and apex 5.2 and 2.5 cm width, wings flat, inflated, strongly nerved; seed oblong, solitary in each follicle, 1.6 × 1.2 cm, pale brown.

Very rare, in deciduous forests of Anantapur especially in Chandrayudu Kona near stream on rocky hills at 300–700 m.. Leaf fall: Jan. New foliage: Apr. Fls.: Apr. Frs.: June–Feb. Vern.: Telibuda, Delibudda.

Chandrayudu Kona (ATP), *CJ & VS*, 14180; Ankalamma Konda (CTR), *DR* 637 (SVU).

INDIA : Tamil Nadu, Orissa.

## KLEINHOVIA Linnaeus

**Kleinhovia hospita** L. Sp. Pl. (ed. 2) 1365. 1763; FBI 1: 364; Beddome, Fl. Sylv. S. India, Anal. Gen. t. 4. f. 4. 1871; Gamble 1: 113 (81); Brandis 92; Flora of India 3: 434. 1993.

Deciduous tree, 6–7 m tall; bark smooth, greyish; young branches tomentose. Leaves thin-coriaceous, broadly ovate, 8–18 × 7–17 cm, glabrous, apex acuminate, margin entire, base cordate, 5–7 nerved; petiole to 15 cm. Flowers pinkish in lax, large, terminal, cymose panicles. Capsule brown, membranous, inflated, pyriform, loculicidal, 5-winged; seeds 1 rarely 2 in each locule, black, rounded, tubercled.

Planted in gardens, compounds, and all university campus areas, for showy flowers. Fls.: Sept.–Oct. Frs.:Dec.–Jan. Vern.: Eng.: The Kleinhovia.
SVU Campus (CTR), *SSR* 18600.

INDIA : Tripura, Orissa, South India — cultivated in most parts.
WORLD : Tropical Asia from Mascarene Isles to Polynesia and East tropical Africa. Native of Molucca.

## PTEROSPERMUM Schreber *nom. cons.*

1. Sepals 8–12 cm long. Capsule large ........................................................ *P. acerifolium*
1. Sepals < 8 cm long. Capsule not large:
   2. Leaves 7–20 × 3.5–7.8 cm, rufous tomentose beneath, base sub-cordate. Bracteoles broadly ovate, lobed; flowers 2.7–5.2 cm long. Capsule angular, pyriform, up to 9 × 4 cm, with rusty tomentum ........................................................................ *P. xylocarpum*
   2. Leaves to 8 × 4 cm, whitish tomentose beneath, base obliquely cuneate. Bracteoles linear; entire, flowers 1.7–2.2 cm long. Capsule terete, ovoid, to 4.5 × 2 cm, with greyish velvety tomentum ........................................................................*P. canescens*

**Pterospermum acerifolium** (L.) Willd. Sp. Pl. 3: 729. 1801; FBI 1: 368; Gamble 1: 109 (77); Brandis 91. Flora of India 3: 448. 1993.

Evergreen tree, up to 12 m tall; bark smooth, greyish-brown; wood red, thin, even and somewhat interlocked-grained; branchlets rusty tomentose. Leaves coriaceous, broadly ovate-oblong to oblong-obovate, 10–25 × 8–15 cm, glabrous, white-tomentose beneath, apex acute to acuminate, margin toothed near apex, base subequally rounded or subcordate, palmately ribbed, reticulations obscure; peltate petiole to 4 cm. flowers white, fragrant, 1–3, large. Capsule dark brown, oblong, woody, 10–12 cm long, 5-angled, tomentose, furfuraceous; seeds numerous, winged, brown, to 5 cm long.

Rare and also cultivated in gardens in Tirupati. It is a moderate shade-bearer and fairly frost hardy, coppices well and produces root suckers well. Fls.:Mar.–June Frs.: Aug.–Dec. Vern.: Masta Kanda.

The petals bruised in honey are reckoned a cure for stomach-ache. The leaves are used as platters and for thatching huts and packing tobacco.

Tirupati, Alipiri garden (CTR), *SSR* 19390.

**Note** : Our authority for including this plant is Elliot, who has reported it from Rayigalla in Palakonda hills and from Parlakimedi (Srikakulam district). Gamble mentions its occurrence in Cuddapah, and Sriharikota forest in Nellore district.

**Pterospermum canescens** Roxb. Fl. Ind. 3: 152. 1832. *P. suberifolium* (L.) Lam. Tabl. Encycl. 3: 136. t. 576. f. 2. 1794; non Willd. 1880: FBI 1: 367. Gamble 1: 110 (78); Brandis 90; Flora of India 3: 453. 1993. *Pentapetes suberifolia* L. Sp. Pl. 678. 1753.

Tree, up to 10 m tall; bark greyish, rough; wood light red; branchlets fulvous tomentose. Leaves coriaceous, obovate-oblong, 5–8 × 2–4 cm, glabrous and dark green above, dense, cream coloured tomentum beneath, apex acuminate, margin entire to irregularly lobed in the upper region, base obliquely cuneate, 3-nerved; secondary nerves 5–7 pairs, inter-secondary nerves obscure; petiole to 0.8 cm, stout. Flowers yellowish-white, fragrant, 2.5–3 cm dia, fragrant, axillary, solitary, pedicel to 2 cm. Capsule ovoid-oblong, terete, to 4.5 × 2 cm, greyish velvety, attenuate at base, apex tapering,

5-valved; seeds 2–4 in each valve, seed part 0.5 cm, wing 1.5 cm long, brown, thin.

Occasional, in moist deciduous forest of Chittoor, Cuddapah and Nellore districts at 400–600 m . Mostly near streams. Fls.: June–July Frs.: Nov.–Apr. Vern.: Lolugu, Narudu, Thada, Pothadi.

Wood little used for carts, gun-stocks and packing cases and as fuel. A paste prepared from the leaves is also used in headache.

Kambakam (CTR), *SSR* 16153; Brahmagundam (CTR), *GVS* 46934 (CAL), Sriharikota (NLR), *YNR* 1489; Venkatagiri (NLR), *ASR* 6731 (VV).

INDIA : Orissa, Maharashtra, Karnataka and Tamil Nadu.
WORLD : Sri Lanka and Myanmar.

**Pterospermum xylocarpum** (Gaertn.) Sant. & Wagh in Bull. Bot. Surv. India 5: 108. 1963; Flora of India 3: 454. 1993. *Velaga xylocarpa* Gaertn. Fruct. 2: 245 t. 133. 1791. *Pterospermum heyneanum* Wall. ex Wt. & Arn. Prodr. Fl. Ind. Orient. 69. 1834; FBI 1: 369. Gamble 1: 106 (77); Brandis 90.

Evergreen tree, up to 12 m tall; wood is light red; branchlets rusty-tomentose. Leaves thick-coriaceous, variable in size and form, oblong to slightly obovate, 7–20 × 3.5–7.8 cm, glabrous above, rusty tomentose beneath, apex acuminate, margin entire to distally dentate, base subcordate obtuse; secondary nerves 4–5 pairs on either side; petiole to 1 cm. Flowers pale white, 2.7–5.2 cm long, fragrant in axillary or terminal clusters. Capsule pyriform, woody, horned, 5-angled, scarcely furrowed, 5–9 × 2.5–4 cm, rusty; seeds 5–10 in each locule, orbicular, compressed, papery, winged.

Common, in dry deciduous forests on the slopes, or along stream banks in shallow ravines at 500–1000 m.. Fls.: Oct.–Dec. Frs.: Jan.–Aug. Vern.: Duddika, Noliki, Lolugu chettu, Thada, Tada. Wood hard and heavy. The leaves are used in leucorrhoea. They are also smoked like tobacco.

Talakona (CTR), *SSR* 13790; Upper Ahobilam (KNL), *TP & SSR* 15203; Pinjara Konda (EG), *SSR* 15233; Srisailam (KNL), *SSR* 15523; Palakonda hills (CDP), *CS* 8241; Penchalakona (NLR), *BSN* 5833 (VV); Kothuru (VSKP), *NPBK* 858 (CAL); Kondapalli (KSN), *KSM* 16093; Rollapenta (PKSM), *RVK* 15892; Appapur (MBNR), *TDCK* 13632.

INDIA : West Bengal, Orissa, Maharashtra, Karnataka and Tamil Nadu.

## STERCULIA Linnaeus

### KEY 1 (Vegetative & Fruiting)

1. Leaves digitate; 5–7-foliolate, glabrous; base acute or tapering, Follicles glabrous, 10–12 cm long, woody; seeds numerous, 1–1.5 cm long—cultivated .............................. *S. foetida*

1. Leaves simple; 3–7-lobed, hairy; base cordate. Follicles hairy, to 5 cm long, thick; seeds 3–6:
   2. Bark white or greenish grey or pale pink, smooth. Leaves shallowly-lobed, apex long caudate-acuminate, velvety beneath. Follicles densely yellowish pubescent, often with stinging hairs ........ ***S. urens***
   2. Bark white with corky warts. Leaves deeply-lobed; apex long narrowly acuminate, villous beneath. Follicles rusty villous, without stinging hairs ........ ***S. villosa***

KEY 2 (Flowering)

1. Flowers in drooping panicles; anthers 10 ........ ***S. villosa***
1. Flowers in erect racemose panicles; anthers 10–15:
   2. Flowers 2–3.5 cm dia., not pubescent, foetid-cultivated ........ ***S. foetida***
   2. Flowers 1–1.8 cm dia., glandular pubescent, not foetid-wild ........ ***S. urens***

**Sterculia foetida** L. Sp. Pl. 1008. 1753; FBI 1: 354; Gamble 1:105 (75); Brandis 80; Flora of India 3: 459. 1993.

Deciduous tree, up to 25 m tall; bark whitish, flaking off; wood yellowish-white to reddish brown, straight-grained, even and coarse textured, branches whorled, horizontal. Leaves digitate, 5–7-foliolate, petioles to 16 cm; leaflets thin-coriaceous, lanceolate or elliptic-oblong, 5–14 × 2.5–5 cm, glabrous, apex narrowly acuminate, margin enitre, base acute or tapering; lateral nerves 15–20 pairs, prominent, parallel; petiolule very short or 0. Flowers yellow, 2–3.5 cm in dia., many in 20 cm long, racemose panicles. Follicles 5, scarlet, boat-shaped, woody, large, 10–12 cm long, nearly glabrous; seeds many, ovoid-ellipsoid, dark grey, 1–1.5 cm, with rudimentary yellow aril.

Occasionally planted as an avenue plant. Leaf fall: Jan.–Feb. New foliage: Apr. Fls.: Mar.–Apr. Frs.: May.–Aug. (ripening following year). Vern.: Jangli Badam.

The seeds are roasted and eaten. The leaves are repellent. Wood is soft and light, is used locally for inferior construction, dugouts, rough packing-cases, tea chests and cheap plywood. The bark yields a weak cordage fibre. The tree exudes a gum, used for book binding and similar purposes.

Tirupati, Alipiri gardens (CTR), *SSR* 19398; Near B.N. Palace, Parlakimidi (SKLM), *GVS* 62372 (MH); Vizag (VSKP), *JSG 16109* (MH).

INDIA : West Bengal, Bihar, Orissa, Maharashtra, Tamil Nadu and Kerala.
WORLD : Sri Lanka, Bangladesh, Myanmar, Eastern tropical Africa, Northern Australia, Pakistan.

**Sterculia urens** Roxb. Pl. Cor. t. 24. 195; FBI 1: 355; Gamble 1: 106 (75); Brandis 80; Flora of India 3: 470. 1993.

Deciduous tree, up to 25 m tall; trunk straight; bark white or greenish-grey, smooth, outer papery, exfoliating in large thin irregular flakes, inner bark fibrous; wood reddish brown, soft, pith with red resin canals. Leaves crowded at the ends of branches, thin-coriaceous, shallowly 3–7- lobed, 15–

20 × 15–24 cm, scabrid above, tomentose beneath; apex long, caudate-acuminate, margin entire, base cordate; petiole to 22 cm. Flowers greenish yellow, 5–9 mm across; panicles clothed with a dense sticky tomentum of glandular stellate hairs. Follcles 5, red, coriaceous, spreading, oblong, or ovoid-oblong, lobes 5 × 2.3 cm, densely, yellow pubescent with stinging hairs; seeds 3–6 in each follicle, 6 × 3 mm, oblong, black, glossy.

Common on the dry steep, rocky slopes in deciduous forests at 600–1300 m, often associated with *Boswellia serrata*. Leaf fall: Dec.–Mar. New foliage from Apr. Fls: Dec.–Mar. Frs.: Apr. onwards dehiscing by June. Vern.: Konda Tamara, Kovila chettu, Thapasi chettu, Tanuku, Yerra poliki, Tabsu, Etta ponaku. Eng.: Gum karaya tree.

Wood soft and used only for toys. Seeds sometimes roasted and eaten. The tree yields a colloidal gum called Katila from the bark throughout the year. This gum extensively used in medicines and in pastes. The gum is one of the least soluble of exudation gums. One of the main uses of gum at present it is a thickening agent in the preparation of printing pastes for the textile industry. The gum is also much used in pharmaceutical and medicinal preparations, such as lozenges, emulsions, lotions, sprays and pastes. It is also employed in the preparation of cosmetics. In the paper industry the gum used as a pulp binder. The bark yields a fibre used in cordage industry. The leaves possess good nutritive value as fodder and are rich in vitamin A. A glue made from the proteins of the seed can be used as an adhesive in making plywood.

Vanipenta (KNL), *SSR* 15503; Laxmapur (NZB), *BR* 6257; Kothakota RF (ATP), *NY* 2110, Pakhal (WGL), *KMS* 11643 (MH); Atmakur (NLR), *ASR* 14642 (VV); Rampa (EG), *MSR* 1608 (MH); Balapalli (CDP), *JLE* 15764 (MH); Godavari (KRN), *GVS* 22472 (MH); Jayantipuram (GTR), *VRK* 6877; Pedamantanala (PKSM), *RVK* 15844; Mannanur (MBNR), *TDCK* 1531.

INDIA : Almost throughout except Himalays.
WORLD : Sri Lanka and Malaysia.

**Sterculia villosa** Roxb. ex DC. Prodr. 1. 483. 1824; FBI 1: 355; Gamble 1: 106 (76); Brandis 81; Flora of India 3: 472. 1993.

Deciduous tree, up to 20 m tall; bark white with corky warts; wood white first then turned into red or crimson, straight or even grained; branches whorled, horizontal, spreading; branchlets with heart shaped scars of fallen leaves; young shoots tomentose with stellate hairs intermixed. Leaves crowded at the ends of branches, sub-orbicular, deeply 5-lobed, lobes ovate-oblong, each lobe up to 20 × 12 cm, glabrescent above and villous beneath, apex long narrowly acuminate, margin entire, nerves 5–7 pairs; petiole 15–35 cm, dilated at tip. Flowers yellow on slender pedicels in long drooping panicles. Follicles oblong, sessile, 2.5–5 × 1–3 cm; spreading, rusty villous, red inside; seeds 3–5 in each follicle, 6–9 × 4–6 mm, oblong, smooth, black.

Occasional, in dry rocky hill slopes of deciduous forests at 600–1000 m.. The tree prefers a light, sandy soil and it is a strong light demander. Leaf fall: Jan.; New foliage: May. Fls.: Jan.–Mar. Frs.: May.–June Vern.: Gugal, Gogal, Kovila, Vakkunara, Kummari poliki. Eng.: The Elephant rope tree.

Wood soft, light and perishable, used chiefly for tea boxes and other light packing cases. It is a good fuel. The inner layer of the bark yields a coarse and strong fibre, used for cordage and making rough bags. Roots eaten by inhabitants during scarcity; seeds roasted and eaten.

Vangasara (VSKP), *SSR* 19669; Rollapenta RF (KNL), *SSR* 15506; Rollapenta (PKSM), *RVK* 15773; Rapur (NLR), *ASR* 6849 (VV).

INDIA : Throughout warmer parts, topical Himalaya from Kumaon eastwards and Andaman and Nicobar Islands.
WORLD : Bhutan, Bangladesh and Myanmar.

## TILIACEAE

1. Leaf margin entire or undulate. Flowers in lax, large, terminal panicles. Capsule 6-winged—Introduced. ........ ***Berrya***
1. Leaf margin not entire or undulate (serrate, crenate or dentate). Flowers in axillary, rarely terminal, short cymes. Drupe 2–4 lobed, (rarely entire only in *G. rothii*) ........ ***Grewia***

### BERRYA Roxburgh

**Berrya cordifolia** (Willd.) Burret in Notizbl. Bot. Garte. Berlin-Dahlem 9: 606. 1926. Flora of India 3: 478. 1993. *Espera cordifolia* Willd. in Ges. Naturf. Freunde Berlin Neue Schriften 3: 450. 1801. *Berrya ammonilla* Roxb. Fl. Ind. 2. 639. 1824; FBI 1: 383; Gamble 1: 122 (87); Brandis 94; Beddome Fl. Sylv. S. India t. 58. 1870.

Deciduous tree, up to 15 m tall; bark pale, smooth, wood dark red. Leaves coriaceous, cordate, 9–20 × 7–15.2 cm, glabrous on both surfaces, apex acuminate, margin entire-undulate, base truncate or subcordate, 5-ribbed; lateral nerves 5–6 pairs above basal nerves, tertiary nerves parallel; petiole to 9 cm. Flowers white, numerous, 1.5 cm dia., pedicellate in terminal, lax, to 1.8 cm long panicles. Capsule oblong or elliptic, 1.8 × 0.8 cm, brown, 6-winged, nerved; seeds 3 mm.

Introduced, planted as an avenue tree. Fls.: Dec.–Jan. Frs.: Apr. Vern.: Cul teak. It is a good timber yielding plant.

Chandragiri (CTR), *SSR* 16187; Chandragiri (CTR), *CAB* 10054 (MH).

INDIA : Peninsula, Andaman.
WORLD : Sri Lanka, Myanmar.

## GREWIA Linnaeus

1. Peduncles much shorter than or as long as petioles.
   2. Flower buds more than 1 cm long; petal base densely appressed pilose. Drupe to 1.5 cm across ........ ***G. flavescens***
   2. Flower buds less than 1 cm long; petal base ciliate. Drupe to 8 mm across.
      3. Peduncles 1–3 together, pedicel ca 2 mm long; petals white. Drupe villous, obscurely 4-lobed. Stipules lanceolate or linear ........ ***G. abutilifolia***
      3. Peduncles 5–7 in a cluster, pedicel 4–10 mm long, petals yellow. Drupe sparsely hirsute, distinctly 2-lobed. Stipules auriculate ........ ***G. tiliifolia***
1. Peduncles much longer than petioles:
   4. Leaves 5–7 nerved:
      5. Leaves pubescent beneath ........ ***G. asiatica***
      5. Leaves tomentose beneath:
         6. Leaves grey beneath. Bark rough, dark. Glands on the petal base 0 or minute. Drupe glaucous-tomentose ........ ***G. orbiculata***
         6. Leaves whitish beneath. Bark smooth, light. Glands conspicuous on petal base. Drupe sparsely pubescent ........ ***G. eriocarpa***
   4. Leaves 3-(4)-nerved:
      7. Leaves glaucous beneath:
         8. Peduncles slender, >1.5 cm long. Petals 3 mm long. Drupes not lobed ...***G. rothii***
         8. Peduncles stout, up to 1.5 cm long. Petals 3.5–6 mm long. Drupe 2-lobed ........ ***G. damine***
      7. Leaves not glaucous beneath:
         9. Flowers bi and unisexual ........ ***G. helicterifolia***
         9. Flowers only bisexual:
            10. Drupe glabrescent when mature ........ ***G. serrulata***
            10. Drupe pilose, tomentose or bristly when mature:
                11. Flower buds ovoid or ovoid oblong; petals reflexed. Drupe rusty pubescent ........***G. bracteata***
                11. Flower buds conical or cylindrical; petals not reflexed. Drupe yellow, wrinkled, bristly with stiff hairs ........ ***G. orientalis***

**Grewia abutilifolia** Vent. ex Juss in Ann. Mus. Natl. Hist. Nat. 4:92. 1804; FBI 1:390; Brandis 99; Flora of India 3: 493. 1993. *Grewia aspera* Roxb. Fl. Ind. 2: 591. 1832; Gamble 1: 119 (85).

Small, deciduous or evergreen tree, up to 6 m tall; young branches stellate hairy, distantly lobed, 7.5–12.7 × 6.5–10 cm; scabrous above, stellate tomentose beneath, apex acute-acuminate, margin irregularly serrate, base subcordate, 5-nerved, petiole to 4.5 cm; Flowers axillary, white, fading yellow, in short umbellate cymes; Drupe brown, subglobose, fleshy, 0.8 cm across, slightly 4-lobed, villous.

Common, in forest clearings and in hilly regions of moist deciduous forests up to 900 m. Fls.: Apr.–Aug. Frs.: Sept.–Feb. Vern.: Pedda Tariki, Pedda Tada, Potucanti, Guvva Tada.

Rollapenta (PKSM), *RVK* 15896; Neridigonda (ADB), *PVP* 9819; Horsely hills (CTR), *BSN* 4031; Maredumilli (EG), *GVS* 24204 (CAL); Venkachinta (VSKP), *GVS* 29587 (CAL); Karad (SKLM), *GVS* 62388 (MH); Polavaram agency (WG), *DCSR* 466 (CAL).

INDIA : Peninsula, Himachal Pradesh, Uttar Pradesh, Bihar, West Bengal, Sikkim, Assam, Manipur, Meghalaya, Tripura, Orissa, Madhya Pradesh, Rajasthan, Gujarath, Maharashtra, Goa, Karnataka, Tamil Nadu and Kerala.

WORLD : Bangladesh, Myanmar, Indonesia and Malaysia.

**Grewia asiatica** L. Mant. Pl. 122. 1767; FBI 1: 386; Brandis 98; Flora of India 3: 494. 1993. *G. subinaequalis* DC. Prodr. 1: 511. 1824; Gamble 1: 118 (84).

Small tree; bark greyish-white or greyish-brown, rough; wood yellowish-white, close grained and elastic; branchlets stellate tomentose. Leaves ovate or sub-orbicular, 2.5–15 × 2–10 cm, scabrous above, white pubescent beneath, apex acuminate, margin coarsely toothed, base obliquely cordate, 5–7-nerved; petiole to 1.8 cm. Flowers yellow in axillary, umbellate cymes, peduncle up to 3.5 cm long; buds oblong-obovoid, ribbed, tomentose. Drupe red or purple, subglobose or globose, 7–12 mm across, indistinctly-lobed, pilose; seeds 1–2.

Rare, in deciduous and semi-evergreen forests. It grows under a wide range of soil and climatic conditions and easily propagated by cuttings. Fls.: May.–Aug. Frs.: Jan.–Sept. Vern.: Nallajana, Peddajana, Phutiki.

Fruits are edible as a desert fruit; it has pleasant flavours and an acid taste and is used largely for the preparation of a beverage, which is popular as a refreshing summer drink. Wood is used for shoulder poles, spear handles, shingles etc. The bark yields fibre, used for making ropes.

Visakhapatnam (VSKP), *CAB* 1716 (MH).

INDIA : Himachal Pradesh, Punjab, Haryana, Delhi, Uttar Pradesh, Bihar, West Bengal, Assam, Orissa, Madhya Pradesh, Rajasthan, Gujarat, Maharashtra, Karnataka and Tamil Nadu.

WORLD : Bangladesh, Sri Lanka.

**Grewia bracteata** Heyne ex Roth, Nov. Pl. Sp. 243. 1821; FBI 1: 389, Brandis 96; Flora of India 3: 495. 1993. *G. obtusa* Wall. ex Dunn in Gamble, Fl. Pres. Madras 1: 117 (83) 1915.

Small tree, up to 8 m tall; branchlets tomentose. Leaves thin-coriaceous, broadly ovate or elliptic-lanceolate, 6–10 × 4–5 cm, glabrous, tufts of hairs on nerve axils beneath, apex acuminate, margin irregularly serrulate, base obtuse, oblique, 3-nerved; petiole 2–6 mm. Flowers white or yellow, 2 cm across in short, 1–3 peduncled, few flowered axillary leaf opposed and terminal, 3 cm long cymes; buds ovoid or ovoid-oblong, 10–15 × 5 mm, tomentose. Drupe rusty, turning orange, broadly ovoid or depressed globose, obscurely or deeply 4-lobed, 1.5–2 cm across.

Occasional, in dry deciduous forests up to 900 m. Fls. & Fls.: Throughout the year.

Bheematheertham (CTR), *DR* 1176 (SVU); Horsely hills (CTR), *CAB* 8319 (MH); Dornal (PKSM), *JSG* 18696 (MH); Balapalli (CDP), *JSG* 11112 (DD); Srisailam (KNL), *JLE* 16832 (MH).

INDIA : South India.
WORLD : Sri Lanka.

**Grewia damine** Gaertn. Fruct 2. 1790; Gamble 1: 118 (84); Flora of India 3: 496. 1993. *G. salvifolia* Heyne ex Roth, Nov. Pl. Sp. 329. 1821; FBI 1: 386. p.p; Brandis 97.

Small tree, up to 6 m tall; wood dark brown; branchlets grey pubescent. Leaves ovate, oblong, 1.8–6.5 × 1–2 cm, few scattered hairs above, densely grey tomentose beneath, apex sub-acute, margin serrulate, base oblique, 3-nerved; petiole to 0.6 cm. Flowers yellow in 3-flowered axillary cymes. Drupe globose, distinctly 2-lobed, 8–10 mm across, rugose, white pubescent.

Common, in scrub jungles and dry deciduous forests. Fls.: July–Sept. Frs.: Aug.–Nov. Vern.: Adavi jama, Adavi pagari, Nara Budama, Chittijana, Konda cipurur, Jana.

Fruits are edible. The wood is used for walking sticks.

Kalasamudramu RF (ATP), *SSR* 13751; Macherla RF (GTR), *VRK* 3983; Madhavaram RF (KNL), *TP & RVR* 1342; Erramalais (KNL), *TP & RVR* 1770; Diguvametta (PKSM), *JLE* 32497 (CAL).

INDIA : Punjab, Haryana, Bihar, West Bengal, Madhya Pradesh, Rajasthan, Gujarat, Maharashtra, Karnataka, Tamil Nadu and Kerala.
WORLD : Nepal, Pakistan and Tropical Africa.

**Grewia eriocarpa** Juss. in Ann. Mus. Natl. Hist. Nat. 4: 93. 1804; Flora of India 3: 497. 1993. *G. elastica* Royle, Ill. Bot. Himal. 1: 104, t, 22. 1834. *G. asiatica* L. var *vestita* (Wall ex Brandis) Masters in Fl. Brit. India 1: 387. 1874; Brandis 99.

Deciduous tree, 8–12 m tall; young branches covered with soft, yellowish pubescence, bark greyish white; wood greyish white to light yellowish-grey, even and straight grained. Leaves obliquely ovate, oblong-ovate or obovate, 7–14 × 5–10 cm, hispid above, whitish tomentose beneath, apex acuminate, margin crenate-serrate, base rounded or sub-cordate, 5–6 nerved; petiole to 1 cm. Flowers yellow, 3–12 mm long, in axillary cymes; peduncle to 1.5 cm long; pedicel to 1 cm long; buds 3 mm across, ovoid to globose, tomentose. Drupe black, globose, 5–10 mm across, obscurely 2–4-lobed, sparsely pubescent.

Rare, in Visakhapatnam district at 200–500 m. Fls.: Feb.-Aug. Frs.: May-Nov. It coppices well. The trees is frost and drought hardy..

The timber is esteemed for its strength and elasticity. It is used for shafts, shoulder poles, oars, tool handles, shinghles, bowls and similar objects. It is a good fuel wood. The branches are lopped for fodder. The bark yields a strong fibre used locally for rope-making.

INDIA : Himachal Pradesh, Punjab, Uttar Pradesh, Bihar, West Bengal, Western Ghats.

**Grewia flavescens** A.L. Juss. in Ann. Mus. Natl. Hist. Nat. 4:91. 1804: Gamble 1: 119 (85); Flora of India 3: 498. 1993; *G. pilosa auct. non* Lam. 1789; FBI 1: 388. Brandis 97. *G. carpinifolia* auct. non A.L. Juss.; Roxb. Fl. Ind. 2: 587. 1832; FBI 1: 387.

Small tree, up to 7 m tall; Leaves thin-coriaceous, elliptic-oblong, ovate-oblong, or oblong, 5–9 × 2.7–4.5 cm, scabrid above, dense tomentose below, apex acute to shortly acuminate, margin irregularly serrate, base truncate or rounded, 3-nerved, petiole to 4 mm. Flowers yellow in short axillary cymes on 1–3 peduncles. Drupe 2–4-lobed, 0.8–1.5 cm across, yellowish brown, stellately hairy; rind crustaceous; stones 1–4.

Common, in scrubs, stream banks and dry deciduous forests at 400–1000 m. Fls.: July–Sept. Frs.: Throughout the year. Vern.: Meda Koava, Bunka Jana, Nelli.

Fruits are edible. Leaves are used as fodder. The flattened, angular branches are used for making baskets.

Tirumala (CTR), *SSR* 13766; Mohammedabad (RR), *MSM & NV* 10427; Lingampet (NZB), *BR* 9298.

INDIA : Delhi, Uttar Pradesh, Bihar, West Bengal, Orissa, Madhya Pradesh,Rajasthan, Gujarat, Maharashtra, Karnataka, Tamil Nadu and Kerala.

WORLD : Tropical Africa.

**Grewia helicterifolia** Wall. ex G. Don, Gen. Hist. 1: 548. 1831; Flora of India 3: 500. 1993. *G. hirsuta* Vahl, var. *helicterifolia* (Wall. ex G. Don) Haines, For. Fl. Chota Nagpur 196. 1910. *G. polygama auct. non* Roxb. 1832, FBI 1: 391; Gamble 1: 119 (83); Brandis 100.

Small tree; branches terete, spreading; branchlets glabrescent. Leaves oblong-lanceolate, 5–10 × 1.5–2.5 cm, hispid or glabrescent above, densely tomentose beneath, apex acute, margin irregularly serrate, base subacute, 3-nerved; petiole very short, 2–3 mm. Flowers white in axillary umbellate cymes; buds ovoid-oblong or subglobose, 5 mm across, densely pubescent; peduncle 5–10 mm long, pedicel to 3 mm long. Drupe brown, obscurely 4-lobed, 0.8–2 cm across, glabrescent or sparsely hirsute.

Rare, in deciduous and evergreen forests of East Godavari and Mahabub Nagar districts. Fls.: May.–Oct. Frs.: Aug.–Feb. Vern: Cinnacipuru, Jibilika.

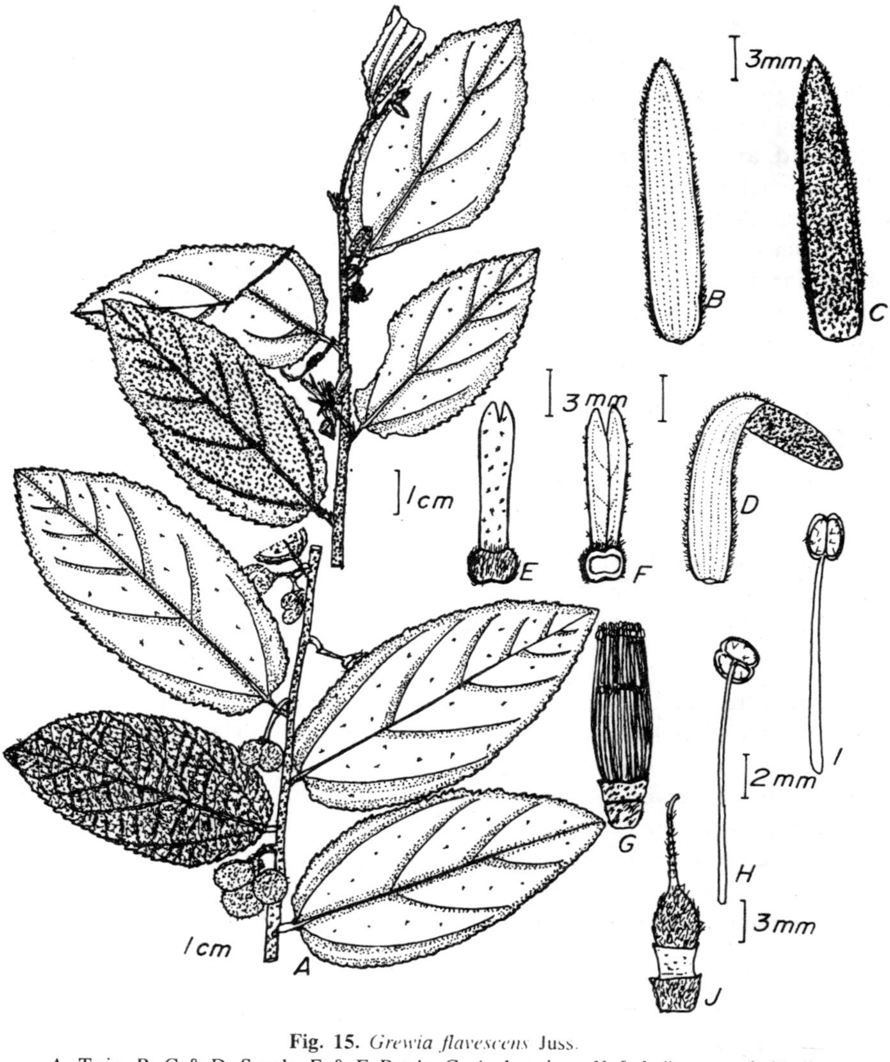


**Fig. 15.** *Grewia flavescens* Juss.
A. Twig, B, C & D. Sepals, E & F. Petals, G. Androecium, H & I. Stamens, J. Pistil.

Gellarayi konda — Mekalametta (EG), *GVS* 7536 (MH).

INDIA : Himachal Pradesh, Punjab, Uttar Pradesh, Bihar, West Bengal, Orissa, Madhya Pradesh, Maharashtra, Karnataka and Tamil Nadu.
WORLD : Bangaldesh, Pakistan, Myanmar, Sri Lanka, Australia.

**Grewia orbiculata** Rottler in Ges. Naturf. Freunde Berlin Neue Schr. 4: 205. 1803. FBI 1: 386; Brandis 100; Flora of India 3: 504. 1993; *G. rotundifolia* A.L. Juss. in Ann. Mus. Natl. Hist. Nat. 4: 92. 1804; Gamble 1: 118 (84).

Small deciduous tree, up to 7 m tall; crooked; bark dark grey, cracked. Leaves thin-coriaceous, ovate-orbicular, 3.5–8.8 cm × 3.6–6.1 cm, pubescent above, grey tomentose beneath, apex acute or obtuse, margin denticulate, base obliquely rotund, 5-nerved, petiole to 1.3 cm; Flowers yellow in ca. 5-flowered, axillary, terminal, leaf-opposed, umbellate, 2.5 cm long cymes on 1–3 peduncles. Drupe subglobose, 5–8 mm across, green, glaucous-tomentose, obscurely 2-lobed.

Common, in dry-deciduous forests and scrub jungles at 300–800 m. Leaf fall: Mar. New foliage: Apr.–May. Fls.: Apr.–May. Vern.: Tegali.

Mamanduru-Thumburatheerthum (CTR), *SSR* 16174; SKU Campus (ATP), *SSR* 15980; Owk RF (KNL), *RVR* 1462; Muchukota RF (ATP), *TP & NY* 217; Sesharayi hills (EG), *VNS* 525 (CAL); Nagarjuna Konda (NLG), *KTH* 9626; Guvvalacheruvu (CDP), *JSG* 15018 (CAL); Narsapur (MDK), *KMS* 8012 (CAL).

INDIA : Bihar, West Bengal, Orissa, Madhya Pradesh, Maharashtra, Karnataka, Tamil Nadu and Kerala.

**Grewia orientalis** L. Sp. Pl. 964. 1753; FBI 1: 384; 3: 506. 1993 excl. Syn. *G. rhamnifolia* Roth Gamble 1: 117 (84); Brandis 96; Flora of India 3: 506. 1993. *G. columnaris* J.E. Smith in Rees, Cycl. 17: 5. 1811; FBI 1: 383.

Small tree, up to 3 m tall; branchlets thinly stellate-hairy. Leaves ovate-elliptic, 4–10 × 2–4.5 cm, glabrescent, rough on both surfaces, apex acute or acuminate, margin crenate-serrate, base rounded or subacute and more or less oblique, 3-nerved, petiole to 7 mm. Flowers white, 3 cm across, in small, axillary 1–3-flowered cymes,; peduncle to 5 mm long, pedicel to 1 cm long. Drupe yellow, subglobose, 1–1.5 cm across, 4-lobed, bristly with stiff hairs, wrinkled.

Common in deciduous and semi-evergreen forests up to 600 m.. Fls.: & Frs.: June–Dec.

Ramallakota RF (KNL), *RVR* 1852; Yadiki RF (ATP), *TP* 484; Macherla RF (GTR), *VRK* 3957; Nellore (NLR), *DBS* s.n. (DD); Bank of Kambakkam tank (CTR), *MCB* 45167 (MH).

INDIA : West Bengal, Maharashtra, Gujarat, Goa, Karnataka, Tamil Nadu and Kerala.

WORLD : Sri Lanka, Malaysia, Indonesia and Tropical Africa.

**Grewia rothii** DC. Prodr. 1: 509. 1857; Gamble 1: 118 (84); Flora of India 3. 507. 1993. *G. excelsa auct. non* Vahl. 790; FBI 1: 355. *p.p.*

Tree, up to 6 m tall; bark black, rough; young branchlets tomentose; Leaves chartaceous, ovate-oblong or ovate-lanceolate, 2.8–5 × 1.2–4.7 cm, glabrous above, densely pubescent and glaucous beneath, apex acute or shortly acuminate, margin serrulate, 3–4 (5) nerved, petiole to 6 mm. Flowers yellow, in axillary, clustered cymes; peduncle 1.7–2 cm long, pedicel 8–10 mm long; buds subglobose, to 5 mm across, tomentose. Drupe globose, 0.8 cm across, entire, tomentose.

Common, in open hilly areas of most districts up to 800 m. Fls.: Apr.–Oct. Frs.: June–Dec. Vern.: Cipuru, Jana Jibilike, Peddacipuru, Putiki. Fruits are edible. The bark gives a fibre used for tieing.

Pothurajupenta (KNL), *SSR* 15988; Neelwai RF (ADB), *TP & GO* 5439; Macherla RF (GTR), *VRK* 3957; Guvvalacheruvu RF (CDP), *R.V. Reddy* 8404; Rampa (EG), *VNS*182 (CAL); Pakhal RF (WGL), *KMS* 13116 (CAL); Karakakonda (VSKP), *CAB* 1701 (CAL); Salur (SKLM), *NPBK* 1043 (CAL); Polavaram Agency (WG), *DCSR* 403 (CAL).

INDIA : Uttar Pradesh, West Bengal, Sikkim, Assam, Orissa, Madhya Pradesh, Karnataka and Tamil Nadu.

WORLD : Bangladesh and Tropical Africa.

**Grewia serrulata** DC. Prodr. 1: 510. 1824; Flora of India 3: 509. 1775. *G. laevigata auct. non* Vahl 1790; FBI 1: 381 incl. Syns. *G. multiflora auct. non* A.L. Juss. 1804; Brandis 97. FBI 1: 388. 1874. *G. disperma auct. non* Rottler ex Sprengel 1825; Gamble 1: 118. *G. glabra* Bl., Bijdr. 115. 1825. *G. barberi* J.R. Drumm. ex Dunn in Gamble Fl. Pres. Madras 1: 118. 1915.

Small evergreen tree, up to 10 m tall; bark dark grey or brown, smooth with slight horizontal stipular ridges; wood yellowish-white or yellowish-green, even and straight grained with streaks of light brown; branchlets sprasely grey stellate-pubescent. Leaves thin-coriaceous, lanceolate, 5–14 × 1.7–4.5 cm, glabrous above, sparsely puberulous along nerves beneath, apex sharply acuminate, margin serrulate, base obliquely acute, 3-nerved, petiole to 6 mm. Flowers white, in 2–3 flowered, umbellate cymes on axillary 2 peduncles. Drupe 2-lobed, druplets globose, 5–15 mm, grey, wrinkled, sparsely hairy, glabrescent when mature; seeds 4.

Common, in moist deciduous and evergreen forests at 500–1200 m. Fls.: June–Oct. Frs.: Dec.–Feb. Vern.: Pegala, Potriki, Thegalle, Allipayaru, Kotrinke.

Wood moderately hard and very flexible. It may be used for turnery and for containers of dry goods, such as cement barrels, rubber boxes and sieve frames. The bark yields a fibre used for cordage. The leaves are lopped for fodder.

Anantagri-Galikonda (VSKP), *TP & EC* 7353; Dharakonda (VSKP), *VNS* 413, 430 (CAL); Rampa (EG), *VNS* 309, 345, 162 (CAL); Araku (VSKP), *NPBK* 10842 (CAL); Sirivaka (Godavary Agency), *CAB* 4996 (CAL); Minumulur (VSKP), *GVS* 28254 (CAL).

INDIA : Punjab, Uttar Pradesh, Bihar, West Bengal, Sikkim, Assam, Meghalaya, Tripura, Orissa. Madhya Pradesh, Gujarat, Maharashtra, Goa, Karnataka, Tamil Nadu, Kerala and Andaman & Nicobar Islands.

WORLD : Pakistan, Nepal, Bhutan, Myanmar, Indo-China, Australia, Malaysia, and Tropical Africa.

**Grewia tiliifolia** Vahl, Symb. Bot. 1: 35. 1790; FBI 1: 386. *p.p.* '*tiliaefolia*'. Gamble 1: 118 (84); Flora of India 3: 511. 1993; Beddome Fl. Sylv. S. India t. 108. 1870. *Grewia leptopetala* Brandis, Indian Trees 100, 1906.

Tree, up to 10 m tall; bark pale brown, smooth; wood reddish-brown to brown with darker streaks, often marked by white spots; young branches stellate-pubescent. Leaves thin coriaceous, 4–15 × 2.5–11 cm, ovate or elliptic-ovate, glabrescent above, sparsely stellate hairy along nerves beneath, apex acute or rounded, margin crenate, base cordate or inequilateral, 3–5 ribbed; petiole to 3.5 cm, stipules auriculate. Flowers yellow in 3–4 flowered cymes on 5–10 axillary peduncles. Drupe black, didymous, globose, 2.5–5 × 7–9 mm, sparsely hirsute.

Common, in moist and dry deciduous, and semi-evergreen forests at 600–1200 m. Fls.: Apr.–June Frs.: Sept.–Oct. Vern.: Tada chettu, Pedda jana, Boddi Bodda, Charachi, Nulli jana.

Stem bark is acrid with a sweetish taste and employed for digestion; bark yields a fibre used for cordage. Fruits edible and has an agreeble acid flavour. The wood has emetic properties, used as an antidote to opium poisoning in a powder form; it is strong and flexible, used in agricultural implements, bent parts of carts, carriages, spokes, horizontal bars etc.

Tirumala (CTR), *SSR* 13712; Seethampet (SKLM), *SSR* 15575; Pothurajupenta (KNL), *SSR* 15991; Gundlabrahmeswaram (KNL), *SSR* 15998; Jaipur (ADB), *GO & PVP* 4943; Palakond hills (CDP), *CS* 7744; Mustapur (RR), *MSM & KH* 10312; Veligonda (NLR), *MSR* 1355 (MH) and 1358 (CAL); Rampa (EG), *MSR* 154 (CAL).

INDIA : Himachal Pradesh, Bihar, West Bengal, Assam, Orissa, Madhya Pradesh, Rajasthan, Gujarat; Maharashtra, Goa, Karnataka, Tamil Nadu and Kerala.

WORLD : Sri Lanka and Tropical East Asia.

## ELAEOCARPACEAE

1. Capsule echinate or covered with dense, stiff spines, flowers solitary ........................ ***Sloanea***
1. Berry or Drupe without any ornamentations or smooth; flowers in fascicles or in racemes:
   2. Flowers 0.6–1.2 cm across in racemes; petals fimbriate or toothed with numerous linear divisions; ovary 3-loculed with 2-ovules in each locule. Leaves coriaceous, base equal sided. ........................ ***Elaeocarpus***

2. Flowers 1.5–3 cm across in pairs or in fasçicles; petals entire; ovary 5-loculed with numerous ovules in each locule. Leaves chartaceous, base oblique-cultivated ... ........................................................................................................ ***Muntingia***

## ELAEOCARPUS Linnaeus

1. Leaf base obtuse. Petals glabrous; anthers not bearded at apex.—Racemes as long as leaves, stamens 20–30 ........................................................................................... ***E. lucidus***
1. Leaf base cuneate. Petals ciliate; anthers bearded at apex:
   2. Leaves lanceolate, elliptic, or elliptic-lanceolate, petiole 0.7–1.5 cm long, without glands at apex. Racemes to 8.5 cm long, puberulous; sepals glabrous along margins; stamens 15. Drupe green, pyrenes strongly rugose, with 3 longitudinal grooves. ... ***E. lanceifolius***
   2. Leaves elliptic-oblong or ovate-oblong, petiole 1.5–4.2 cm long with 2 glands near at apex. Racemes to 12 cm long, pubescent; sepals densely along margins; stamens 40–50. Drupe greenish yellow, pyrenes prominently rugose with 2-longitudinal grooves ... ........................................................................................................***E. tectorius***

**Elaeocarpus lanceifolius** Roxb. [Hort. Beng. 42. 1814, nom. nud.], Fl. Ind. 2:5. 1832 '*lanceaefolius*'; FBI 1: 402, Brandis 103. Flora of India 3: 543. 1993. *E. serrulatus* Benth. in Hook., J. Bot. 3: 363. 1852, non Roxb. 1814.

Tree, up to 20 m tall; wood soft, light brown; branchlets pubescent. Leaves thinly coriaceous-chartaceous, lanceolate, elliptic, elliptic-lanceolate, 7–15 × 2–5 cm, glabrous above, thinly or sparsely pubescent and glandular beneath, apex acuminate, margin distantly crenate-serrate, base much narrowed, lateral nerves 6–12 pairs, united by a conspicuous intramarginal vein; petiole 7–15 mm. Flowers white, 6–9 mm across, puberulous in 5–8.5 cm long, stiff racemes. Drupe green, oblong-ovoid to ellipsoid, 2.5–3 cm long, pyrenes oblong, with 3 longitudinal grooves, unilocular, seed one.

Rare in moist deciduous and evergreen forests above 1200 m altitute. Fls.: Mar.–Aug. Frs.: Oct.–Dec.

Fruits are edible. Wood is employed in making tea boxes, charcoal and also in house building. Nuts used in rosaries, necklace and bracelets. Rare, in Visakhapatnam district (Gamble).

INDIA : West Bengal (Darjeeling), Sikkim, Assam, Meghalaya, Nagaland, Manipur and Karnataka.

WORLD : Nepal, Bhutan, Myanmar, South China, Taiwan, Hong Kong and Indonesia (Java).

**Elaeocarpus lucidus** Roxb. Fl. Ind. 2: 600. 1832; FBI 1: 403; Gamble 1: 124 (88). Flora of India 3: 543. 1993.

Tree to 15 m tall; bark black, smooth; branchlets pubescent. Leaves thick coriaceous, ovate-lanceolate, elliptic, 7–11 × 2.3–4.1 cm, glabrous, slightly puberulous beneath, apex acute-acuminate, margin serrate, base obtuse, acute or rounded; secondary nerves 6–8 pairs, petiole 1.3–1.8 cm, with 2 glands near apex. Flowers white, 6–8 mm across in 6–12 cm long, axillary racemes. Drupe red, to 2 cm dia, oblong, ovoid, grooved, unilocular; seed 1, brownish.

Rare, in moist deciduous and semi-evergreen forests between 1000–1500 m. Rare in Visakhapatnam, Vijayanagaram and Srikakulam districts. Fls.: Feb.–May.. Frs.: July–Sept.

Galikonda (VSKP), *SSR* 15927.

INDIA : West Bengal, Assam and Meghalaya.
WORLD : Bangladesh.

**Elaeocarpus tectorius** (Lour.) Poir. in Lam. Encycl. Suppl. 2. 704. 1812. Flora of India 3: 559. 1993; *Crespedum tectorius* Lour. Fl. Cochinch 336. 1790. *Elaeocarpus oblongus auct. non* Gaertn. 1788; FBI 1.403; Gamble 1: 124 (88). Brandis 103, *E. robustus* Roxb. Fl. Ind. 2: 597. 1832; FBI 1: 402; Brandis 103.

Evergreen tree, up to 25 m tall; with aerial roots at base in swampy places; bark blackish, slightly rough; wood hard, brown, then yellowish; branchlets yellow pubescent. Leaves coriaceous, elliptic-oblong, 12–23 × 5.7–8.3 cm, green above, brownish and glandular beneath, apex acuminate, margin repand-serrate or crenate-serrate with a small glandular point at each of the crenatures, base broadly cuneate or rounded; secondary nerves 10–12 pairs, nerves raised beneath; petiole to 4.2 cm with 2 glands near its apex. Flowers white in 12 cm long, pubescent racemes. Drupe, ellipsoid or oblong or ovoid 2–3.2 × 1.8–2.5 cm, pyrenes 2–3, prominently rugose with 2-longitudinal grooves, each pyrene 1-seeded.

In moist deciduous, semi-evergreen and evergreen forests between 1200–1500 m, and it occurs along the streams in the forests of Visakhapatnam district.. Fls.: Apr.–May. Frs.: Aug.–Oct. Old leaves turn red when falling.

Kokkarampalli forest (VSKP), *SSR* 15963; Minumuluru (VSKP), *GVS* 28250, 30006; Paderu (VSKP), *GVS* 42596 (MH).

INDIA : Bihar, West Bengal, Sikkim, Assam, Arunachal Pradesh, Meghalaya, Nagaland, Manipur, Mizoram, Tripura, Orissa, Maharashtra, Karnataka, Tamil Nadu, Kerala and Andaman & Nicobar Islands.
WORLD : Nepal, Bangladesh, Sri Lanka, Malay peninsula, Sumatra, Indo-China.

## MUNTINGIA Linnaeus

**Muntingia calabura** L. Sp. Pl. 509. 1753; Flora of India 3: 570. 1993.

Small tree, up to 6 m tall; branches spreading, branchlets densely villous, glandular pubescent. Leaves chartaceous, oblong-lanceolate, 6–10 × 2–4 cm, villous above, softly woolly below, apex acuminate, margin serrate, base obtuse, oblique, 3–5 nerved; petiole to 4 mm. Flowers white, 1.5–3 cm across, in supra axillary pairs or in fascicles. Berry red, globose, 1–1.5 cm across, pulp juicy, sweet; seeds numerous, obovoid-ellipsoid, minute, embedded in pulp.

Introduced and cultivated. It is drought resistent and can be propagated by cuttings. Fls. & Frs.: Throughout the year. The pulpy fruits are edible and can be made into jams. The tough fibre of the inner bark is used for making ropes and cordage. Infusion of the leaves is used as tea, and of flowers for headache and incipient colds.

Tirumala hills (CTR), *SSR* 15281.

INDIA : Cultivated in all the states.
WORLD : Native of south America. West Indies and from Southern Mexico to Peru, Northern Argentina and Brazil.

## SLOANEA Linnaeus

**Sloanea sterculiacea** (Benth.) Rehder & Wilson in Sarg. Pl. Wilson 2: 363. 1916. *p.p.* Flora of India 3: 566. 1993. *Echinocarpus sterculiaceus* Benth in J. Linn. Soc. 5 suppl. 2: 72. 1861; FBI 1: 400; Brandis 101.

Large, evergreen tree, up to 25 m tall; often buttressed. Leaves elliptic or ovate, 11–22 × 4.5–10 cm, glabrous above, softly tomentose beneath, apex short acuminate or cuspidate, margin minutely serrulate, base cuneate or sub-cordate; secondary nerves 7–9 pairs; petiole to 4 cm, thick, puberulous, slightly dilated; Flowers whitish in axillary, solitary or few-flowered fascicles. Capsule red, globose, covered with straight, sharp, subulate hispid, persistent spines, 6 cm across, 4–5-valved; seeds numerous.

In moist deciduous forests, above 1000 m altitude. Fls.: Sept. Frs. Dec.–May.

Sapparla (VSKP), *SSR* 15965; Minumuluru (VSKP), *GVS* 30043.

INDIA : West Bengal, Sikkim, Assam, Arunachal Pradesh and Meghalaya.
WORLD : Bhutan, Myanmar, Nepal, Bangladesh.

## ERYTHROXYLACEAE
### ERYTHROXYLUM P. Browne

**Erythroxylum monogynum** Roxb. Pl. Cor. t. 88. 1798; FBI 1: 414. Gamble 1: 127 (90); Brandis 107; Flora of India 3: 590. 1993. *Erythroxylum indicum* (DC.) Beddome Fl. Sylv. S. India t. 81. 1871.

Small tree, up to 8 m tall; glabrous, much branched; bark brown, rough, lenticellate; wood dark brown, twigs often with deciduous scales at base. Leaves coriaceous, elliptic-obovate, 2–5.6 × 1.3–2.8 cm, shining above, pale glaucous beneath, apex obtuse, margin entire, base cuneate; petiole to 7 mm. Flowers pale green with white disk, scented, 1–6, in axillary fascicles. Drupe red, oblong, to 9 × 3 mm, furrowed with persistent calyx, seed one.

Very common, in dry hilly regions and evergreen forests up to 910 m. Fls. and Frs. throughout the year. Vern.: Pagadamu chettu, Devadaru, Adavi Goranta, Gatrinta, Gaderi. The dried leaves are cooked and eaten during famine. Sometimes the leaves are also used in treating stomachic and as diphoretic and diuretic. Wood very hard with pleasant resinous smell; used in turnery and to adulterate sandal paste. It is an excellent fuel. The wood tar is a preservative for paste. Fruits edible.

Tirumala hills (CTR), *SSR* 13768; Kalasamudram RF (ATP), *TP & EC* 2831; Peapaly (KNL), *RVR* 3180; Palkonda RF (CDP), *KNS* 2394 (DD); Jaipur RF (ADB), *GO & PVP* 4950; Mellavagu (GTR), *VRK* 5702; Diguvametta (PKSM), *CAB* 8091 (MH); Pakhal (WGL), *ANH* 15950 (MH); Mannannur (MBNR), *TDCK* 13640; Somasila (NLR), *ASR* 4224 (VV).

INDIA : Maharashtra, Karnataka and Tamil Nadu.
WORLD : Sri Lanka.

## AVERRHOACEAE
### AVERRHOA Linnaeus

**Averrhoa carambola** L. Sp. Pl. 428. 1753; FBI 1: 439; Gamble 1: 133 (95).

Evergreen tree, up to 8 m tall; bark smooth; branches drooping, branchlets pubescent. Leaves to 15 cm long; imparipinnate; petiole 1.5–2 cm; leaflets 5–11, ovate-oblong, elliptic or lanceolate, 2–6.7 × 0.7–4.5 cm, glabrous above, glaucous and pubescent beneath, apex acuminate, margin entire, base oblique, acute; petiolule to 1.5 cm, stout, pubescent. Flowers pink or purple, 8 mm long in axillary panicles, often from the old wood. Berry yellow or golden-brown, oblong, 5–7 cm long, 3–5-ridged or lobed; seeds arillate, ovoid, compressed, light brown.

Handsome ornamental tree, cultivated in gardens. Fls.: June–Sept. Frs.: Oct.–Nov. Eng.: Caramba, Blimbing, Country Goosberry.

Fruits edible, cooked or pickled. The leaves, roots and fruits have antiscorbutic properties and are used as cooling medicines. The unripe fruit is astringent and used as an acid in dyeing and for removing iron mould. The timber is used in buildings and furniture.

Tirumala (CTR), *SSR* 15976.

INDIA : Cultivated throughout the tropics.
WORLD : Probably native of Central and East Java.

## RUTACEAE

1. Leaves 1-foliolate:
   2. Petiole winged, rarely not winged. Stamens numerous. Fruit up to 8 cm in dia.—Cultivated ........ ***Citrus***
   2. Petiole not winged. Stamens 8–10. Fruit up to 2 cm in dia.:
      3. Flowers polygamous in long peduncled corymbs. Plants unarmed ........ ***Acronychia***
      3. Flowers bisexual in short peduncled corymbs, umbels or racemes. Plants armed:
         4. Staminal filaments connate at base and forming a staminal tube. Leaves oblong-lanceolate ........ ***Atalantia***
         4. Staminal filaments free. Leaves broadly ovate or elliptic ........ ***Pamburus***
1. Leaves 3-multifoliolate:
   5. Plants armed:
      6. Petiole and rachis not winged. Stamens numerous ........ ***Aegle***
      6. Petiole and rachis winged (except *Z. rhetsa* and *Z. ovalifolium*). Stamens up to 12:
         7. Stamens 3–5. Fruit dehiscent. Leaves alternate ........ ***Zanthoxylum***
         7. Stamens 8–12. Fruit indehiscent. Leaves opposite:
            8. Ovules 1 or 2 in each cell. Flowers bisexual ........ ***Naringi***
            8. Ovules more than 2 in each cell. Flowers polygamous ........ ***Limonia***
   5. Plants unarmed:
      9. Fruit dehiscent, 3–4-valved. Leaflets opposite.—Leaflets 3. ........ ***Euodia***
      9. Fruit indehiscent berry. Leaflets alternate:
         10. Style short, persistent.—Leaflets 3–5. ........ ***Glycosmis***
         10. Style articulate at the top of the ovary, deciduous:
            11. Petals usually valvate.—Leaflets 10–13 pairs ........ ***Micromelum***
            11 Petals imbricate:
               12. Filaments dilated below, stamens 8–10, anthers rather large.—Leaflets 5–7 ........ ***Clausena***
               12. Filaments subulate, stamens 10, anthers small ........ ***Murraya***

### ACRONYCHIA J. R. Forster & J.G.A. Forster *nom. cons.*

**Acronychia pedunculata** (L.) Miq. Fl. Ind. Bat. Suppl. 532. 1861. *Jambolifera pedunculata* L. Sp. Pl. 349. 1753. *Acronychia laurifolia* Bl. Gew. Buiten 3. 63. 1823. & Bijdr. 245. 1825; FBI 1: 498; Gamble 1: 152 (108); Brandis 116. *A. barberi* Gamble in Kew Bull. 1915 & Fl. Pres. Madras 1: 152 (108). 1915.

Small evergreen tree, up to 7 m tall; bark pale, smooth; wood greyish-white, smooth; young branches glabrous, stout. Leaves elliptic-lanceolate or elliptic-oblong, 9.7–14 × 3–4.6 cm, glabrous, glossy above, apex obtusely acuminate, margin entire, base obtuse or acute; secondary nerves 16 pairs, irregular and finely reticulate; petiole to 2.2 cm. Flowers yellowish-green or white, fragrant, polygamous, axillary in long stout pedunculate corymbs. Drupe aromatic, globose, slightly lobed, 0.7–1.2 cm dia., pubescent, rough with immersed glands, tip bluntly spiny; seeds 3–5.

In hills of Visakhapatnam forest above 1000 m. Fls.: June–Aug. Frs.: Nov.–Dec. Wood little used.

Sunkarimetta (VSKP), *NPBK* 10928 (CAL & MH); Minumuluru (VSKP), *GVS* 28252 (MH).

INDIA : Assam, Orissa, Chittagong, Tamil Nadu, Sikkim, Khasi hill, Bengal.

WORLD : Sri Lanka, Myanmar, S.W. China, Taiwan, Malaysia to S.E. Australia, New Caledonia.

## AEGLE Correa

**Aegle marmelos** (L.) Cor. in Trans. Linn. Soc. London 5: 223. 1800; FBI 1: 516; Gamble 1: 161 (115); Brandis 119; Beddome Fl. Sylv. S. India t. 161. 1871. *Crateva marmelos* L. Sp. Pl. 444. 1753.

Armed deciduous tree, up to 13 m tall; bark grey, exfoliating with longitudinal irregular flakes; wood yellowish-white; branchlets pubescent; spines axillary, single, straight and strong. Leaves 3(5)-foliolate, petiole to 4.7 cm; leaflets chartaceous, ovate-elliptic, 3.2–7.4 × 1.8–3.8 cm, upper one larger, side ones are smaller, glabrous, apex obtuse to acuminate, margin crenulate, base acute; petiolule to 3 mm. Flowers greenish-white in 10 cm long, axillary panicles. Fruit greenish-yellow, ovoid or globose, 8 × 6 cm, rind woody, softer, which becomes reddish and hard in drying; seeds numerous, oblong, compressed, 8 × 4 mm, embedded in aromatic pulp.

Common, in dry deciduous forests at 300–800 m. Often planted in the plains and grown in backyards and in temples. It is found typically on stiff, dry, clayey and alluvial soils. Leaf fall: Jan. New foliage: Feb. Fls.: Mar.–May. Frs.: Appear immediately but ripen in summer of following year. Vern.: Maredu, Bilvamu, Sriphala. Eng.: Bael tree.

Leaves are used as sacred offerings to the deity Siva. Fruit pulp used in diarrhoea and dysentry and also used in preparation of sherbet. Wood hard with strong aromatic scent when freshly cut, used as pestles of oil and sugar mills, naves of wheels and agricultural implements. Being a sacred tree, rarely cut for timber. Fruit pulp is highly nutritious, aromatic, cooling and laxative and contains mucilage, pectin, sugar, tannin, volatile oil and marmelosin. Leaves contain aeglin and aegelinin. Bark contains coumarin. Baked unripe fruit is a powerful astringent and stomachic. Leaves are analgesic and root bark is febrifuge.

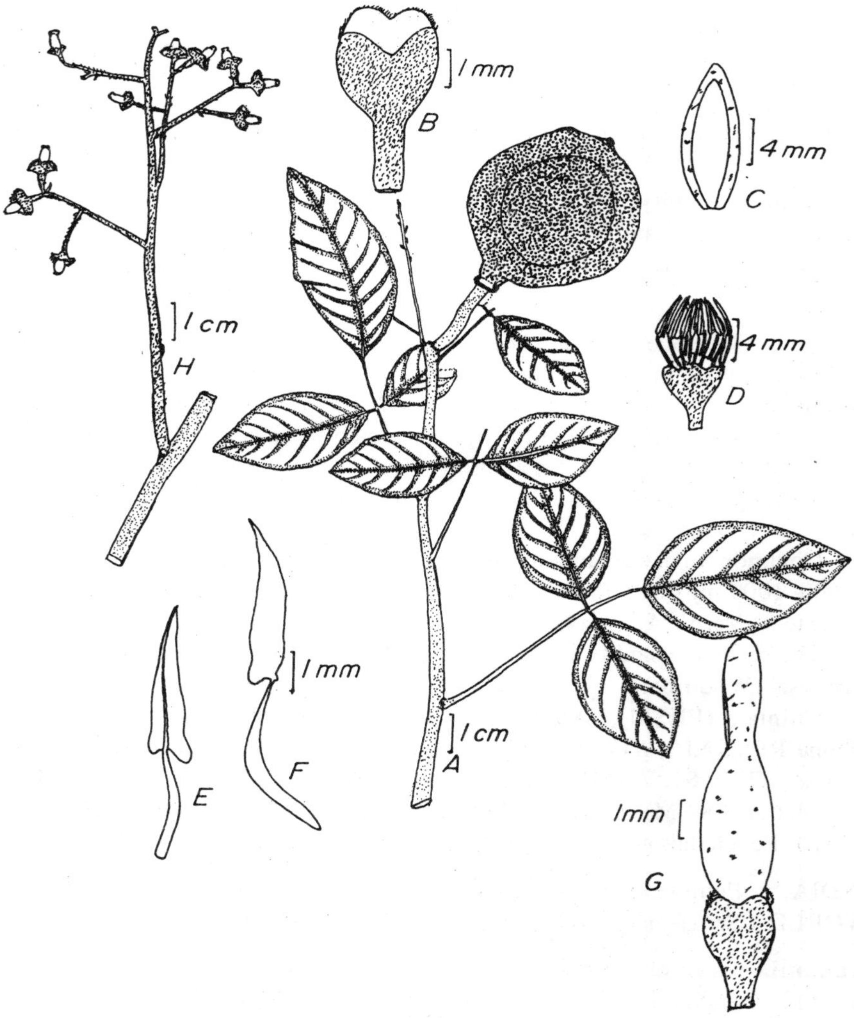


**Fig. 16.** *Aegle marmelos* (L.) Correa
A. Twig, B. Calyx, C. Petal, D. Androecium, E & F. Stamens G. Pistil, H. Flowering twig.

Rampur (ADP), *SSR* 19321; Galikonda (VSKP), *SSR* 19636; Jayanthipuram (GTR), *VRK* 6973; Pathur (MDK), *TP & CP* 12129; Mudheli RF (NZB), *BR & KH* 7130; Diguvametta (PKSM), *RVK* 157; Ramayapatnam (NLR), *BSN* 5163 (VV); Gokavaram (KSN), *AWL*, 1213(DD); Anantagiri RF (VSKP), *KMS* 2293 (DD); Mannanur (MBNR), *TDCK* 13635; Chandragiri (CTR), *DR* 538 (SVU).

INDIA : Throughout.
WORLD : Myanmar, Indochina, Sub Himalayan tract, Baluchistan.

## **ATALANTIA** Correa

1. Flowers in corymbs; pedicel slender, to 1.3 cm long. Leaf tip obtuse. Fruit 2 × 2 cm. ........ .......... ***A. monophylla***
1. Flowers in elongate racemes; pedicel stout, to 2 mm long. Leaf tip acute. Fruit 1 × 0.8 cm. .......... ***A. racemosa***

**Atalantia monophylla** (L.) Cor. in Ann. Mus. Natl. Hist. Nat. 6. 383. 1805; FBI 1: 511; Gamble 1: 159 (113); Brandis 121; Beddome Fl. Sylv. S. India Anal. Gen. 56. t. 701. 5. 1870. *Limonia monophylla* L. Mant. Pl. 237. 1771.

Densely foliaceous, armed tree, up to 10 m tall; bark grey, smooth; wood yellowish, close-grained; branchlets rigid. Leaves coriaceous, laceolate or elliptic, 3.8–6.8 × 1.8–3.6 cm, bright green, glabrous on both surfaces, apex obtuse, notched or 2-lobed at the tip, margin entire, base obtuse; petiole to 5 mm. Flowers cream in 2 cm long, axillary racemose umbels or corymbs. Berry orange coloured, globose, 2 × 2 cm, rind thick, epicarp pitted; seeds 4.

Common, in dry deciduous and thorny scrub forests in almost all districts at 600–1000 m. Fls. with two peaks during Sept.–Dec. and Feb.–Mar. Fruit persistent. Vern.: Adavi nimma, Karu nimma, Erra Munukudu.

The seeds are said to yield a sweet smelling oil, used in rheumatism and paralysis. Wood hard.

Chintala (PKSM), *SSR* 16126; Chitvel RF (CDP), *SSR* 16145; North Dhone RF (KNL), *RVR* 3151; Macherla RF (GTR), *VRK* 3868; Kondapalli (KSN), *CAB* 8127 (MH); Baramkonda (VSKP), *KCJB* 17159 (MH); Kodagandi (SKLM), *KCJB* 17276 (MH); Bhadrachalam (KMM), *JSG* 15871 (CAL); Penchalakona (NLR), *ASR* 2867 (VV).

INDIA : Peninsula, Khasia hills.
WORLD : Sri Lanka to Indochina, Malaysia, Vietnam.

**Atalantia racemosa** Wt. & Arn. Prodr. Fl. Ind. Orient. 91. 1831; FBI 1: 512; Gamble 1: 159 (114); Brandis 121; Beddome Fl. Sylv. S. India, Anal. Gen. 46. 1870.

Armed tree, up to 12 m tall; crown compact, young branches glabrous, lenticellate. Spines axillary. Leaves thin-coriaceous, elliptic or ovate-lanceolate, 3.5–9 × 1.5–4 cm, glabrous, apex acute with slightly emarginate, margin entire, often crenulate, base cuneate or subacute; secondary nerves 10–12 pairs, nerves prominent beneath; petiole to 1 cm. Flowers cream coloured, fragrant 1 cm long, pedicellate (3 mm) on 4.5 cm long, axillary, elongate racemes. Berry 1 × 0.8 cm.

Very rare, in dry deciduous forests at 700–1000 m. Fls.: Dec.–Feb. Frs.: Mar.–May. Vern.: Kada nimma.

Talakona (CTR), *DR* 2034 (SVU); Cuddapah hills (CDP), *CBDM* 15795 (MH).

INDIA : Peninsula.
WORLD : Sri Lanka.

## CITRUS Linnaeus

1. Petiole not winged and without clear articulation to the lamina.—Fruit large, pericarp thick, pulp scanty ........ ***C. medica***
1. Petiole winged or margined with clear articulation to the lamina:
   2. Petiole narrowly margined. Flower buds tinged with reddish outside; stamens usually more than 5 times as number of petals. ........ ***C. limon***
   2. Petiole with broad or narrow winged. Flower buds white, not tinged; stamens usually 4–5 times as number of petals:
      3. Flowers and fruits small, very sour. ........ ***C. aurantifolia***
      3. Flowers and fruits larger, sweet or slightly sour:
         4. Leaves rather broadly winged:
            5. Leaf base subcordate, sparsely pubescent on twigs and under surface of leaves, Fruits 10–20 cm in dia. ........ ***C. grandis***
            5. Leaf base other than subcordate, glabrous on twigs and under surface of leaves. Fruit 4–6 cm in dia. ........ ***C. aurantium***
         4. Leaves narrowly winged:
            6. Leaves to 8 × 4 cm. Pericarp loose and easily separable from remaining fruit segments. ........ ***C. reticulata***
            6. Leaves to 1.5 × 8 cm. Pericarp tightly adherent to the remaining fruit segment. ........ ***C. sinensis***.

**Citrus aurantifolia** (Christon. & Panz.) Swingle in J. Wash. Acad. Sci. 3: 465. 1913. *Limonia aurantifolia* Christon. & Panz. Pflanzeur. Syst. 1: 618. 1777. *Citrus medica* L. var.. *acida* (Roxb.) Hook. f. Fl. Brit. India 1: 515. 1875; Gamble 1: 161 (151).

Small tree, up to 5 m tall; branches many; spines short and sharp. Leaflets small, ovate-elliptic, 4–8 × 2–5 cm, margin crenulate; petiole narrowly winged. Flowers white, small, 8–12 mm long, 1–7 in axillary fascicles or cymes. Fruit green or tinged with yellow when ripe, ovate or globose, 3.5–6 cm in dia., often with apical papillae; pericarp thin, smooth, pulp yellowish-green; seeds small, ovate.

Occasionally planted. Vern.: Nimma. Eng.: Lime.

Fruits acidic, aromatic, used for making fresh juice and flavouring many foods; lime oil is prepared from the peel, citric acid is made from the fruit. It is a good source of vitamin C. The important commercial products of lime are lime juice, lime oil and calcium citrate. Dried lime peel powder and lime sediment obtained when lime juice is clarified, are utilized for cleaning metalware. Lime peel is also used as cattle feed. Lime is an appetizer, stomachic, artiscorbutic and anthelmintic. It checks biliousness. Salted lime peel is recommended for indigestion.

INDIA : Wild in the warm valleys of the Himalayas; different varieties cultivated all over India.
WORLD : Most parts of the tropics; probably a native of Malayasia.

**Citrus aurantium** L. Sp. Pl. 782. 1753; FBI 1: 575; Gamble 1: 162 (115); Brandis 123.

Tree, up to 10 m tall; thorns slender. Leaflets ovate or elliptic, up to 10 × 7 cm, margin crenulate; petiole 2–3 cm, rather broadly winged. Flowers white, large, very fragrant, with 5–12 per cent staminate only. Fruit orange, subglobose, somewhat flattened; peel thick, rough, strongly aromatic, often bright orange-red at maturity; pulp very sour and bitter; central core usually hollow; seeds numerous with predominant glucoside aurantamarin.

Cultivated largely in Guntur. Occasionally planted in Tirumala (Thammanna *et al.*, 1994). Vern.: Mallika narangi. Eng.: Sour, Bitter, Seville, Begarade orange.

Fruits are too bitter to use as fresh fruits, but are highly regarded for making marmalade, they are used in flavouring and liqueurs (Curacao). Leaves, flowers, and fruits yield volatile bigarade oil used in perfumes. The timber is hard, used in cabinet making and turnery.

INDIA : Cultivated.
WORLD : Probably a native of S. China and Indochina.

**Citrus grandis** (L.) Osbeck. Dagb. Ostind. Resa 98. 1759. *C. aurantium* L. var.. *grandis* L. Sp. Pl. 783. 1753. *C. decumana* L. Syst. ed. 12. 508. 1767; Gamble 1: 762 (115).

Tree, 5–15 m tall; spreading; crown round; young branches pubescent. Leaves ovate-elliptic, 5–20 × 2–12 cm, midvein often pubescent beneath, base subcordate; petiole broadly winged. Flowers whitish or cream, large, 3–7 cm in dia., solitary or in clusters. Fruits globose or pear-shaped, 10–20 cm in dia., often yellowish when ripe; peel thick with few large oil glands; segments 11–14, pale yellow to orange, with sweetish juice; seeds yellowish, large, ridged, with predominant glucoside naringin.

Occasionally planted (Thammanna *et al.*, 1994). Vern.: Pampala masam. Eng.: Shaddock, Pummelo, Forbidden fruit.

It is highly esteemed in the East as a desert fruit; the strong membrane (endocarp) round the segments peeled off and the large pulp vesicles, which fall apart easily are eaten.

INDIA : Cultivated throughout on a smaller scale.
WORLD : Probably a native of Thailand and Malaysia and spread from there to China, India and Persia; then it reached Europe as a curiosity.

**Citrus limon** (L.) Burm. f. Fl. Ind. 173. 1768. *Citrus medica* L. var. *limon* L. Sp. Pl. 782. 1753. *Citrus medica* L. var *limoninum* (Risso) Hook. f. in Flora of British India 1: 515, Gamble 1: 162 (115).

Small tree, 3–4 m tall; thorns stiff, stout. Leaves ovate, serrate, 5–10 × 3–6 cm; petiole slightly winged, articulated with lamina. Flowers white, axillary, solitary or clustered, 3.8–5 cm in dia. Fruit oblong or ovoid, bright yellow with terminal nipple; pericarp thick, adherent, prominently gland dotted; seeds ovoid. Prominent glucoside hesperidin.

Planted. Fls: In all seasons. Vern.: Nimma.

Lemons are not eaten as fresh fruits, but are widely used in the preparation of lemonade, squashes and for culinary and confectionary purposes as a flavouring and garnish. They are used in cosmetics and for the production of lemon oil, citric acid and pectin.

SKU Campus (ATP), *SSR* 15599.

Native of S.E. Asia, now widely cultivated.

**Citrus reticulata** Blanco Fl. Filip. 610. 1837. *C. chysocarpa* Lush. in Indian For. 36: 343. 1910. *C. crenatifolia* Lush. in Indian For. 36: 343. 1910.

Evergreen small tree, 2–8 m tall; rarely armed; branches slender. Leaflets narrowly ovate, elliptic or lanceolate, 4–8 × 1.5–4 cm, glabrous, glossy, dark green above, yellowish green beneath, margin crenate; petiole narrowly, slightly winged or margined. Flowers white, 1.5–2.5 cm in dia. Fruit yellow or orange-red, depressed globose, 5–8 cm in dia., segments 10–14, readily separating from each other and form the smooth, loose, thin skin; pulp sweet and juicy, orange in colour; seeds small.

Cultivated in hilly areas. Vern.: Kamala pandu. Eng.: Mandarin orange, Loose-skinned orange. Fruit is eaten as a desert fruit and it gives orange juice. It is the most valued commercial orange of India. The rind of the fruit yield an oil, which is rich in limonen.

INDIA : Widely cultivated.
WORLD : Probably native of S.E. Asia.

**Citrus sinensis** (L.) Osbeck. Dagb. Ostind. Resa 41. 1775. *C. aurantium* L. var. *sinensis* L. Sp. Pl. 782. 1753. *C. aurantium* L. var. *aurantium p.p.*

Evergreen tree, 6–12 m tall; crown compact; wood yellowish, close grained; branchlets angular when young; spines straight, axillary, stout. Leaves ovate or elliptic-ovate, 5–15 × 2–8 cm, glabrous, dark green above, margin entire or slightly serrate; petiole 1–2.5 cm long, narrowly winged, articulated. Flowers white, fragrant 2–3 cm in dia. in axillary, solitary or short racemes. Fruit orange, sub-globose with solid core, 4–12 cm in dia; peel 0.5 cm thick, smooth, tightly adherent, segments 10–14; pulp sweet, orange; seed 0–many, obovoid.

Cultivated; widely grown. It is eaten as desert fruit. Vern.: Battavinarina, Buddasini, Narangi, Satghud, Sini. Eng.: Sweet orange, Tight skinned orange.

The fruit is flavourous, sweet, juicy and nutritious. It is highly esteemed as a desert fruit and is extensively used for flavouring pastry products and in making marmalades. It is a rich source of antiscorbutic vitamin. Orange purifies blood, allays thirst in fevers, cures catarrh and improves appetite. The juice is useful in bilious affections and stops bilious diarrhoea. Fresh rind is rubbed on the face as a remedy for acne. The dried outer portion of the rind possesses stomachic and tonic properties. The wood hard, employed in turning, engraving and cabinet making. Walking sticks are made from the branches.

INDIA : Widely cultivated.
WORLD : Probably native of Southern China and Indochina cultivated widely in subtropical regions.

**Citrus medica** L. Sp. Pl. 782. 1753; FBI 1: 514; Gamble 1: 161(115); Brandis 123.

Small tree, 4–8 m tall; spines stout; wood white, fine grained. Leaves elliptic, oblong or ovate-oblong, 8–18 × 3–8 cm, glabrous, apex obtuse, margin serrate or crenate, base rounded or cuneate; petiole 8–10 mm, slightly winged, not distinctly articulated with lamina. Flowers white, often unisexual, ca. 1.5 cm long. Fruit ovoid or oblong or obovoid, yellow, 10–20 cm long; pulp greenish, sour, scanty; seeds small, white. Predominant glucoside is hesperidin.

Rare, in Visakhapatnam and Prakasam districts up to 1000 m in thickets along streams. Vern.: Lungamu.

The fruits are used mainly for pickling. The juice is refrigerent and astringent. Preserved rind is used as a remedy for dysentry. Candied peel of citron is a commercial product. The peel and pulp are made into marmalades and other preserves. Wood fairly hard and rather heavy used for making agricultural implements. The branches are used as walking sticks.

Korai Road (VSKP), *GVS* 22631 (MH).

INDIA : Considered to be a native tree.
WORLD : Myanmar, cultivated throughout the world moist regions.

## CLAUSENA N. Burmann

**Clausena heptaphylla** (Roxb.) Wt. & Arn. Prodr. Fl. Ind. Orient. 95. 1834; FBI 1: 504; Gamble 1. 155 (110); Brandis 115. *Amyris heptaphylla* Roxb. Fl. Ind. 2: 248. 1832.

A small tree, strongly aromatic; branchlets glabrous. Leaves 5–7-foliolate, 15–35 cm long; petiole to 8 cm, cylindric, slender; leaflets membranous, or chartaceous, lanceolate to obovate, 7.5–11 × 3.5–6.3 cm, glabrous, yellowish when dry with immersed glands, pale beneath, apex acuminate or emarginate, margin obscurely crenate, base oblique, obtuse, round; secondary nerves 8-pairs, slender, conspicuous; petiolule to 2 cm. Flowers yellowish-green or yellowish-white, 4 to 5 mm across in subterminal or axillary, to 25 cm long thyrsoid panicles; peduncle to 6.5 cm. Berry reddish-white or pale yellow, oblong, 1.2 cm long, glandular.

Occasional, in the hills of Visakhapatnam and East Godavari districts up to 1500 m. Fls.: Apr.-May. Frs.: June–July.

Maddiveedu (EG), *NRR & TRS* 86116; Galikonda (VSKP), *GVS* 19630 (MH); Paderu (VSKP), *GVS* 30003 (MH); Sunkarimetta (VSKP), *NPBK* 10928 (CAL).

INDIA : Assam, ascending to 1200 m; Eastern Ghats up to 1500 m; Western Ghats.

WORLD : Bangladesh.

## EUODIA J.R. Forster & J.G.A. Forster

**Euodia lunu-ankenda** (Gaertn). Merr. in Philipp. J. Sci. 7: 378. 1912. var. ***lunu-ankenda*** (*Evodia lunar-ankenda)*; Gamble 1: 148 (105) (Evodia). *Fagara lunu-ankenda* Gaertn. Fruct. 1. 334. t. 68. f. 9. 1788. *Euodia roxburghiana* Benth. Fl. Hongs. 59. 1861; FBI 1: 487; Brandis 112. *E. triphylla* (Wt.) Beddome Fl. Sylv. S. India, Anal. Gen. 41. t. 62. 1870.

Evergreen tree, up to 15 m tall; bark grey, smooth, corky; wood greyish-white, close-grained; young branches puberulous. Leaves 3-foliolate, rachis to 12 cm; petiole to 10 cm, terete; leaflets sub-coriaceous, elliptic or obovate, 4–18 × 3–8 cm, glabrous, shining above, paler beneath, closely pellucid dotted, apex narrowly, shortly, abruptly acuminate, margin entire, base obtuse or acute; secondary nerves 14–18 pairs, horizontal, slender, not prominent; petiolule to 8 mm. Flowers cream, fragrant, 4 mm across, unisexual, 4-merous in dense axillary cymes; panicles divaricate, spreading, to 14 × 5 mm, peduncle to 10 cm. Fruits light brown, globose, cocci 2–4, wrinkled, 4–6 mm; seeds black, oblong, 3.5 × 2 mm, shining, muricate.

Rare, in Rampa hills of East Godavari at above 900 m. Fls.: July–Aug. Frs.: Throughout the year.

Wood moderately hard. It is used for only planks, rafters, match splints and plywood. The leaves are sour and aromatic when crushed, they are said to be used for flavouring curries. The plant is used in fever.

Rampa (EG), *JSG* 15995 (CAL).

INDIA : Tamil Nadu.
WORLD : Sri Lanka, Malay Peninsula.

## GLYCOSMIS Correa

**Glycosmis pentaphylla** (Retz.) DC. Prodr. 1: 538. 1824. Tanaka in J. Indian Bot. Soc. 16: 229. 1937; FBI 4: 499; *p.p.*Brandis 112; *Limonia pentaphylla* Retz. Obs. 5: 24. 1789. *Glycosmis cochinchinensis* sensu Gamble 1: 153 (109). 1918.

Small, evergreen tree, up to 6 m tall; young shoots glabrous. Leaves 5–7 foliolate, petiole to 1.5 cm; leaflets variable, elliptic-oblong, 5–12.5 × 2.5–6 cm, alternate and subopposite, glabrous, green above, paler beneath, apex acute, or shortly acuminate, margin entire, base acute; secondary nerves 7–10 pairs; petiolule to 5 mm. Flowers white, small, 7 mm across, in small clusters on erect, axillary, 8 cm long panicles. Berry cream coloured, globose, 1 cm in dia., smooth, apiculate.

Common, in moist, dry deciduous forests at 600–1000 m. Fls.: Most of the year especially Sept.–Feb.

Fruits are edible. Leaf paste is externally applied for poisonous bites.

Galikonda (VSKP), *SSR* 19630; Mamanduru-Thumburatheertham (CTR), *SSR* 16171; Anantagiri (VSKP), *NPBK* 10983 (MH, CAL); Korada RF (SKLM), *GVS* 62389 (CAL, MH).

INDIA : Tamil Nadu.
WORLD : Sri Lanka, S.E. Asia and W. Malaysia.
**Note** : It is a most variable species.

## LIMONIA Linnaeus

**Limonia acidissima** L. Sp. Pl. (ed. 2) 554. 1762. *Feronia elephatum* Cor. in Trans. Linn. Soc. London 5: 224. 1800; FBI 1: 516; Gamble 1: 160 (114); Brandis 119; Beddome Fl. Sylv. S. India t. 121. 1871.

Deciduous, armed tree, up to 12 m tall; crown spreading; bark blackish, deeply fissured; wood yellowish-grey to greyish-white; branchlets densely tomentose. Leaves 1–3 in a cluster, imparipinnate, 3–9 foliolate, 8 cm long, rachis flattened, narrowly winged, petiole to 3 cm; leaflets thin-coriaceous, obovate, 3.8 × 2 cm, glabrous, bright green, apex obtuse, margin entire to faintly crenulate; base narrow-acute, venation reticulate and arching type; petiolule 0. Flowers cream coloured, polygamous in axillary or terminal short racemes. Berry small, globose 1.5 cm in dia, woody, indehiscent; seeds numerous, embedded in pulp.

Common, in open dry forests and also in open fields up to 800 m, often cultivated along the road sides. Leaf fall: Dec. New foliage: Jan.–Mar. Fls.: Apr. Frs.: May onwards. Vern.: Velaga, Pushpa-phalamu. Eng.: Wood apple.

The pulp of the fruit is much eaten. Fruit and bark used for digestive troubles. Trunk and branch yield gum, used as substitute for gum arabic. Wood is used for house building, naves of wheels and oil crushers. It can be used as patterned wood for shoe lasts, pen holders, rulers, and similar articles, also used for Agricultural implements.

Bairluti (KNL), *SSR* 15515; Anantapur town (ATP), *NY* 801; Palakonda hills (CDP), *CS* 9318; Chintala (PKSM), *RVK* 15761; Mamidipally RF (NZB), *BR & KH* 9022; Chodavaram (EG), *MRS* 1457 (CAL), Kodimiarampalli (NLR), *MSR* 1245 (CAL); Punyagiri (VSKP), *GVS* 23322 (MH); Komaticheruvu (CTR), *GVS* 45873; Valagapalli (KSN), *AWL* 1229 (DD); Mannanur (MBNR), *TDCK* 15366; Venkatagiri (NLR), *ASR* 2322 (VV).

INDIA : Throughout.
WORLD : Pakistan, Sri Lanka, Java.

## MICROMELUM Blume

**Micromelum pubescens** Blume Bijdr. 1: 137. 1825; FBI 1: 501. Gamble 1: 154 (109); Brandis 114.

A small evergreen tree, up to 8 m tall; wood yellowish-white, close grained; young branches pubescent, greyish brown, raised with minute glands. Leaves imparipinnate, to 30 cm long, petiole to 8 cm; leaflets alternate, or opposite, 10–13 pairs, ovate-lanceolate, elliptic, glabrous above, pubescent beneath, apex abruptly acuminate or caudate-acuminate, margin wavy, or slightly dentate, base oblique, rounded, petiolule to 3 mm. Flowers pale green, scented (sweet smell), 0.8–1.4 cm dia. in 20 cm long, corymbose, terminal panicles. Berry yellow or scarlet, ovoid, foetid, 1.4 cm long, glabrous; seeds 1–2, cotyledons folded and crumpled.

Rare, in evergreen forests of Visakhapatnam and Kurnool districts. Fls.: Jan.-Mar. Frs.: May–July.

Nallamalais (KNL) without collector's name 8219 & 89220 (MH).

INDIA : Sikkim, Assam, Khasi and Naga hills.
WORLD : Sri Lanka, Malay Peninsula, China, Nepal, ascending to 2000 m.

## MURRAYA J.G. Koenig ex Linnaeus

1. Leaflets 5–7, not pungent; rachis glabrous. Flowers in short corymbs. Fruit red when ripen. .......... ***M. paniculata***
1. Leaflets 11–25, very pungent; rachis pubescent. Flowers in long corymbose panicles. Fruit purplish black when ripe. .......... ***M. koenigii***

**Murraya koenigii** (L.) Spreng. Syst. Veg. 2: 315. 1826; FBI 1: 503; Gamble 1: 156 (111); Brandis 113; Beddome Fl. Sylv. S. India Anal. Gen. 44. 1870. *Bergera koenigii* L. Mant. Pl. 565. 1771.

Aromatic, deciduous tree, up to 7 m tall; bark dark brown; wood greyish-white, a close grained; branchlets green to grey, finely pubescent, but glabrate. Leaves imparipinnate, rachis ca. 17 cm long, petiole to 4 cm; leaflets 11–25, oblique, elliptic-ovate, 1.5–3.8 × 1.4–2 cm, glabrous and dark green above, pubescent beneath with black dots, apex acute, or retuse, margin irregularly crenate-dentate, base asymmetrical; petiolule 3 mm. Flowers white in many-flowered, dense, corymbose cymes. Fruit subglobose, 1 cm across, purplish black when ripe, apiculate, rough with glands, pulp whitish; seeds 1–2, green.

It is cultivated for its aromatic leaves but also found wild in sun exposed slopes, open dry forests and also in open fields up to 1200 m. New foliage: Mar. Fls. with two peaks during Mar.–May and July–Aug. Frs. Throughout the year. Vern.: Kalima, Karivepaku, Karivemu.

Aromatic leaves used for curry flavouring and also used in indigestion. Fruits are edible, yield yellow volatile oil with pepper like taste. The leaves, root and bark are considered tonic, stomachic and carminative. Leaves are taken internally in dysentry and diarrhoea, and also for checking vomiting; they are applied externally to bruises and eruptions. Wood hard, used for Agricultural implements.

Itukalapalli (ATP), *NY* 170; V.P. South (GTR), *VRK* 3931; Rollamadugu (CDP), *JSG* 10802 (DD); Nulakamaddi (EG), *GVS* 24538; Krishnapuram (KSN), *CAB* 1974 (MH); Lumbasingi (VSKP), *CAB* 11814 (MH); Kudichintalabailu (MBNR), *TDCK* 15370; Udayagiri (NLR), *ASR* 500 (VV).

INDIA : Throughout.
WORLD : Sri Lanka, Myanmar, Indo-China, S. China.

**Murraya paniculata** (L.) Jack. in Malayan Misc. 1(5): 31.1820. *Chalkas paniculata* L. Mant. 1: 63. 1767. *Murraya exotica* L. Mant. Pl. 563. 1771. ' *Murraea*', FBI 1: 502; Gamble 1: 156 (111); Brandis 118; Beddome Fl. Sylv. S. India Anal. Gen. 44. t. 7. f. 2. 1870.

Densely foliaceous, evergreen tree, up to 5 m tall; crown spreading, trunk short; bark greyish-white, smooth, fragrant; wood white, close grained; young branchlets pubescent. Leaves imparipinnate to 15 cm long, petiole to 8 mm; leaflets 5–7, ovate, elliptic-rhomboid, 2–6 × 2–3 cm, glabrous, shining, gland dotted, apex obtusely acuminate, apically notched, margin entire, base oblique to acute; secondary nerves 5–8 pairs; petiolule to 2.3 mm. Flowers few, white in terminal or axillary corymbs. Fruit red or deep orange, ovoid; 1–2-seeded.

Hills, on the floor of the forest in shade especially in the ravines, along the slopes and near streams in the hilly areas at 500–900 m, often cultivated in gardens and makes an excellent hedge. Fls. with two peaks during Mar.–May and Sept.–Oct. Frs.: Throughout the year. Vern.: Golimi, Naga Golugu, Gana Penu, Nara Gongi.

Wood very hard resembling bore wood; used for tool handles, cabinet work and walking sticks. It is suitable for mathematical instruments, pen holders, brushes etc. The leaves are stimulent and astringent and yield a dark coloured volatile oil with pleasant odour. Twigs are used for cleansing teeth.

Horsely hills (CTR), *SSR* 13795; N.U. Campus (GTR), *VRK* 6991; Venkatagiri (NLR), *ASR* 6872 (VV).

INDIA : Outer Himalayas, up to 1350 m, Bihar, S. India, cultivated throughout.

WORLD : Native of South-East Asia. Sri Lanka, Myanmar, Indo-China. Malaysia to Australia.

## **NARINGI** Adanson

1. Leaves 3-foliolate, margin subentire, glands not conspicuous. Flowers in 6 cm long panicles. Berry to 2.5 cm dia., seeds 1.5 × 1 cm. .......... ***N. alata***
1. Leaves 5–7-foliolate, margin crenulate, glands very conspicuous. Flowers in 4 cm long subumbellate racemes. Berry to 8 mm dia; seeds 5 × 3.5 mm. .......... ***N. crenulata***

**Naringi alata** (Wall. ex Wt. & Arn.) Ellis in Bull. Bot. Surv. Ind. 22: 193. 1980. *Limonia alata* Wall. ex Wt. & Arn. Prodr. Fl. Ind. Orient. 92. 1834; FBI 1: 508; Gamble 1: 157 (112); Brandis 119. *Pleiospermum alatum* (Wall. ex Wt. & Arn.) Swingle in Journ. Wash. Acad. Sci. 6: 427. 1916; Beddome Fl. Sylv. S. India Anal. Gen. 45. t. 7. f. 4. 1870. *Hesperethusa alata* (Wt. & Arn) Alston in Trimen Handb. Fl. Ceyl. Suppl. 38. 1931.

Armed tree, up to 6 m tall; wood hard and close grained. Leaves trifoliate, petiole to 3.6 cm, leaflets coriaceous, obovate, central one 6.2 × 2.8, lateral ones 3–4 × 1.4–2 cm, apex obtuse, margin entire, rarely faintly crenulate, base cuneate, petiolule 0. Flowers white, 4 cm across in terminal, 6 cm long racemes. Berry globose, to 1.8 cm; seeds ca. 5, to 1.5 × 1 cm.

Occasional, in dry deciduous forests of Nallamalais, Visakhapatnam and Srikakulam district to 1200 m. Fls.: Feb.-Apr. Frs.: May–Sept. Vern.: Konda nimma, Munukudu, Nalla Munukundu.

Chintala forest (PKSM), *SSR* 16130; Loyapalli (GTR), *VRK* 3847, 3981; Dornal (KNL), *JSG* 17687;. Anantagiri (VSKP), *NPBK* 1830 (CAL); Donubai (SKLM), *GVS* 62458 (CAL); Rapur (NLR), *ASR* 4834 (VV).

INDIA : South India.
WORLD : Sri Lanka.

**Naringi crenulata** (Roxb.) Nicolson in Saldanha & Nicolson, Fl. Hassan Dist. 387. 1976. *Limonia crenulata* Roxb. Pl. Cor. t. 86. 1709; Gamble 1: 157 (112). *Limonia acidissima auct. non* L. 1762; FBI 1: 507; Brandis 118.

Armed, deciduous tree, up to 15 m tall; crown columnar; bark greyish, rough; wood yellowish white; branchlets glabrous. Leaves 3–5 in a cluster, imparipinnate; petiole and rachis articulated, narrowly winged, 3.2 cm long; leaflets 3 pairs, thin-coriaceous, glandular, ovate-elliptic, 2–7 × 1–3 cm, villous along nerves above, glabrous beneath, apex acute to obtuse, margin crenulate, base cuneate, petiolule 0; Flowers white, 1 cm across in 3 cm long, subumbellate racemes. Berry globose, 1 × 0.5 cm, bluish black; seeds ovoid, 3 to 4, 5 × 3.5 mm.

Common, in the scrubs and the hills of dry deciduous forests at 600–900 m. Leaf fall: Feb. New foliage: Apr. Fls.: Apr.–June. Frs.: July-Oct. Vern.: Torri velaga.

Wood very hard. The roots possess purgative and sudorific properties; dried fruit is tonic and stomachic and useful in malignant and persistent fevers and as an antidote for fever.

Tirumala hills (CTR), *SSR* 13769; Donubai (SKLM), *SSR* 15584; Eturnagaram (WGL), *SSR* 18536; Dumakonda RF (EG), *GVS* 68586 (MH); Araku (VSKP), *DSRJ* 95892 (MH); Diguvametta (PKSM), *CAB* 8108 (MH); Anantagiri (VSKP), *GVS* 21736; Bheemram (ADB), *GO* 5181; Peccheruvu (KNL), *RVR & PVP* 2146; Appapur (MBNR), *TDCK* 13637; Udayagiri (NLR), ASR 4510 (VV).

INDIA : Throughtout.
WORLD : Sri Lanka, Pakistan, Bangladesh, Myanmar, Indochina, S.W. China.

## PAMBURUS Swingle

**Pamburus missionis** (Wt.) Swingle in J. Wash. Acad. Sci. 6. 338. 1916. *Limonia missionis* Wt. Ic. t. 175. 1837. *Atalantia missionis* (Wt.) Oliver in J. Linn. Soc. Bot. 5 (Suppl. 2): 25. 1861; FBI 1: 513; Gamble 1: 160 (114); Brandis 121; Beddome, Fl. Sylv. S. India Anal. Gen. 46. 1870.

Densely foliaceous, thorny, evergreen tree, up to 10 m tall; crown compact; wood yellowish white; branchlets glabrous. Spines axillary, straight. Leaves thick coriaceous, broadly ovate or elliptic, 5–11 × 2–5 cm, glabrous, dark green, grey when young, apex obtuse or rounded, margin entire or slightly crenulate, base acute, nerves obscure; petiole to 1.2 cm. Flowers cream, numerous in 2.5–5 cm long, axillary racemes. Berry compressed globose, 2.5 cm in dia., green, rind thick; seeds numerous, globose, embedded in glutinous mucilage.

Rare, in mixed forests and along stream banks at 600–700 m. Fls.: Feb.–Apr. Frs.: May-June. Vern.: Adavi nimma, Adavikitchili.

Tirumala hills (CTR), *SSR* 13721; Cuddapah hills (CDP), *CBDM* 8505 (MH); Sriharikota (NLR), *ASR* 2315 (VV).

INDIA : South India
WORLD : Sri Lanka.

## ZANTHOXYLUM Linnaeus

1. Leaflets 8–32. Trees middle sized ........ *Z. rhetsa*
1. Leaflets 3–7. Trees small:
   2. Leaflets 3–7, narrowly elliptic-lanceolate, rachis and petiole winged. Flowers yellow, petals 0. Seeds rugose ........ *Z. armatum*
   2. Leaflets 3, broadly elliptic-oblong to lanceolate; rachis and petiole not winged. Flowers white, petals present. Seeds smooth ........ *Z. ovalifolium*

**Zanthoxylum armatum** DC. Prodr. 1: 727. 1824. *Z. alatum* Roxb. Fl. Ind. 3. 768. 1832; FBI 1: 493; Gamble 1: 150 (106); Brandis 116.

Densely foliaceous tree, up to 7 m tall; with highly pungent aromatic smell and taste; bark brown, smooth; wood white, close grained; prickles shining, strong, broad and flat on the branchlets, petioles and midrib those on older branches often on a conical corky base. Leaves imparipinnate, rachis glabrous, winged; petiole winged to 3.3 cm; leaflets 3–7, narrowly elliptic-lanceolate, 2.8–10 × 1.2 cm, thick green with few scattered glands, apex obtusely acuminate, margin crenulate, glandular, base oblique or cuneate. Flowers yellow, dioecious in lax, sparingly branched, axillary panicles. Fruit pale red, ripe carpels 1–2, often 3, 4–5 mm dia, somewhat compressed, tubercled; seed black, solitary, rugose.

Occasional, in moist deciduous forests of Visakapatnam at 800–1300 m. Fls.: Jan.–Mar. Frs.: June–Dec. Vern.: Konda Kasimi, Trade or Popular name; Wing leaf prickly ash.

Bark used for intoxicating fish. Wood is hard and used in tooth aches, and carpels as a condiment.

Galikonda (VSKP), *SSR* 15917; Anantagiri-Galikonda (VSKP), *TP & EC* 7315, *DAM* 6501; Araku (VSKP), *NPBK* 10826 (MH); Tanjavanam (VSKP), *GVS* 47301 (MH, CAL); T. Bunglow-Anatagiri (VSKP), *GVS* 19502 (MH).

INDIA : Eastern Ghats, Himalays, up to 2000 m.
WORLD : Pakistan, Japan, Taiwan, Philippines.

**Zanthoxylum ovalifolium** Wt.. Ill. 1: 169. 1840; FBI 1: 492. incl. var. *sepiarium*; Gamble 1: 150 (107); Brandis 118.

Small tree, up to 8 m tall; lenticellate, armed with short, straight or slightly curved, parallel prickles from thickened bases. Leaves 3-foliolate, petiole to 2 cm; leaflets coriaceous, obliquely elliptic-oblong to lanceolate, 6–11 × 2–5.5 cm, glabrous, apex bluntly acuminate with emarginate, margin distantly crenate, base obliquely acute, lateral nerves numerous, slender, parallel; petiolule short or 0. Flowers white, 5 mm across in dense paniculate cymes with alternate branches. Fruit solitary, spherical, 5 mm dia., slightly pitted, glabrous; seed black, globose, smooth, shining.

Occasional, in the forests of Chittoor district up to 1350 m. Fls. and Frs.: Nov.–Mar. Vern.: Rijayi.

Murthivari Cheruvukatta (CTR), *GVS* 45928 (MH); Shikarline (CTR), *GVS* 45988 (MH).

**Zanthoxylum rhetsa** (Roxb.) DC. Prodr. 1. 728. 1824; FBI 1: 495; Gamble 1: 150 (107); Brandis 118. *Fagara rhetsa* Roxb. Fl. Ind. 1. 438. 1820. *Zanthoxylum budrunga* (Roxb.) DC. Prodr. 1: 728. 1824; FBI 1: 493; Gamble 1: 150 (107); Brandis 118.

A middle-sized, deciduous tree, trunk prickly, stem resembles that of *Bombax*; bark corky, deeply cracked, pith large septate; wood yellowish-grey, hard. Leaves crowded at the ends of stout branchlets, imparipinnate, 30–50 cm long, petiole to 11 cm; leaflets 4–16 pairs, opposite, ovate-oblong or lanceolate, 6–11 × 3.5–4.5 cm, glabrous, apex caudate-acuminate, margin entire, base very oblique, the upper side of each leaflet rounded at the base, the lower side narrowed into the costa, secondary nerves 8–12 pairs; petiolule to 2 mm. Flowers yellow, 4-merous, 4 mm across, on 20 cm long cymose panicles. Fruits solitary, spherical, 6 mm in dia., tuberculate; seeds globose, bluish-black, smooth, shining, aromatic and tasting of pepper.

Occasional, in moist evergreen forests of East Godavari and Visakhapatnam districts at 1000 m. Fls.: Mar.-June. Frs.: Aug. Vern.: Rhetsa-man.

Rampa hills (EG), *JSG* 7950 (MH); Maredumilli (EG), 24295 (MH).

INDIA : Himalayas; W. Peninsula, up to 300 m; Sikkim, Bihar.
WORLD : Sri Lanka, Myanmar, Thailand, S. Vietnam, Malaysia, Java, Philippines, Mollucca, S. Papua.

## SIMAROUBACEAE

1. Leaflets 8–14 pairs, margin irregularly toothed. Flowers yellowish, polygamous, 5-merous. Fruit winged. .......... ***Ailanthus***
1. Leaflets 1–3 pairs, margin entire. Flowers greenish-white, unisexual, 4-merous. Fruit not winged. .......... ***Picrasma***

## AILANTHUS Desfontaines *nom. cons.*

**Ailanthus excelsa** Roxb. Pl. Cor. t. 23. 1795; FBI 1: 518; Gamble 1: 163 (116); Brandis 125.

Densely foliaceous, deciduous tree, up to 25 m tall; bark whitish, rough; branchlets hoary tomentose with persistent leaf scars. Leaves imparipinnate, rachis to 45 cm long; petiole to 20 cm; leaflets 8–14 pairs, foetid, elliptic to ovate-lanceolate, 7–8 × 4–5 cm, glabrous above, softly pubescent beneath, apex acuminate, margin irregularly toothed, base inequilateral, petiolule to 2 cm. Flowers yellowish, polygamous in 40 cm long, axilllary or terminal panicles. Samara yellowish, oblong, flat, 4–6 × 1 cm, narrowed or pointed at both ends, twisted at the base, wings with prominent nerves; seed one in the middle of the samara, compressed.

Common, generally in the plains around villages and old forts and broken ground, rarely in the forests up to 700 m. It is able to adopt itself in a variety of soils. It prefers well drained soil and avoids clayey and waterlogged soils. It is a fast growing tree. Leaf fall: Feb.–Mar. New foliage: Apr.–June. Fls.: Jan.–Mar. Frs.: May. Vern.: Pedda Manu, Pedda vepa. Eng.: Tree of heaven.

Stem bark given for cold and cough. The wood is soft, light, but fairly strong and it is used especially for making catamarans and small boats. It is also used for veneers and plywood; packing cases, sword handles, spear-sheaths, toys and poor quality matchwood. The bark is bitter, astringent, febrifuge and anthelmintic. The tree yields an inferior quality of gum. Leaves and twigs suitable for fodder as cattle, sheep and goats.

Lower Ahobilam (KNL), *SSR* 15294; Kadam RF (ADB), *GO & DAM* 5195; Domakonda (NZB), *BR* 9683; Visakhapatnam (VSKP), *JSG* 15749 (MH); Rampa (EG), *VNS* 21 (CAL); Polavaram Agency (EG), *DCSR* 255 (CAL).

INDIA : Throughout.
WORLD : Tropical Africa, Sri Lanka, Myanmar, Australia and Pakistan.

## PICRASMA Blume

**Picrasma javanica** Bl Bijdr. 5: 248. 1825; FBI 1: 520; Basak in Fasc. Fl. India 4: 13. 1980; Brandis 127.

An evergreen, small tree; branchlets glabrous, lenticellate, smooth, grey or brown. Leaves imparipinnate, rachis 7 cm long, petiole to 4 cm; leaflets 3–7, coriaceous, opposite, lanceolate to elliptic-oblong, odd leaflet obovate, 4.5–11 × 2–4.7 cm, glabrous, shining and grey above, brown or yellowish brown below, apex shortly, obtusely acuminate, margin entire, base oblique, rounded or cuneate; secondary nerves 4–6 pairs, midrib crested above, raised below, petiolule to 6 mm. Flowers greenish-white, tetramerous, 1 mm across, unisexual, pedicellate in 18 cm long, cymose panicles; peduncle to 5 cm. Drupe white, 7 cm dia., fleshy.

Occasional, in Visakhapatnam district to 1000 m. Fls. Mar.-Apr. Frs.: July-Oct.

Ramajelladu waterfalls (VSKP), *GVS* 44394 (MH); G. Madugula. Thanjavanam (VSKP), *GVS* 47311 (MH).

INDIA : Sikkim, Khasi hills, Andamans, Arunachal Pradesh, Assam, Manipur, Meghalaya, Tripura, West-Bengel.

WORLD : China, Java, Nepal, Bangladesh, Myanmar, Malaysia, Cambodia, Thailand, Laos, Vietnam, Indonesia, Phillippines.

## BALANITACEAE
### **BALANITES** Delile *nom. cons.*.

**Balanites aegyptiaca** (L.) Del. Descr. Egypt. Hist. Nat. 2: 221. t. 28. f. 1. 1813. *Ximenia aegyptiaca* L. Sp. Pl. 1194. 1753. *Balanites roxburghii* Planch. in Ann. Sci. Nat. Bot. Ser. 4. 4. 258. t. 2. 1854; FBI 1: 522; Gamble 1: 164 (117); Brandis 124; Beddome Fl. Sylv. S. India 2: 50 in part and Anal. Gen. t. 8 f. 1871.

Small armed tree, up to 8 m tall; bark yellowish or cinerous; wood yellowish-white; branchlets brown, lenticellate; spines sharp, up to 3.5 cm. Leaves 2-foliolate, petiole to 5 mm; leaflets thin-coriaceous, oblanceolate-elliptic-obovate, 1.5–3 × 5–1.5 cm, puberulous, dotted with brown coloured glands, apex obtuse, apiculate, margin entire, base acute; petiolule 0. Flowers greenish in axillary pseudo-umbels. Drupe ovoid-oblong, 4.5–8.6 × 1.3–6 cm, woody, green, rotund, fleshy, slightly 4–5- angled; seed solitary, embedded in pulp.

Introduced, common in dry forests of most districts plains to 500 m. Leaf fall: Feb. New foliage: Mar. Fls.: Mar.–Apr. Frs.: May.

SKU Campus (ATP), *SSR* 15549; Chintala (PKSM), *SSR* 16129; Tadwai (WGL), *SSR & KSM* 18547; North Dhone RF (KNL), *TP & RVR* 1311; Gopalapatnam (ADB), *GO* 4401; Macherla RF (GTR), *VRK* 5890; Talakona (CTR), *GVS* 46002 (MH); Nattavaram (VSKP), *CAB* 8639 (MH).

WORLD : Dry localities of Africa, Sri Lanka and Myanmar.

## OCHNACEAE
### **OCHNA** Linnaeus

1. Flowers solitrary or 2–3 in cymes. Leaves < 5 cm long. Drupe 1–3-lobed. ... ***O. lanceolata***
1. Flowers many in corymbose racemes or umbels. Leaves > 5 cm long. Drupe to 10-lobed. ... ............ ***O. obtusata***

**Ochna lanceolata** Spreng. Syst. Veg. 2. 592. 1825; Kanis in Blumea 16: 24. 1968. *O. wightiana* Wall. ex Wt. & Arn. Prodr. Fl. Ind. Orient. 1: 152. 1634; FBI 1: 524; Gamble 1: 166 (119). Brandis 129. Beddome Fl. Sylv. S. India 51. 1870. *O. heyneana* Wt. & Arn. Prodr. Fl. Ind. Orient. 1: 152. 1834; Gamble 1: 166 (119).

Semi-evergreen tree, up to 8 m tall; branchlets glabrous. Leaves thin-coriaceous, lanceolate, 4–4.8 × 1–2 cm, glabrous, apex apiculate, margin serrate, base obtuse; petiole to 2 mm. Flowers yellow, 1.2–2 cm across, solitary or 2–3-flowered cymes. Drupe 1–3-lobed, 1 × 0.7 cm; seed one in each lobe, to 5 mm.

Occasional, in very few districts at 500–700 m. Fls.: Mar. Frs.: Apr.

Cuddpah (CDP), *CBDM* 8710 (MH); Achampet range (MBNR), *SJRC* 503 (MH); Nagarjuna Konda (NLG), *KTH* 9731 (CAL).

INDIA : Peninsula.
WORLD : Sri Lanka.

**Ochna obtusata** DC. in Ann. Mus. Nat. 1. Hist. Nat. 17. 481. 1811. *O. squarrosa auct. non* L. 1753. FBI 1: 523; Gamble 1: 165 (118); Beddome Fl. Sylv. S. India 50. t. 8. f. 3. 1870.

1. Leaves glaucous below. Flowers in racemes; stigma lobed. .............................. var. ***gamblei***
1. Leaves green below. Flowers in umbellate panicles; stigma simple. ............... var. ***obtusata***

**Ochna obtusata** DC. var. ***gamblei*** (King ex Brandis) Kanis in Blumea 16: 34. 1968. *O. gamblei* King ex Brandis. Indian Trees 128. 1906. *O. beddomei* Gamble Fl. Pres. Madras 1: 166 (188). 1915, nom. & Bull. Misc. Inform. 1916. 34. 1916 descr.

Small tree, up to 10 m tall; bark smooth, greyish-white; wood reddish; branches stiff, erect. Leaves tufted on branchlets, coriaceous, elliptic or elliptic-oblong, 6–14 × 3–5 cm, glaucous, whitish above, pale beneath, apex obtuse, margin entire or serrate, base cuneate, often cordate; secondary nerves numerous and irregularly nerved; petiole to 1 mm. Flowers yellow, dense, many-flowered racemes at the ends of woody leafless branchlets. Drupe 10-lobed, druplet black with persistent calyces; seeds one in each lobe.

Common, on the full sun-exposed dry rocky hills, of dry deciduous forests at 600–900 m. Leaf fall: Feb.-Mar. New foliage: Apr. Fls.: Mar.–Apr. Frs.: May onwards. Vern.: Chitti Movi, Kukka Movi. It is a fire resistant and good coppicer.

Guvvalacheruvu RF (CDP), *RV Reddy* 8181; Kalasamudram RF (ATP), *EC* 2979; Palakonda hills (CDP), *CS* 9328; Horsely hills (CTR), *CECF* 4358 (CAL); Murrem Konda (NLR), *MSR* 1269 (CAL); Krishna (KSN), *CBDM* 8678 (MH); Appapur (MBNR), *TDCK* 15375.

INDIA : Central and Peninsula.

**Ochna obtusata** DC. var. ***obtusata***; Kanis in Blumea 16: 32. 1968. *O. squarrosa auct. non* L. 1753; FBI 1: 523; Gamble 1: 165 (118); Brandis 128.

Deciduous tree, up to 8 m tall; bark brown, smooth; wood reddish-brown; branchlets pubescent. Leaves membranous or chartaceous, green, oblong-lanceolate, elliptic-obovate, 5.5–11.2 × 2.5–4.8 cm, glabrous, apex acute, margin finely serrulate; secondary nerves numerous, slender, parallel, petiole to 3 mm. Flowers bright yellow, fragrant in corymbose racemes. Fruit of to 10-lobed drupes, 1.5 × 1 cm, oblong-ovoid, sessile, black, surrounded by persistent calyx; seeds one in each lobe.

Occasional, in rocky ravines and cool rocky slopes of deciduous forests at 500–900 m. Leaf fall: Feb.–Mar. New foliage: Apr. Fls.: Apr.–May; Frs. May onwards. Vern.: Sunari, Tammi, Erra Juvvi, Kukkamovi, Erra Jammi.

Simhachalam (VSKP), *SSR* 15565; Simhachalam (VSKP), *DAM* 6543; Sriharikota (NLR), *JSG* 12811 (DD); Thungathampara (SKLM), *GVS* 6237; (CAL); Godavary, *JSG* 8665.

INDIA : Peninsula, West Bengal, Bihar, Madhya Pradesh.
WORLD : Nepal, Bangladesh, Sri Lanka.

## BURSERACEAE

### KEY 1 (Vegetative)

1. Leaf rachis winged —cultivated ........ ***Bursera***
1. Leaf rachis not winged:
    2. Branchlets and leaves glabrous:
        3. Leaflets 2–4 pairs ........ ***Commiphora***
        3. Leaflets 5–16 pairs ........ ***Boswellia***
    2. Branchlets, rachis and leaves tomentose or pubescent:
        4. Leaflets sub-opposite; rachis and lower surface of leaflets with brownish pubescence; secondary nerves 10–15 pairs, 0.7 cm apart; petiolule 0 or 2 mm ........ ***Garuga***
        4. Leaflets opposite; rachis and both surfaces of leaflets with whitish tomentum, secondary nerves 5–9 pairs, 1.2 cm apart; petiolule to 1 cm ........ ***Protium***

### KEY 2 (Flowering and Fruiting)

1. Drupe trigonous; seeds winged ........ ***Boswellia***
1. Drupe globose-oblong; seeds not winged:
    2. Flowers minute in short axillary racemes ........ ***Bursera***
    2. Flowers not minute in long axillary racemose panicles:
        3. Disc annular, usually lobed. Flowers aggregated or dense in a racemose panicles ........ ***Protium***
        3. Disc cup shaped; Flowers in lax racemose panicles:
            4. Disc adnate to the calyx. Drupe oblong, yellowish when ripe ........ ***Garuga***
            4. Disc free from calyx. Drupe globose-subglobose, green or tinged red when ripe........ ***Commiphora***

## BOSWELLIA Roxburgh ex Colebrooke

1. Leaflets 2–6 pairs, ovate-oblong, slightly glaucous beneath. Flowers greenish white; panicles much branched. ............................................................................................ *B. ovalifoliolata*
1. Leaflets 8–15 pairs, oblong-lanceolate, green beneath. Flowers pinkish white; panicles not much branched or look like simple racemes ................................................... *B. serrata*

**Boswellia ovalifoliolata** Bal. et Henry in Journ. Bombay Nat. Hist. Soc. 58: 546–548. 1961.

Deciduous tree, up to 10 m tall; bark greyish, smooth, exfoliating in thin flakes; branchlets red coloured, glabrous. Leaves imparipinnate, about 18 cm long, petiole to 3.5 cm; leaflets 5–13, coriaceous, ovate or elliptic ovate, 1.5–3.6 × 1–2.5 cm, glabrous, slightly glaucous beneath, apex obtuse, margin entire, base inequilateral or rounded; secondary nerves 8–10 pairs, reddish, sessile. Flowers greenish white, in monochasial cymes on 15 cm long, much branched panicles. Drupe 3-gonous, pyrenes 3, 1.5 × 0.8 cm, bony; seeds winged and compressed.

Rare, in Seshachalam hills of Chittoor and Cuddapah districts at 600–900 m. Leaf fall: Dec.–Feb. New foliage: Apr.–May. Fls.: Dec.–Feb. Frs.: Apr.–June. Vern.: Konda sambrani.

Chitvel RF (CDP), *SSR* 16143; Tirupati (CTR), *KMS* 7336 (MH); Nallamalais (KNL), *CBDM* 8776 (MH).

**Note** : It is a new species reported by Balakrishnan and Henry from Tirupati hills of Chittoor district. Ellis in his "Flora of Nallamalais" said that he could not trace the new species from Nallamalais of Kurnool district because Beddome might have collected it slightly further South of Nallamalais in Cuddapah and Chittoor districts of Andhra Pradaesh, which form type locality of new species —the major collecting centres of Beddome.

**Boswellia serrata** Roxb. ex Colebr. in Asiat. Res. 9. 379. t. 5. 1807; FBI 1: 525; Gamble 1: 168 (120); Brandis 120. *B. glabra* Roxb. Pl. Cor. t. 207. 1811; Gamble 1: 168 (120); Beddome Fl. Sylv. S. India t. 124. 1871. *B. serrata* Roxb. var. *glabra* (Roxb.) Bennet in Hook. f. Fl. Brit. India 1: 528. 1875.

Deciduous tree, up to 16 m tall; crown spreading and flat; bark greenish, ashy grey, exfoliating in thin flakes; wood brownish; branchlets puberulous. Leaves apically clustered, imparipinnate to 18.5 cm long; petiole to 3.5 cm; leaflets 14–26, thin-coriaceous, oblong-lanceolate, 2.8–6 × 1.5–2 cm, apex obtuse or subacute, margin entire or crenate (one side serrate, another side crenate) base obliquely acute; secondary nerves ca. 16 pairs, not prominent or conspicuous, petiolule sessile. Flowers pinkish-white in 10 cm long, little

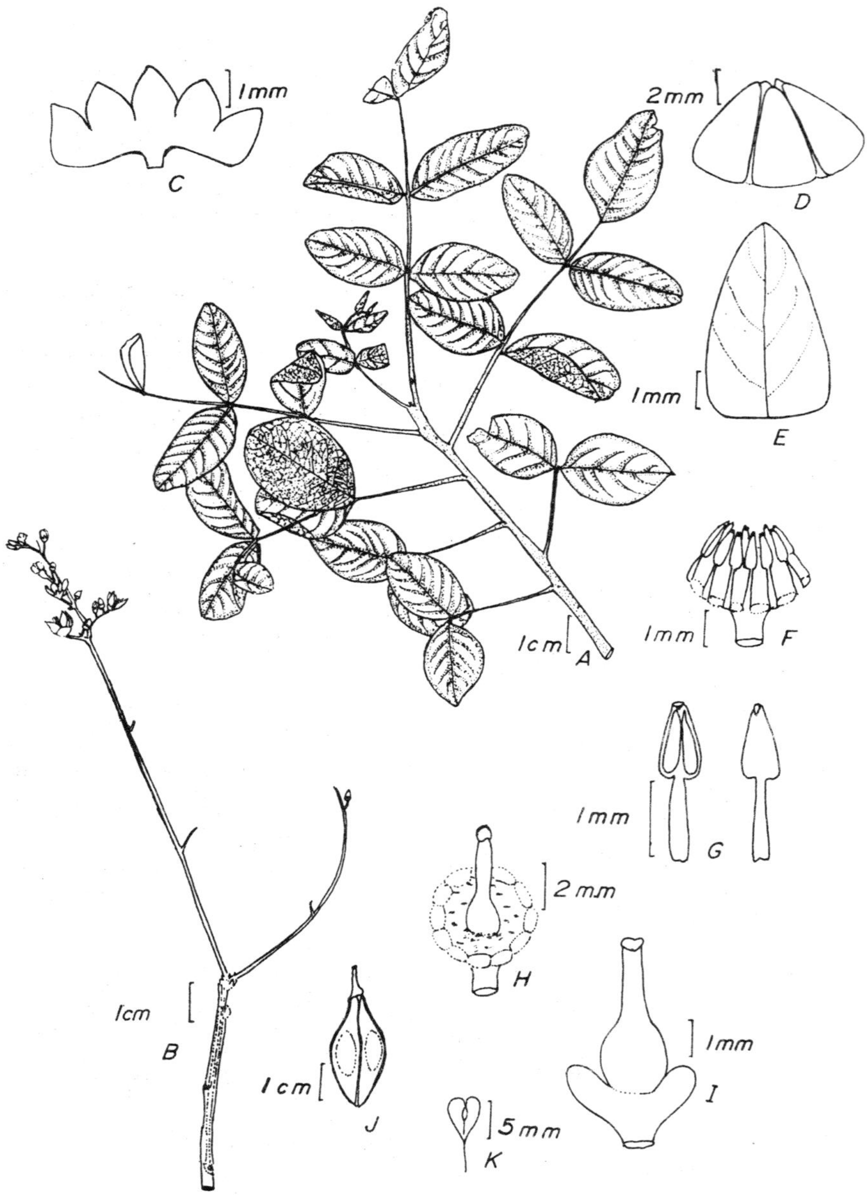


**Fig. 17.** *Boswellia ovalifoliolata* Bal. & Henry
A. Twig, B. Inflorescence branch, C. Calyx, D. Corolla, E. Petal, F. Androecium, G. Stamens, H. Pistil with disc, I. Pistil, J. Capsule, K. Seed.

branched, axillary panicles. Drupe trigonous, 1.6 × 1.1 cm, brown, pyrenes 3, each 1-seeded, enclosed in heart shaped pyrenes attached to the inner angle.

Common, on dry hill slopes of dry deciduous forests in gravelly, poor soil at 800–1200 m. Stem brittle. Leaf fall: Dec. New foliage: Jan. Fls.: Feb.–Mar. Frs.: May–June. Vern.: Guggilam, Anduga, Dhapamu.

Wood is used for charcoal. This tree yields a golden yellow coloured gum resin, which is used in treating skin diseases and rheumatism. It is prepared easily through cuttings.

Vanipenta (KNL), *SSR* 15298; Manchippa (NZB), *SSR* 19359; Kottakota RF (ATP), *TP & NY* 2112; Macherla RF (GTR), *VRK* 3884; Rollapenta (PKSM), *RVK* 15891; Godavary (KNR), *GVS* 22298 (MH); Pakhal (WGL), *ANH* 15951; Lankamalais (CDP), *CECF* 4738; Appapur (MBNR), *TDCK* 15369.

INDIA : North-west, Central and Peninsular region.

## **BURSERA** N.J. Jacquin ex Linnaeus

**Bursera penicellata** (Sesse & Moc. ex DC.) Engl. Bot. Jahrb. Syst. 44. 1881; McVaugh & Rzed. in Kew Bull. 18: 334. 1965. *Elaphrium penicillatum* Sesse & Moc. ex DC. Prodr. 1: 724. 1824.

Much branched, deciduous, medium sized tree; branchlets pubescent. Leaves clustered, imparipinnate, 7–8 cm long, rachis narrowly winged, petiole to 4 cm; leaflets 7, thin-coriaceous, or elliptic, obovate, glabrous above, villous beneath especially on the nerves, margin dentate, base oblique, subacute; secondary nerves 8–10 pairs; petiolule 0. Flowers cream, polygamous, 4.5 mm across in short axillary racemes. Berry reddish, globose, 9 mm in dia, pericarp fleshy with oil glands, apex acute; seed 1.7 × 6 mm.

Introduced. Leaf fall: Jan. New foliage: Apr. Fls.: Apr.–May. Frs.: May–Aug. Trade name. Mexican linaloe, Indian Lavender. Essential oil is extracted from seeds.

Mohammedabad forest (RR), *MSM & NV* 10416.

INDIA : Introduced in several parts for the source of Linaloe oils.
WORLD : Native of Mexico.

## **COMMIPHORA** N.J. Jacquin

1. Branchlets unarmed. Flowers in lax, dichotomous panicles. Drupe green, apex entire. Leaf apex caudate acuminate, margin entire ........ ***C. caudata***
1. Branchlets ending with sharp spines. Flowers in fascicles. Drupe tinged red, apiculate. Leaf apex obtuse, margin crenate ........ ***C. berryi***

**Commiphora berryi** (Arn.) Engler in DC. Monog. Phan. 4: 17. 1883; Gamble 1: 170 (122). *Balsamodendron berryi* Arn. in Ann. Nat. Hist. Serv. 13: 86. 1839; FBI 1: 529; Brandis 133; Beddome Fl. Sylv. S. India t. 126. 1871.

Deciduous, small tree, 3–6 m tall; much branched; branchlets spiny-tipped at right angles to the branches. Leaves 3-foliolate, rarely 4-foliolate, petioles slender to 3.5 cm; leaflets chartaceous, obovate, terminal one to 4 × 2 cm, lateral ones 3 × 1 cm, glabrous, glossy above, apex obtuse, margin crenate, base cuneate, or subacute; petiolule 0. Flowers 8 mm across, subsessile, solitary or in axillary clusters. Drupes tinged red, subglobose, 1 × 0.7 cm, apiculate.

Rare, in dry deciduous forests commonly grown as hedge plant at 400–600 m. Fls.: Feb.–Mar. Frs.: Apr. Leaffall: Feb. New foliage: Aug. onwards. It yields a fragrant gum-resin.

Near Bhongiri (NLG), *SSR* 19685; Vijayawada (KSN), *PVK* 5653; Nuzividu (KSN), *PVK* 5051; Karempudi (GTR), *CAB* 17443 (MH).

INDIA : Tamil Nadu.

**Commiphora caudata** (Wt. & Arn.) Engl. in DC. Monog. 4. 27. 1883; Gamble 1: 171 (122). *Protium caudatum* Wt. & Arn. Prodr. Fl. Ind. Orient. 176. 1834; FBI 1: 530; Brandis 132; Beddome Fl. Sylv. S. India t. 125. 1871.

Deciduous tree, up to 15 m tall; bark greenish-grey, smooth, papery, flakes-off in large pieces; wood greyish; branchlets glabrous. Leaves imparipinnate, to 25 cm long, petiole to 9 cm; leaflets 3–7, chartaceous, broad ovate to lanceolate, 5–7.5 × 3–4 cm, glabrous, glossy above, subglaucous below, apex caudate, margin entire, base attenuate or subacute; petiolule to 2 mm. Flowers polygamous, cream to reddish, scented in 12 cm long, axillary, paniculate dichasial cymes. Drupe globose or subglobose, 1 × 0.5 cm, green, fleshy; seed solitary, black with 4 radiating wings.

Very common, in the hill slopes of dry deciduous forests, usually near streams 400–900 m. Leaf-fall: Feb. New foliage: Apr. reddish, turning green. Fls.: Apr.–May. Frs.: Throughout the year. Vern.: Konda regu, Konda mamidi, Metta mamidi, Konda Ragi. Eng.: Mango.

Fruits are pickled. The bark and leaves have the odour of mangoes and the oleo-resin obtained from the tree is sometimes used as incense.

Tirumala hills (CTR), *SSR* 13701; Avagondapalem (ATP), *SSR* 18579; Kalasamudram (ATP), *EC* 2957; Palakonda hills (CDP), *CS* 7660; Upper Ahobilam (KNL), *TP & RVR* 2722; Kondavedu (GTR), *CAB* 17498 (MH); Maredumilli (VSKP), *GVS* 68527 (MH); Marripalem forest (PKSM), *RVK* 18768; Bourapur (MBNR), *TDCK* 13652.

INDIA : Peninsula.
WORLD : Sri Lanka.

## GARUGA Roxburgh

**Garuga pinnata** Roxb. Pl. Cor. t. 208. 1819; FBI 1: 528. Gamble 1: 169 (121); Brandis 131; Beddome Fl. Sylv. S. India t. 118. 1871.

Densely foliaceous, deciduous tree, up to 25 m tall; bark thick and reddish brown, peeling off in flakes; branchlets pubescent. Leaves crowded at the ends of branches, imparipinnate, to 35 cm long; petiole to 10 cm; leaflets 13–19 pairs, chartaceous with characteristic fruit like reddish galls, oblong, 3–15 × 1.5–5 cm, slightly puberulous above, brownish pubescent beneath, apex abruptly caudate or acuminate, margin crenate-serrate, hairy, base oblique or sub-acute; secondary nerves 10–15 pairs; petiolule 0 or 2 mm. Flowers pale yellow in 15 cm long much branched, axillary panicles. Drupes globose, 1.5 × 1 cm, yellowish green, pendulous, pyrenes 3, rugose, horned; seed one per pyrene.

Common, in the hills of deciduous forests in full sun at 450–800 m. It is good coppicer. Leaf fall: Feb.–Mar. New foliage: Apr.–May. Fls.: Mar.–Apr. Frs.: June–Aug. Vern.: Garuga, Garugu, Garga. Wood is of fair quality. The fruit is edible.

Upper Ahobilam (KNL), *SSR* 15289; Rollapenta (KNL), *SSR*, 15514; Donubai (SKLM), *SSR* 15581; Jaipur RF (ADB), *GO* 4491; Mudheli RF (NZB), *BR & KH* 7132.

INDIA : Throughout.
WORLD : S.E. continental Asia, Bangaladesh, Myanmar, S. China, Thailand, Indo-China.

## PROTIUM N.L. Burmann *nom. cons.*

**Protium serratum** (Colebr.) Engler in DC. Monogr. Phan. 4: 88. 1883; Gamble 1: 171. *Bursera serrata* Colebr. in Trans. Linn. Soc. 15: 361, t. 4. 1827; FBI 1: 530; Brandis 132.

Large evergreen tree, up to 20 m tall; bark papery, grey, slightly cracked; wood reddish, close grained; young branches terete, striate, puberulous. Leaves compound, imparipinnate, to 17 cm long, leaflets 7, elliptic-oblong, 4.5–11 × 2.5–5.3, glabrous above, pubescent beneath, apex abruptly and obtusely acuminate, margin crenate-serrate; secondary nerves 5–9 pairs, petiolule to 1 cm, rachis brown tomentose. Flowers green, 2 mm long in 8.5 cm long, axillary panicles. Drupes globose, 8 mm across, red.

Northern circars, Deccan in ravines and along streams at 500–1000 m. Fls.: Mar.–Apr. Frs.: May. Vern.: Gandru chettu, Chitrek, Chitraka chettu.

Donubai (SKLM), *SSR* 15581; Vantlamamidi (VSKP), *SSR* 15955; Gundlabrahmeswaram (KNL), *JLE* 25849 (MH); Maredumilli (EG), *GVS* 24224 (MH); Lumbasingi (VSKP), *CAB* 11815 (MH).

INDIA : Assam, Orissa, Chota Nagapur.
WORLD : Myanmar, Bangladesh.

## MELIACEAE

### KEY 1 (Vegetative & Flowering)

1. Flowers 4-merous. Leaves even-pinnate; leaflets obovate ........ ***Xylocarpus***
1. Flowers 5-merous. Leaves odd-pinnate; leaflets other than obovate (in *Aphanamixis* terminal one obovate):
  2. Petals 1.5 cm long ........ ***Chukrasia***
  2. Petals < 1 cm long:
    3. Peduncle to 15 cm long; pedicels articulate ........ ***Trichilia***
    3. Peduncle to 10 cm long; pedicels not articulate:
      4. Leaves 2–3-pinnate ........ ***Melia***
      4. Leaves 1-pinnate or 3-foliolate:
        5. Petals 3–4:
          6. Leaflets 4–8 pairs, glabrous. Inflorescence as long as or nearly as long as leaves; petals 3 ........ ***Aphanamixis***
          6. Leaflets 2–3 pairs, lepidote. Inflorescence much shorter than leaves; petals 4 ........ ***Amoora***
        5. Petals 5:
          7. Staminal filaments free ........ ***Toona***
          7. Staminal filaments united:
            8. Leaflets with obtuse-retuse apex ........ ***Soymida***
            8. Leaflets with acuminate, acute or rounded tip:
              9. Anthers apiculate ........ ***Azadirachta***
              9. Anthers not apiculate:
                10. Ovary 2 or 3-celled:
                  11. Leaves 3-foliolate. Flowers 8 mm across, bisexual ........ ***Walsura***
                  11. Leaves 5–7-foliolate. Flowers 5 mm across; polygamodioecious.—Staminal tube apically entire ..... ***Aglaia***
                10. Ovary 5-celled:
                  12. Leaflets toothed. Staminal filaments connate below and not formed into a tube. Trees small. ........ ***Cipadessa***
                  12. Leaflets entire, lower part crenate. Staminal filaments united throughout into a tube. Trees medium-large sized. ........ ***Swietenia***

### KEY 2 (Fruiting)

1. Fruit capsule:
  2. Capsule 2-valved ........ ***Trichilia***
  2. Capsule 3–5-valved:
    3. Capsule 1.5–3.5 cm long:
      4. Seeds numerous, winged at both ends, without aril ........ ***Toona***
      4. Seeds 1–3, not winged, with a scarlet aril:
        5. Fruit yellowish, 3–5 cm in dia.; seeds oblong-subglobose ........ ***Aphanamixis***
        5. Fruit grey to green, 1.2 cm in dia.; seeds ellipsoid ........ ***Amoora***
    3. Capsule 4–16 cm long:

6. Capsule very thick, woody.—Capsule 9–16 cm long, seeds to 6 cm long ... ........ ***Swietenia***
6. Capsule not very thick and woody:
    7. Capsule irregularly sub-globose, 7–10 cm in dia., seed angular. Mangrove trees ........ ***Xylocarpus***
    7. Capsule globose-obovoid, 4–6 cm in dia., seeds not angular. Terrestrial trees:
        8. Seed 2 × 0.5 cm, thin, broadly winged at base ........ ***Chukrasia***
        8. Seed 4 × 1.5 cm, compressed, winged at both ends ........ ***Soymida***

1. Fruit not capsule:
    9. Fruit drupe:
        10. Drupe 5-furrowed, seeds 5 ........ ***Cipadessa***
        10. Drupe not furrowed, seed one:
            11. Drupe oblong-ovoid ........ ***Azadirachta***
            11. Drupe globose ........ ***Melia***
    9. Fruit berry:
        12. Berry 2 cm long, seeds ovoid ........ ***Walsura***
        12. Berry 1 cm long, seeds ellipsoid ........ ***Aglaia***

## AGLAIA Loureiro *nom. cons.*

**Aglaia elaeagnoidea** (Juss.) Benth. var ***beddomei*** (Gamble) K.K.N. Nair in J. Bombay Nat. Hist. Soc. 78: 426. 1981; Beddome Fl. Sylv. S. India t. 130. Anal. Gen. 55. 1871. *A. roxburghiana* (Wt. & Arn.) Miq. var. *beddomei* Gamble, Fl. Pres. Madras 1: 180 (129). *A. roxburghiana* sensu Bedd. Fl. Sylv. S. India t. 130. A. 1871. non (Wt. & Arn.) Miq. 1868.

Evergreen tree, up to 12 m tall; bark light brown, smooth; wood red, hard; branchlets and inflorescence densely covered with minute brown peltate scales. Leaves odd-pinnate, rachis to 10 cm, petiole to 5 cm; leaflets 5–7, coriaceous, elliptic-oblong or lanceolate, 6–12.5 × 1.8–4 cm, glabrescent, drying grey, apex acute, obtuse or shortly acuminate, margin entire, base oblique, acute or cuneate; secondary nerves 13–15 pairs, not prominent, petiolule 5 mm. Flowers yellowish, polygamo-dioecious, globular in supra axillary, 15 cm long panicles. Berry globose, 2 cm dia., orange yellow, velvety tomentose; seed one, ellipsoid, brown with white aril.

Rare, in moist deciduous forests in valleys 500–1200 m. Fls.: Nov.–Dec. Frs.: May. Vern.: Yerra-Adugu.

Tirumala (CTR), *SSR* 13734, 15261; Upper Ahobilam (KNL), *TP & SSR* 15204; Minumuluru (VSKP), *SSR* 15953; Rampa (EG), *VNS* 482 & *MSR* 1596 (CAL).

INDIA : Bihar.

## AMOORA Roxburgh

**Amoora canarana** (Turez.) Hiern in Hook. f. Fl. Brit. India 1: 560. 1875; Gamble 1: 182 (130). *Oraoma canarana* Turez. Bull. Soc. Imp. Naturalistes Moscou 31: 411. 1858.

Evergreen tree, up to 15 m tall; latex milky, branchlets with stellate scales, rusty. Leaves odd-pinnate, 15–32 cm long, petiole to 10 cm; leaflets 2–3 pairs, opposite or alternate, thin-coriaceous, ovate-oblong, 7.5–13 × 3.5–4.5 cm, glabrous above, lepidote beneath, apex acute or acuminate, margin entire or undulate, base cuneate or narrowed; secondary nerves 9–11 pairs; petiole 0.8–1.2 cm. Flowers greenish-yellow, polygamo-dioecious, 5 mm across, in divaricately branched axillary panicles; panicles much shorter than the leaves; disc obscure. Capsule grey to greenish, subglobose, 1.5 × 1 cm, 3-valved; seeds 1 or 2, elliposid.

Rare, in Visakhapatnam district on high hills at 800–1200 m. Fls.: Mar.–May. Frs.: May–Aug.

Narsipatnam-Lumbasingi Ghat Road (VSKP), *HSP* 20709 (BLAT) & SKW 2673 (BLAT).

INDIA : Peninsula from Canara Southwards, Tamil Nadu, Maharashtra.

## APHANAMIXIS Blume

**Aphanamixis polystachya** (Wall.) Parker in Indian For. 57: 486. 1931. *Aglaia polystachya* Wall. in Roxb. Fl. Ind. 2. 429. 1824. *Amoora rohituka* (Roxb.) Wt. & Arn. Prodr. Fl. Ind. Orient. 119. 1834; FBI 1: 559; Gamble 1: 181 (130); Brandis 141; Beddome Fl. Sylv. S. India t. 132. 1871. *Andersonia rohituka* Roxb. Fl. Ind. 2 : 213. 1832.

Evergreen tree, 10–20 m tall; crown spreading; trunk short; bark smooth; wood red; branchlets stout, young shoots silky pubescent. Leaves large, imparipinnate, 15–40 cm long, petiole to 4 cm; leaflets 9–17, elliptic-oblong, terminal one obovate, 9–18 × 4–7 cm, glabrous, apex obtuse or obtusely acuminate, margin entire, base oblique, obtuse or acute; secondary nerves 8–12 pairs; petiolule to 2 mm. Flowers white, small, sessile; male flowers numerous, erect, 4 mm in axillary, spreading panicles; female flowers larger than male in axillary or supra axillary, 45 cm long, solitary spikes. Capsule yellow, 3-valved, globose, 3.5 cm in dia., pericarp coriaceous, glabrescent with accrescent calyx; seeds 1–3, sub-globose, oblong, aril scarlet, polished, apiculate, 1.6 cm long.

In the semi-evergreen forests of Visakhapatnam district above 900 m. It is typical of moist localities occurring in ravines and banks of streams. Fls.: Aug.–Sept. Frs.: May–June.

Peda valasa (VSKP), *SSR* 19662; Visakhapatnam (VSKP), Coll. 9166 (MH).

INDIA : Assam, Orissa, Avadh, Andamans
WORLD : Sri Lanka, Myanmar, Bangaladesh, Malay Peninsula, Indonesia, Phillippines.

## AZADIRACHTA A. Jussieu

**Azadirachta indica** A. Juss. in Mus. Nat. 19. 221. 2. f. 5. 1830; Gamble 1: 177 (127); Brandis 139. *Melia azadirachta* L. Sp. Pl. 385. 1753; FBI 1: 544; Beddome Fl. Sylv. S. India t. 14. 1869.

Semi-evergreen tree, up to 15 m tall; bark brown with vertical furrows; wood red; branchlets glabrous and lenticellate. Leaves bipinnate, paripinnate to 2.8 cm long, petiole to 7 cm; leaflets 6–10 pairs, coriaceous, sub-opposite, obliquely-lanceolate, falcate, 6–7.5 × 1.5–2.6 cm, shining, apex acuminate, margin serrate, base oblique, petiolule to 2 mm. Flowers white, fragrant in 10–20 cm long axillary panicles. Drupe yellow, oblong-ovoid, 1.6 × 6 cm, glabrous, endocarp woody; seed solitary.

Common, generally grows wild in plains, road sides and rare in forests up to 500 m. It grows on a variety of soils from sandy to clayey to black cotton soils. It grows best where drainage is good and where sub-soil water level is fairly high and good coppicer. Never leafless. Fls.: Mar.–Apr. Frs.: Ripen from May–July. Vern.: Yapa, Vepa, Vemu.

It is a very important and multipurpose tree. The bark is bitter and nearly every part is used medicinally in intermittent fever and as a tonic. The ripe fruits are largely collected for the oil, which is used in parasitic skin diseases and also applied to hair for destroying lice. Internally it is anthelmintic. The twigs are used as tooth brushes. Wood yields good timber. Leaves with turmeric powder is ground and applied all over the body to cure chicken pox. A sacred tree of the Hindus.

SKU Campus (ATP), *SSR* 15979; North Dhone RF (KNL), *TP & RVR* 1309; Sirpur RF (ADB), *GO & PVP* 4809; Macherla RF (GTR), *VRK* 3952; Narsingi (MDK), *BR & CP* 11527; Mudheli forest guest house (NZB), *TP & BR* 6361.

INDIA : Cultivated and naturalized throughout.
WORLD : Native of India and China, and it is cultivated and naturalized also in Pakistan and Malaysia.

## CHUKRASIA **A.H.L. Jussieu**

**Chukrasia tabularis** Adr. Juss. in Mem. Mus. Natl. Hist. Nat.19: 251. 21. 1850; FBI 1: 568 "*Chikrasia*", Gamble 1: 186 (131); Brandis 144; Beddome Fl. Sylv. S. India t. 9. 1869.

A very large, densely foliaceous, deciduous tree, up to 30 m tall; bark reddish brown, narrowly longitudinally fissured, with short horizontal cracks; wood yellowish-brown to reddish-brown resembling Mahogany; branchlets lenticellate. Leaves even-pinnate to 27 cm, petiole to 5 cm; leaflets 10–16, coriaceous, ovate or ovate-lanceolate, 4.5–12 × 2.5–5.5 cm; glabrous above,

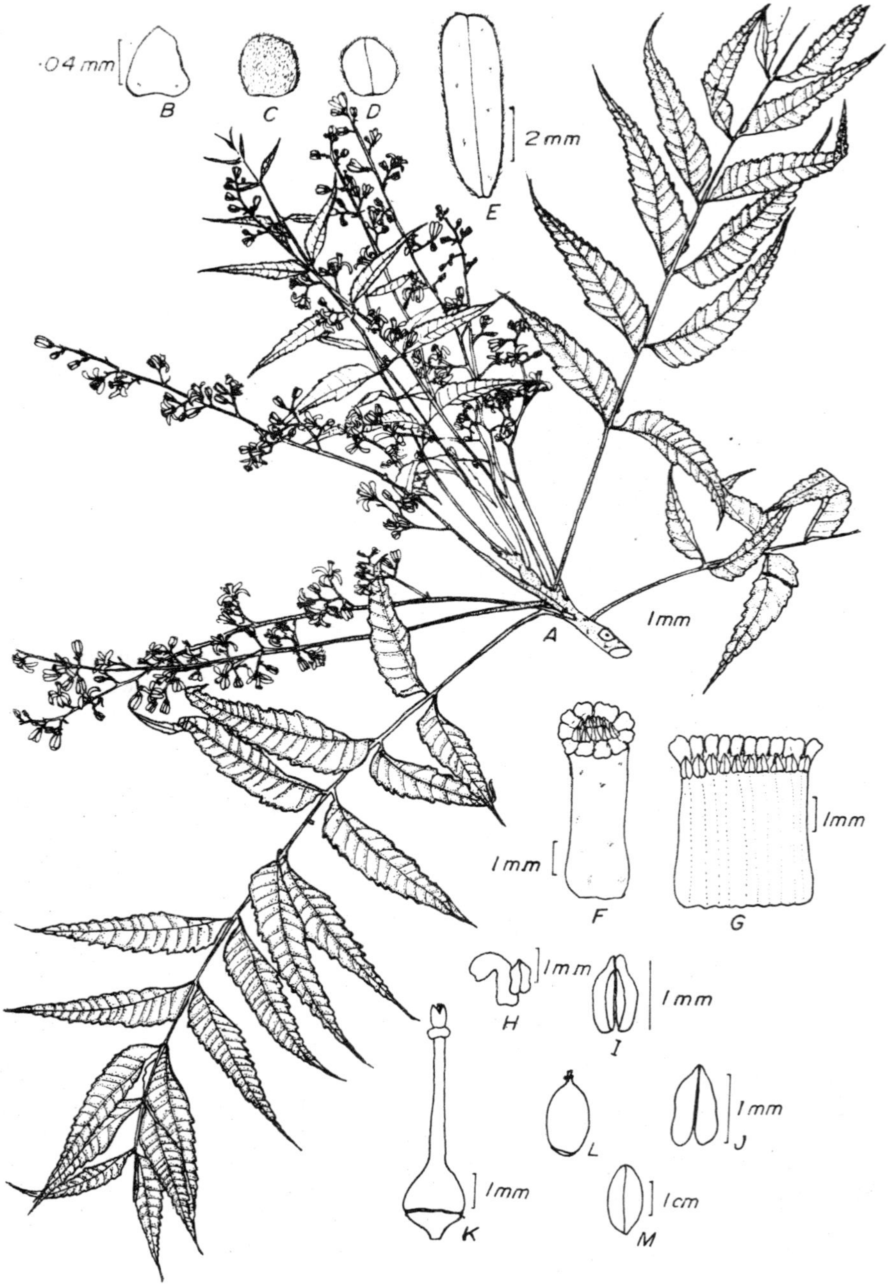


**Fig. 18.** *Azadirachta indicia* A. Juss.
A. Twig, B–D. Sepals, E. Petal, F & G. Staminal column whole and spread out, H. Staminal column lobe with anther, I & J. Anthers, K. Pistil, L. Fruit, M. Seed.

**Fig. 19.** *Chukrasia tabularis* A. Juss.
A. Twig, B. Bract, C. Bracteole, D & E. Calyx, F. Petal, G & H. Staminal column whole and spread out, I. Stamen, J. Pistil.

rusty beneath, apex acuminate, margin entire, base oblique; petiolule to 1 cm. Flowers yellowish-red in 25 cm long terminal panicles. Capsule globose-obovoid, 4 cm long, deep dark brown, 3–4- valved, woody, septifragal, smooth; seeds numerous, winged at base, flat, closely packed, 1.8–2 × 0.5 cm, brown.

Rare, in mixed deciduous forests at 600–1000 m.

Leaf fall: Feb.–Mar. Fls.: Apr.–June. Frs.: Aug–Sept., dehiscing by next Feb. Vern.: Konda Vepa.

It is one of the tallest among forest trees in the top canopy. Wood is a valuable timber, takes a fine polish and is suitable for furniture, and is useful for railway carriages, panelling, building purposes and for carving. Bark is a powerful astringent.

Talakona (CTR), *SSR* 13725; Tirumala (CTR), *DR* 1532 (SVU), Cuddapah (CDP), *CBDM* 9400 (MH).

INDIA : Orissa, (W. Peninsula).
WORLD : Sri Lanka, Pakistan, Bangaladesh, Malaysia, Myanmar, China.

## CIPADESSA Blume

**Cipadessa baccifera** (Roth) Miq. in Ann. Mus. Bot. Lugduno. Batavum 4: 6. 1868; Gamble 1: 176 (126). *Melia baccifera* Roth, Nov. Pl. Sp. 215. 1821. *Cipadessa fruticosa* Bl., Bijdr. 162. 1825; FBI 1: 545; Brandis 137. *Melia rothii* A. Juss. in Mem. Mus. Natl. Nat. 19: 222 t. B. f. 6. 1830.

Deciduous small tree, up to 7 m tall; wood red; branchlets tomentose. Leaves imparipinnate, 15.5–20 cm long, petiole to 3 cm; leaflets ca. 4 pairs, thin-coriaceous, oblong-lanceolate, 3.2–8.2 × 2.2–3.5 cm, glabrous above, pilose below, apex acuminate, margin proximally entire, distally serrate, base cuneate; petiolule to 4 mm. Flowers white in axillary panicles. Fruit globose, 6–8 mm in dia, scarlet, 5-furrowed; seeds 5, angular.

Common, in scrub and open mixed forests above 600 m. Fls.: Mar.–June; Frs.: Throughout the year. Vern.: Chedubira, Chenda Bara, Rana Beri, Turaka Vepa, Hanumantha Beera.

The wood is used for fuel. Green twigs burn with a crackling noise.

Tirumala (CTR), *SSR* 13777; Horsely hills (CTR), *SSR* 13793; Pecheruvu RF (KNL), *RVR* 2153; Araku valley (VSKP), *GVS* 21552 (MH); Udayagiri (NLR), *MSR* 1485 (CAL); Polavaram Agency (WG), *DCSR* 224 (CAL).

INDIA : Peninsula.
WORLD : Sri Lanka, Nepal, Myanmar, Thailand to Vietnam, China, Malaysia.

## MELIA Linnaeus

1. Trees small or middle sized. Leaflet margin serrate. Flowers bluish-purple or lilac; anthers included. Drupe 1.5 cm long, smooth when dry; seed one. .............................. ***M. azedarach***

1. Trees large. Leaflet margin crenate-serrate. Flowers greenish white; anthers exerted. Drupe 2.5–3 cm long, compressed with longitudinal ridges when dry; seeds 3 or 4 ........ ***M. dubia***

**Melia azedarach** L. Sp. Pl. 384. 2753; FBI 1. 544; Gamble 1: 176 (126); Brandis 140; Beddome Fl. Sylv. S. India t. 13. 1869.

Deciduous tree, up to 10 m tall; crown spreading; bark thick and deeply fluted, smooth in young; wood light red, straight grained and coarse textured. Young branches sparsely clothed with deciduous stellate hairs. Leaves tri-pinnate, rachis 30–40 cm long, petiole to 15.2 cm, pinnae 3–5 paired; leaflets 5–9 in each pinnae, chartaceous, ovate or lanceolate, 4.5–7 × 1.5–3 cm, glabrescent, apex acuminate, margin serrate, base oblique. Flowers bluish purple or lilac, fragrant in 20 cm long, axillary panicles. Drupe yellow, elliposid-globose, 1.5 cm in dia.; seed 1 yellowish brown, hard with a natural perforation through the centre, pendulous.

It is generally grown as an avenue tree and occasionally found running wild to 800 m. Recommended for plantations on wide range of soils, specifically loamy laterite soils. Leaf fall: Dec.–Jan. New foliage: Apr.–May. Fls. Apr.–June (occasionally, Fls: Jan.), Frs.: Ripen in Nov.–Dec. Vern.: Turaka Vepa, Kali Vepa.

Wood useful for furniture making but little used. It is used as fuel wood. The fruits are poisonous and used in leprosy. Foliage lopped for fodder. Nuts hung on the doors are believed to avert epidemics. The plant used in removing vile influences in cattle. The leaves, bark and fruits are having insect-repellant properties. It yields gum similar to that of neem and wood apple.

Reddipalli farm (ATP), *NY* 271; Veldurthi (KNL), *RVR* 1737; V.P. South (GTR), *VRK* 3940; Kodimial (KNR), *GVS* 20093(MH); Talantrakota (SKLM), *GVS* 62327 (MH); Nattavaram (VSKP), *CAB* 1869 (MH); Hindi Konda (CDP), *JSG* 2100 (MH); Mombojipally (MDK), *BR & CP* 11630; Mamidipalli (NZB), *BR & KH* 9007; Bellampalli (ADB), *PVP* 9444; Tirumala (CTR), *DR* 2447 (SVU).

INDIA : Commonly cultivated.
WORLD : Native in Western Himalaya and naturalized in China, Myanmar, Pakistan, Iran, Turkey.

**Melia dubia** Cav. Diss. 7: 364. 1789; FBI 1: 545. *M. composita* Willd. Sp. Pl. 2: 559. 1799; Gamble 1: 176 (126); Brandis 140. Beddome Fl. Sylv. S. India t. 12. 1869. *M. superba* Roxb. Fl. Ind. 2: 396. *M. robusta* Roxb. Fl. Ind. 2: 397. 1832.

Densely foliaceous, majestic deciduous tree, up to 25 m tall; spread of the crown is more than its height, bark dark brown, smooth, exfoliating in thin, narrow strips; wood reddish-white, young branches pinkish-tomentose. Leaves tri-pinnate, pinnae 3 to 4 paired, each pinnae ca. 40 cm long, alternate, petiole to 16 cm; leaflets 5–7 (each pinnae), thin-coriaceous, ovate-lanceolate, 4–7 × 2–3 cm, thinly puberulous when young, then glabrous, apex acuminate, margin crenate-serrate, base obtuse, slightly oblique, petiolule to 3 mm. Flowers greenish-white, fragrant in ca. 15 cm long axillary panicles. Drupe ovoid-elliposid, to 2.5–3 × 0.8 cm, yellowish, pendulous, smooth; endocarp woody; seeds 3–4 smooth, pointed, pendulous.

Hills above 750 m on the slopes of moist deciduous forests, especially in Nallamalais. Leaf fall: Jan. New foliage: Feb. Fls.: Feb.–Mar. Frs.: Nov.–Jan. Vern.: Munnuthi-Karaka. It is one of the fast growing tree.

Wood soft, useful for tea-boxes, packing cases, cigar boxes, ceiling planks, pencils, match boxes, splints and agricultural implements. It is suitable for musical instruments, tea boxes and ply board. It serves as a good fuel wood. Fruits are bitter and anthelmintic.

Pangidi (KNL), *TP & SSR* 13216; Muthyalu forest (SKLM), *SSR* 15908; Rollapenta (KNL), *JLE* 42282 (MH); Horsely hills (CTR), *DR* 1080 (SVU).

INDIA : Peninsula.
WORLD : Sri Lanka, Malay Peninsula, Australia, Tropical Asia, Angola.

## **SOYMIDA** A.H.L. Jussieu

**Soymida febrifuga** (Roxb.) Adr. Juss. in. Mem. Mus. Natl. Hist. Nat. 19: 251. 1830; FBI 1: 557; Gamble 1: 185 (133); Brandis 144; Beddome Fl. Sylv. S. India t. 8. 1869. *Swietenia febrifuga* Roxb. Pl. Cor. t. 17. 1795.

Deciduous tree, up to 20 m tall; trunk straight, symmetrical; bark dark brown, rough with oblong flakes; wood dark reddish brown, bright red when fresh cut; branchlets wrinkled with persistent leaf scars. Leaves paripinnate, rachis to 30 cm long; petiole red to 7 cm; leaflets 6 pairs, opposite, thick-coriaceous, ovate-oblong, 8.5–13 × 6–7.5 cm, glabrous, apex obtuse or retuse, margin entire, base oblique or truncate; secondary nerves 8–12 pairs, conspicuous beneath; petiolule to 2 cm. Flowers small, greenish-white in 10–20 cm long axillary or terminal panicles. Capsule globose-obovoid, woody, to 6 × 3.5 cm, brown, septifragally 5-valved; valves separating from the dissepiments, which remain attached to thick spongy axis; seeds numerous, oblong, 4 × 1.5 cm, compressed, winged at both ends with a soft felty covering.

Common, in dry deciduous forest 400–1000 m. Leaf fall:Feb.–Mar. Recommended for plantations on lime soils, black cotton soils, dry stony hills of laterite and kankar. New foliage: Apr.–May. Fls.: Feb.–Apr. Frs.: May–June dehiscing by July. Vern.: Somi, Somigi.

The wood is extremely hard, very durable and is used in furniture making; bark is bitter. It coppices well.

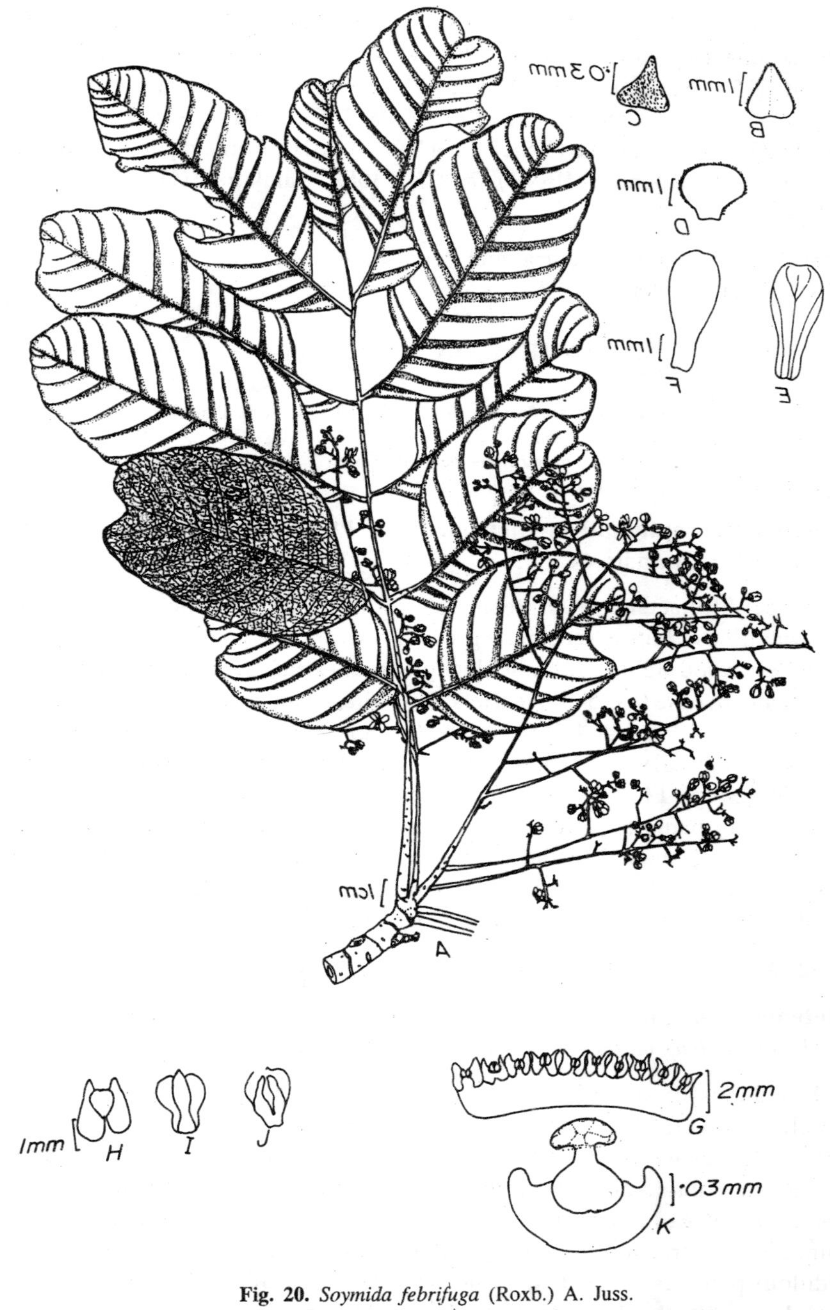


**Fig. 20.** *Soymida febrifuga* (Roxb.) A. Juss.
A. Twig, B. Bract, C. Bracteole, D. Sepal, E & F. Petals, G. Staminal Column spread out, H, I. J. Anthers, K. Pistil.

Bairluti (KNL), *SSR* 15516; Mamandur (CTR), *SSR* 1679; Vengalammacheruvu RF (ATP), *TP & NY* 2116; Ichampalli RF (ADB), *GO & DAM* 5155; Narsapur RF (MDK), *CP* 11574; Malleswaram (Godavary), *JSG* 16056 (MH); Cheedipalem (EG), *GVS* 24523 (MH); Krishna forests, *CBDM* 9325 (MH); Karaka (VSKP), *CAB* 9318 (MH); Ramayapatnam (NLR), *JSG* 12689; Yerlapenta (MBNR), *TDCK* 13689; Talakona (CTR), *DR* 2440 (SVU).

INDIA : North-West, Central and Peninsulae.
WORLD : Sri Lanka.

## SWIETENIA N.J. Jacquin

1. Leaflets ovate-lanceolate, glabrous beneath, up to 19 × 4.5 cm. Panicles to 13 cm long; petals not hooded. Capsule globose, .................................................... ***S. macrophylla***
1. Leaflets oblong-lanceolate, puberulous beneath, up to 7 × 2 cm. Panicles to 8 cm long; petals hooded. Capsule oblong or lanceolate, ................ ........................... ***S. mahagoni***

**Swietenia macrophylla** King in Hook. Icon. Pl. t. 1550. 1886; Madras trees 128. 1911; Brandis 44.

Densely foliaceous, evergreen tree, up to 20 m tall; trunk straight and cylindrical, base buttressed; bark greyish-brown, fairly smooth, flaking off in patches; wood light red; branchlets glabrous. Leaf rachis to 20.5 cm long, paripinnate, petiole to 8.5 cm; leaflets 4–5 pairs, coriaceous, ovate-lanceolate, sub-falcate, 8–19 × 2.2–4.5 cm; glabrous, apex acuminate, margin slightly wavy and irregularly wavy, base oblique, acute; secondary nerves 8–12 pairs; petiolule 0.5 cm. Flowers greenish, 1 cm across, fragrant, in 9–13 cm long supra-axillary panicles. Capsule brown, globose, 16 cm long; seeds winged.

Introduced, planted in the gardens and road sides. Leaf-fall: Jan. New foliage: Mar. Fls.: April. Frs.: Ripened following year. Eng.: Colombian, Mexican, Braziliiam, Peruvian Mahogany tree.

WORLD : Native of Honduras, now widely cultivated in the tropics.

**Swietenia mahagoni** (L.) Jacq. Enum. Pl. Carib. 20. 1760; Gamble 1: 187 (134). *Cedrus mahagoni* L. Syst. nat. (ed. 10) 2: 940. 1759.

Deciduous or semi-deciduous tree, up to 25 m tall; bark grey-black, tubercled; wood reddish-brown, medium-fine textured, straight to interlocked grained. Leaves to 15 cm, petiole to 3 cm, Leaflets 4 or 5 pairs, coriaceous, oblong-lanceolate, slightly falcate, 4.5–7 × 1–2 cm, glabrous above, puberulous beneath, apex acuminate, margin entire, base oblique; petiolule 5 mm. Flowers cream or greenish yellow, 8 mm across in 8 cm long, axillary pendulous panicles. Capsule dark brown, oblong or lanceolate, to 13 × 7 cm, apex obtuse, 5-valved; seeds winged, 6 × 1.5 cm, numerous,

Occasionally planted. Fls.: Apr. Frs.: Dehiscing by Jan. Vern.: Mahaagoni chettu.

Timber very durable, suitable for radio, television cabinets and other high-grade work. It yields gum from the trunk.

Horsely hills (CTR), *DR* 1123 (SVU).

INDIA : Cultivated throughout.
WORLD : Native of West Indies and Honduras.

## TOONA (Endlicher) Roemer

**Toona ciliata** Roemer Syn. Hesper. 139. 1845. *Cedrela toona* Roxb. ex Rottler Ges. Naturf. Freunde Berlin Neue Schriften 4: 198. 1803; FBI 1: 568; Gamble 1: 186 (133); Brandis 145; Beddome Fl. Sylv. S. India t. 10. 1869.

Densely foliaceous, deciduous tree, up to 15 m tall; crown spreading; branches forming shady head; bark reddish-brown; wood even-grained, red; branchlets tomentose. Leaves pari-pinnate, rachis 30–50 cm long, petiole to 5 cm long; leaflets 4–15 pairs, thin-coriaceous, sub-opposite, ovate-lanceolate or lanceolate, 7–12.5 × 3–5 cm, glabrous, apex acuminate, margin entire to slightly undulate, base obliquely acute; secondary nerves 8–9 pairs; petiolule to 1.5 cm. Flowers white, honey scented in 25 cm long, drooping, terminal panicles. Capsule oblong, 2.5 × 1.2 cm, dark grey with small, white tubercles, 5-valved, woody, having a thick central 5-angled column; seeds numerous (25–30), wings brown, membranous, on both ends, to 1.6 cm, seed part 5 mm dia.

Rare, in Viskhapatnam district in valleys and moist forest localities above 1000 m. Leaf fall: Dec. New foliage: Feb., red, rapidly turning to green. Fls.: Mar.–Apr. Frs.: Seeds ripen May–July, the empty capsules remaining on the tree for months. Vern.: Gali Manu.

Wood useful for building and furniture purposes, cigar boxes, shuttles, door panels and carvings. This tree near the houses is considered as a bad omen.

Chintapalli (VSKP), *SSR & KSM* 16139; Anantagiri (VSKP), *SSR* 19623; Matsygundam (VSKP), *GVS* 47360 and 47361 (CAL).

## TRICHILIA P. Browne *nom. cons.*

**Trichilia connaroides** (Wt. & Arn.) Bentvelzen in Acta Bot. Neerl. 11: 13. 1962. *Zanthoxylum connaroides* Wt. & Arn. Prodr. Fl. Ind. Orient. 148. 1834. "*Zanthoxylon*" *Heynea trijuga* Roxb. ex Sims. Bot. Mag. 41. t. 1738. 1815; Brandis 134 non *Trichilia trijuga* Vell. 1825; FBI 1: 565; Gamble 1: 183 (131). *Heynea affinis* A. Juss. in Mem. Mus. Natl. Hist. Nat. 19: 235; 275. 1830 non *Trichilia affinis* A. Juss. 1829.

Evergreen tree, up to 10 m tall; bark brown, cracked; wood grey; branchlets reddish tomentose, lenticellate. Leaves imparipinnate, rachis to 20 cm long, petiole to 8 cm; leaflets 7–9, thin-coriaceous, oblong-lanceolate, 4.5–10 × 1.5–4 cm; pale, glabrous above and softly pubescent or glaucous beneath, apex obtusely acuminate, margin entire, base rounded; petiolule to 1.2 cm. Flowers white, small in ca. 15 cm long corymbosely branched panicles. Capsule subglobose, 0.8–1.4 cm long, red; valves 2, thin, coriaceous; seed 1 or rarely 2, pendulous with thin, white, fleshy aril.

Common, in moist valleys and rare in dry deciduous forests, from Godavary to Visakhapatnam district at 800–1200 m. Fls.: Mar.–May. Frs.: Aug.–Nov. Vern.: Gore Gudda, Garugu.

The wood does not seem to be used.

Minumulur (VSKP), *SSR* 15941; Gudem (VSKP), *SSR* 19654; Lumbasingi (VSKP), *CAB* 11817 (MH); Araku valley (VSKP), *GVS* 19664 (MH).

WORLD : Malaysia.

## **WALSURA** Roxburgh

**Walsura trifoliata** (A. Juss.) Harms in Engler & Prantl., Pflanzenf. ed. 2. 19B; 119, 177. 1940 ('trifolia'). *Heynea trifoliata* A. Juss. in Mem. Mus. Natl. Hist. Nat. 19: 235. 1830. *Walsura piscida* Roxb. Fl. Ind. 2: 387. 1832; FBI 1: 564; Gamble 1: 183 (131); Brandis 136, Beddome Fl. Sylv. S. India Anal. Gen. 56. t. 1887.

Densely foliaceous, evergreen tree, up to 12 m tall; bark grey, slightly cracked; wood pink then yellowish; branchlets tomentose. Leaves thick-coriaceous, 3-foliolate, rachis to 7.5 cm, petiole 3–8 cm, leaflets elliptic or oblong, or ovate-lanceolate, 5–9 × 2.6–4.2 cm, glabrous, shining above, pale glaucous beneath, apex rounded, margin entire, base cuneate; secondary nerves 6–8 pairs. not conspicuous; petiolule to 9 mm. Flowers cream-coloured, small, in 9 cm long, axillary or terminal panicles. Fruit obovoid, 1.6 × 1 cm, bright orange-yellow, velvety tomentose, slightly apiculate with persistent calyx; seeds often 1 rarely 2, ovoid, 1.3 × 0.6 cm.

Occasional, on rocky hill slopes of moist deciduous forests at 600–700 m. Fls.: Mar.–Apr. Frs.: June–Aug. Vern.: Walsuri, Yerra-valsura, Erra-valudu.

Oil extracted from wood upon heating is used for itches.

Tirumala-papavinasanam (CTR), *SSR* 13723; Mamandur-Tirumala (CTR), *SSR* 16181; Sidduliah Kona (NLR), *SSR* 19378; Balapalle (CDP), *JLE* 15708 (MH); Madalamma Konda (VSKP), *KCJB* 70671 & 9091 (MH).

INDIA : Peninsula.
WORLD : Sri Lanka.

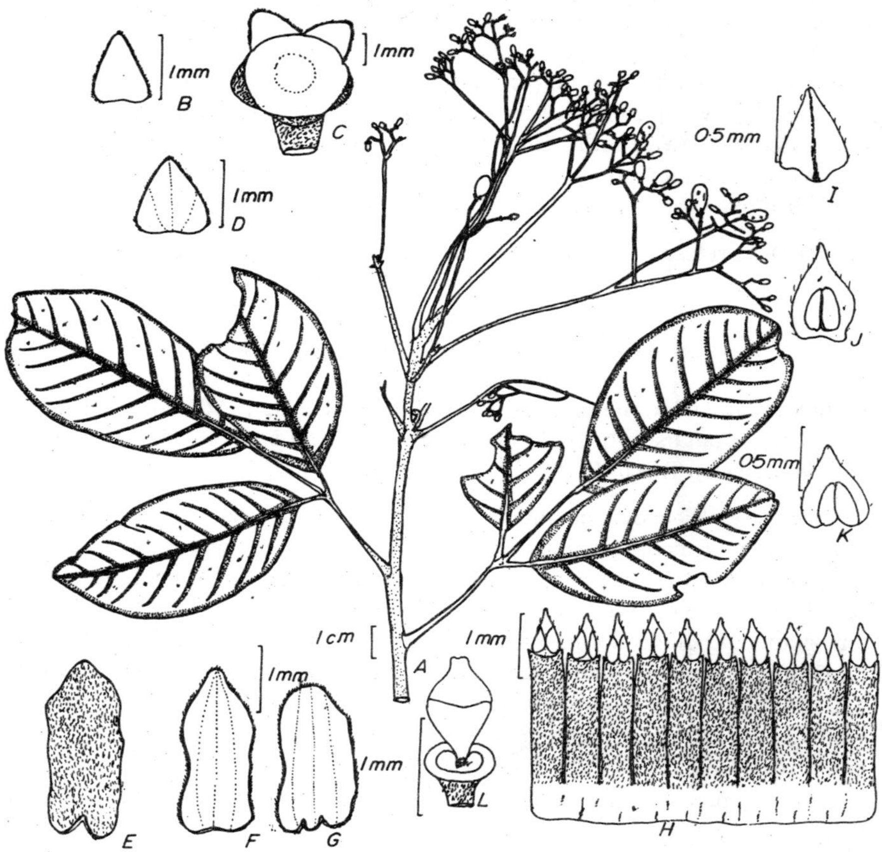


**Fig. 21.** *Walsura trifolia* (A. Juss.) Harms
A. Twig, B. Bract, C. Calyx, D. Sepal, E–G. Petals, H. Staminal column spread out, I,J, & K. Anthers, L. Pistil.

## XYLOCARPUS Koenig

**Xylocarpus granatum** Koenig in Natur. forscher 20: 2. 1784. *X. obovatus* (Bl.) Adr. Juss. Mem. Mus. Hist. Nat. 19: 244. t. 9. f. 22. 1830; Gamble 1: 185 (132). *Carapa obovata* Bl., Bijdr. 179. 1825; Brandis 140. *C. moluccensis auct. non* Lamk. 1784, FBI. 1: 567; Beddome Fl. Sylv. S. India t. 136. 1871.

Tree, up to 8 m tall; bark grey, smooth; wood reddish-brown; branchlets sparsely lepidote. Leaves paripinnate, rachis 7.5–15 cm long, petiole to 6 cm; leaflets 1–2 pairs, thick-coriaceous, obovate, 5–8 × 2.8–3.6 cm, puberulous, apex obtuse or round, margin entire, base cuneate, subequilateral; petiolule to 2 mm. Flowers white, few, in axillary, cymose panicles. Capsule irregularly sub-globose, 7–10 cm in dia., 4-valved, woody; seeds 6–12, large, angular, yellowish, with a thick orange coloured aril.

Rare, in swamp forests of Krishna delta. In Nizampatnam mangroves it is associated with *Avicennia* populations. Fls.: Aug.–Oct. Frs.: Apr.–May.

Wood hard and valuable.

Nizampatnam (GTR), *VRK* 6913.

WORLD : Sri Lanka, tropical Africa, Malaysia, Archipelago, N. Australia.

## FLINDERSIACEAE

### **CHLOROXYLON** A. de Candolle

**Chloroxylon swietenia** DC. Prodr. 1: 625. 1824; FBI 1: 569. Gamble 1: 152 (101); Brandis 146. Beddome Fl. Sylv. S. India t. 11. 1869. *Swietenia chloroxylon* Roxb. Pl. Cor. t. 64. 1796 & Fl. Ind. 2: 400. 1832.

Densely foliaceous, deciduous tree, up to 16 m tall; bark thick, yellowish, corky, spongy, deeply and characteristically fissured; wood yellow, close grained; branchlets pubescent. Leaves paripinnate, rachis to 22 cm long, petiole to 3.7 cm; leaflets 10–16 pairs, thin-coriaceous, elliptic or lanceolate, 2.3 × 7 cm, glaucous, gland dotted on lower surfaces, apex obtuse, margin entire, base oblique, unequal sided; petiolule to 2 mm. Flowers white in terminal and axillary 10 cm long panicles. Capsule oblong, to 2.8 × 1.2 cm, 3-loculicidal, walls grey with white glandular spots, glabrous; seeds ca. 15, oblong, 1.8 × 5 cm, brown, apically broad winged, compressed.

Common, in open and hills of dry deciduous forests at 300–1000 m. Recommended for plantations in very poor soils and areas subjected to high temperature, good drainage and soil containing large proportion of sand and gravel. It is a god coppicer, and fire resistant. Leaf fall: Feb.–Mar. New foliage: May. Fls: Mar.–Apr. Frs.: May–Aug. Vern.: Billudu, Billu.

Wood hard, valuable and has great demand for agricultural implements. It also serves as a good fuel, as a consequence this is over exploited.

Kalasamudram (ATP), *SSR* 13749; Kusimi forest (SKLM), *SSR* 15902; Peddda chamala (PKSM), *SSR* 16115; Manchippa (NZB), *SSR* 19336; Cement Nagar RF (KNL), *TP & RVR* 1823; Macherla RF (GTR), *VRK* 3829; Velagalapalli (KSN), *AWL* 1220 (DD); Narsapur RF (MDK), *CP* 1580; Guvvalacheruvu (CDP), *RV Reddy* 8185; Panapakam (CTR), *GVS* 31871; Pakhal (WGL), *ANH* 15956 (MH); Appapur (MBNR), *TDCK* 1539.

INDIA : Peninsula India.

WORLD : Sri Lanka, S.E. Asia, Madagascar, Endemic to S.W. India, Sri Lanka

**Note** : According to Gandhi (1976) this genus with its capsular fruits and alate seeds, was included in Meliaceae by Bentham & Hooker although it has glandular-punctate leaves. Some place this genus in the Flindersiaceae.

## CELASTRACEAE

1. Fruit a dehiscent capsule:
   2. Leaves opposite. Flowers on long pedunculate solitary cymes ...................... ***Euonymus***
   2. Leaves alternate. Flowers on short pedunculate divaricate cymes ................. ***Maytenus***
1. Fruit an indehiscent capsule:
   3. Leaves ovate-oblong or obovate, margin crenulate-serrulate, to 11.5 × 4.7 cm. Flowers yellowish-green in axillary dichotomously branched, corymbose cymes. Drupe 1 × 0.4 cm. ........................................................................................................ ***Cassine***
   3. Leaves elliptic-obovate, margin entire; to 6 × 2.5 cm. Flowers-greenish in lateral cymes. Drupe 6 × 4 mm. .............................................................................***Pleurostylia***

### **CASSINE** Linnaeus

**Cassine glauca** (Rottlb.) Kuntze in Rev. Gen. Pl. 1: 114.1891; Ding Hou in Steenis Fl. Males. Ser. 1.6: 284–286. 1962. *Mangifera glauca* Rottlb. in Nye. Saml. Kongel. Danske Vidensk. Selesk. Skr. 2: 534. t. 4. f. 1. 1783. *Elaeodendron glaucum* (Vahl.) Pers. Syn. 1: 241. 1806; FBI 1: 623; Gamble 1: 211 (152); Brandis 164, *E. roxburghii* Wt. & Arn. Prodr. Fl. Ind. Orient. 157. 1834; Beddome Fl. Sylv. S. India t. 148. 1870.

Evergreen, densely foliaceous tree, up to 15 m tall; crown hemispherical; bark grey, nearly smooth, exfoliating in very small scales, relatively thin, flesh coloured when cut; wood light brown, close grained; branches reddish, branchlets glabrous. Leaves thin-coriaceous, variable, ovate-oblong, or obovate, 6–11.5 × 2.5–4.7 cm, glabrous, apex acute, margin crenulate-serrulate, base acute; nerves prominent, petiole to 2.4 cm. Flowers yellowish-green, to 8 mm across in axillary, dichotomously branched, corymbose cymes. Drupe oblong, smooth, 1 × 0.4 cm, yellowish green, apiculate; seed 1.

Common, in deciduous forests in most forest districts up to 1000 m. Fls.: With a peak during Mar.–Apr. (Plains to 500 m) and July–Sept. (hill above 500 m). Frs.: Throughout the year. Vern.: Nirija, Noorijia, Noridi, Neridi, Nerinja, Bhutan-Kusamu.

Leaf powder is used for headaches; decoction of the bark is given for cholera.

Papi hills (WG), *SSR* 15256; Kasipentla (CTR), *SSR* 15981; Ichampalli (ADB), *GO & DAM* 5153; Ballapalli (GTR), *KCJB* 17334 (MH); Anantagiri (VSKP), *NPBK* 10959 (MH); Diguvametta (PKSM), *RVK* 15720.

INDIA : Throughout.
WORLD : Sri Lanka, Myanmar, Indo-China, Thailand, Indonesia, Malay. Philippines.

## EUONYMUS Linnaeus

**Euonymus glaber** Roxb. Fl. Ind. 1: 628. 1820; FBI 1: 609; Brandis 158.

Tree, up to 10 m tall; bark pale, nearly smooth. Leaves chartaceous, elliptic-oblong, oblong-lanceolate or obovate, 5–10 × 2–4 cm, glabrous, glossy, paler beneath, apex shortly acuminate, margin serrate, base acute, petiole 6–10 mm; stipules small, with fimbriate tips. Flowers white, 7–10 mm across, pedicellate, in long-peduncled, 10–12-flowered, axillary, solitary cymes. Capsule red, 4-lobed, loculicidal, 1.5–1.6 cm across.

Occasional, near streams in the hilly forests of Vijayanagaram and Srikakulam districts (R.S. Rao & Hara Sreeramulu, 1986). Fls.: Apr.–May. Frs.: June–Aug.

INDIA : East Bengal.
WORLD : Bhutan, Myanmar.

## MAYTENUS Molina emend Bosch

1. Leaves lanceolate-elliptic; secondary nerves 6–10 pairs. Cymes sessile. Capsule globose to obovoid, 3-gonous ............ ***M. rufa***
1. Leaves elliptic to elliptic-ovate; secondary nerves to 15 pairs. Cymes short pedunculate. Capsule turbinate, 3-lobed. ............ ***M. bailadillana***

**Maytenus bailadillana** (Narayan & Mooney) Raju & Babu, in Bull. Bot. Surv. India 10: 348. 1968. *Gymnosporia bailadillana* Narayana & Mooney in J. Indian Bot. Soc. 20: 40. 1941.

Small tree, up to 7 m tall; branches with few, straight axillary thorns, rarely unarmed; branchlets dark brown, ribbed, glabrous. Leaves elliptic to elliptic-ovate, 3–7 × 1.8–5 cm, glabrous, shining above, greenish-brown beneath, apex acute to shortly, obtusely acuminate, margin minutely glandular crenulate, base cuneate; secondary nerves 15 pairs, petiole 5–7.5 mm. Flowers white in short-peduncled, axillary simple or dichotomous cymes. Capsule turbinate, 3-lobed, obtrigonous, 6 mm long, loculicidal, slightly depressed, more or less striate dorsally; seeds 2 in each cell, ovoid, shining, aril completely covering the young seed, splitting irregularly when mature and hanging loosely at the base of the seed.

Rare, along water courses on rocky, dry ground above the bank up to 1000 m. Fls.: Nov.–Dec. Frs.: Jan.–Mar.

Sunkarimetta (VSKP), *HSP* 20551; *SKW* 2563 (BLAT).

INDIA : Madhya Pradesh, Orissa.

**Maytenus rufa** (Wall.) Hara in J. Jap. Bot. 40. 327. 1965. *Celastrus rufa* Wall. in Roxb. Fl. Ind. 2. 397. 1824. *Gymnosporia rufa* (Wall.) Lawson in Hook. f. Fl. Brit. India 1: 271. 1875; Gamble 1: 209 (151); Brandis 163.

Small tree, branches slender, thorns axillary on many of the branches. Leaves sub-coriaceous, lanceolate-elliptic, 6.7–9.2 × 2.5–3.5 cm, glabrous, brownish-yellow when drying, apex obtusely acute, margin serrate, base acute-cuneate; secondary nerves 6–10 pairs; petiole to 0..8 cm Flowers white in axillary divaricate cymes. Capsule pale yellow, globose to obovoid, 3-gonous, 0.8 cm in dia; seeds 2 per each locule, black.

Common, in Visakhapatnam district up to 1000 m. Fls.: Apr.–June. Frs.: July–Aug.

Minumuluru (VSKP), *GVS* 30010 (MH).

## PLEUROSTYLIA R. Wight et Arnott

**Pleurostylia opposita** (Wall.) Alston in Trimen. Handb. Fl. Ceylon 6 (Suppl.): 48. 1931. *Celastrus opposita* Wall. in Roxb. Fl. Ind. 2. 398. 1824. *Pleurostylia wightii* Wt. & Arn. Prodr. Fl. Ind. Orient. 157. 1834; FBI 1: 617; Gamble 1: 211 (152); Brandis 165.

Densely foliaceous, semi-evergreen tree, up to 8 m tall; crown compact; bark grey, fissured (tessellated) in small squares; wood light red or greyish; branchlets glabrous. Leaves decussate, thin-coriaceous, elliptic-obovate, to 6 × 2.5 cm, grabrous, apex obtuse, margin entire, base attenuate; petiole to 4 mm. Flowers minute, greenish on lateral cymes (below the leaves). Drupe indehiscent, 6 × 4 mm, yellowish; one-seeded, ovoid, 3 mm.

Occasional in dry deciduous forests and in wide variety of habitats from full sun-exposed slopes to the interior forests up to 1300 m. Leaf fall: Jan.–Feb. New foliage: Mar. Fls. and Frs.: Apr.–Aug. Vern.: Piyari, Kuntichinta.

Wood hard, and useful for making combs.

Tirumala hills (CTR), *SSR* 13717; Penchalakona (NLR), *SSR* 19359; Talakona (CTR), *GVS* 45995; Cuddapah hills (CDP), *JSG* 10157 (MH); Yeleswaram (EG), *JSG* 16057; Chingora RF (VSKP), *KCJB* 17197 (DD).

WORLD : Sri Lanka, Madagascar, Mauritius.

# HIPPOCRATACEAE

## SALACIA Linnaeus *nom. cons.*

**Salacia chinensis** L. Mant. Pl. 293. 1771; Gmel. in L. Nat.Reg. Veg. ed. 13.1.107. 1791; *S. prinoides* DC. Prodr. 1: 571. 1824; FBI 1: 626; Gamble 1: 215(154); Brandis 166.

Small tree or straggling shrub; branches divaricate; branchlets angular, glabrous, twigs ridged from the decurrent bases of the deciduous stipules. Leaves coriaceous, elliptic-oblong, 4–8 × 1.5–2.6 cm, glabrous, apex shortly acuminate, margin serate-crenate or entire, base subacute to obtuse; petiole 6–10 mm; stipules deciduous. Flowers yellowish, fragrant, 6.5 mm across in

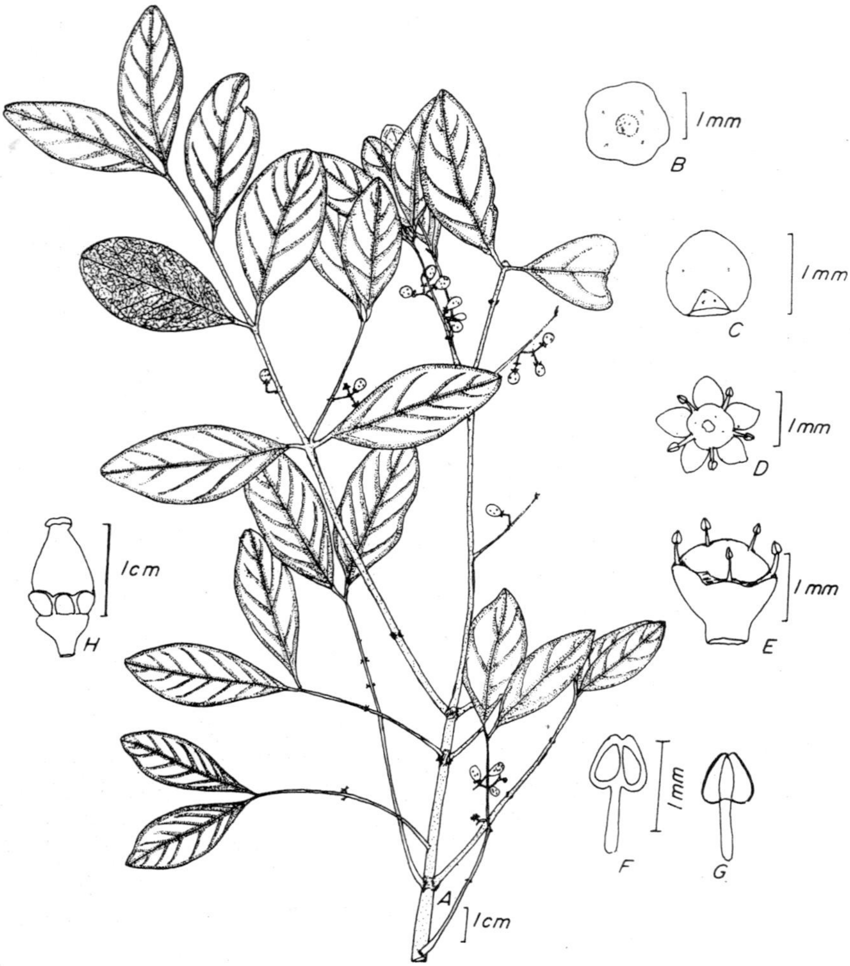


**Fig. 22.** *Pleurostylia opposita* (Wall.) Alston
A. Twig, B. Calyx, C. Petal, D. Petals with Androecium, E. Sepals with androecium and disc, F & G. Stamens, H. Pistil.

1 cm long umbels from tubercles of leafless axils. Berry red, subglobose, 1 × 0.5 cm, fleshy, minutely apiculate; seed 1, light brown, 6–7 mm in dia., subglobose.

Frequent, in semi-evergreen forests to 800 m. Fls.: Feb.–Apr. Frs.: Oct. Vern. Narani.

Palakonda hills (VSKP), *JSG* 13898 (DD); Kadagandi RF (VSKP), *KCJB* 17274 (MH); Annavaram (EG), *CAB* 12595 (MH).

INDIA : Eastern and Western Peninsula, Assam, West Bengal, Bihar, Andaman & Nicobar.

WORLD : Sri Lanka, Java, Philippines, Myanmar, Thailand, Indo-China, China, throughout Malaysia.

## SIPHONODONTACEAE
### **SIPHONODON** Griffith

**Siphonodon celastrineus** Griff. in Calc. Journ. Nat. Hist. 4: 247. t. 14. 1843; FBI 1: 629; Brandis 165.

An evergreen, small tree; bark grey, slightly rough; wood white; branchlets glabrous. Leaves coriaceous, elliptic-oblong, 10–14 × 4.5–6.5 cm, glabrous, shining above, apex emarginate or obtuse, margin crenate-serrate, base obtuse to acute; secondary nerves 8–10 pairs, slender, not prominent; petiolule 0.5 cm. Flowers yellowish, 6–8 mm across. Fruit yellowish, pyriform, hard, fleshy, 2.2–5 × 2.5–4 cm in dia, with numerous woody 1-seeded pyrenes; seeds brown, 1.4 × 0.4 cm.

Occasional, in Visakhapatnam and East Godavary districts at 700–1200 m. Fls.: Apr.–June. Frs.: Feb.

Dharakonda gedda (VSKP), *GVS* 47420 (MH); Rampa (EG), *MSR* 1558 (CAL).

INDIA : Eastern India, Bihar, Andamans, Sikkim.
WORLD : Bhutan, Myanmar, Phillippines, Java, Thailand.

## RHAMNACEAE

1. Trees unarmed. Leaves pinnately nerved. Flowers 4-merous. Fruit fleshy, berry like drupe; pyrenes 2–4. .......... ***Rhamnus***.
1. Trees armed, rarely unarmed. Leaves 3–5-nerved from base. Flowers 5-merous. Fruit fleshy drupe with 1–4 seeded stone .......... ***Ziziphus***

### **RHAMNUS** Linnaeus

**Rhamnus virgatus** Roxb. Fl. Ind. 1: 604. 1820; Gamble 1: 222 (159); Brandis 172. *R. davurica* Pall. var. *hirsutus* (Wt. & Arn.) Lawson in Hook. f. FBI 1: 639. 1875. *R. hirsutus* Wt. & Arn. Prodr. Fl. Ind. Orient. 165. 1834.

Small tree, up to 3 m tall; much branched; generally with a stout spine, bark smooth. Leaves membranous, frequently fascicled on arrested branchlets, rhomboid, ovate-lanceolate or lanceolate, 0.5–6 × 0.3–2.9 cm, sub-opposite, glabrous, slightly pubescent beneath when young, apex shortly obtusely acuminate, margin finely serrate, base round; secondary nerves 4-pairs; petiole to 1.1 cm. Flowers greenish, unisexual, numerous, minute in axillary fascicles; pedicles filiform. Drupe obovate-orbicular, 4–6 mm, 0.4 cm in dia., faintly 2-lobed with mucronate tip, wrinkled, glabrous above, pubescent beneath; seed grooved, obovoid.

Rare, in Visakhapatnam district above 900 m. Fls.: Mar.-Apr. Frs.: May–June.

Sunkarimetta (VSKP), *GVS* 22650 (CAL); Minumuluru (VSKP), *GVS* 30008 (MH); Paderu (VSKP), *SKW* 8563–65 (BLAT).

INDIA : Throughout the hilly areas.
WORLD : Afghanistan, Bhutan, Myanmar, China, Japan, Pakistan.

## **ZIZIPHYS** P. Miller

### KEY 1 (Vegetative)

1. Plants unarmed ............ *Z. glabrata*
1. Plants armed:
   2. Leaves 5–10 × 4.5–6 cm; petiole to 1.2 cm, prickles both recurved. ............ *Z. xylopyrus*
   2. Leaves 2–4 × 1.2–2.6 cm; petiole to 5 mm, prickles one is straight, another one recurved. ............ *Z. mauritiana*

### KEY 2 (Flowering & Fruiting)

1. Flowers in divaricate, elongated cymes to 3 cm long; peduncle to 1.5 cm; petals spathulate; styles 3, distinct; stigma 3-fid. Drupe tomentose, 2.5 cm across ............ *Z. xylopyrus*
1. Flowers fascicled in condensed cymes; peduncle 0.5–0.6 cm; petals other than above; styles 2, connate up to middle, stigma 2-fid. Drupe rugose, < 2 cm across:
   2. Calyx tube pubescent; petals obovate; stamens 3 mm long. Drupe globose-obovoid............ *Z. glabrata*
   2. Calyx tube villous; petals ovate; stamens 1 mm long. Drupe oblong-globose ............ *Z. mauritiana*

**Ziziphus glabrata** Heyne ex Roth, Nov. Pl. Sp. 159. 1821; FBI 1: 633. *Z. trinervia* Roxb. Fl. Ind. 1: 606. 1832; Gamble 1: 220 (158); Brandis 171; Beddome Fl. Sylv. S. India, Anal. Gen. t. 68. 1870.

Unarmed tree, up to 8 m tall; wood brown; branchlets tomentose. Leaves elliptic, elliptic-oblong or ovate, 3–7.5 × 2.3–5 cm, glabrous, white woolly-tomentose beneath when young, apex acute or subobtuse, often mucronate, margin serrate, base oblique, obtuse or acute; 3-nerved; petiole to 1.2 cm. Flowers greenish-yellow in axillary short pedunculate cymes. Drupe globose or obovoid, 1.2 cm dia., yellow when ripe, apiculate; seeds 1 or 2.

In dry deciduous forests of almost all the districts at 500–800 m. Fls.: Mar–Sept. Frs.: Sept.–Jan. Vern.: Goti, Kakupala.

Wood hard, but not as good as that of *Z. mauritiana* Lam.

Ankusapuram RF (ADB), *GO* 4370; Macherla RF (GTR), 3833; Razoli (KNL), *RVR* 3146; Nasrullabad (NZB), *TP & BR* 6165; Bhuthappagudi-Kekati RF (ATP), *TP & NY* 795.

INDIA : Peninsula, N.E. India.
WORLD : Bhutan.

**Ziziphus mauritiana** Lam. Encycl. 3: 319. 1789. *Z. jujuba* (L.) Gaertn. Fruct. 1: 203. 1788. non. Mill. 1768; FBI 1: 632. 1875; Gamble 1: 219 (157); Brandis 169; Beddome, Fl. Sylv. S. India, Anal. Gen. 68. t. 1871.

Deciduous, much branched, thorny tree, up to 10 m tall; crown rounded; branches drooping; bark nearly black, rough; wood reddish brown; young parts rusty tomentose. Leaves chartaceous, ovate-elliptic or suborbicular, 2–4 × 1.2–2.6 cm, apex rounded, margin fairly denticulate, base rounded, acute or slightly inequilateral, glabrous and shining above, densely white tomentose beneath, strongly 3-nerved; petiole to 5 mm. Flowers greenish-yellow in small, axillary, clusters. Drupe oblong-globose, 1 × 0.8 cm, reddish brown, fleshy, smooth with a single stone; seed one, tubercled and irregularly furrowed.

Wild in scrub and dry deciduous forests and cultivated in and around villages and then run wild on waste lands to 700 m. It is extensively cultivated for its edible fruits. Though specifically recommended for sandy loams, gravelly alluvial soils, it can be planted in laterites, black cotton and moderately saline soils having good drainage. Leaf fall: Nov.–Dec. New foliage: Mar.–Apr. Fls.: Apr.–Oct. Frs.: Oct.–Mar. Vern.: Reni, Regu, Badari, Ganga regi.

Wood tough and strong. Useful for many purposes and is a source of good fuel.

Tutigunta (WG), *SSR* 15253; SKU Campus (ATP), *TP & NY* 149; Palakonda hills (CDP), *CS* 7776; Macherla RF (GTR), *VRK* 3919; Veldurthi RF (KNL), *RVR* 1754; Kamareddy (NZB), *BR* 9271; Pochammarlu (MDK), *BR & CP* 11482; Chandoor (NZB), *BR & GO* 9206; Kalyani dam (CTR), *DR* 692 (SVU).

INDIA : Throughout the greater part of India.
WORLD : Myanmar.

**Ziziphuys xylopyrus** (Retz.) Willd. in Sp. Pl. 1: 1104. 1798; FBI 1: 634; Gamble 1: 220 (158); Brandis 171; Beddome Fl. Sylv. S. India Anal. Gen. 68. 1870. *Rhamnus xylopyrus* Retz. Obs. Bot. 2: 11. 1781.

Unarmed or sparsely armed deciduous tree, up to 12 m tall; bark grey or brown with thick oblong scales; wood reddish-brown; branchlets tomentose, thorns paired or solitary. Leaves elliptic-ovate or orbicular, 5–10 × 4.5–6 cm, glabrous above, covered with white or yellowish tomentum beneath, apex rounded, margin serrulate, base sub-cordate or obtuse or asymmetrical, 3-nerved; petiole to 1.2 cm. Flowers greenish-yellow in 4–5 cm long, axillary, divariate cymes. Drupe globose, 2.5 cm across, deep green, white tomentose, hard and woody; seeds 3, oblong.

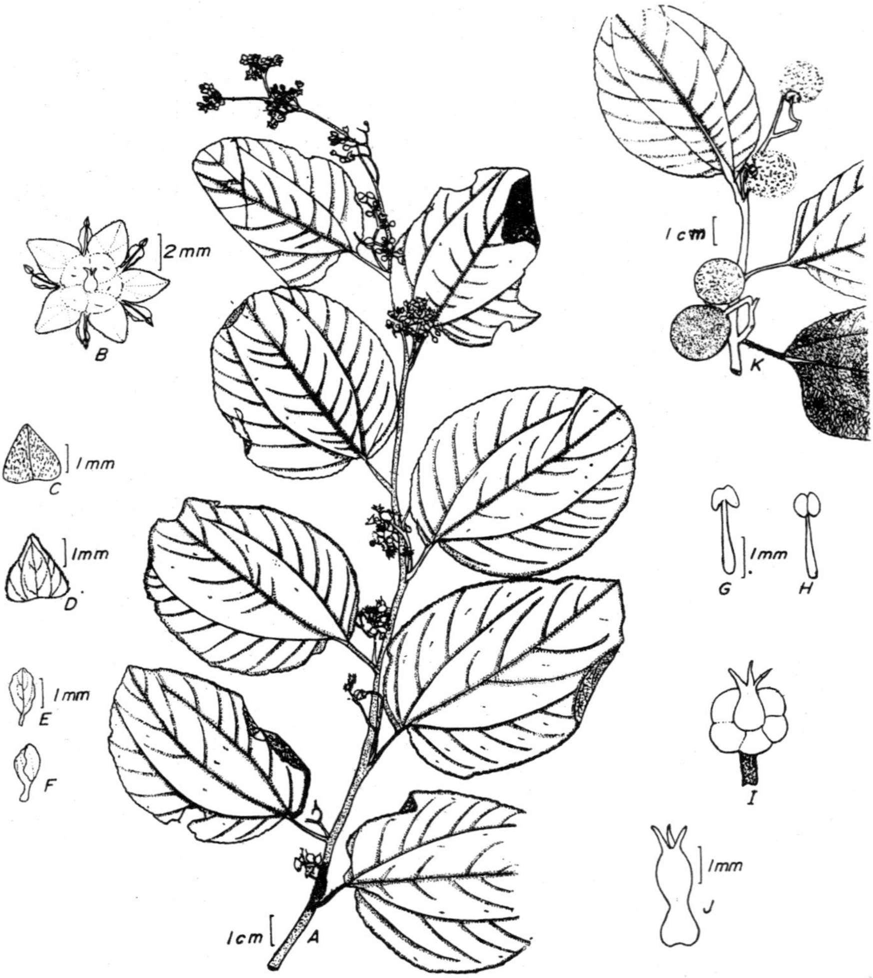


**Fig. 23.** *Zizyphus xylopyrus* (Retz.) Willd.
A. Twig, B. flower, C & D. Sepals, E & F. Petals, G & H. Stamens, I. Disc with pistil, J. Pistil, K. Fruiting twig.

Fairly common, in dry deciduous forests and almost in all districts at 400–1000 m. Leaf fall: Mar. New foliage: Apr.–May. Fls.: Apr.–July. Frs.: Ripens in Jan. Vern.: Gol, Gotti chettu, Gotiki. Eng.: Golden silk cotton tree. Wood hard.

Tirumala (CTR), *SSR* 13716 & 15267; Mothugudem (KMM), *SSR* 15240; Eturnagaram (WGL), *SSR* 18546; Racherla RF (KNL), *RVR* 1376; Macherla RF (GTR), *VRK* 3913; Bourapur (MBNR), *TDCK* 16813; Panasagonda (SKLM), *GVS* 62354 (CAL).

INDIA : Throughout.
WORLD : Sri Lanka and Nepal.

## VITACEAE

### CISSUS Linnaeus

**Cissus vitiginea** L. Sp. Pl. 117. 1753; Gamble 1: 234 (167). *Vitis linnaei* Wall. ex Wt. & Arn. Prodr. Fl. Ind. Orient. 126. 1834; FBI 1: 649; Brandis 176.

Small tree, 4–6 m tall; trunk ca. 60 cm dia; bark greyish, suberous and fissured; some times climbing, tendrils stout, leaf-opposed; branchlets densely pubescent. Leaves thick-coriaceous, orbicular-ovate or broadly cordate, 3–5-lobed, 5–12 × 7–13 cm, puberulous above, densely curly pubescent beneath, apex acuminate, margin dentate, base cordate, 3–7-nerved; secondary nerves 4–5 pairs above basal ones; petiole to 1.3 cm, pubescent. Flowers white, 4-merous, 4 mm across in leaf-opposed, umbellate cymes. Drupe pale purple, ellipsoid, 0.8 × 0.5 cm; seed one, tessellate.

Common, in scrub jungles and deciduous forests on rocky soils at 200–600 m. Fls.: June–July. Frs.: July–Nov. Vern.: Adavi draksha, Gummudu.

Kalasamudram RF (ATP), *SSR* 13747; Eswaramala RF (ATP), *TP & NY* 838; Owk RF (KNL), *RVR* 1475; Diguvametta (PKSM), *JLE* 42193 (CAL); Tirumala (CTR), *DR* 1506 (SVU).

INDIA : Peninsula.
WORLD : Sri Lanka.
**Note** : Since Linnaeus publication, the climbing habit of this taxon has been emphasised by all workers including Planchon (i.e. 1883). Gamble for the first time, has pointed out of arboreal habit of this taxon. It has been observed that the plants have arboreal habit also in Kalasamudram during our field study. Ellis & Swaminathan have also reported such type of arboreal habit from Bairani-Diguvametta, Nallamalais (in JBNHS 72: 230. 1975).

## LEEACEAE

### LEEA Linnaeus *nom. cons.*

**Leea indica** (Burm. f.) Merr. in Philipp. J. Sci. 14: 245. 1919; Ridsdale in Blumea 22. 95. 1974. *Staphylea indica* Burm. f. Fl. Ind. 75. t. 23. f. 2. 1768. *Leea sambucina* (L.) Willd. Sp. Pl. 1177. 1798; FBI 1: 666; Gamble 1: 240 (172); Brandis 179. *Aquilica sambucina* L. Mant. Pl. 2: 211. 1771.

Small tree, up to 6 m tall; branchlets puberulous. Leaves 2–3-pinnate, to 30 cm long, petiole 4–10 cm; pinnae 3-paired; odd pinnate; leaflets 2–5 pairs, oblong or elliptic-oblong, 7.5–15 × 2–3.5 cm, glabrous above, white pubescent beneath, apex subacute or obtuse, margin distantly serrate, base cordate; secondary nerves 8–10 pairs, prominent beneath, arched; petiolule 0.8–1.2 cm; stipules adnate to petiole, deciduous. Flowers white, ca. 6 mm across, in lateral leaf-opposed, 15 cm long, corymbose, divaricate cymes. Berry black, globose, 7.5–10 mm dia., depressed; pyrenes 5 or 6, 1-seeded.

Rare, in North coastal, Prakasam and Nellore districts at 700–1300 m. Fls. & Frs.: Throughout the year.

Anantagiri (VSKP), *SSR* 19625; Malleswara Kona (NLR), *BSN* 5293 (VV); Borra caves (VSKP), *GVS* 44267 (MH); Bison hill (EG), *CAB* 5071 (MH); Polavaram (WG), *DCSR* 421 (CAL).

INDIA : Throughout.
WORLD : Nepal, Bangladesh, Myanmar, Sri Lanka, Thailand, Cambodia, Laos, N. & S. Vietnam, China, Malaysia.

## SAPINDACEAE

### KEY 1 (Leaf)

1. Leaves 3-foliolate; margin serrate ........ ***Allophylus***
1. Leaves pinnate; margin entire or undulate:
    2. Leaves imparipinnate; leaflets > 3 pairs:
        3. Rachis winged; leaflets 12 pairs—cultivated ........ ***Filicium***
        3. Rachis not winged; leaflets 4–6 pairs ........ ***Erioglossum***
    2. Leaves paripinnate; leaflets 2–3 pairs:
        4. Leaves pubescent beneath, apex emarginate-retuse; petiole to 4.5 cm long ... ***Sapindus***
        4. Leaves glabrous on both surfaces, apex other than above; petiole 5–10 cm long:
            5. Leaflets sub-opposite, red when young, apex obtuse or rotund; secondary nerves plaited below; petiolule 0 ........ ***Schleichera***
            5. Leaflets alternate, not red at any stage, apex obtuse or rounded, secondary nerves not plaited; petiolule to 5 mm long ........ ***Lepisanthes***

### KEY 2 (Flowering)

1. Corolla absent ........ ***Schleichera***
1. Corolla present:
    2. Flowers in simple racemes ........ ***Allophyllus***
    2. Flowers in racemiform panicles:
        3. Flowers regular:
            4. Petals hooded, orbicular, greenish ........ ***Filicium***
            4. Petals clawed, obovate, white ........ ***Sapindus***
        3. Flowers irregular (Zygomorphic):
            5. Petals white; panicles axilllary, or lateral, spreading ........ ***Lepisanthes***
            5. Petals white or pinkish; panicles terminal, erect ........ ***Erioglossum***

## KEY 3 (Fruiting)

1. Drupe lobed:
   2. Drupe lobes to 6 mm across, orange red ........ ***Allophyllus***
   2. Drupe lobes 1–1.5 cm across, other than above:
      3. Drupe deeply lobed; purplish black ........ ***Erioglossum***
      3. Drupe not deeply lobed; brownish ........ ***Sapindus***
1. Drupe not lobed:
   4. Drupe 3-gonous, yellow tomentose ........ ***Lepisanthes***
   4. Drupe ovoid or sub-globose, glabrous:
      5. Drupe spiny-tipped or pointed, 2–2.5 × 1.5 cm ........ ***Sehleichera***
      5. Drupe apiculate, 1.0 cm across ........ ***Filicium***

## ALLOPHYLUS Linnaeus

**Allophylus cobbe** (L.) Raeusch. Nomencl. Bot. ed. 3. 108. 1797; FBI 1: 673; Leenhouts in Blumea 15: 322. 1967. *Rhus cobbe* L. Sp. Pl. 267. 1753. *Allophylus serratus* (Roxb.) Kurz. in Journ. Asiat. Soc. 44(2): 185. 1876; Radlk. in Engler, Pflanzenr. 986. 562. 1932; Leenhouts in Blumea 15: 351. 1967; S.K. Mukerjee in Indian For. 98. 494. 1972. *Allophylus cobbe* (L.) Raeusch. forma *serratus* (Roxb.) Heirn. in Hook. f. FBI 1: 673. 1875. *Allophylus rheedii* (Wt.) Radlk. in Engler & Prantl. Pflanzenf. 3(5): 313. 1895; Gamble 2: 246 (176); Brandis 185.

Tree, up to 8 m tall; bark grey or purple spotted; branches yellowish brown, lenticellate. Leaves trifoliate, subcoriaceous, elliptic-lanceolate, 5.3–12 × 2.4–5.9 cm, terminal one larger, laterals are smaller and inequilateral, softly pubescent, apex acuminate, margin serrate, base cuneate to attenuate, petiole to 12 cm, petiolule to 2 mm. Flowers yellowish-white, 2–2.5 cm across in 15 cm long, axillary racemes. Fruit globose, 5 × 3 mm, red when ripe, smooth glabrous; seed one.

Occasional, in slopes of dry deciduous forests at above 700 m. Fls.: Aug.–Dec. Frs.: Dec.–Jan. falling off early. Vern.: Erra avalu, Guna gutti, Sallikunkudu.

Tirumala (CTR), *SSR* 13763; Sriharikota (NLR), *YNR* 2067; Srungavarapukota (VSKP), *NPBK* 11026 (MH); Balapalli (CDP), *JLE* 14918 (MH); Jaddingi stream (EG), *GVS* 68660 (MH).

INDIA : Peninsula.
WORLD : Sri Lanka.

## ERIOGLOSSUM Blume

**Erioglossum rubiginosum** Bl. Bijdr. 229. 1825; FBI 1: 672; Gamble 1: 248 (177); Brandis 192.

Small, evergreen tree; branchlets rusty tomentose. Leaves paripinnate, rachis tomentose, petiole to 11.5 cm, leaflets 4–6 pairs, sub-opposite, elliptic-oblong, 6.5–13 × 3–4.8 cm, glabrous and greyish above, brownish tomentose beneath especially on the nerves, apex shortly acuminate, margin entire, base oblique; secondary nerves 8–10 pairs, not quite uniting with the marginal nerve; petiolule 3 mm. Flowers white or pinkish, fragrant, 6 mm long, polygamo-dioecious, tomentose, clustered on the racemiform branches of a 40 cm long terminal panicles. Fruit of 1–3 purplish black, fleshy, basally connate, oblong, 1.3 cm long carpels.

Rare, in evergreen forests of Visakhapatnam above 800 m. Fls. Feb.–Apr. Frs.: May. Vern.: Ishirashi.

Fruits sweet, edible.Wood is strong durable and is useful for furniture.

Vankachintala near Minumuluru (VSKP), *GVS* 47339 (MH); Maredumilli (EG), *GVS* 2728 (MH); Palakonda hills (VSKP), *JSG* 14035 (MH); Punyagiri (VSKP), *GVS* 23308 (MH).

INDIA : Orissa, Assam, Sikkim, Andaman & Nicobars.
WORLD : Myanmar.

## **FILICIUM** Thwaites ex Hook.

**Filicium decipiens** (Wt. & Arn.) Thw. Enum. Pl. Zeyl. 408. 1864; Bedd. Fl. Sylv. S. India t. 129. 1871; FBI 1: 539; Gamble 1: 249 (178). *Rhus decipiens* Wt. & Arn. Podr. 172. 1834.

Densely foliaceous, evergreen tree to 15 m tall; crown compact, hemispherical and equal to the height of the tree, branchlets covered with scales. Leaves imparipinnate, rachis 20–30 cm long, broadly winged; petiole to 6 cm; leaflets 12 pairs, subcoriaceous, 5.5–12 × 1–2 cm, lanceolate, glabrous, glossy, punctate, apex acute, acute-retuse, margin entire, base attenuate; petiolule 0. Flowers greenish, 4 mm across in 20 cm long panicles. Drupe dark purple, subglobose, 1 cm across, apiculate; seeds 1–2.

Occasionally planted in gardens. Fls.: Oct.–Jan.

Alipiri Nursery (CTR), *SSR* 19399.

INDIA : Peninsula.
WORLD : Sri Lanka.

## **LEPISANTES** Blume

**Lepisanthes tetraphylla** (Vahl) Radlk. in Sitzungsber. Math.Phys. Cl. Koenigl. Bayer Akad. Wiss. Muenchen 8: 276. 1878; Gamble 1: 247 (176); Brandis 189. *Sapindus tetraphylla* Vahl, Symb. Bot. 3: 54. 1794. *Hemigyrosa deficiens* (Wt. & Arn.) Bedd. Fl. Sylv. S. India t. 151. 1871; 6. 231 1872; FBI 1: 671. *H. canescens* (Roxb) Bl. in Rumphia 3: 166.

1849; FBI 1: 671. *Lepisanthes deficiens* (Wt. & Arn.) Radlk. in Sitzungsber. Math. Phys. Cl. Boenigl. Bayer Akad. Wiss. Muenchen 8: 276. 1878; Gamble 1: 247 (176).

Evergreen tree, up to 12 m tall; bark grey, fairly rough; wood white; branchlets conspicuously striate when young, glaucous, puberulous. Leaves paripinnate, rachis to 6 cm, petiole to 10 cm; leaflets 2-pairs, thick-coriaceous, variable, narrowly elliptic-oblong, 3.2–11 × 1.1–3.3 cm with cuneate base in mature trees; broadly elliptic-oblong, 10–20 × 6–9 cm, base oblique with cuneate in young trees, glabrous, apex obtuse, or rounded; margin entire; secondary nerves 8–9 pairs, midvein raised beneath, nerves and nervules prominent either side, petiolule 4–5 mm, stout, yellow pubescent in young. Flowers white, irregular, 4 mm across, pubescent; racemiform, panicles 5–10 cm long, dense flowered, fascicled, lateral or axillary from the old wood. Drupe yellow, tomentose or velvety, 2 × 1.5 cm, 3-gonous, woody, coriaceous; seed 1, oblong, exarillate.

Common, in evergreen, and dry forests at 500–1000 m. Fls.: Mar.–Apr. Frs.: Apr.–May. Vern. Korivi, Konda Kunkundu, Salikunkudu.

Wood hard and used as good fuel. Occasionally used by the natives for building purposes.

Penchalakona (NLR), *SSR* 19365; Sidduliah Kona (NLR), *SSR* 19373; Punyagiri (VSKP), *GVS* 23306; Chandragiri (CTR), Coll. 10022 (MH); Sunnipenta (KNL), *JLE* 22769.

INDIA : Western Peninsula, Tamil Nadu.
WORLD : Sri Lanka, Myanmar, Thailand, N. & S. Vietnam, Cambodia, Malay Peninsula, Martinique.

## SAPINDUS Linnaeus

**Sapindus emarginatus** Vahl, Symb. Bot. 3: 54. 1794; Gamble 1: 250 (178); Beddome Fl. Sylv. S. India t. 154. *S. trifoliata* sensu Hiern. in Hook. f. Fl. Brit. India 1: 682. 1875; *p.p.* non L. 1753.

Densely foliaceous, deciduous tree, up to 12 m tall; bark grey with rough scales; wood yellow with streaks of lighter and darker colour; branchlets tomentose. Leaves even-pinnate, 8–12 cm long; petiole to 4.2 cm; leaflets 3-pairs, thick-coriaceous, broadly elliptic, oblong-obovate, 3–13 × 2–10 cm, glabrous above, pubescent beneath, apex emarginate-retuse, margin entire, base cuneate-obtuse; petiolule 0. Flowers brownish-white in 10 cm long rusty tomentose panicles. Drupe deep-brown, ovoid, 3-lobed, to 1 × 2 cm, fleshy, pubescent in young; rind wrinkled, formed by dried pulp; seed one in each locule, enclosed by black, smooth, hard endocarp.

Common, in plains, occasionally cultivated, often in forests. The tree grows well in almost all kinds of soils, but loamy, clayey, and black cotton soils are more favourable. Leaf fall: Jan.–Feb. New foliage: Mar. Fls. with two peaks during Feb.–Apr. (lower altitudes) and Aug.–Dec. (higher altitudes). Frs.: Throughout the year. Vern.: Kunkudu, Ritta chettu, Kukudu. Eng.: Soap-nut tree.

Dried fruit juice is used as detergent and subsitute for soap. Fruits possess emitic, tonic, astringent, and anthelmintic properties and are used in treatment of asthma, colic due to indigestion, diarrhoea etc. Fruit as well as root bark are used as fish poison. Wood hard, heavy and not durable, splits and cracks on exposure, not much used except for carts and for making small articles.

Srisailam (KNL), *SSR* 15525; Bhasar (ADB), *GO & PVP* 4639; Pochammarlu (MDK), *BR & CP* 11499; Kalasamudram RF (ATP), *NY* 681; Ethipothala (GTR), *VRK* 3921; Gokaram (KSN), *AWR* 1202; Bairani-Diguvametta (PKSM), *KNS* 3001 (DD); Mannanur (MBNR), *TDCK* 15398.

INDIA :Karnataka, Maharashtra, Tamil Nadu; cultivated in Bihar, Orissa, and West Bengal.
WORLD :Sri Lanka.

## **SCHLEICHERA** C.L. Willdenow *nom. cons.*.

**Schleichera oleosa** (Lour.) Oken. Allg. Naturgesch 3(2): 1341. 1841. *Pistacia oleosa* Lour. Fl. Cochinch. 2. 615. 1790. *Schleichera trijuga* Willd. Sp. Pl. 4: 1096. 1805; FBI 1: 681; Gamble 1: 248 (177); Brandis 189; Beddome Fl. Sylv. S. India t. 11. 1871.

Densely foliaceous, deciduous tree, up to 20 m tall; bark grey, exfoliating in small irregular flakes; wood reddish-brown, branches greyish; young shoots silky. Leaves paripinnate, to 15 cm long, petiole to 6.5 cm; leaflets 2–3 pairs, sub-opposite, coriaceous, oblong, elliptic, 5–12 × 2.5–5 cm, glabrous, apex acute to acuminate, obtuse or rotund, margin undulate, base cuneate, petiolule 0. Flowers greenish-yellow, small in 8–12 cm long, much branched axillary racemes. Drupe ovoid, to 2.5 × 1.5 cm, acute, tipped with persistent sytle, dry indehiscent; seed 1.2 cm long, brown, smooth, enclosed in a succulent aril having an acid taste.

Common, in mixed deciduous forests near streams at above 600 m. It is a shade bearer and drought hardy. It prefers sandy or sandy loam soils. It is a good coppicer and produces suckers freely. Leaf fall: Jan. New foliage: Mar. Pinkish when very young. Fls.: Mar.–Apr. Frs.: June–July and quickly falls on the ground. Vern.: Puska, Kodali-pulusu, Madaka pulusu, Sagada-puska, Mavita-vitiki, Boshi chettu, Pusugu, Pulusari-Kosangi, Pulla Kaya, Karaka booshi. Eng.: The lac tree, Macassar oil tree, Ceylon Oak.

It is considered to be the best for lac rearing. Ripe fruits possess whitish pulp with pleasant, acidic taste often eaten during summer, and also pickled; seeds also eaten raw or roasted. Flowers are said to yield a dye. Bark is

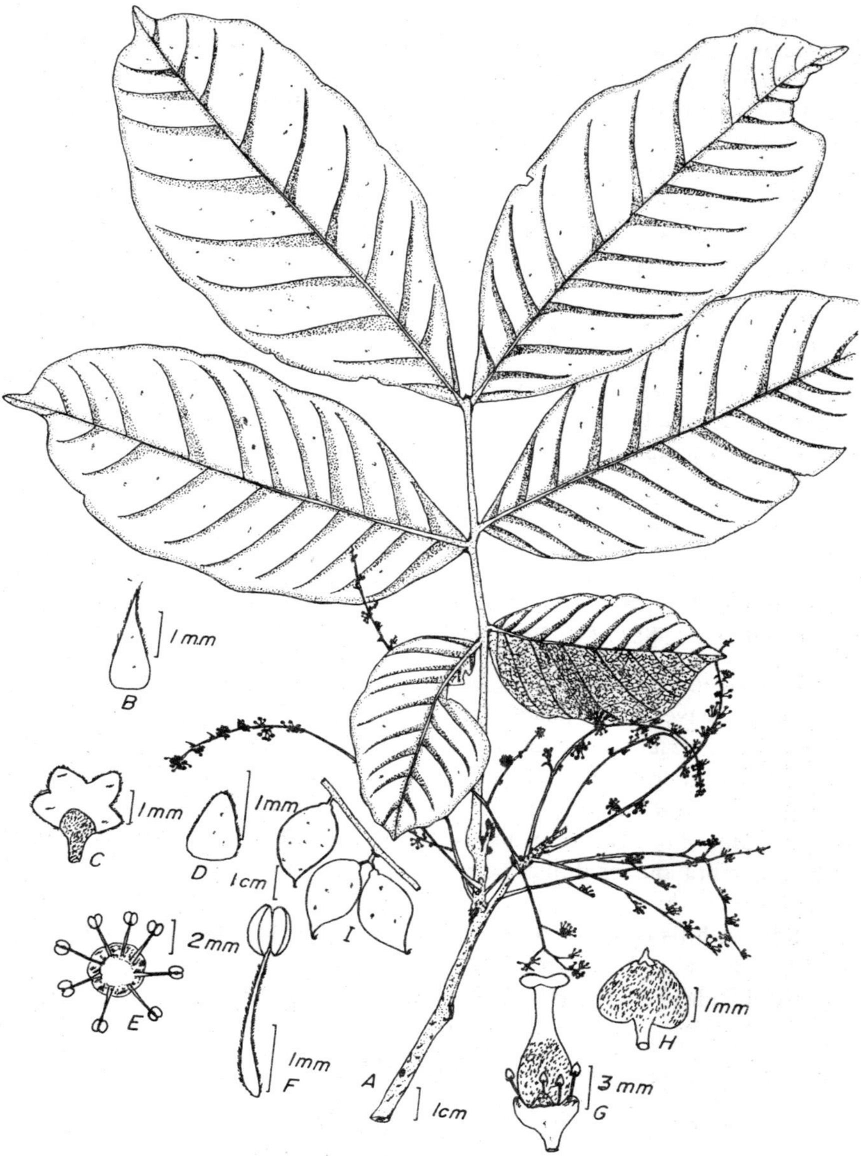


**Fig. 24.** *Schleichera oleosa* (Lour.) Oken.
A. Twig, B. Bract, C. Calyx, D. Calyx lobe, E. Disc with stamens, F. Stamen, G. Pistil with staminodes, H. Pistillode, I. Fruits.

astringent and used in external application for itch, back pain, inflammation and ulcers. A maceration or infusion of the bark is prescribed in malaria. Wood very durable, used for rice-pounders in oil and other mills, carts. The timber is suitable where hardness, bending, strength and toughness are required. It makes a good fuel and excellent charcoal. The seed yields oil and much valued for cooking and other purposes.

Galikonda (VSKP), *SSR* 15921; Vantla mamidi (VSKP), *SSR* 15945; Navabkota forest (ATP), *SSR* 18562; Manchippa (NZB), *SSR* 19337; Sidduliah Kona (NLR), *SSR* 19362.

INDIA : Throughout.
WORLD : Myanmar, Sri Lanka, Java.

The trees are exploited very badly, because of its multiple economic importance. Seed germination is very poor so it requires immediate conservation method.

## ACERACEAE
### **ACER** Linnaeus

**Acer niveum** Bl. Cent. Plant. Nov. Rumph. 3: 193. t. 167. 1847; FBI 1: 693; Gamble 1: 242 (173); Brandis 181.

Evergreen tree, up to 30 m tall; bark smooth having streaks, branchlets lenticular. Leaves ovate-lanceolate, 5–20 × 2.5–8.5 cm, glabrous, dark green above, white glaucous beneath, apex acute-acuminate, margin entire, base rounded or semicordate, trinerved from base but not distinct; secondary nerves 5–8 pairs, irregular, petiole to 5.5 cm. Flowers yellowish-green on 8 cm long terminal lax cymose panicles. Fruit a double samara, parallel, brown; wings membranous or thin papery, nerved, to 3.5 × 1.5 cm; seed part 1 × 7 cm.

Rare in hills of moist deciduous forests of Visakhapatnam at 950–1200 m. Fls.: Oct. Frs.: Nov.–Dec.

Galikonda (VSKP), *SSR* 15925; Minumuluru (VSKP), *GVS* 47350 (CAL).

INDIA : Assam.
WORLD : Sumatra, Java.

## SABIACEAE
### **MELIOSMA** Blume

1. Leaves simple .......... ***M. simplicifolia***
1. Leaves pinnate .......... ***M. pinnata***

**Meliosma pinnata** (Roxb.) Maxim. Bull. Sci. Acad. Imp. Sci. Saint-Ptersborgh (Ser. 3) 12. 64 in Obs. 1867; FBI 2: 6; Gamble 1: 256 (182)

Brandis 195. *Millingtonia pinnata* Roxb. Fl. Ind. 1: 103. 1820.

Medium sized tree; branches smooth, glabrous; branchlets lenticellate. Leaves pinnate, 15–40 cm long; leaflets 6–12 pairs, oblong or oblong-lanceolate, glabrous, apex acuminate, margin serrate; petiole cylindric. Flowers white, minute in terminal, puberulous panicles. Drupe subglobose, stone rugose.

Very rare, in Madgole hills of Visakhapatnam district at 800–1000 m (A. Lushington, cf. Gamble).

INDIA : Sikkim, Assam, Khasi hills.
WORLD : Laos, Vietnam, China, Taiwan, S. Korea, Japan, Philippines, Malay Peninsula, Malaysia.

**Meliosma simplicifolia** (Roxb.) Walp. Rep. 1: 423. 1842; sub sp. ***simplicifolia***; FBI 2: 5.; Gamble 1: 256 (182); Brandis 194. *Millingtonia simplicifolia* Roxb. Pl. Cor. t. 254. 1820.

Small evergreen tree, wood reddish-brown; branchlets pubescent. Leaves membranous, oblanceolate or obovate, 13–22 × 3.5–7 cm, glabrous, brownish above, pale beneath, apex obtuse, margin serrate, base cuneate-attenuate; secondary nerves to 15 pairs, arching, prominent, tertiary nerves parallel; petiole to 1.7 cm. Flowers yellowish-white, minute, sessile, small in rusty pubescent, 20 cm long panicles. Drupe small 4 mm in dia., bilobed, stone deeply rugose, segments oblong, indehiscent; seed oblong, exalbuminous.

Rare, in hills of Visakhapatnam distrct above 1000 m. Fls.: Dec.–Jan. Frs.: Dec.–Apr.

Sunkarimetta (VSKP), *GVS* 21677 (MH); Chintapalli (VSKP), *JLE* 37132 (MH); Minumuluru (VSKP), *GVS* 47314 (CAL).

## ANACARDIACEAE

### KEY 1 (Vegetative)

1. Leaves compound:
   2. Leaves trifoliate ........ ***Rhus***
   2. Leaves odd-pinnate; leaflets 3–5 pairs:
      3. Leaflets ovate-lanceolate, apex abruptly acuminate; secondary nerves 6–8 pairs, curved upwards and not forming intramarginal vein ........ ***Lannea***
      3. Leaflets oblong or elliptic-oblong, apex caudate-acuminate; secondary nerves 20–30 pairs, nearly parallel, slender, joined by strong intramarginal vein.-mango smelled when scratched........ ***Spondias***
1. Leaves simple
   4. Leaves slightly glaucous beneath ........ ***Nothopegia***
   4. Leaves green beneath:
      5. Inter-secondary nerves absent, basal secondaries retroflexed; apex obtuse:
         6. Secondary veins branched; ashy-grey pubescent beneath; petiole thick to 3 cm long ........ ***Semecarpus***

6. Secondary veins not branched, glabrous on both surfaces; petiole slender to 2 cm long ........................................ ***Anacardium***

5. Inter-secondary nerves present, all secondaries curved towards margin or upwards, apex not as above:

7. Secondary nerves branched, straight, horizontal towards margins; apex obtuse retuse, margin entire ........................................ ***Buchanania***

7. Secondary nerves not branched and straight, curved upwards; apex acute or shortly, obtusely acuminate, margin slightly undulate ........................................ ***Mangifera***

## KEY 2 (Flowering)

1. Flowers bisexual — style lateral ........................................ ***Buchanania***
1. Flowers polygamous, polygamodioecious or unisexual:
   2. Sepals and petals 4 each: — style apical.
      3. Stamens 4; racemes short erect ........................................ ***Nothopegia***
      3. Stamens 8; racemes long, pendulous ........................................ ***Lannea***
   2. Sepals and petals 5 each:
      4. All stamens not fertile, staminodes 1(2), style 1, lateral:
         5. Flowers 1 cm across; petals greenish with red stripes ........................................ ***Anacardium***
         5. Flowers 3–5 mm across; petals cream, base with yellow glandular ridges... ........................................ ***Mangifera***
      4. All stamens fertile; styles > 2, apical.
         6. Flowers yellow or yellowish-white:
            7. Stamens 5; styles 3, ovary 1-celled ........................................ ***Rhus***
            7. Stamens 10; styles 5, ovary 5-celled ........................................ ***Spondias***
         6. Flowers greenish-styles 3, ovary 1-celled ........................................ ***Semcarpus***

## KEY 3 (Fruiting)

1. Drupe on fleshy hypocarp (enlarged receptacle):
   2. Drupe oblong-ovoid; hypocarp orange, to 2 × 2.2 cm ........................................ ***Semecarpus***
   2. Drupe reniform, hypocarp yellowish, or reddish, , 1.5–8 × 2.5–3 cm. ........................................ ***Anacardium***
1. Drupe not on fleshy hypocarp:
   3. Drupe to 7 mm long:
      4. Drupe apiculate, striate, pericarp glandular ........................................ ***Nothopegia***
      4. Drupe not apiculate, smooth, pericarp not glandular ........................................ ***Rhus***
   3. Drupe above 8 mm long:
      5. Drupe to 1 cm long:
         6. Drupe globular, not compressed ........................................ ***Buchanania***
         6. Drupe oblong-ovoid, compressed ........................................ ***Lannea***
      5. Drupe above 3 cm long:
         7. Drupe ellipsoid, furrowed ........................................ ***Spondias***
         7. Drupe ovoid-oblong, not furrowed ........................................ ***Mangifera***

## **ANACARDIUM** Linnaeus

**Anacardium occidentale** L. Sp. Pl. 383. 1753; FBI 2: 20; Gamble 1: 260 (185); Brandis 204; Beddome Fl. Sylv. S. India t. 163. 1871.

Evergreen tree, up to 10 m tall; trunk short and thick; crown compact-hemispherical; bark rough; wood reddish-brown; branchlets glabrous. Leaves

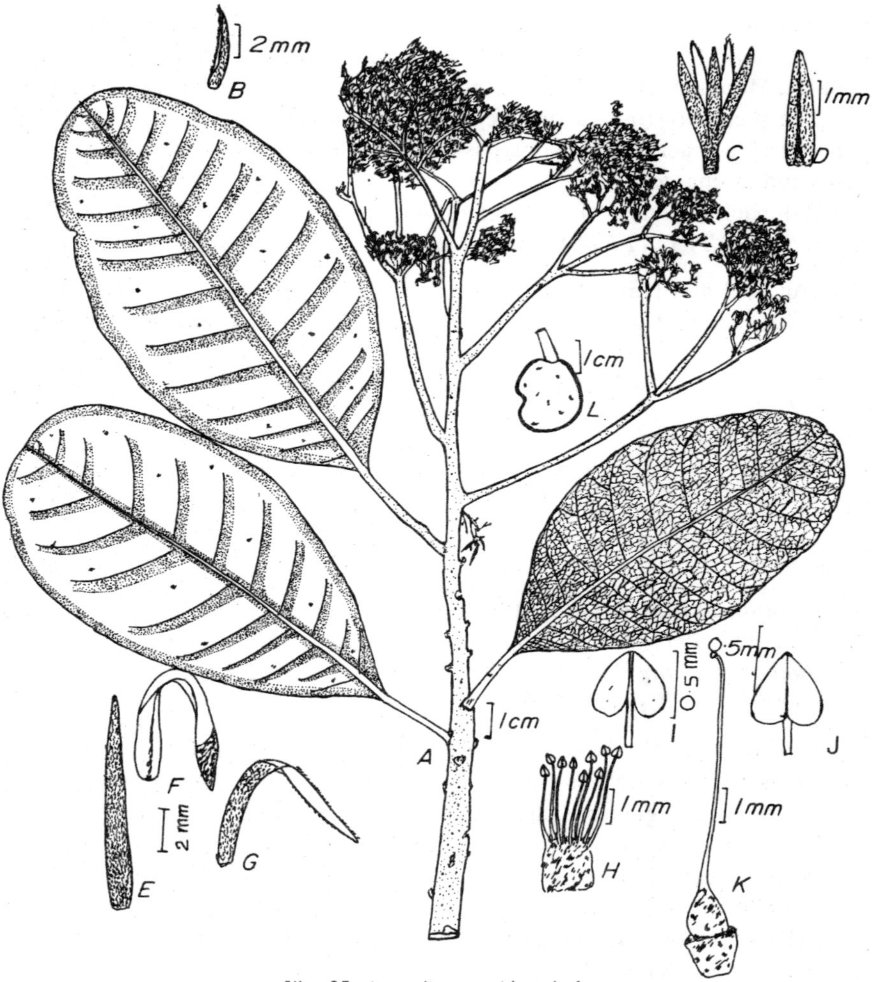


**Fig. 25.** *Anacardium occidentale* L.
A. Twig, B. Bract, C. Calyx, D. Sepal, E, F & G. Petals, H. Androecium, I & J. Stamens, K. Pistil.

thick-coriaceous, obovate, 11–13.5 × 5.5–7.6 cm, glabrous, apex rounded, margin entire, base cuneate; finely reticulately veined, secondary nerves 10–12 pairs; petiole to 2 cm. Flowers greenish with red streaks, 1 cm across, in ca. 15 cm long, axilllary or terminal panicles. Drupe reniform, to 3 cm long, greenish-grey, hypocarp pyriform, yellowish or reddish, fleshy, formed by the enlarged disk; seeds reniform, ascending, 1.5–2 × 1 cm.

Introduced, but naturalised now in the coastal areas, extensively being planted by clearings, scrub jungles and natural forests in the plains of Visakhapatnam and Srikakulam districts. The tree can be grown on any type of soil except heavy clay, alkaline soils and swampy areas. Recommended for bare sandy soils, hard laterite soils and deep red loamy soils. It avoids brakish soils. Fls.: Jan.–Mar. Frs.: Apr.–May. Vern.: Jidi mamidi, Mokka mamidi, Munta mamidi. Eng.: Cashew nut.

The seeds are roasted or eaten raw. The pericarp of the fruit yields on roasting a black acrid oil, hypocarp is fleshy and edible. Wood used for packing cases and as charcoal.

Rampachodavaram (EG), *SSR* 15249; Kurnool (KNL), *RVR* 3802; S.K. University (ATP), *NY* 1249; Gummada (SKLM), *GVS* 62414 (MH).

INDIA : Hotter parts especially near the sea.
WORLD : Exotic from America, widely cultivated in the tropics and Malaysia.

## BUCHANANIA K.P.J. Sprengel

1. Leaves pubescent to villous beneath, broadly oblong-ovate; secondary nerves parallel; petiole stout, to 1.5 cm. Panicles terminal and axillary, pubescent, stout, dense; Drupe oblique .......... ***B. lanzan***
1. Leaves glabrous on both surfaces, narrowly oblong- elliptic; secondary nerves curved; petiole not stout, to 2.7 cm. Panicles axillary, glabrous, not stout, sparsely branched. Drupe not oblique .......... ***B. axillaris***

**Buchanania axillaris** (Desr.) Ramam. in Saldanha & Nicolson, Fl. Hassan Dist. 374. 1976. *Mangifera axillaris* Desr. in Lam. Encycl. 3: 697. 1792. *Buchanania angustifolia* Roxb. Pl. Cor. t. 262. 1820; FBI 2. 23; Gamble 1: 259 (184); Brandis 205.

Tree, 8–10 m tall; bark grey, rough, deeply fissured; wood greyish-brown; branchlets glabrous. Leaves scarcely coriaceous, narrowly oblong-elliptic, 5.5–13.8 × 2.5–4.2 cm, glabrous, apex retuse, margin entire, base acute; secondary nerves 13–15 pairs; petiole to 2.7 cm. Flowers cream, 6 mm across in 10 cm long, slightly branched, axillary panicles. Drupe globose, 1.5 cm in dia., black, glabrous; seed one, gibbose.

Hills on the dry localities of deciduous forests at 500–900 m. Fls.: Apr.–June. Frs.: Dec. Vern.: Sarapappu, Karnel, Pedda Mori, Morlisara, Pedda Moralli.

The kernal of the seeds is edible and used in the preparation of sweets. Wood hard but is of no value.

Tirumala hills (CTR), *SSR* 13733, 13779; Eturnagaram (WGL), *SSR & KSM* 15533; Kalasamudram RF (ATP), *EC* 2918and *EC* 2996; Balugram RF (KNL), *RVR* 2935; Rayachoti range (CDP), *KNS* 2351 (DD); Rollapenta (PKSM), *RVK* 15894; Farahabad (MBNR), *TDCK* 15345.

INDIA : Peninsula.
WORLD : Sri Lanka.

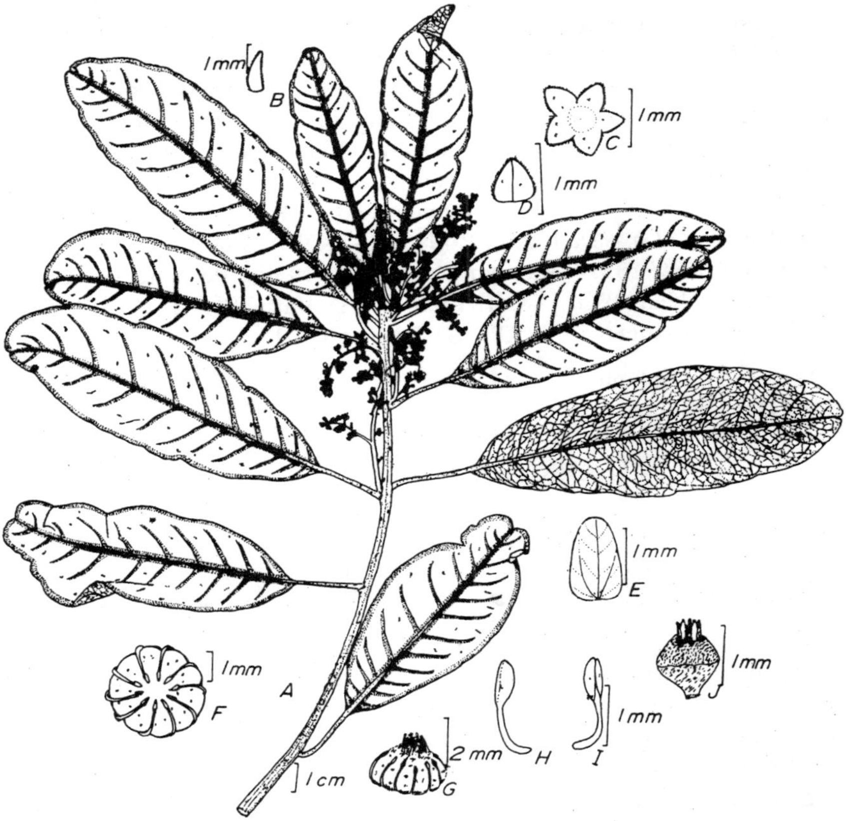


Fig. 26. *Buchanania axillaris* (Desr.) Ramam.
A. Twig, B. Bract, C. Calyx, D. Sepal, E. Petal, F. Stamens with disc, G. Stamens with pistil, H & I. Stamens, J. Pistil.

**Buchanania lanzan** Sprengel in J. Bot. (Schrader) 2: 234. 1800; Gamble 1: 258 (184). *B. latifolia* Roxb. Fl. Ind. 2: 385. 1832; FBI 2: 23; Brandis 205; Beddome Fl. Sylv. S. India t. 165. 1871.

Evergreen tree, up to 10 m tall; trunk straight; bark dark grey, fissured; wood greyish-brown, branchlets clothed with silky hairs. Leaves thick-coriaceous, broadly oblong-ovate, 12–20 × 5.3–9.7 cm, glabrescent above, slightly villous beneath, apex obtuse or retuse, margin entire, base acute-round; secondary nerves 10–18 pairs, parallel, nerves impressed above; petiole stout, to 1.5 cm. Flowers creamish in 15 cm long, axillary and terminal, villous panicles. Drupe globose, oblique, 0.8 × 1.3 cm dia., black, stone hard, 2-valved; seeds 1 or 2, biconvex.

Occasional, on high hills of deciduous forests, mostly in eroded ravine lands at 700–1400 m. It avoids water logged areas, but occurs locally in clayey soils. A good species for afforesting bare hill tops. It has poor coppicing capacity and produces root suckers sparingly. Fls.: Jan.–Mar. Frs.: Apr.–May and remain on the tree for quite a long time. Vern.: Sarapappu, Mori, Morli, Chinna Morali, Jaru mamidi. The fruit is largely eaten and also used in the preparation of sweets. Wood is not much used.

Kakavada (EG), *SSR* 15223; Rollapenta (KNL), *SSR* 15510; Anantagiri (RR), *SSR & KSM* 18514; Rampur (ADB), *SSR* 19306; Pedachema (PKSM), *RVR* 15856; Pakhal (WGL), ANH 15953 (MH); Barmakonda (VSKP), *KCJR* 17175 (MH); Anantagiri (VSKP), *GVS* 19555 (MH).

INDIA : Throughout.
WORLD : Myanmar.

## **LANNEA** A. Richard *nom. cons.*

**Lannea coromandelica** (Houtt.) Merr. in J. Arnold Arbor. 19: 353. 1938. *Dialium coromandelicum* Houtt. Nat. Hist. Ser. 2. 2. 39. t. 2. 1794. *Odina wodier* Roxb. Fl. Ind. 2: 293. 1837; FBI 2: 29; Gamble 1: 263 (187); Brandis 200; Beddome Fl. Sylv. S. India t. 123. 1871.

Deciduous tree, up to 15 m tall; bark grey, smooth in young trees, rough and dark in old trees; wood bright crimson, streaked or flushed pale pink or white; branchlets rusty-tomentose. Leaves imparipinnate, rachis to 17 cm long, petiole 6–10 cm; leaflets 4 pairs, chartaceous, ovate-lanceloate, 2–9 × 2.5–4 cm, glabrous, shining, apex abruptly acuminate, margin entire, base oblique; secondary nerves, 6–8 pairs; petiolule 5 mm. Flowers unisexual, pale yellow, in 20–35 cm long, pendulous, terminal racemes. Drupe oblong-ovoid, 1 × 0.4 cm, red, smooth, compressed, epicarp thin, fleshy and a large stone seated on enlarged calyx; stone irregularly pitted, hard.

Common, in open deciduous forests at 200–1200 m and always found associated with *Boswellia serrata* planted as an avenue tree. Recommended for afforesting deep alluvial soils with good drainage. It is a strong light demander, fire resistant and good coppicer. It is one of the first trees to lose and one of the last to regain its leaves, being leafless November–May. Fls.: Mar.–Apr. Frs.: May–July and persist for long time. Vern.: Gumpena. Oddi, Dumpidi.

Wood fairly good but not much used. It yields gum in considerable quantity.

Vanipenta (KNL), *SSR* 15300; Navabkota (ATP), *SSR* 18565; Kunthala water falls (ADB), *GO & DAM* 5114; Guvvalacheruvu (CDP), *RV Reddy* 8151; Vinukonda RF (GTR), *VRK* 6876; Gangapur RF (MDK), *BR & CP* 11438; Mothe RF (NZB), *BR & GO* 9067; Lothuvagu (RR), *MSM & KH* 1063;

Diguvametta (PKSM), *RVR* 15716; Jalupenta (MBNR), *TDCK* 15311; Tirumala (CTR), *DR* 1303 (SVU).

INDIA : Throughout.
WORLD : Sri Lanka, Thailand, Myanmar, Indo-China, China, in Malaysia.

## MANGIFERA Linnaeus

**Mangifera indica** L. Sp. Pl. 200. 1753; FBI 2: 13; Gamble 1: 259 (185); Brandis 206; Beddome Fl. Sylv. S. India t. 162. 1871.

Large, spreading, evergreen tree, up to 25 m tall; crown dome-shaped, heavy, trunk straight; bark dark grey, rough; wood grey, often streaked; branchlets glabrous. Leaves crowded at the ends of branches, coriaceous, oblong-lanceolate, 5.5–15 × 3–4.7 cm, apex glabrous and shining, emits aromatic odour when crushed, apex acute to shortly or obtusely acuminate or subobtuse, margin entire to undulate, base cuneate; secondary nerves ca. 20 pairs; petiole to 3 cm. Flowers whitish in 25 cm long many-flowered pubescent panicles. Drupe ovoid or obliquely pyriform, 5 to 15 cm long, fleshy, yellowish when ripe, laterally compressed with a protruberance on one side near the tip; stone compressed, longitudinally furrowed, fibrous, very hard; seed one, ovoid, oblique.

In moist forests, wild in the ravines and along rocky hills. It is cultivated throughout. Fls.: Jan.–March. Frs.: Ripens June–July. Vern. Mamidi. Eng. Mango.

Fruits are edible and raw fruits are made into preserves and pickles. Wood moderately hard used for furniture, planking, packing cases, boats, match boxes, window frames, floor and ceiling boards and other purposes. It is not durable when exposed, but is quite durable under water.

Rampachodavaram (EG), *SSR* 15250; Kalasamudram RF (ATP), *NY* 397; Maddileti stream (KNL), *RVR* 3077; V.P. South (GTR), *VRK* 3952.

INDIA : Cultivated; Sikkim, the Nambar forest Assam, the Khasi hills, in ravines on the higher hills of the Satpura range in Khandeisi and along the Western Ghats. Except northern parts of Punjab.
WORLD : Indigenous in Myanmar. Cultivated in Southern China, Malaysia, Indonesia, Warmer parts of Australia, Philippines, Hawai and West Indies, Madagascar and along the coast of tropical Africa. A variety of sweet fruits from different hybrids of the species are usually cultivated.

The wild *Mangifera* species is quite different in its size, morphology and nature of fruits etc. from those of the commonly cultivated varieties. The plant is restricted to sholas, along the streams, dominating the entire tree species. One of the tallest with huge buttressed trunks. Though it is not of great importance as a forest species; the numerous shady groves throughout the state have certainly some climatic influence as they must affect the amount of rainfall.

The mango is the most popular crop and the choisest fruit of India and occupies a prominent place among the best fruits of the world. It is grown as plantation crop and also in orchards, but more often in homeyards, field borders and roadside avenues.

The mango fruit is one of the most highly prized desert fruits of the tropics. It has a rich, luscious, aromatic flavour and a delicious taste. Sucrose, glucose, and fructose are the principal carbohydrates, it is a rich source of vitamin A, and potassium. Ripe mango fruit is considered invigorating refreshing and flattening. The juice along with aromatics, is recommended as a restorative tonic. It contains vitamin A and C is useful in heat apoplexy. The bark yields a colouring matter which produces beautiful, thick and light yellow shades on cotton, silk and wool. The stem exudes a gum resin used as a substitute for gum arabic.

## NOTHOPEGIA Blume *nom. cons.*

1. Branchlets glabrous. Petiole twisted. Racemes paniculate. Drupe with thin pericarp ........ ***N. racemosa***
1. Branchlets pubescent. Petiole slender. Racemes simple. Drupe with thick pericarp ........ ***N. heyneana***

**Nothopegia heyneana** (Hook. f.) Gamble Fl. Pres. Madras 1: 265 (189). *N. colebrookiana* (Wt.) Bl. var. *heyneana* Hook. f. F. Brit. India 2: 40. 1876; Brandis 207.

Evergreen tree, up to 12 m tall; bark smooth, greyish; branchlets rusty pubescent. Leaves thinly coriaceous, variable, narrowly oblong or linear-elliptic, 9–15 × 2–4 cm, glabrous, green and glossy above, slightly glaucous beneath, very young leaves red, apex obtusely acute to shortly acuminate, margin entire, base acute; secondary nerves 15–25 pairs, close, meeting at the margin, secondaries and tertiaries prominent beneath; petiole to 13 mm. Flowers white, 2.5 mm long, polygamous; female flowers sessile or subsessile in rufous pubescent, 1.5–2 cm long racemes; male racemes much branched and slightly longer than female raceme. Drupe red and then dark purple or blue when ripe, globose, to 0.7 × 1.2 cm, torus slightly enlarged, longitudinally striate, large, tipped with persistent style; seed one, transversely oblong, laterally attached by a short thick funicle.

Rare, near streams of valleys of evergreen forests above 1000 m. Fls.: Mar.–Apr. (May). Frs.: May–June. Fruit pulp sweet, edible and contains resins and dyes (gives black colour to the clothes).

Galikonda (VSKP), *SSR* 19627 and 19631; Dharakondagedda (VSKP), *GVS* 47418 (MH).

INDIA : Maharashtra, Orissa and Tamil Nadu.

**Nothopegia racemosa** (Dalz.) Ramam. in Saldanha & Nicholson Fl. Hassan Dist. 327. 1976. *Glycycarpus racemosus* Dalz. in J. Asiat. Soc. Bombay 3: 69. 1849; Hooker's Icon Pl. t. 842. 1852. *Nothopegia dalzellii* Gamble Fl. Pres. Madras 265. 1918 & 1: 189. 1957 (repr. ed.). *N. colebrookeana* sensu Hook. f. Fl. Brit. India 2: 40. 1876 p.p non (Wt.) Bl. 1850. Small tree.

Specimen is not available, but it is reported from Dharakonda of Visakhapatnam by Subba Rao & Kumari (1981).

INDIA : Western Ghats, Eastern Ghats.
WORLD : Sri Lanka.

### **RHUS** Linnaeus

1. Armed tree. Leaves small, margin sinuate-toothed .................................... ***R. mysorensis***
1. Unarmed tree. Leaves large, margin entire or slightly undulate .................... ***R. paniculata***

**Rhus mysorensis** Don, Gen. Hist. 2: 74. 1832; FBI 2: 9. 1876; Gamble 1: 264 (188); Brandis 197; Beddome Fl. Sylv. S. India, Anal. Gen. 78. t. 11. f.3. 1821.

Small tree; aromatic, bark brown, thin, rough; wood reddish-brown, darkening on exposure, interlocked grained; branchlets woolly at apices, spines to 1.5 cm. Leaves 3-foliolate, petiole to 1.7 cm; lateral leaflets obovate, 1.5–2 × 1–1.3 cm; terminal one oblanceolate, 2–4 × 1–2 cm, coriaceus, apex obtuse, apiculate, margin sinuate-toothed, base cuneate; petiolule 0. Flowers yellow, polygamo-dioecious, 2 mm across in 15 cm long terminal panicles. Drupe globose, 5 mm, tinged red.

Dry stony lands and scrub jungles to 600 m. The tree rapidly grows on waste lands and it is also used for fencing. Fls.: Nov.–Jan. Frs.: Dec.–Jan. Vern.: Kadapurugudu, Sitha Sundhare.

Fruits are sourish-bitter in taste and are eaten either cooked or roasted. Leaves are eaten and also used as fodder. Bark gives brown colour to leather. Leaves are used for tanning. Wood hard and heavy used as fuel, suitable for turnery, other utility aritcles and small tool handles.

Palakonda RF (CDP), *DAM* 4692; Cuddapah, *EKK* 80388 (DD); Ketaki RF (ATP), *TP & NY* 792; Kalyani dam (CTR), *DR* 2450 (SVU).

INDIA : N.W. India through to the Peninsula.

**Rhus paniculata** Wall. ex Hook f. Fl. Brit. India. 2: 10. 1876; Gamble 1: 264 (188); Brandis 197.

Small, unarmed tree; wood reddish-brown; branchlets glabrous. Leaves palmately 3-foliolate, petiole 2.5–3.8 cm, leaflets coriaceous-obovate or oblanceolate, apical one 7.5–12.5 long, side ones 2.5–7.5 cm long, glabrous, apex obtuse, margin entire or slightly undulate, base contracted; secondary nerves numerous, spreading, parallel, prominent beneath; petiolule 0. Flowers small, polygamous, shortly pedicellate; panicles elongate, slender, axillary and terminal, branchlets recurved. Drupe orbicular, 5 mm in dia, depressed.

Rare, in dry hilly regions of Rumpa of East Godavari district at 500–1000 m. Fls. & Frs. June–Aug.

Jahmaniri hills (EG), *JSG* 16086.

WORLD : Bhutan.

### **SEMECARPUS** Linnaeus f.

**Semecarpus anacardium** L.f. Suppl. Pl. 182. 1781; FBI 2: 30; Gamble 2: 266 (190); Brandis 207; Beddome Fl. Sylv. S. India t. 166. 1871.

Deciduous tree, up to 13 m tall; trunk short, thick; bark brown, rough, exfoliating in very irregular patches; wood brownish-grey, often with yellow streaks; branchlets sericeous. Leaves coriaceous, oblong, elliptic-obovate, 9–19 × 6–8.6 cm, glabrous above, greyish-white tomentum beneath, apex obtuse, rotund, margin entire, cartilaginous, base obtuse-cuneate; secondary nerves 15–25 pairs, horizontal; petiole to 3 cm. Flowers greenish-white in 18 cm long terminal panicles. Drupe oblong-ovoid to 3.5 cm, compressed, shining black when ripe, hypocarp orange and fleshy and equal to the Drupe.

Occasional, in dry deciduous forests at 600–1100 m. Moderate shade bearer and coppices well. Leaf fall: Feb.–Mar. New foliage: May. Fls.: May–June. Frs.: Dec.–Mar. Vern.: Nalla Jeedi, Simidi Jeeri. Eng.: Marking nut, Oriental cashew.

The fruit juice with the addition of lime water makes an excellent marking ink for clothes, and the hypocarp is edible when ripe; it is slightly astringent. Kernal of the fruit (inside the hard shell) has an almond like taste and is edible, but is likely to cause cutaneous eruptions. The kernal oil is used as a wood preservative. Wood is of little use, contains an irritating juice.The wood is suitable for cheap and light furniture, packing boxes and is fairly good for match-boxes and splits. Bark exudes a brownish gum.

Talakona (CTR), *SSR* 13726; Kakavada (EG), *SSR* 15224; Tirumala hills (CTR), *SSR* 15263, Rampur (ADB), *SSR* 19319; Mulayaguda-Borra caves (VSKP), *TP & EC* 7401; Ramannapenta RF (KNL), *RVR & PVP* 2375.

INDIA : Throughout hotter parts and tropical outer Himalaya.
WORLD : Northern Australia, Malayan Archipelago.

**SPONDIAS** Linnaeus

**Spondias pinnata** (L.f.) Kurz. Perelim. Rep. For. & Veg. Pegu Append. A. 44 & B. 42. 1875. *Mangifera pinnata* L.f. Suppl. Pl. 156. 1781. *Spondias mangifera* Willd. Sp. Pl. 2: 751. 1799; FBI 2: 42; Gamble 1: 261 (186); Brandis 201. Beddome Fl. Sylv. S. India t. 169. 1872. *S. acuminata* Roxb. Fl. Ind. 2: 453. 1832; FBI 2: 41; Gamble 1: 261 (186).

Deciduous tree, up to 15 m tall; bark white, smooth, aromatic; wood grey, soft; branches nearly horizontal; branchlets stout, glabrous. Leaves coriaceous, usually crowded at the ends of branchlets, imparipinnate, 17–25 cm, petiole to 7 cm; leaflets 3–5 pairs, coriaceous, oblong or elliptic-oblong, glabrous, apex caudate-acuminate, margin entire, base rounded, slightly oblique, lateral nerves 20–30 pairs, very close, parallel and forming an intramarginal nerve, petiolule to 5 mm. Flowers small, yellowish white, polygamous in axillary or terminal, 23 cm long panicles. Drupe yellowish, ellipsoid, 3.8 × 2.3 cm; rough with irregular furrows.

Cultivated, it occurs on the hills of Tirumala. Leaf fall: Feb.-Mar. New foliage: Apr., red to green. One of the first trees to shed its leaves and one of the last to renew them, often not until the rains have set. Fls.:Feb.-Mar. Frs.: Ripening June onwards. Vern.: Adavi mamidi, Konda mamidi, Ampilepi chettu.

The leaves having sweet mango smell. Fruit is acidic generally eaten as condiment and made into chutney. Wood is soft and light and of little or no value. Bark used in dysentry. Decoction of leaves applied locally for earache.

Tirumala gardens (CTR), *SSR* 15977; 16188; Cuddapah RF, *AMR* 3830 (VV).

INDIA : Throughout, wild and cultivated.

MORINGACEAE

**MORINGA** Adanson

1. Leaves 3-pinnate; leaflets to 2.5 × 1.5 cm. Flowers white. Capsule 9-ribbed. .................. ***M. pterigosperma***
1. Leaves 2-pinnate; leaflets to 5 × 3 cm. Flowers purplish-white. Capsule sharply 3-angled ... ***M. concanensis***

**Moringa concanensis** Nimmo ex Gibs. in Dalz. & Gibs. Bombay Fl. 311. 1861; FBI 2: 45; Gamble 1: 270 (192); Brandis 210.

Deciduous tree, up to 15 m tall; bark thick, grey, fissured; branchlets warty, tomentose. Leaves bipinnate to tripinnate, very long, to 40 cm, petiole to 8 cm, pulvinate, leaflets 9–13, chartaceous, elliptic-obovate, 2.5–5 × 0.8–3 cm, glaucous beneath, apex emarginate to slightly apiculate, margin entire, base oblique; petiolule to 3 mm. Flowers purplish-white in large divaricate panicles. Capsule elongate, 15–40 cm long, pointed, sharply 3-angled; seeds numerous, with thick membranous hyaline wings at both ends.

Hills in the deciduous forests in full sun at 200–900 m.. Leaf fall: Jan. New foliage: Apr. Fls.: Feb.–Apr. Frs.: Apr. onwards, dehiscing by September. Vern.: Karu Munaga, Konda Munaga, Adavi Munaga. Fruits are not edible. Stem brittle , wood soft.

Panyam (KNL), *EC* 7430; Srisailam (KNL), *BSN* 4776; Bheemram (ADB), *TP & GO* 5455; Anakapalli (VSKP), *JSG* 21792 (DD); Molachintapalli (MBNR), *TDCK* 15315.

INDIA :Peninsula
WORLD :Pakistan.

**Moringa pterigosperma** Gaertn. Fruct. 2: 314. 1791; FBI 2: 45; Beddome Fl. Sylv. S. t. 80. 1870. *Moringa oleifera auct. non* Lam. 1: 398. 1785; Gamble 1: 270 (192); Brandis 209.

Tree, up to 8 m tall; trunk grey-white with longitudinal wrinkles; bark corky, thick, grey, deeply fissured. Leaves tri-pinnate, ca. 40 cm long, petiole to 6.5 cm; leaflets small, ovate, elliptic-obovate, to 2.5 × 1.5 cm, base obtuse; secondary nerves indistinct; petiolule slender, 3 mm. Flowers white, strongly honey scented in 25 cm long panicles. Capsule elongate, to 30 cm long, green, pendulous, 9-ribbed, loculicidally dehiscent; seeds 2–3 cm long, woody, rounded-trigonous, wings broad.

Often found wild in the forests to 1100 m, but very commonly cultivated in gardens and house compounds. It grows on variety of soils, prefers alluvial soils with sufficient moist. It is easily propagated by cuttings. Fls.: Jan.–Apr. Frs.: May–July. Vern.: Munaga, Mulaga, Eng.: Drumstick, Horse radish.

Leaves, flowers, and fruits eaten as vegetable. The roots have the flavour of horse radish and the seeds are eaten in curries and yields a valuable oil. Wood is used as floater.

Maddileti stream (KNL), *RVR* 3061; Reddipalli farm (ATP), *NY* 260; Ethipothala (GTR), *VRK* 3901.

INDIA : Cultivated throughout. Perhaps indigenous in the sub-Himalayan tracts.
WORLD : Widely cultivated in tropical countries.

## FABACEAE

### KEY 1 (Vegetative)

1. Leaves palmately compound, trifoliolate:
    2. Leaflet margin shallowly crenate, base not ribbed ........ ***Ougeinia***
    2. Leaflet margin entire, sinuate-lobed, base 3–5 ribbed:
        3. Leaflet apex obtuse or retuse ........ ***Butea***

3. Leaflet apex obtuse, acute, sub-acute or acuminate .................................. ***Erythrina***

1. Leaves pinnately compound, 3-many-foliolate.:
   4. Leaflets alternate:
      5. Leaflets large ........................................................................ ***Pterocarpus***
      5. Leaflets small ........................................................................ ***Dalbergia***
   4. Leaflets opposite or sub-opposite:
      6. Leaflets 5–9:
         7. Leaflets up to 12 cm long, apex acuminate ........................................ ***Pongamia***
         7. Leaflets up to 3 cm long, apex acute ................................................ ***Millettia***
      6. Leaflets > 10:
         8. Leaf rachis to 30 cm long; leaflets oblong, apex emarginate................ ***Sesbania***
         8. Leaf rachis to 20 cm long; leaflets elliptic or elliptic-lanceolate, apex acuminate:
            9. Bark deeply fissured, spongy. Leaves sericeous .............................. ***Mundulea***
            9. Bark slightly fissured, not spongy. Leaves not sericeous ............. ***Gliricidia***

## KEY 2 (Flowering)

1. Flowers large > 2.5 cm long:
   2. Flowers whitish or cream coloured ................................................... ***Sesbania***
   2. Flowers red or scarlet:
      3. Petals unequal ........................................................................ ***Erythrina***
      3. Petals equal ........................................................................ ***Butea***
1. Flowers small < 2.5 cm long:
   4. Flowers yellow ........................................................................ ***Pterocarpus***
   4. Flowers white, lilac, purplish or pinkish:
      5. Flowers in fascicles ........................................................................ ***Ougeinia***
      5. Flowers in racemes or panicles:
         6. Flowers to 1 cm across:
            7. Calyx lobes minute ........................................................ ***Pongamia***
            7. Calyx lobes long, predominant:
               8. Flowers in panicles.................................................... ***Dalbergia***
               8. Flowers in racemes.................................................... ***Millettia***
         6. Flowers 1.5–2.5 cm across:
            9. Racemes to 20 cm long ........................................................ ***Gliricidia***
            9. Racemes to 10 cm long ........................................................ ***Mundulea***

## KEY 3 (Fruiting)

1. Pod orbicular, compressed with broad wing ........................................ ***Pterocarpus***
1. Pod elongate or obliquely oblong:
   2. Pod turgid ........................................................................ ***Erythrina***
   2. Pod flattened, compressed:
      3. Pod 25–30 × 0.5 cm, tetragonous ........................................................ ***Sesbania***
      3. Pod less than 20 cm long:
         4. Pod oblong follicle, base flat and wing like, apex 1-seeded, rarely 2 ........ ***Butea***
         4. Pod other than above:
            5. Pod slightly constricted between the seeds ...................................... ***Ougeinia***
            5. Pod not constricted:
               6. Pod woody, obliquely oblong; seed large ................................ ***Pongamia***
               6. Pod not woody, linearly oblong; seeds small:
                  7. Pod greyish-brown or blackish, samaroid like, 1–4 seeded ... ***Dalbergia***
                  7. Pod greenish or yellowish, not samaroid, 5–12 seeded:

8. Pod densely velvety with golden-brown hairs, sutures thick ............ ***Mundulea***
8. Pod glabrous, sutures not thickened .................................. ***Gliricidia***

## BUTEA Roxburgh ex Willdenow *nom. cons.*

**Butea monosperma** (Lam.) Taub. in Engler & Prantl. Pflanzenf. 2(3): 366. 1894. *Erythrina monosperma* Lam. Encyl. 1: 391. 1785. *Butea frondosa* Koen. ex Roxb. in Asiat. Res. 3: 369. 1792; FBI 2: 194; Gamble 1: 357 (252); Beddome Fl. Sylv. S. India t. 176. 1872.

Deciduous tree, up to 12 m tall; trunk crooked and irregular; bark grey or dark-brown, rough; wood grey; branches nodose; branchlets tawny tomentose. Leaves tri-foliolate, petiole to 14 cm; leaflets coriaceous, laterals rhomboid, obliquely rounded, 12.5 × 12 cm; terminal leaflets broadly ovate, to 16.5 × 15 cm, apex obtuse or retuse, margin undulate, base cuneate in terminals, obliquely rounded in lateral leaflets, 3–5 ribbed; petiolule to 7 mm, stout. Flowers, 3.5–5 cm long, showy, scarlet with dark brown velvety calyces in 30–35 cm long rigid racemes. Pod velvety brown, to 18 × 0.5 cm; base wing like, indehiscent without seeds, apical part dehiscent with seeds; sutures thickened; seeds 1 or 2, 3.3 cm long, oval, compressed.

It is very common in Telangana region, especially in Adilabad forests, occasionally in Coastal and Rayalaseema regions of mixed dry deciduous forests, being found in open grass lands and scrub forests. Recommended for afforestation in water logged areas, tank bund, black cotton soil, saline-alkaline and badly drained ground. It is a light demander, drought hardy, good coppicer and develops root suckers freely and survives in degraded forests. Leaf fall: Jan. New foliage Mar. Fls.: Feb.–Apr. Frs.: May. Vern.: Moduga chettu; Eng.: Flame of the forest, Bastard Teak.

It is a good host for lac insect and thus produces huge quantity of lac every year. The dye obtained from the lac and flowers is used as colouring agent. The leaves serve the purpose of cups and plates and also as fodder. Fibre from the root suckers are used in making ropes. Wood soft and durable in water is used to some extent as fire wood. It has little value as timber. Seeds are oily and protinaceous, thus susceptible to the attack by a large number of fungi during storage. Wood yields a red 'Kino' gum.

Manchippa (NZB), *SSR & KSM* 19334, Rollapenta (KNL), *KSM & SSR* 14296; Gundumala RF (ATP), *NY* 1222; Laxmapur RF (NZB), *TP & BR* 6254; Gangapur RF (MDK), *CP* 11429; Tummalagunta (CTR), *YYR* 2716; Peddachema (PKSM), *RVK* 13847.

INDIA : Throughout except Jammu & Kashmir and Arunachal Pradesh.
WORLD : Pakistan, Sri Lanka, Nepal and Myanmar.

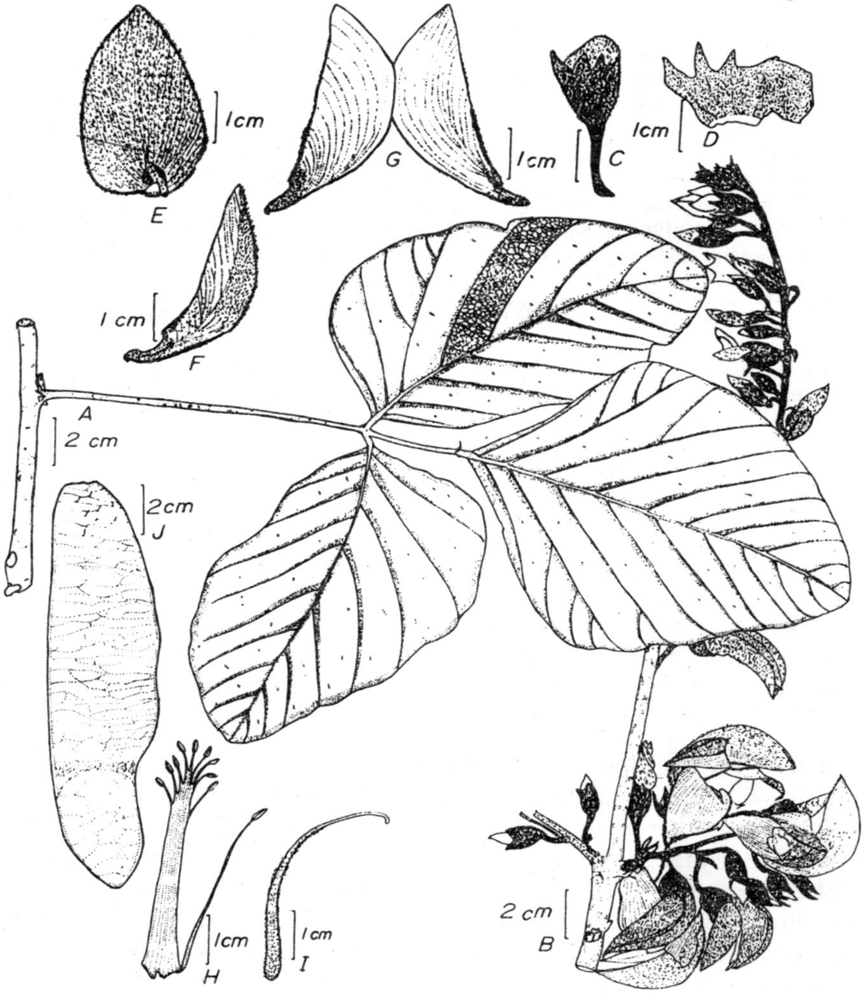


**Fig. 27.** *Butea monosperma* (Lam.) Taub.
A. Vegetative twig, B. Flowering twig, C. Calyx, D. Calyx split open, E. Standard petal, F. Wing petal, G. Keel petals, H. Staminal column split open, I. Pistil, J. Pod.

## **DALBERGIA** Linnaeus f. *nom. cons.*

### KEY 1 (Vegetative)

1. Plants armed with thorns; branches spreading, drooping or straggling ....... ***D. melanoxylon***
1. Plants unarmed; branches not drooping or straggling:
    2. Leaflets 3–9; 5–10 × 3–7.5 cm:
        3. Leaflets 7–9, terminal leaflets obovate, apex acute; secondary nerves not distinct. ........................................................................................................ ***D. sissoides***

3. Leaflets 3–7, terminal leaflets elliptic, apex other than acute; secondary nerves distinct:
    4. Leaflet apex cuspidate-acuminate, broadly elliptic or ovate; leaf rachis zig-zag ... ..... *D. sissoo*
    4. Leaflet apex retuse or obtuse, orbicular-elliptic; leaf rachis straight ...*D. latifolia*
2. Leaflets 9-above, 1.8–4.8 × 1–2.8 cm:
    5. Leaflets oblong; secondary nerves irregular, not parallel, curved towards margin; petiolule thick, to 3 mm; black when dry ..... *D. paniculata*
    5. Leaflets elliptic-oblong; secondary nerves regular, parallel, straight and curved towards apex; petiolule slender, to 5 mm; greyish brown when dry ..... *D. lanceolaria*

## KEY 2 (Flowering)

1. Stamens 9 or 10:
    2. Flowers yellowish-white; style short, thick ..... *D. sissoo*
    2. Flowers white; style elongate, curved:
        3. Flowers in subterminal, congested, corymbose panicles ..... *D. latifolia*
        3. Flowers in loose, terminal, single panicles ..... *D. sissoides*
1. Stamens 5+5:
    4. Flowers bluish or purplish in puberulous lax panicles ..... *D. lanceolaria*
    4. Flowers whitish in tomentose, congested panicles:
        5. Calyx rusty tomentose to 7 mm long, standard petal 1 cm long, ovary glabrous ..... ..... *D. paniculata*
        5. Calyx yellowish tomentose to 4 mm long, standard petal 0.6 cm long; ovary puberulous ..... *D. melanoxylon*

**Dalbergia lanceolaria** L.f. Suppl. Pl. 316. 1781; FBI 2: 235; Gamble 1: 383 (270); Brandis 236. *D. frondosa* Roxb. Fl. Ind. 3: 226. 1332.

Large, deciduous, densely foliaceous tree, up to 25 m tall; bark smooth, grey; wood white or yellowish white, straight-grained and medium coarse textured; branchlets glabrous. Leaves imparipinnate, rachis to 21 cm long, petiole to 3 cm long; leaflets 11–17, chartaceous, elliptic-oblong, 1.8–4.8 × 1–2.8 cm, sericeous when young, glabrous later, apex obtuse, margin entire, base cuneate-obtuse; secondary nerves regular, parallel, straight and curved towards apex; petiolule to 5 mm. Flowers pale pink in short unilateral racemes (silky) in terminal and axillary large panicles. Pod purplish-brown, lanceolate, 9 × 2 cm, apex obtuse, base acute; stipe to 1 cm, seeds 1–3.

Common, in deciduous slopes, 600–1400 m. Fls.: July. Frs.: Sept.–Jan. Vern. Patchari, Yerra patsari, Nagalla patchari, Pasarganni.

The decoction of the bark is used in dyspepsia. The seed oil is applied in rheumatism. Wood moderately hard and of no great value, used for tool handles and agricultural implements.

Tirumala (CTR), *SSR* 13705, 15280; Donubai (SKLM), *SSR & KSM* 15590; Anantagiri (RR), *SSR & KSM* 18513; Eturnagaram (WGL), *SSR* 18551; Sidduliah kona (NLR), *SSR* 19366; Palakonda hills (CDP), *CS* 9352; Zaheerabad (MDK), *CP* 11588; Kundram RF (ADB), *TP & GO* 5457; Racherla (KNL), *RVR* 1368; Kalasamudram (ATP), *EC* 2971; Macherla RF (GTR), *VRK* 3881.

INDIA : Throughout from tropical Himalaya Southwards.
WORLD : Sri Lanka, Myanmar.

**Dalbergia latifolia** Roxb. Pl. Cor. t. 113. 1799 & Fl. Ind. 3:221. 1832; FBI 2: 231; Gamble 1: 383 (270); Brandis 236; Beddome Fl. Sylv. S. India 24. 1869.

Large deciduous tree, nearly evergreen, up to 20 m tall; crown rounded; trunk straight, cylindrical; bark light grey, rough; wood dark purple with black streaks; branchlets glabrous. Leaves imparipinnate to 20 cm long; petiole to 6 cm; leaflets 5–7, alternate, coriaceous, orbicular-elliptic, 5–10 × 3.2–7.5 cm, glabrous above, puberulous beneath, apex retuse or obtuse, margin entire, base subacute-obtuse; petiolule to 8 mm. Flowers whitish in short, axillary, much branched panicles on old wood. Pod oblong-lanceolate, abruptly narrowed to a stipe, flat, 4–8 × 1.5–2 cm, glabrous; seeds 1–3.

Occasional, in mixed deciduous forest at 600–1200 m. It prefers good, deep loam or clayey soil containing lime. It is a drought resistant, good coppicer and produces root suckers. Leaf fall: Dec.–Jan. New foliage: Mar.–Apr. Fls.: Jan.–Mar., rarely Aug.–Sept. Frs.: Jan.–Feb. Vern.: Jittegi, Virugudi-cheva, Iridi, Irugudi. Eng: Black wood, Rose wood.

Leaves are used as fodder. Bark contains tannins. Oil from the seed is used as antidote for poison. Wood extremely hard and valuable, especially useful for calico-printing blocks, Mathematical instruments, and screws and also used for gun carriage wheels, knives, felloes, handles, shelves, decorative carriage parts, temple cars, boat knees, well construction, agricultural implements and combs.

Tirumala (CTR), *SSR* 13778; Penchalakona (NLR), SSR 19358; Tummala bailu (PKSM), *RVR* 15880.

INDIA : Andaman islands, Peninsula, Bihar, Gujarat, Rajasthan, Sikkim, Uttar Pradesh and West Bengal.
WORLD : Nepal and Malaysia.

**Dalbergia melanoxylon** Guill. et Perr. Fl. Seneg. Tent. 227. t. 53. 1832; Brandis 235; *D. stocksii* Benth., in J. Linn. Soc. 4: Suppl. 42. 1860; FBI 2: 234.

Tree, up to 12 m tall; bark white, fairly rough; wood dark purple to brownish-black, close and fine grained; branches spreading, drooping or straggling. Leaves imparipinnate, rachis to 8 cm long with scattered pubescence; petiole to 1.5 cm; leaflets 7–15, sub-coriaceous, 1.4–7.6 cm, puberulous, glaucous, apex obtuse-shallowly emarginate; margin entire, base obtuse, rounded or cuneate; secondary nerves 4–5 pairs, irregular, inconspicuous; petiolule to 1.5 mm. Flowers whitish, numerous and sub-secundly arranged in axillary or terminal congested panicles. Pod greyish. broadly elliptic-lanceolate, 2–4 cm long; 1–2 seeded.

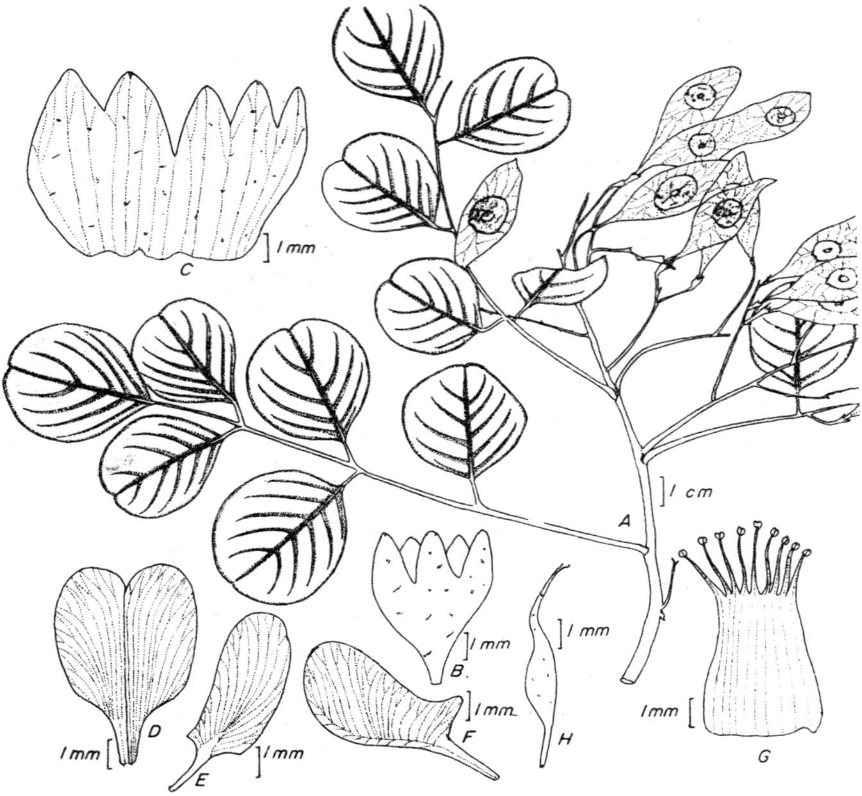


**Fig. 28.** *Dalbergia latifolia* Roxb.
A. Twig, B. Calyx, C. Calyx split open, D, E & F. Standard, Wing, & Keel petals, G. Staminal column spread out, H. Pistil.

Introduced. It is planted by forest department in Telangana region. Fls.: June–Aug.

It is commonly known as African black wood, Senegal or Sudan ebony, Chinese black wood etc. It gives much valuable hard timber and mainly used for making costly surgical tool handles, walking sticks, paper cutters, combs, hair pins, fancy articles and musical instruments etc. and it takes beautiful polish.

Narsapur (MDK), *SSR, KSM & CP* 16196.

INDIA : Naturalised in West India — often cultivated in Bombay, Madras and Bengal.

WORLD : Native of tropical Africa.

**Dalbergia paniculata** Roxb. Pl. Cor. t. 114. 1799 & Fl. Ind. 3:221. 1832; FBI 2: 236; Gamble 1: 383 (270); Brandis 236; Beddome Fl. Sylv. S. India Anal. Gen. 88. 1871.

Large, densely foliaceous, deciduous tree, up to 15 m tall; trunk irregularly fluted; bark smooth, grey; wood greyish-white, rather soft, no heartwood, its structure most remarked, entirely different from that of other arborescent Dalbergias. Leaves imparipinnate, rachis to 10 cm, petiole to 2 cm; leaflets 9–13, turn black when drying, alternate, chartaceous, oblong, 1.4–3 × 1–2 cm, apex obtuse, margin entire, base cuneate-obtuse; petiolule 3 mm. Flowers whitish in terminal or axillary panicles. Pod green, shortly stipitate, lanceolate, apex acute, faintly nerved; seeds 1 or 2.

Hills, on the dry deciduous slopes to 800 m. Occasionally in the foot scrub jungles. Leaf fall: Feb. New foliage: Apr. Fls.: Apr.–June; Frs.: Nov.–Jan. Vern.: Sopera, Porla-patchari, Patsaru.

The wood is not very durable, it is reported to be used for building purposes. The wood is also employed in musical instruments.

Tirumala (CTR), *SSR* 13713; Rampur (ADB), *SSR & KSM* 19810; Vantlamamidi forest (VSKP), *KSM & SSR* 16016; Bhimnagar RF (NZB), *TP & BR* 6209; Narsapur (MDK), *TP & CP* 11968; Mannanur (MBNR), *TDCK* 16803.

INDIA : Peninsula, Gujarat, Rajasthan, Uttar Pradesh.
WORLD : Myanmar.

**Dalbergia sissoides** Graham ex Wt. & Arn. Prodr. Fl. Ind. Orient. 265. 1834; Gamble 1: 381 (270); Brandis 233. *D. latifolia* var. *sissoides* (Graham ex Wt. & Arn.) Baker in Hook. f. Fl. Brit. India 2: 257. 1876. *D. latifolia* sensu Wt. Ic. t. 1153. 1846 non Roxb. 1798.

Deciduous tree, up to 10 m tall; crown less compact. Leaves imparipinnate, rachis 15–20 cm long, petiole to 3 cm; leaflets 7–9, lanceolate to elliptic-oblong, terminal one slightly glaucous, apex acute, margin entire, base cuneate; secondary nerves 8 pairs, parallel, not distinct; petiolule to 3 mm. Flowers white in lax panicles. Pod stipitate, acute at apex, 9 × 1.5 cm, reticulately veined; seeds 1–3.

Rare, in dry deciduous forests of Medak district. Fls.: Mar.–May. Frs.: July–Aug. Narsapur (MDK), *KMS* 7961 (CAL).

**Dalbergia sissoo** Roxb. Fl. Ind. 3: 223. 1832; FBI 2: 231; Gamble 1:383 (270); Brandis 233; Beddome Fl. Sylv. S. India t. 25. 1869.

Densely foliaceous, semi-evergreen or deciduous tree, up to 15 m tall; bark thick, grey, exfoliating in narrow longitudinal strips. Leaves imparipinnate, rachis to 8 cm long, petiole to 3 cm long; leaflets 3–5, alternate, broadly elliptic or ovate, pubescent when young, glabrous when full

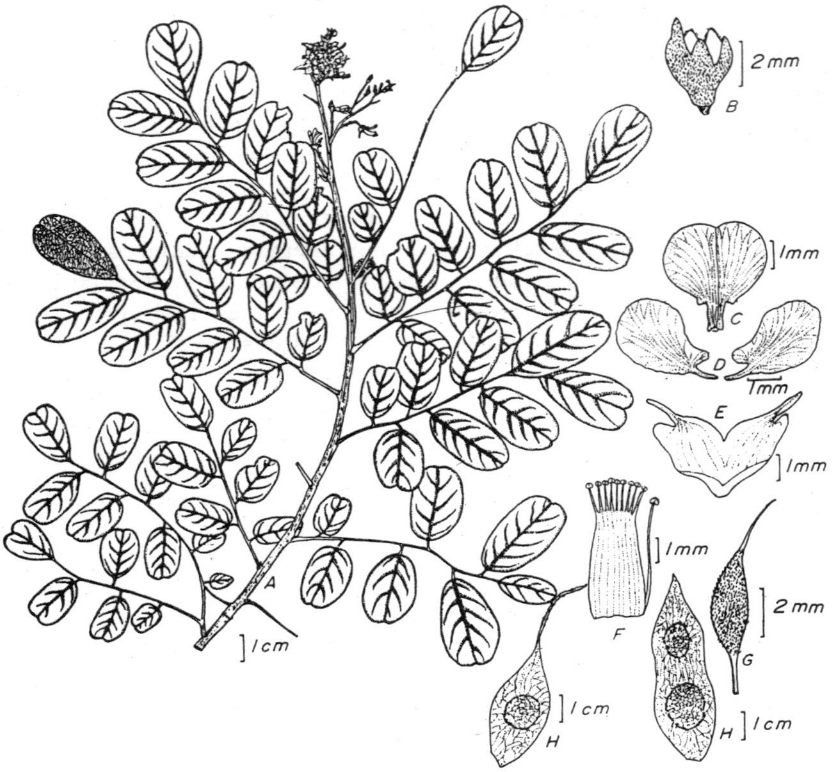


**Fig. 29.** *Dalbergia paniculata* Roxb.
A. Twig, B. Calyx, C. Standard, D. wings, E. Keels, F. Staminal column split open, G. Pistil, H pod.

grown, apex cuspidate-acuminate, margin entire, base narrowed or cuneate; petiolule to 5 mm. Flowers yellowish-white in short axillary panicles. Pod strap shaped, 5–7.5 cm long, pale brown, glabrous, indehiscent, base cuneate, apex obtuse; seeds 1–4, kidney shaped, thin, flat, light brown.

Plains, on dry slopes. It is frequently planted on road sides. Recommended for plantations on sandy and gravelly alluvium soils on beds of river (partial). It is fairly drought resistant and frost hardy. Strong light demander. It coppices vigorously and reproduces vegetatively by root suckers. Leaf fall: Feb. New foliage: Apr. Fls.: Apr. Frs.: Mar. onwards. Vern.: Sissoo.

Wood is valuable and used for furniture and ordinary purposes. Leaf twigs are used as cattle fodder. The leaves are bitter and stimulant; decoction of the leaves is said to be useful in gonorrhea.

SKU Campus (ATP), *SSR* 13798; Mahanandi (KNL), *KSM & SSR 16002;* Kalasamudram RF (ATP), *EC* 2899; Mombojipally (MDK), *BR* 11648; Palakonda hills (CDP), *CS* 9313; Hajipur nursery (NZB), *TP & BR* 6237; Asifabad (ADB), GO 4478; Nallapadu (GTR), *VRK* 6760.

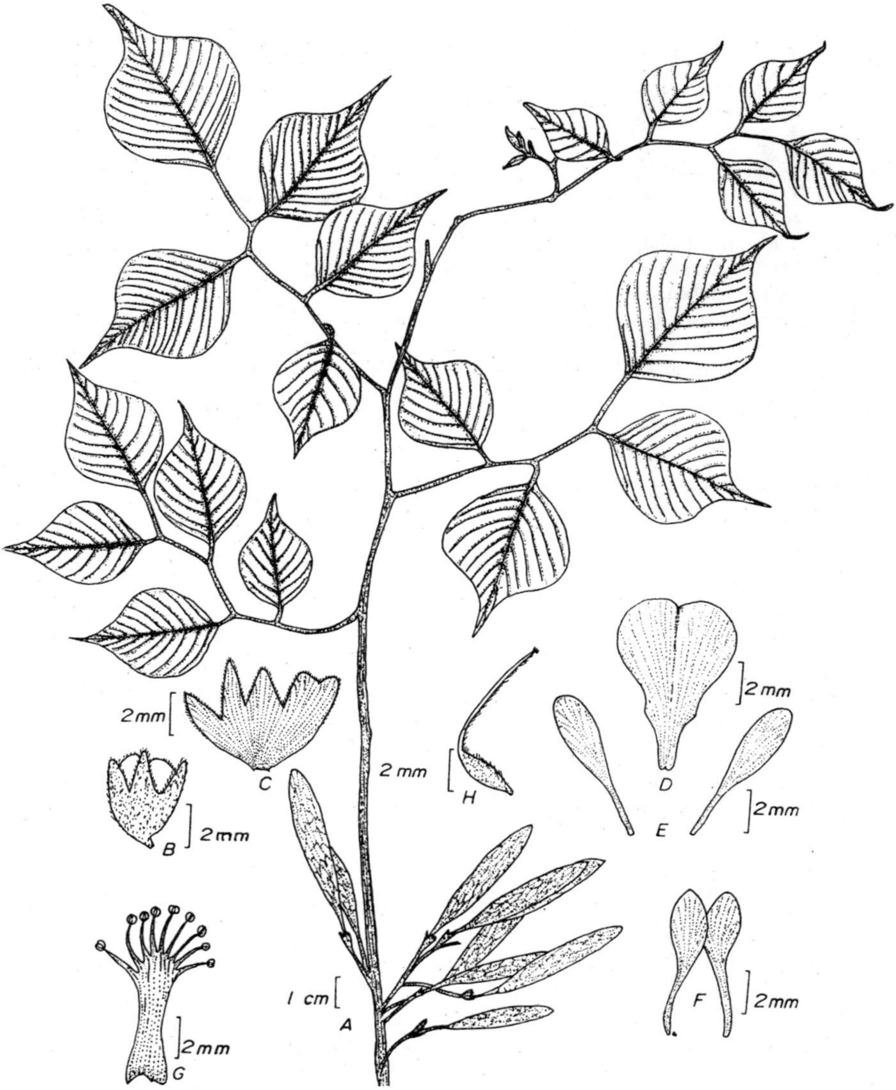


**Fig. 30.** *Dalbergia sissoo* Roxb.
A. Twig, B. Calyx, C. Calyx split open, D, E, & F. Standard, wing and keel petals, G. Staminal column spread out, H. Pistil.

INDIA : Throughout plains.

WORLD : Bangladesh, Bhutan, Myanmar, Afghanistan, Pakistan, Iran, Iraq, Sikkim.

**Notes** : Thothathri (1987) has discussed on the nativity of *D. sissoo* and has concluded that it is indigenous to the gravelly river beds of the sub-Himalayan tract extending from Assam to Punjab, and this agrees with the conclusion of Prain (1904) also.

## ERYTHRINA Linnaeus

1. Calyx spathaceous:
   2. Trees unarmed, leaflets apex acuminate; wing and keel petals equal, one fourth the length of standard. ........ *E. variegata*
   2. Trees armed, leaflets apex sub-acute, wing much smaller than keel petals, keels half the length of standard. ........ *E. stricta*
1. Calyx not spathaceous:
   3. Calyx irregularly splitting into 2 or more lips; standard petal broadly deltoid ... *E. fusca*
   3. Calyx regularly splitting into 2 lips; standard petal oblong or oblong-lanceolate:
      4. Leaflets sub-coriaceous, densely tomentose beneath, apex acute, margin sinuate-lobed. ........ *E. suberosa*
      4. Leaflets membranous, glabrous on both surfaces, apex acuminate, margin entire. ... ........ *E. subumbrans*

**Erythrina fusca** Lour. Fl. Cochinch. 427. 1790. *E. ovalifolia* Roxb. Fl. Ind. 3: 254. 1832; FBI 2: 189; Brandis 227.

Tree, up to 10 m tall; bark pale brown, thick, splitting longitudinally; wood soft, prickles 1–10 mm long, straight, stout, pale to black, on pyramidal corky tubercles; branchlets greenish-grey, tomentose. Leaves pinnately 3-foliolate; rachis 3–7 cm long, prickly, petiole 5 cm long; leaflets coriaceous, elliptic-oblong, or ovate, 4.5–8 × 3–5 cm, glabrescent, pale beneath, apex obtuse, margin entire, base rounded; secondary nerves 6–8, finely reticulate; petiolule 5 mm; stipels gland like. Flowers crimson, to 5 cm long, in terminal, 12–25 cm long, lax racemes. Pod torulose on one side, 7–20 cm long, tomentose, base attenuate, long apiculate; seeds 2–11, ca. 13 × 6 mm, oblong-ovoid, black.

Rare, along streams in Srikakulam district (Rama Rao *et al*, 1987). Fl.: Feb.–Mar. Fr.: Mar.–May. Vern.: Thetta baditha, Seed paste mildly heated and applied for bone fractures by Jatapus.

Donubai and Nettiparthi (SKLM), *NRR* 79476 (MH).

INDIA : Assam, Orissa, Bihar, West Bengal.
WORLD : Bangladesh, Myanmar, Malaysia, Java, Philippines and Sri Lanka.

**Erythrina stricta** Roxb. Fl. Ind. 3: 251. 1832; FBI 2: 189; Gamble 1: 354 (250); Brandis 227; Beddome Fl. Sylv. S. India t. 175. 1872.

Deciduous tree, 15–20 m tall; bark greenish, corky, exfoliating in thin papery flakes; wood white; branchlets pubescent, armed with sharp conical prickles. Leaves 3-foliolate, rachis 4–7 cm long, petiole to 17 cm, very sparsely prickled, leaflets thin coriaceous, terminal ones ovate-rhomboid, lateral ones strongly rhomboid, 9.5–16 × 8.5–15.5 cm, glabrous, apex subacute, margin entire, base truncate or subcordate; secondary nerves 5–8 pairs, petiolule to 7 mm. Flowers bright scarlet, 3 cm long, parallel, 2–3 in fascicles on rachis; racemes horizontal, brown pubescent, terminal; peduncle woody, to 15 cm long. Pod brownish, slightly torulose, ca.15 cm long; stipe slender; seed 3–4, oblong, 1–1.2 cm long.

Occasional, in deciduous forests, up to 1000 m. Leaf fall: Dec. New foliage: Feb. Fls.: Jan.–Feb. Frs.: Mar.–May.. Vern: Mullu Moduga.

It is used as fence plant. The wood is soft and light, but tough fairly durable. The bark in powdered form is used in biliousness, fever, rheumatism, itch, asthma, leprosy, epilepsy. The flowers are said to be given as an antidote to poison.

Bodhan (NZB), *BR* 7270; Manjeera barriage (MDK), *BR & CP* 14095; V.P. South (GTR), *VRK* 3936.

INDIA : Western Peninsula, Arunachal Pradesh, Assam, Mizoram, Orissa, Rajasthan, Sikkim.

WORLD : Nepal, Butan, Bangladesh, East Tibet, Myanmar, Vietnam, Laos, China, Thailand.

**Erythrina suberosa** Roxb. Fl. Ind. 3: 253. 1832; FBI 2: 189; Gamble 1: 354 (250); Brandis 227; Beddome Fl. Sylv. S. India Anal. Gen. 87. 1871.

Deciduous tree, up to 20 m tall; trunk straight and cylindrical; bark corky, thick, deeply cracked; wood white, straight grained and coarse textured; branchlets tomentose, armed with stout conical, brown prickles. Leaves trifoliolate; rachis to 5.5 cm; petiole to 11 cm; leaflets thin-coriaceous, rhomboid-ovate, 10.5–13.5 × 9–15 cm, glabrous and green above, densely tomentose and paler beneath, apex obtuse or acute, margin sinuate-lobed, base truncate; secondary nerves 5–6 pairs; petiolule to 3 mm, pubescent. Flowers scarlet, to 3.8 cm long, densely capitate in 5–10 cm long racemes. Pod stipitate, subterete, torulose, 12.5–15 × 1 cm; seeds 2–5.

Rare, in dry forests, up to 1000 m. It coppices well. Leaf fall: Feb.–Mar. New foliage: July. Fl.: Mar.–May. Fr.: Apr.–June Vern.: Mullu Moduga, Barijama, Barjapa.

Leaves are used as cattle fodder and re-applied externally for relieving pain in joints. The bark yields a pale yellow fibre, suitable for cordage. Seeds may be eaten after boiling and roasting, it contains a fixed oil. The wood is used for rafters, canoes and catamarans.

Madharam beat RF (ADB), *TP & GO* 5449; Mannanur (MBNR), *S.T. Ramachandrachari* 490 (MH).

INDIA : Peninsula, tropical Himalayas from Jammu & Kashmir to Sikkim, Bihar, Delhi and Punjab.

WORLD : Nepal, Bhutan, Myanmar, Indochina.

**Erythrina subumbrans** (Hassk.) Merr. in Phillipp. J.Sci. 5:113.1910. *Hyphaphorus subumbrans* Hassk. Hort. Bogor. Desc. 198. 1858. *Erythrina lithosperma* Bl. ex Miq. Fl. Ind. Bat. 1(2): 209. 1859; FBI 2:190; Brandis 227.

Tall tree; bark white; trunk and branches armed with straight sharp prickles; foliage dark green; young branches puberulous. leaves 3-foliolate, leaflets membranous, ovate, 15×11 cm, glabrous, apex acuminate; margin entire; stipellary glands large, oblong. Flowers red in racemes. Pod recurved, 10–12.5 cm long; basal portion 7.5 cm long, flat, seedless, indehiscent; upper portion shorter and narrower, dehiscent with seeds; seeds 1–4, large, dark purple.

Introduced. Fls. & Frs. Aug.–Mar.

Minimuluru(VSKP), *GVS* 29599(MH); Rajampet (CDP), *KCJB* 8322 (MH); ICRISAT (MDK); *LJGM* 2699 (CAL).

INDIA : Western Ghats, Rajasthan.
WORLD : Malaysia.

**Erythrina variegata** L.Herb. Amboin. 10. 1754; var *orientalis* Merr. Interpr. Herb. Amboin. 276. 1917. *E. indica* Lam. Encycl 2: 391. 1786; FBI 2:188; Gamble 1:353(249); Brandis 220; Beddome Fl. Sylv. S. India, Anal. Gen 87. 1871.

Deciduous tree, up to 13 m tall; unarmed; bark yellowish or greenish grey, smooth; wood white; young branches pubescent, prickled, which fall when a few years old. Leaves 3-foliolate, rachis 20–30 cm; petiole 15–20 cm; leaflets thin-coriaceous, ovate-rhomboid, 5–15 × 5.5–17 cm, glabrous, apex acuminate, margin entire, base truncate, obtuse; secondary nerves 5–6 pairs, petiolule to 8 mm. Flowers scarlet, 2.5 cm across in 5–15 cm long, dense racemes. Pod torulose, 20–35 × 2.5–3 cm, deep brown, stipitate; seeds 7–12, 1.5–1.8 cm, dark purple.

Common, often planted as a wind-break, and also wild. Fls.:Mar. Fr. Apr. –July, Vern., Badisa, Bedita, Baridamu. Eng. : Indian Coral tree. It is valuable as shade tree.

Near Vantlamamidi (VSKP), *KSM* & *SSR* 16015. Mokarlabad (RR), *MSM* & *KH* 10261.

INDIA : Throughout, often cultivated.
WORLD : Sri Lanka, Myanmar, Polynesia, Malaysia.

## **GLIRICIDIA** Humboldt, Bonpland et Kunth

**Gliricidia sepium** (Jacq.) Kunth ex Walp. Rep. 1:679.1842. *Robinia sepium* Jacq. Enum. Pl. Carib. 28. 1760. *Gliricidia maculata* (Steud.) Kunth, Nov. Gen. Sp. 6:393.1824

Deciduous tree, up to 10 m tall; trunk short; bark greyish; branchlets lenticellate, pubescent. Leaves imparipinnate, rachis to 17 cm, petiole to 4 cm; leaflets 6–7 pairs, thin-coriaceous, lanceolate-elliptic, 2.5–7 × 1.6–2.5 cm, glabrous, green above, greyish beneath after drying, apex acute to slightly acuminate, margin entire, base oblique and obtuse; secondary nerves 8–10 pairs; petiolule to 5 mm. Flowers pinkish-white, 2 cm across in axillary and terminal, 20 cm long racemes. Pod linear-oblong, yellowish, to 15 cm long, stipitate; seeds 5–7.

Introduced. It is quick growing tree, planted in the gardens, roadsides as an avenue plant. Fls: Feb.–Apr. Fr. Apr.-May. Vern.; Seema kanuga Eng: The spotted Gliricidia. It is cultivated in hedges for green manure.

Macha reddy (NZB), *BR & KH* 9052; Owk (KNL), *RVR* 3056; SKU. Campus (ATP), *NY* 998; V.P. South (GTR), *VRK* 3942., Tirupati (CTR), *DR* 1329. (SVU)

INDIA : Planted throughout.
WORLD : Native in South America.

## MILLETTIA Wight & Arnott *nom. cons.*

**Millettia peguensis** Ali in Kew Bull. 21: 489. 1918. *M. ovalifolia* sensu Kurz. J. Asiat. Soc. Bengal 42:68. 1873; non *Pongamia ovalifolia* Wt. & Arn. 1834.

Moderate-sized tree. Leaflets 3 pairs, ovate to 3 × 3 cm, glabrous, apex acute. Flowers pinkish, small in axillary and terminal recemes. Pods compressed, 6.5 cm long; seeds 2–4, beaked.

Planted in gardens and road sides for its elegant foliage and flowers. It is grown in Siripuram of Visakhapatnam district. Fls. Jan.–Mar. Frs. Apr.–Aug.

INDIA : Introduced in Gardens.
WORLD : A native of Myanmar; Thailand.

## MUNDULEA (AP. de Candolle) Bentham

**Mundulea sericea** (Willd.) A. Chev. in Compt. Rend. Acad. Sci. Par. 180:1521. 1925. *Cytisus sericeus* Willd., Sp. Pl. 3:1121. 1802. *Mundulea suberosa* (DC.) Benth. in Miq. Pl. Jungh. 248.1852; FBI 2: 110; Gamble 1: 314(222); Beddome Fl. Sylv. S. India 85. Anal. Gen t. 12. f 2. 1869. *Tephrosia suberosa* DC. Prodr. 2:249; Brandis 219.

Deciduous tree, up to 6 m, tall; bark thick, pale yellow, spongy, deeply fissured; wood close-grained, yellow; branchlets silky tomentose. Leaves imparipinnate, rachis to 13 cm long; leaflets 8–10 pairs,coriaceous, sub-opposite, elliptic-lanceolate, 2–6 × 1.2–3 cm, glabrous and bright green above, silky beneath, apex acuminate, apiculate, margin entire, base obtuse; petiolule 3 mm. Flowers pinkish-violet in close terminal racemes. Pod linear, flattened, 3–8 × 0.5–1cm; yellowish, densely velvety with short golden-brown hairs, thickened at the sutures; seeds 10–12, yellowish-brown.

Occasional, clumps on fully sun-exposed slopes. It coppices well and regarded as useful plant for reafforesting bare hills and rocky hills of dry deciduous forests. Fls. May.–June. Frs.: Through the year. Vern: Verri Billu, Neela Murri, Tella Billu.

Seeds are to be used for poisioning fish. Plant possesses insecticidal and piscicidal properties. All parts of the plant, particularly the bark and seeds are toxic and used as dust, it is toxic to caterpillers, cockroaches, pulse beetles and house flies.

Mannanur RF (MBNR), *SSR & KSM* 16023; Gundumala RF (ATP), *TP & NY* 712; Bairanidokka (KNL), JLE 42240(MH); Guramkonda (CDP), *JSG* 13028(MH); Vinukonda (GTR), *KCJB* 17372(MH); Rajupalli (VSKP), *CAB* 1554 (MH); Tirumala (CTR), *DR* 1453. (SVU).

INDIA : Western Peninsula.
WORLD : Sri Lanka, Madagascar, South Africa.

## **OUGEINIA** Bentham

**Ougeinia oojeinensis** (Roxb.) Hochr. in Bull. Soc. Bot. Geneve 13 & 14:51. 1909. *Dalbergia oojeinensis* Benth. in Miq., Pl. Jungh. 216. 1852; FBI 2:161; Gamble; 340 (240); Brandis 224; Beddome Fl. Sylv. S. India t. 26. 1869.

Deciduous, often gregarious tree, up to 13 m tall; trunk short, crooked (irregularly shaped); bark dark brown, deeply cracked (regular vertical and transverse cracks); wood light-brown and close grained; branches slender. Leaves trifoliate, petiole 6.4 cm; leaflets stipellate, rigid-coriaceous, obliquely ovate, 9.2–13 × 6.5–11 cm, glabrous above, slightly pubescent beneath, apex retuse to pointed, margin shallowly crenate, base cordate; secondary nerves 8 pairs, petiolule to 5 mm. Flowers rose, in short fascicled racemes from the nodes of old branches. Pod distinctly 2–5 jointed, flat, 5–8 cm long, reticulately veined; seeds 2 to 5, smooth, brown when ripe.

Frequently occurs along the Godavari and in Visakhapatnam, Telangana districts and more or less scarce in other forests up to 1300 m. It is recommended for plantation on land slips, river banks and other exposed places. It is a good coppicer and produces abundant root suckers. Fls.: Feb.–Mar. to May. Frs. May.–June Vern.: Dargu, Tella-moduga, Modi-Mudsu, Vandanam, Chikkudu.

Wood hard, tough and durable, useful for carts, implements, picker arms and other articles when strength and toughness are necessary and makes excellent furniture. Bark exudes astringent red gum and yields a bast fibre useful for cordage. Leafy twigs are used as cattle fodder.

Jalalpur (NZB), *BR & CP* 9653; Rampur (ADB), *SSR & KSM* 19304; Gangapur RF (MDK), *BR & CP* 11426; Yarlagadda (EG), *VNS* 359 (CAL);

INDIA : Bihar, Karnataka, Kerala, Madhya Pradesh, Orissa, Tamil Nadu.
WORLD : Nepal.

## **PONGAMIA** Ventenat *nom. cons.*

**Pongamia pinnata** (L.) Pierre For. Fl. Cochinch. t. 385. 1899; Thoth. in Bull. Bot. Surv. India 3:418. 1962. *Cytisus pinnatus* L. Sp. Pl. 741. 1753. *Pongamia glabra* Vent. Jard. Malm. t. 28. 1803; FBI 2:240; Gamble 1: 385 (272); Brandis 243; Beddome Fl. Sylv. S. India t 177. 1872. *Derris indica* (Lam.) Bennet in J. Bombay nat. hist. Soc. 68: 303. 1971.

Densely foliaceous, evergreen tree, up to 18 m tall; crown spreading; trunk short; bark greyish-green, tubercular; wood yellowish-white; branches spreading. Leaves imparipinnate, reachis up to 18 cm long, petiole 4–5 cm; leaflets 5–9, opposite, thin-coriaceaous, ovate or ovate-oblong, 5.5–12 × 5.3–7 cm, glabrous, shining, apex acute to shortly acuminate, margin entire, base cuneate; secondary nerves 6-pairs; petiolule to 1 cm. Flowers pinkish-white in 15–18 cm long racemes. Pod obliquely-oblong or elliptic, 4–5 × 1.5–2.5 cm, apex with short decurved mucro; seed 1, rarely 2, elliptic or reniform, 1.8–2 cm long, compressed, wrinkled, reddish-brown.

Common, along river banks, and near sea coasts; occasionally in deep forests and also in plains up to 1000 m. It is adapting itself to diverse conditions. It grows best in fairly moist situations on porous and well drained soil, and it is good coppicer and produces root suckers. In some areas it is deciduous in May. Fls.: May–July. Frs.: Dec.–May. Vern.: Ganuga, Kagu, Kanuga. Eng.: Pongam oil tree. It is a shade bearer and is considered to be a good tree for planting in pastures. It resists drought well, moderatley tolerant of salinity, so it is used for afforestating the forests and waste lands.

The dried flowers are used in dyeing. Seeds are used as fish poison and give an oil used for burning and skin diseases. Wood tough, fibrous, not durable and it is easily attacked by insects and commonly used as fire wood.

Papampet (EG), *KSM & SSR* 14251; SKU Campus (ATP), *NY* 549; Madhavaram RF (KNL), *TP & RVR* 1341; Vijayapuri South (GTR), *VRK* 3940; Peddachema (PKSM), *RVK* 15710, Mannanur (MBNR), *TDCK* 13608.

INDIA : Throughout.
WORLD : Pakistan, Sri Lanka, Bangaldesh, Malaysia, Australia, Polynesia.

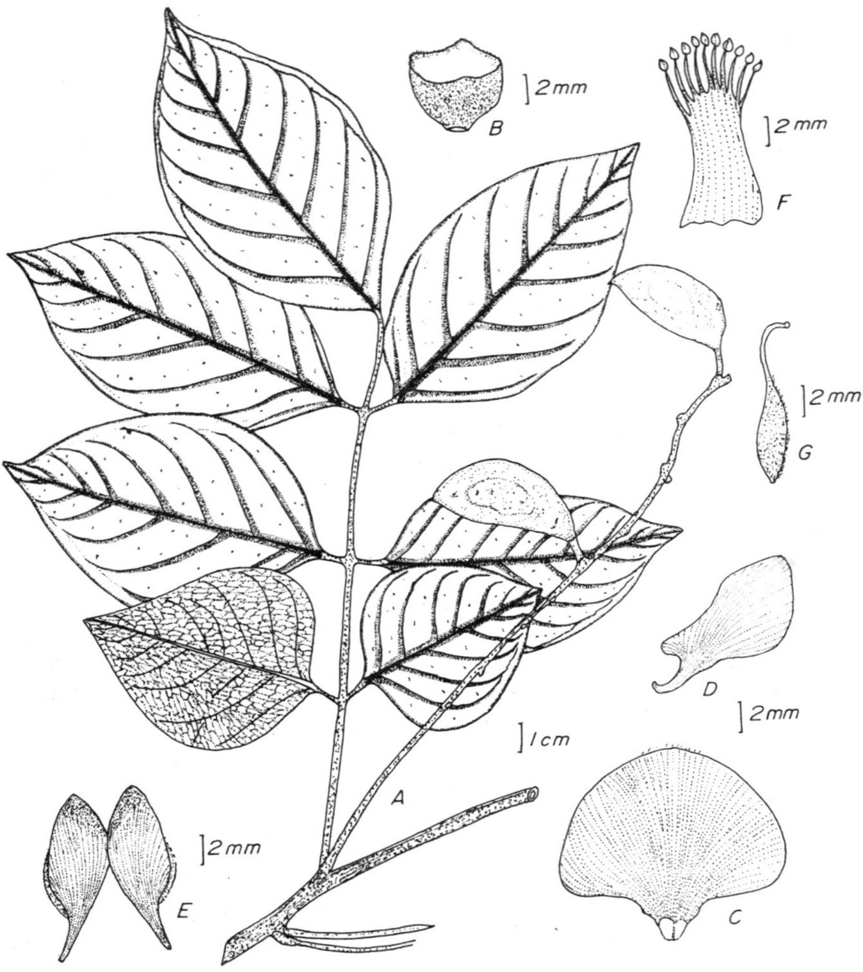


**Fig. 31.** *Pongamia pinnata* (L.) Pierre
A. Twig, B. Calyx, C,D & E. Standard, wing & keel petals, F. Staminal column spread out, G. Pistil.

## PTEROCARPUS N.J. Jacquin

1. Bark blackish-brown, divided into rectangular plates by deep and vertical and horizontal cracks. Leaflets 3, elliptic or orbicular-elliptic, shiny; lateral nerves not very conspicuous, curved towards margin; apex sligtly emarginate. Flowers in short racemes; calyx greyish tomentose, deeply lobed, apex acuminate. Pod with narrow wings. ................ ***P. santalinus***
1. Bark grey, longitudinally fissured, exfoliating in irregular scales. Leaflets 5–7, elliptic-oblong, not shiny, lateral nerves conspicuous, curved towards apex; apex deeply emarginate, sometimes acute to shortly acuminate. Flowers in long panicles; calyx brownish pubescent, shortly lobed, apex acute. Pod with broad wings. ........................................ ***P. marsupium***

**Pterocarpus marsupium** Roxb. Pl. Cor. t. 116. 1799; FBI 2: 239; Gamble 1: 385 (271); Brandis 240; Beddome Fl. Sylv. S. India t. 21. 1869.

Large deciduous tree, up to 25 m tall; bark thick, grey, longitudinally fissured, exfoliating in small irregular scales; wood close grained, yellowish brown; red juice exudes from the bark on incisions. Leaves odd-pinnate, to 15 cm; petiole to 5 cm; leaflets 5–7, coriaceous, elliptic-oblong or ovate, 5–12 × 2.2–7.5 cm, glabrous, shining, apex emarginate or even bilobed, margin entire, base rounded to truncate; secondary nerves numerous, close, prominent, petiolule to 8 mm. Flowers golden yellow in 17 cm long, axillary panicles. Pod orbicular, flat, stipitate, dark reddish-brown, 4 cm dia., wings broad.

Occasional, in hilly slopes of dry and moist deciduous forests up to 1500 m. It grows better on well-drained alluvial and sandy loam to loamy soils. Leaf fall: Feb.–Mar. New foliage: May.–June. Fls.: June–Oct. Frs.: Dec.–Mar. and remain on the tree, up to end of May. Vern.: Yerra Yegisa, Yegisa, Pedda Yegi.

The tree yields a 'kino' gum. The bark is used as an astringent and in toothache. The flowers are used in fever. The bruised leaves used in external application for boils, sores and skin diseases. The leaves are used as fodder.

Pothu Rajupenta (KNL), *TP & SSR* 15219; Kakavada (EG), *KSM & SSR* 14236; Chintala (PKSM), *KSM & SSR* 16027; Tirumala (CTR), *SSR* 15277; Narsapur (MDK), *SSR* 18504; Avagondapalem (ATP), *KSM & SSR* 19119; Bheemaram (ADB), *GO & DAM* 5180; Yerrakunta (MBNR), *TDCK* 16820; Chintapalli (VSKP), *GVS* 32829 (MH).

INDIA : Peninsula, Bihar, Uttar Pradesh.
WORLD : Sri Lanka.

**Pterocarpus santalinus** L.f. Suppl. Pl. 318. 1781; FBI 2:239; Gamble 1: 385 (274); Brandis 24; Beddome Fl. Sylv. S. India t. 22. 1869. *Ligonium santalinum* (L.f.) O. Kuntze, Rev. Gen. Pl. 1: 193. 1891.

Large deciduous tree, up to 20 m tall; bark blackish brown, divided into rectangular plates by deep vertical and horizontal cracks, exuding deep red juice when cut; wood dark red to almost black. Leaves 3-foliolate, petiole to 4 cm long; leaflets thick-coriaceous, elliptic or orbicular-elliptic, 4.5–10 × 3.2–6 cm, shiny, glabrous, apex obtuse with slightly emarginate, margin entire, base rounded; petiolule to 6 mm; secondary nerves 25–35 pairs. Flowers yellow in short axillary racemes. Pod obliquely orbicular, gradually narrowed into short stalk or stipe, 3–4 cm in dia., brown when ripe; seeds 1 or 2, 1–1.5 cm long, dolabriform, reddish brown.

Rare, it has very restricted natural range and grows typically on dry, hilly often rocky ground of Seshachalam hill ranges (Cuddapah and Chittoor districts and in some parts of Nellore district) at 500–800 m. Recommended

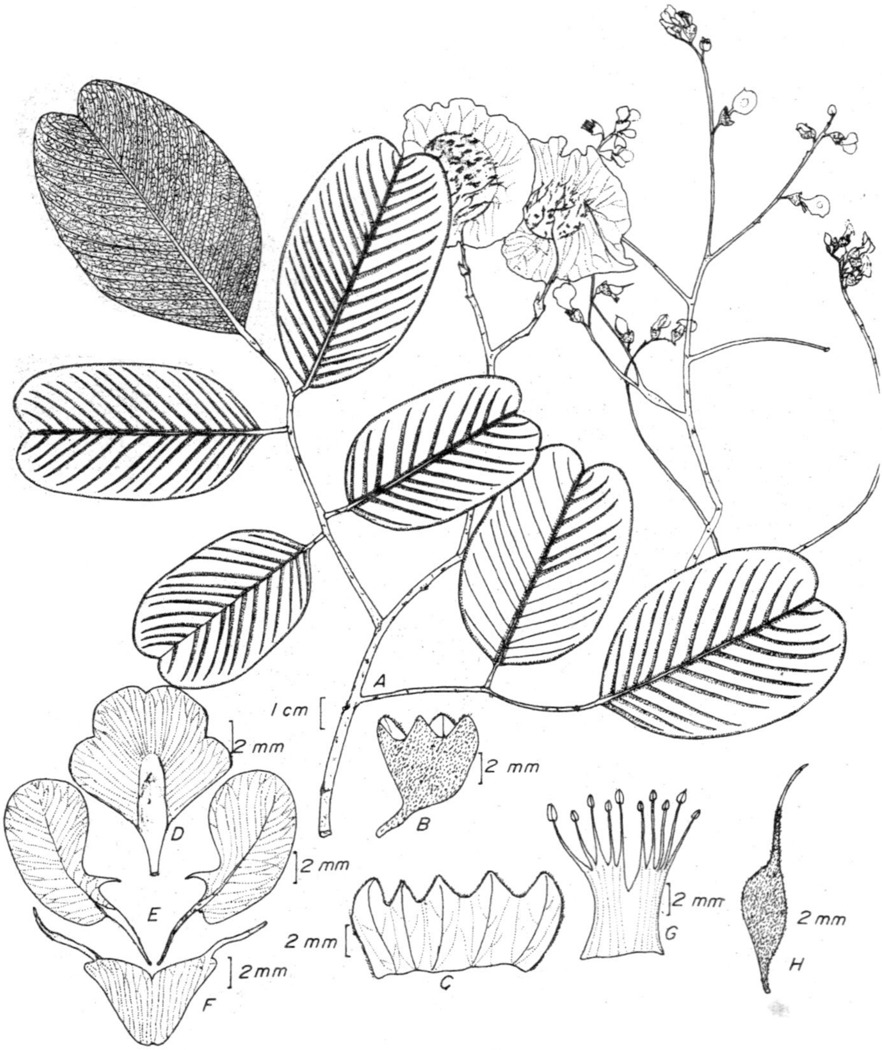


**Fig. 32.** *Pterocarpus marsupium* Roxb.
A. Twig, B. Calyx, C. Calyx split open, D, E & F. Standard, wing and keel petals; G. Staminal column spread out, H. Pistil.

for plantation on laterite loam, quartazite shale and lime stone. It requires perfect drainage. It is a strong light demander, and it coppices well and produces root suckers freely. Leaf fall: Jan. to the middle of Mar. New foliage: Leaves appearing nearly Apr. Fls.: Apr.–June Frs.: Feb.–Mar. (following year). Vern.: Chandanam, Rakthachandanamu, Yerra chandanam. Eng.: Red sander tree.

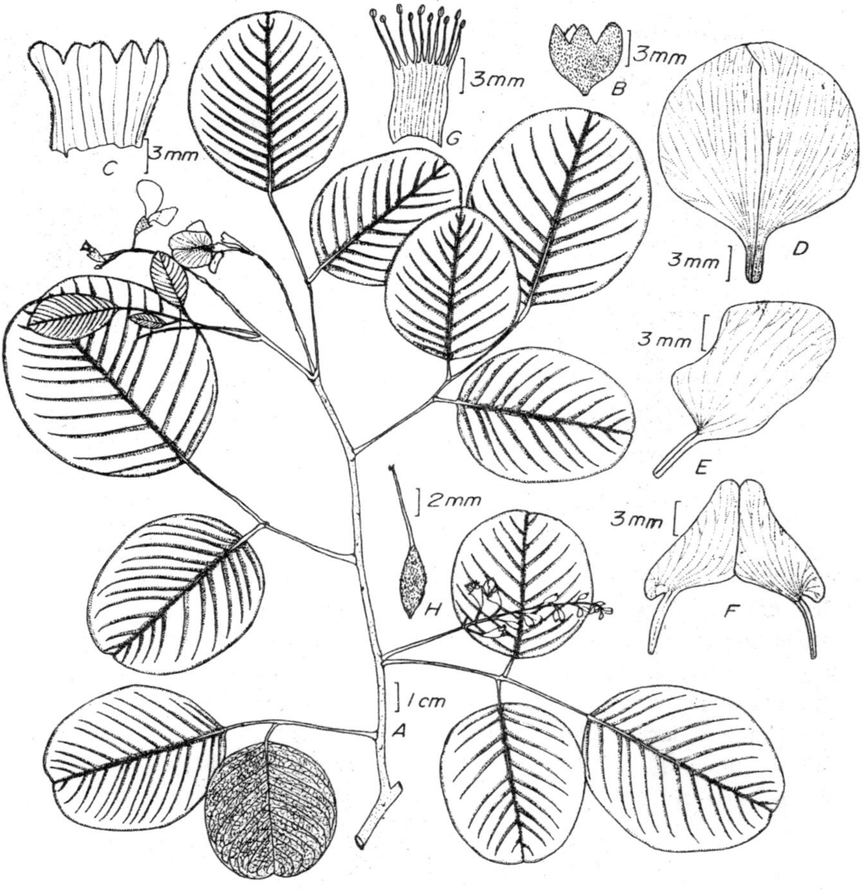


**Fig. 33.** *Pterocarpus santalinus* DC.
A. Twig, B. Calyx, C. Calyx split open, D, E & F. Standard, wing & keel petals, G. Staminal column spread out, H. Pistil.

Wood extremely hard, used for making traditional musical instruments called Shamisen. To a large extent it is exported to Japan, Malaysia, Sri Lanka and Germany. The choice of the Japanese is more for sound wavy grained logs and it is also used as dye. It is considered as astringent and diaphoretic. A paste of the wood is applied externally to give a cooling effect during inflammations and headache.

Talakona (CTR), *SSR* 13732; A.U. campus (VSKP), *TP & EC* 7324; Sidduliah kona (NLR), *SSR & KSM* 19377; Rollamadugu (CDP), *JSG* 11148 (CAL); Kodur (CDP), *B.H. Anandathirtha Rao* 107361 (DD).

INDIA : It is endemic to Andhra Pradesh, Tamil Nadu and Kerala. Cultivated in other parts, especially in Maharashtra.

**Note** : The interesting endemic tree is threatened due to indiscriminate felling and smuggling. Considering its slow regeneration and growth, it is not exactly a renewable resource in terms of time-scale. Conservation of this species in its natural habitat and long-term reforestation might only save this from depletion.

## SESBANIA Scopoli *nom. cons.*

**Sesbania grandiflora** (L.) Poir. in Lam. Encycl. 7: 127. 1806; FBI 2: 115; Gamble 1: 323 (228); Brandis 219. Beddome Fl. Sylv. S. India 86. 1871. *Robinia grandiflora* L. Sp. Pl. 722. 1753.

Tree, up to 10 m tall; trunk straight, cylindrical; bark grey, deeply furrowed; branchlets pubescent. Leaves paripinnate, rachis to 30 cm long; petiole to 2.2 cm, leaflets 18–30 pairs, thick, chartaceous, oblong, 1.5–3.6 × 8–12 cm, glaucous, apex obtuse or slighly emarginate, margin entire, base obtuse or acute; petiolule to 3 mm. Flowers white very large, in lax, 2–4 flowered racemes. Pod pendulous, up to 30 × 0.5 cm compressed, four angled, curved, margin thick; seeds ca. 30, pale.

It is commonly seen growing on the bunds between the fields and in backyard vegetable gardens. Recommended for afforestation on eroded, grassy waste lands and black cotton soils. The tree is short-lived, quick growing, and is harvested on a 3 year rotation. It serves as an effective wind breaker for banana plantation. Fls.: Dec.–Jan. Frs.: Apr.–May. Vern.: Tella-suminta, Avisi, Aushika, Agise, Patta, Suka Nasam, Bakku-Pushapamu. Eng.: Swamp pea, Sesban.

The young leaves are eaten as leafy vegetable and also used as good fodder. The juice of the flowers is said to improve the sight when squeezed into the eyes. It is also a popular remedy for catarrh. Bark yields a good fibre and gum. The juice of bark is used by fishermen for toughening their nets. Tender leaves, pods and flowers are eaten as vegetable.

Cumbumcheruvu (PKSM), *KSM* 16009; Near Panapakam forest (CTR), *GVS* 46337 (CAL).

INDIA : Cultivated throughout.

WORLD : Cultivated and naturalized in tropical Africa and Asia, probably a native of Indonesia.

## CAESALPINIACEAE

1. Calyx petaloid, corolla absent:
   2. Flowers greenish-cream in axillary or terminal lax panicles .......... ***Hardwickia***
   2. Flowers yellow-orange or red in lateral dense corymbs .......... ***Saraca***

1. Calyx sepaloid, corolla present:
    3. Pod moniliform, stipules spinescent. Rachis flattened with very small leaflets ............... .......... ***Parkinsonia***
    3. Pod not moniliform, stipules not spinescent, Rachis rounded with large leaflets:
        4. Leaf bilobed or leaflets adnate at their length up to middle or above the middle ... .......... ***Bauhinia***
        4. Leaf compound; 1–2 pinnate:
            5. Flowers large > 6 cm across .......... ***Delonix***
            5. Flowers to 4 cm arcoss:
                6. Stamens 3 .......... ***Tamarindus***
                6. Stamens 3–10
                    7. Inflorescence condensed heads. Leaflet apex caudate-acuminate ... .......... ***Brownea***
                    7. Inflorescence not condensed heads. Leaflet apex not as above:
                        8. Inflorescence rusty tomentose .......... ***Peltophorum***
                        8. Inflorescence glabrous or puberulous:
                            9. Pods twisted .......... ***Caesalpinia***
                            9. Pods terete or flattened .......... ***Cassia***

## **BAUHINIA** Linnaeus

### KEY 1 (Leaf)

1. Leaflets connate almost to the apex .......... ***B. semla***
1. Leaflets connate up to above the middle:
    2. Leaves broader than long, 3 or 7–9-nerved:
        3. Leaves pale green beneath .......... ***B. tomentosa***
        3. Leaves glaucous beneath:
            4. Leaflets suborbicular, base semicordate, leaf notch with mucro, 4.5–13 × 8–15 cm .......... ***B. malabarica***
            4. Leaflets oblong, base rounded or truncate; leaf notch without mucro, 2–5.5 × 3.5–7.3. cm .......... ***B. racemosa***
    2. Leaves as broad as long, 11 — above nerved.
        5. Leaflets connate about two-thirds of their length, ovate, rounded at apex ...... .......... ***B. variegata***
        5. Leaflets connate about half way up, oblong, obtuse or subacute at apex ... ***B. purpurea***

### KEY 2 (Flowering)

1. Flowers small, up to 2 cm long:
    2. Fertile stamens 3. Leaflets connate throughout their length .......... ***B. semla***
    2. Fertile stamens 10. Leaflets connate up to above the middle:
        3. Flowers pale yellow in corymbs; calyx 5-cleft. Pod 25–42 cm long, straight, flattened, parallel veined .......... ***B. malabarica***
        3. Flowers whitish in racemes; calyx spathaceous. Pod 12–25 cm long, falcate, turgid, reticulately veined .......... ***B. racemosa***
1. Flowers large, up to 10 cm long:
    4. Flowers yellow; calyx spathaceous, tube short; fertile stamens 10 .......... ***B. tomentosa***
    4. Flowers other than yellow; calyx not spathaceous, tube elongate; fertile stamens 3–5:
        5. Flower buds not angular; one petal is darker and variegated. Leaflets ovate, with rounded tip .......... ***B. variegata***

5. Flower buds angular; all petals are similar in colour. Leaflets oblong with obtuse tip ............ ***B. purpurea***

**Bauhinia malabarica** Roxb. Fl. Ind. 2:321.1832; FBI 2:217; Gamble 1:407(288); Brandis 256. *Piliostigma malabaricum* (Roxb) Benth. in Miq. Pl. Jungh 261.1852.

Evergreen tree, up to 15 m tall; crown bushy, diffused; bark brown-grey, nearly smooth with linear oblong flakes when old; wood reddish-brown with darker patches; branchlets pubescent-tomentose. Leaves sub-orbicular, 4.5–13 × 8–15 cm, leaflets connate to three fourths of their length, coriaceous, dark green and glabrous above, glaucous and glabrescent or sometimes minutely pubescent beneath, apex rounded, margin entire, base semi-cordate, nerves 7–9 from base; petiole to 3.5 cm. Flowers pale yellow, sub regular on slender pedicels in axillary, tomentose, sub-corymbose racemes. Pod oblong, 25–42 × 2 cm, greyish-reddish brown, flattened, reticulately veined, nearly straight, somewhat turgid; seeds 20–30, globose-ovoid, flattened, dark brown, polished.

Rare, moist localities and drier districts, loving the vicinity of streams. Fls.: Aug.–Nov. Frs. Jan.–Mar.. Vern. Pulishinta.

Wood hard, but is of little value or only used for fuel. The leaves are eaten which give pleasant taste (due to acidic contents).

Eturnagaram (WGL), *SSR* 18540. Anantagiri-Galikonda (VSKP), *TP & EC* 7400; Kammarapadu-Polavaram (WG), *DCSR* 187 (CAL); Maredumilli to Kakur (EG), *GVS* 67587 (MH); Rolla penta (KNL), *JLE* 42234. (CAL)

INDIA : Throughtout
WORLD : Nepal, Bhutan, Myanmar, Thailand, Laos, Cambodia, Vietnam, Philippines, Indonesia, commonly cultivated in tropics.

**Bauhinia purpurea** L.Sp. Pl. 375. 1753; FBI 2:284; Gamble 1:407 (258); Brandis 258; Beddome Fl. Sylv.S. India 92. *Phanera purpurea* (L.) Benth. in Miq. Pl. Jungh. 262. 1852.

Densely foliaceous tree, up to 12 m tall; bark grey to brown, smooth, thin-coriaceous with numerous lenticels; wood brown; branchlets warty, glabrous. Leaves oblong or sub orbicular, 6–12 × 6–12.5 cm, leaflets connate about half-way up sometimes overlapping, glabrous, apex obtuse or subacute, margin entire, base cordate, 11-nerved; petiole to 3.2 cm. Flowers large, purple, showy in terminal panicles. Pod linear-oblong, 30–40 × 1–2 cm, greenish-purple, flat, apiculate, glabrous, narrowed at base; seeds 12–15, sub-orbicular, 1.6 × 1.2 cm, dark-brown, smooth, glabrous.

Plains and hills of dry deciduous forests, chiefly along water courses, and also often cultivated in gardens up to 1000 m. It is frost hardy and grows on variety of soils. Fls. : Oct.–Dec. Frs.: Jan.–Mar. Vern.: Devakasia, Peddari, kanchanam, Bodanta. Eng: Camel's foot tree.

Wood hard and durable and used for agricultural implements. The bark gives fibre. The leaves are eaten as vegetables and also used as fodder. Stem yields gum.

Pangidi (KNR), *SSR* 15215; Borracaves (VSKP), *PRS* 15450; N.U. Campus (GTR), *VRK* 6994; Anantapur town (ATP), *NY* 805; Araku - Anjodigadda (VSKP), *TP & EC* 7438; Pampampet (EG), *PRS* 15419; Barmakonda (SKLM), *KCJB* 17155 (MH); Gundla kamma river, Gundlabrahmeswaram range (KNL), *KNS* 3047(DD); Pullalacheruvu (PKSM), *RVK* 18301; Farahabad (MBNR), *TDCK* 16829.

INDIA : Throughout, often cultivated.
WORLD : Nepal, Bhutan, S.E. Asia, China, Cultivated.

**Bauhinia racemosa** Lam. Encycl. 1:390. 1875; FBI 2:276; Gamble 1: 407 (288); Brandis 256; Beddome Fl. Sylv. S. India t. 182. 1872. *Piliostigma racemosa* (Lam.) Benth. in Miq. Pl. Jungh. 262. 1852.

Deciduous tree, up to 12 m tall; crown compact; bark dark grey, rough; wood brown; branchlets densely-tomentose. Leaves thin-coriaceous, oblong, 2–5.5 × 3.5–7.3 cm, leaflets connate to two-thirds of their lengths, glabrous above, slightly pubescent and pale glaucous beneath, apex obtuse, margin entire, base rounded or truncate, 9-nerved; petiole to 1.6 cm. Flowers whitish, small in 12 cm long, terminal panicles. Pod oblong, 12–25 × 1.8–2.5 cm, brown, woody, turgid, curved falcate, indehiscent; seeds 10–18, compressed, dark reddish brown or black, glabrous.

Foot hills of dry deciduous forests of most districts at 500–1000 m. Leaf fall: Feb. New foliage: Mar. Fls.: Mar.–June Frs.; Nov.–Feb. and remain on the tree for several months Vern.; Are, Pacchare.

The inner bark gives a good fibre for rough ropes. Wood is only used for fuel. The fruit is used in cough, asthma and diarrhoea and the bark is astringent in dysentry.

Bairluti (KNL), *SSR* 15518; Chintala (PKSM), *SSR* 16117; Bala Raju palli (CDP), *SSR* 16154; Narsapur (MDK), *SSR & CP* 16193; Kalasamudram RF (ATP), *EC* 2925; Madhavarm RF (KNL), *RVR* 1318; Sattenapalli RF (ADB), *TP & PVP*, 4091; Borra caves (VSKP), *PRS* 15478; Kakavada (EG), *PRS* 15408; Sangam (NLR), *MSR* 1145 (CAL); Pakhal (WGL), *KMS* 13163(CAL); Gokavaram (KSN), *AWL* 1214 (DD); Marripalem (PKSM), *RVK* 15769; Molachintapally (MBNR), *TDCK* 15322;. Kailasakona (CTR), *DR* 2093 (SVU).

INDIA : Bihar, Sikkim, West Bengal, Peninsula.
WORLD : Nepal, Bhutan, Bangladesh, Thailand, Indonesia, Vietnam, Laos, Cambodia.

**Bauhinia semla** Wunderlin in Taxon 25:362. 1976. *B. retusa* Ham. ex Roxb. Fl. Ind. 2: 322. 1832; FBI 2: 279; Gamble 1:407 (288); Brandis 257.

Deciduous tree, up to 12 m tall; bark dark grey-brown; wood red with darker patches and streaks; branchlets long, slender, pendulous. Leaves 12 × 15.5 cm; leaflets connate almost to the apex, rigidly coriaceous, glabrous, apex slightly emarginate, margin entire, base cordate, 9-nerved; petiole to 3 cm. Flowers white, numerous in corymbose racemes. Pod sub-oblong, 15 × 4 cm, straight, flattened, gradually widening to an obtuse tip; seeds 6–8.

Slopes and hills of moist and dry deciduous forests at 500–1000 m. Leaf fall: Dec.–Jan. New foliage: May.–June. Fls.: Sept–Dec. Frs.: Feb.–Mar. Frs.: Feb. Vern.: Nirpa, Goddari. The bark yields a fibre and gum, which is used in sweet meats. It is frost hardy. Wood hard and is of good quality.

Anantagiri (VSKP), *SSR* 19629; Gundaram RF (N2B), *BR* & *KH* 9676; Anantagiri-Galikonda (VSKP), *TP* & *EC* 7390;

INDIA : Bihar, Himachal Pradesh, Orissa, Pubjab, Uttar Pradesh.
WORLD : Pakistan, Nepal.

**Bauhinia tomentosa** L. Sp. Pl. 375. 1753; FBI 2: 275, Gamble 1: 406 (288); Brandis 256; Beddome Fl. Sylv. S. India 92. 1871.

Small tree, up to 5 m tall; wood black; branchlets pubescent. Leaves orbicular, 3–6 × 3–6.3 cm, leaflets connate above half their length, chartaceous, glabrous above, tomentose beneath, apex obtuse, margin entire, base cuneate, 3-nerved, petiole to 2 cm. Flowers yellow, large in racemes. Pod oblong, to 9 × 1.5 cm, brown, compressed, reticulate, puberulous, apex horned, base subacute-cuneate, dehiscent; seed 8–12, 8 mm across, horizontal.

Common, in coastal districts especially Krishna and Guntur; Often planted for ornamental. Fls. July–Aug. Frs.; Jan.–Feb. Vern.: Kanchini. Wood tough.

Tirupati (CTR), *SSR* 15559; Madhavaram RF (KNL), *RVR* 1665; Kondapalli (KSN), *CAB* 4563 (CAL); Upper Ahobilam (KNL), *JLE* 16865 (MH).

INDIA : Central and southern region.
WORLD : Africa, Asia.

**Bauhinia variegata** L. Sp. Pl. 375. 1753; FBI 2: 284. Gamble 1:406 (288); Brandis 258; Beddome Fl. Sylv. S. India 92.1877. *B. candida* Ait. Hort. Kew ed. 1,2. 49. 1789.

Deciduous tree, up to 13 m tall; bark grey, longitudinally cracked with numerous lenticels; wood dark-coloured with irregular masses near the centre; branchlets brown, glabrescent. Leaves ovate, to 8.9 × 10 cm, leaflets connate about two-thirds, thin-coriaceous, glabrous above, grey glaucous and pubescent beneath, apex obtuse or rounded, margin entire, base cordate, 11–15-nerved; petiole to 3.2 cm. Flowers large, white or pink or purple with one petal variegated yellow in short racemes. Pod linear-oblong, to 15 × 1.8 cm, brown, flat, stalked, horned, dehiscent.

Occasional, naturally occurs in dry rocky hills and very beautiful in the forest, often cultivated. Fls.: Feb.–Mar. Frs.; Apr.–May. Vern.: Mandara, Madapaku, Devakanchanam.

The leaves and flower buds are sometimes eaten as vegetable. Bark yields a fibre and it is used for tanning and dyeing and also to remove intestinal worms; decoction of bark applied in ulcer, itching, skin diseases, diarrhoea, wood is only used for fuel.

Tirumala (CTR), *SSR* 15278; Pahrabad (MBNR), *SSR* 15535; Upper Ahobilam (KNL), *PRS* 15441; Anantapur (ATP), *KCJB* 83237 (MH); Aklaspur (KNR), *GVS* 22496 (MH); Anantagiri (VSKP), *NPBK* 836 (CAL).

INDIA : Throughout
WORLD : Bhutan, Myanmar, China, Nepal—probably a native of China, widely cultivated in tropics.

## BROWNEA Jacquin

**Brownea coccinea** Jacq. Enum. Pl. Carib. 26. 1760; Benthal Tr. Cal. 180. 1946.

Tree, 8–10 m tall; crown spreading, round; branchlets glabrous. Leaves paripinnate, rachis stout, striate, up to 40 cm long; petiole very short to 1 cm; leaflets 8–10 pairs, sub-opposite, elliptic-oblong, (lower most ovate), upper ones are longer than lower ones, 6.5–18 × 2.5–5 cm, glabrous, apex caudate-acuminate, margin entire, base oblique, obtuse, (sub-cordate in lower most ones) with a small gland to one side on base of the midrib; secondary nerves 10–12 pairs, jointed near the margin and forming loops, mid vein raised; petiolule stout, 4–6 mm long, rusty tomentose. Flowers scarlet red, conspicuous, in condensed heads on older branchlets, bract very conspicuous, broadly ovate to linear, puberulent; floral sheath rusty tomentose.

Cultivated, occasionaly grown in Andhra University campus for its showy flowers. Fls.: May.

Andhra University campus (VSKP), *SSR* 19664.

WORLD : Native of South America (Panama).

## Caesalpinia Linnaeus

**Caesalpinia coriaria** (Jacq.) Willd. Sp. Pl. 2:532. 1799; Gamble 1: 394 (277). *Poinciana coriaria* Jacq. Select. Strip. Amer. Hist. 123. t. 175. 1763.

Evergreen, unarmed tree, 8–10 m tall; crown dense and low spreading; branchlets warty. Leaves bipinnate, to 10 cm long, pinnae 5–8 pairs, 2.5–5 cm long, petiole to 2 cm; leaflets ca. 25 pairs, close, cordate or linear-oblong, 3–7 × 1 mm, glabrous, apex obtuse-acute, margin entire, base truncate; petiolule 0. Flowers small to 8 mm across, whitish-yellow, sweet-scented in racemose panicles, pale to blackish brown. Pod flat, oblong, spirally twisted, glabrous, 3 × 2 cm, glossy, smooth; seeds oblong, flattened.

Planted for ornamental purpose along road sides and gardens, and also naturalised. Fls.: Sept.–Nov. Frs. through the year. Vern: Dividivi.

Pods are rich in tannins. The pods and bark are said to be antiperiodic. A decoction of the pods is used in the treatment of bleeding piles.

Madgole (VSKP), *GVS* 28125 (MH); Anantagiri (VSKP), *NPBK* 816 & *GVS* 19487 (CAL & MH).

INDIA : Cultivated, naturalised in Gujarat, Karnataka, Kerala, Maharashtra.

WORLD : Native of tropical America.

## **CASSIA** Linnaeus

1. Flowers yellow. Pod dehiscent or indehiscent:
   2. Inflorescence pendent. Three lower most stamens with slender filaments at least 5 times the length of the ovate anthers. Pod indehiscent, long (30–60 cm), cylindric.... ***C. fistula***
   2. Inflorescence erect. Lower most stamens mostly with the oblong anthers longer than the filaments. Pod dehiscent, not very long (15–25 cm), and not cylindric:
      3. Leaves oblong-lanceolate. Pod turgid—quadrangular-cultivated ......... ***C. spectabilis***
      3. Leaves other than above. Pod flat:
         4. Foliar glands absent; leaflets 7–15 pairs, Flowers in corymbose panicles. Medium sized tree. .......... ***C. siamea***
         4. Foliar glands present; leaflets 4–6 pairs. Flowers in corymbose racemes. Small tree. .......... ***C. surattensis***
1. Flowers reddish or pinkish. Pod indehiscent.
   5. Bracts absent. Pod transversely wrinkled. .......... ***C. grandis***
   5. Bracts present and conspicuous. Pod without transverse wrinkles:
      6. Sepals red:
         7. Leaflets up to 14 pairs. .......... ***C. javanica***
         7. Leaflets up to 20 pairs. .......... ***C. renigera***
      6. Sepals green:
         8. Leaflets hairy, apex emarginate with apicule. Flowers in corymbiform racemes. .......... ***C. roxburghii***
         8. Leaflets glabrous, apex acute or pointed, Flowers in simple racemes. ... ***C. nodosa***

**Cassia fistula** L. Sp. Pl. 377. 1753; FBI 2:261; Gamble 1:400 (283); Brandis 253; Beddome Fl. Sylv. S. India 91. 1871. *C. rhombifolia* Roxb. Fl. Ind. 3: 334. 1832.

Deciduous tree, up to 10 m tall; bark smooth, rough when old, greenish grey or pale; wood reddish brown when fresh cut, darkening on exposure; branchlets glabrous. Leaves paripinnate, rachis 20–40 cm long, petiole 60 cm, leaflets 4–5 pairs, thick-coriaceous, opposite, elliptic-ovate, ovate-oblong or ovate-lanceolate, 5–16 × 3–7.5 cm, bright green and glabrous above, pale and silvery-pubescent when young, apex acute or gradually tapering, margin entire, base cuneate; secondary nerves numerous, close; petiolule to 8 mm. Flowers bright yellow, large, in 20–45 cm long, pendent, lax racemes. Pod cylindrical, smooth, pendulous, 30–60 cm long, dark browinsh-black, shining, indehiscent; seeds 40–100, immersed in dark coloured sweetish pulp, about 1 cm in diameter, flattened.

Common, in dry deciduous forests of all districts, up to 800 m. Frequently cultivated in gardens and avenues. Recommended for plantations in poor shallow granite, sand-stone soil and laterite soil. It is a very good coppicer. Leaf fall: Mar.-Apr. New foliage. Apr.–May. Fls.; Apr.–May. Frs. Pod commence ripening in Dec. and continue from Jan. to Mar.–Apr. Vern.: Rela.. Eng.: Golden-shower, Indian laburnum, Pudding-pipe tree, Golden-rain tree.

The pulp of the ripe-fruit is a strong purgative, the flowers eaten. The wood is strong, hard and useful for agricultural work. It also gives a good firewood and charcoal.

Upper Ahobilam (KNL), *SSR* 15208; Simhachalam (VSKP), *PRS* 15462; Devapur RF (ADB), *GO* and *PVP* 245; Gundumala RF (ATP), *TP* & *NY* 744; Palakonda hills (CDP), *CS* 9349; Macherla RF (GTR), *VRK* 4000; Pocharam sanctuary (MDK) *BR* & *CP* 11611; Kubgal RF (NZB), *TP* and *BR* 6201; Mahommadabad (RR) *MSM* and *KH* 10230; Lankapakala (EG) *PRS* 15407; Papikonda RF (WG), *DCSR* 506 (CAL); Diguvametta (PKSM), *RVK* 15719; Rala Mannanur (MBNR), *TDCK* 13611, Tirupati (CTR), *DR* 1387 (SVU).

INDIA : Throughout drier parts, often cultivated.
WORLD : Myanmar, Sri Lanka and Malaysia.

**Cassia grandis** L.f. Suppl. Pl. 230. 1781; Gowda and Krishnas., Pl. Wealth Lalbagh 35. 1968.

Medium-sized tree; foliage green. Leaflets 10–20 pairs, oblong, glabrous above, tomentose beneath, abruptly rounded at both ends. Flowers axillary, 15 cm long, on lateral racemes. Pod cylindrical, compressed, smooth, 7.5–12 cm long, transversely wrinkled.

Cultivated. Eng.: Pink shower. Fls. Feb.–Apr. wood is used for making small agriculatural implements. Bitter fruit pulp is used as purgative.

INDIA : Cultivated, Karnataka, Madhya Pradesh, Maharashtra, Tamil Nadu, West Bengal.
WORLD : Native of Tropical America.

**Cassia javanica** L. Sp. Pl. 379. 1753.

Tree, up to 12 m tall; crown broad, spreading; stipules large to 1.4 cm long. Leaves tripinnate. Leaflets 6–14 pairs, oblong, ovate, 2.5–5 × 1–2.5 cm, hairy beneath, glabrous above, apex rounded, margin entire, petiolule short. Flowers pink, fading to white, larger; sepals brown to reddish.

Cultivated in gardens. Fls. May.–July. Eng.: Apple-blossom.
Tirumala gardens (CTR), *SSR* 15978.

INDIA : Cultivated, naturalised in some parts.
WORLD : Native of Java. Indonesia, Malaysia, Philippines—elsewhere cultivated.

**Cassia nodosa** Buch.-Ham. ex Roxb. Fl. Ind. 2: 336. 1832; FBI 2: 261; Brandis 253. *Cassia javanica* L. var. *agnes* de Wit. in Webbia 11: 220. 1955.

Medium-sized, evergreen tree; bark reddish brown or grey coloured; young branchlets silky. Leaves ca. 15 cm long, leaflets 6–12 pairs, thinly coriaceous, elliptic-ovate or oblong, glabrous and glossy above, slightly puberulous, paler beneath, apex acute or pointed, margin entire, base oblique; petiolule 6 mm. Flowers showy, pink, fading to dull white, dense in erect, pedunculate, ca. 15 cm long racemes from the scars of fallen leaves; bracts narrow-lanceolate, silky, persistent. Pod cylindrical, 30–45 cm long with transverse partitions.

Cultivated. Fls.: May.–June Eng.: Pink and white shower.

INDIA : Andaman Islands, Assam, Gujarat, Manipur, Tripura-Cultivated.
WORLD : Myanmar, Bhutan, Bangaldesh, Thailand, Malaysia.

**Cassia renigera** Wall. ex Benth. Trans. Linn. Soc. 27. 518. 1871; Brandis 254.

Deciduous tree, up to 10 m tall; branchlets drooping, softly pubescent. Leaflets 8–20 pairs, 5 cm long, membranous, oblong, strap-shaped, slightly oblique, softly pubescent on both surfaces, shining. Stipules prominent, persistent, large, reniform, very oblique, 2.5 cm, across, cuspidate. Flowers large, pinkish, fading to rose-white, scented in short, erect, axillary racemes; rachis velvety. Pod terete, cylindrical, hard, 30–50 × 2.5 cm , smooth.

Cultivated in gardens for its showy flowers. Fls.: Apr.–June. Wood is very hard and well suited for wheel making and handles for tools.

INDIA : Cultivated in Delhi, Maharashtra, Madhya Pradesh, Tamil Nadu, Tripura, Uttar Pradesh and in some other states.
WORLD : Native of Myanmar, Malay Peninsula.

**Cassia roxburghii** DC. Prodr. 2: 489. 1825; Beddeme Fl. Sylv. S. India t. 180. 1872. *Cathartocarpus roxburghii* (DC.) Loud. Hort. Brit. 167. 1830. *Cassia marginata* Roxb. Fl. Ind. 2: 262. 1878; auct non Willd. FBI 2: 262; Gamble 1: 401 (284). Brandis 253.

Tree, up to 10 m tall; bark greyish-brown, rough; wood light brown, branches long, spreading and hanging; branchlets tomentose. Leaves 15–27 cm long, petiole to 0.9 cm; leaflets 10–20 pairs, bifarous, coriaceous, upper ones oblong, lower ones ovate, 2–3 × 1–1.12 cm, younger ones densely yellow pubescent, mature ones silky beneath, glabrous above, apex emarginate with apicule, margin entire, base subacute, inequilateral; petiolule to 2 mm. Flowers rose, 2.5 cm across; in short axillary racemes or the terminal ones panicled. Pod terete or cylindrical, 15–30 × 2 cm, dark brown, pendulous, indehiscent, pulp spongy with transverse partitions in between numerous seeds.

Cultivated. Planted on road sides, gardens and also grown as an avenue plant. Fls.: May–June. The wood is very hard, and well adopted for turning naves of wheels and handles of tools are made of it.

INDIA : Cultivated throughout.
WORLD : Native of Sri Lanka, Myanmar, Malaysia.

**Cassia siamea** Lam. Encycl. 1: 648. 1789; FBI 2: 264; Gamble 1: 402 (285); Brandis 254. *C. florida* Vahl, Symb. Bot. 3: 57. 1794; Beddome Fl. Sylv. S. India t. 179. 1872.

Evergreen tree, up to 12 m tall; bark grey, rough, longitudinally fissured; wood dark brown; branchlets glabrescent, lenticellate. Leaves paripinnate, 10–15 cm long, petiole to 3 cm; leaflets 7–15 pairs, chartaceous, elliptic-oblong, 2.3–5.5 × 1.1–2.2 cm, glabrous above, pilose below, apex obtuse, tipped with small sharp mucro, margin entire, base obtuse, petiolule to 4 mm. Flowers pale yellow in 20–30 cm long, axillary and terminal corymbose panicles. Pod oblong, flat, 15–25 × 0.8–1.5 cm long, brown when ripe, minutely velvety, thickened at sutures; seeds 20–30, dark greenish.

Planted as an avenue tree and then naturalized. Recommended for covering the denuded areas, hill slopes, flat terrain provided the drainage is good. Fls.: Sept.–Dec., but the flowering period is comparatively long. Frs.: Mar.–Apr. Vern.: Seema tangedu, Niala tangedu.

Wood hard and durable but useless. This species is one of the hosts of the lac insect.

Narsapur (MDK), *SSR & CP* 18505; Adilabad town (ADB), *GO & PVP* 4251; Sriharikota (NLR), *BSN* 2072; N.U. Campus (GTR), *VRK* 6986; Mohammedabad (RR), *MSM & KH* 10209; Biknoor (NZB), *BR & CPR* 7249; Simhachalam (VSKP), *PRS* 15460.

INDIA : Cultivated and naturalized throughout.
WORLD : Sri Lanka, Myanmar, Malaysia, Thailand, Indonesia, Taiwan, Phillippines, widely cultivated in tropics.

**Cassia spectabilis** DC. Cat. Pl. Hort. Monsp. 90. 1813; Sen & Naskar. Bull. Bot. Surv. India 7: 39. 965.

Tree, to 15 m tall; branchlets softly pubescent; Leaves 7.5 cm long; leaflets 8–15 pairs, oblong-lanceolate, pubescent beneath; Flowers bright yellow, 3.7 cm across, 30 cm long racemes. Fruit cylindrical, 15 cm long.

Cultivated in Tirumala.

INDIA : Cultivated throughout.
WORLD : Tropical America.

**Cassia surattensis** Burm. f. Fl. Ind. 97. 1768; de Wit. in Webbia 11: 269. 1955. *C. glauca* Lam. Encycl. 1: 647. 1785; FBI 2: 265; Gamble 1: 403 (285); Brandis 255; Beddome Fl. Sylv. S. India 91. 1871.

Small tree, branchlets sparsely pubescent. Leaves 12–15 cm long, petiole to 3 cm, eglandular; leaflets 4–6 pairs, chartaceous, oblong-elliptic, 2–5 × 1–3 cm, glaucous beneath, apex obtuse or subacute, margin entire, base cuneate, rachis with clavate glands in between the 3 basal pairs of leaflets, petiolule to 1.5 mm. Flowers yellow, 4 cm across, in axillary to 5 cm long, corymbose racemes. Pod flat, thin, to 15 × 2 cm, strongly nerved, compressed; seeds brown, oblong-ellipsoid, 7 mm long, beaked, smooth.

Cultivated. Rare, in forests of Chittoor district. Fls.: & Frs.: Aug.–Dec. Bark and leaves are used in diabetes and gonorrhoea.

Horsely hills (CTR), *DR* 1978 (SVU).

INDIA : Throughout
WORLD : Native of South-East Asia, Malay Peninsula, Malaysia, Australia. Cultivated in many countries.

## **DELONIX** Rafinesque

1. Flowers bright scarlet, 10–12 cm across, petals thick, claw long; stamens as long as or shorter than petals, not spreading. Pod 30–80 cm long, pinnae to 20 pairs ............***D. regia***
1. Flowers cream, then yellow, 4–5 cm across, petals thin,claw short; stamens exerted, longer than petals, spreading. Pod 10–12 cm long, pinnae to 12 pairs ..............................***D. elata***

**Delonix elata** (L.) Gamble Fl. Pres. Madras 396. 1919 & 1: 280. 1957 (repr. ed.). *Poinciana elata* L. Cent. Pl. 2: 16. 1756; FBI 2: 260; Brandis 249; Beddome Fl. Sylv. S. India t. 178. 1872.

Deciduous tree, up to 12 m tall; crown spreading; bark greyish; wood yellowish-white, close and even grained; branchlets warty, puberulous. Leaves 15–22 cm long, pinnae 6–12 pairs, petiole to 3.5 cm; leaflets 10–22 pairs, chartaceous, 0.5–1 × 0.1–0.3 cm, puberulous, apex obtuse, apiculate, margin entire, base cuneate, sessile. Flowers yellowish-white in terminal, corymbiform racemes. Pod linear or oblanceolate 12 × 2.5 cm, pale to dark brown, apices attenuate, horned, sutures grooved, glabrous, reticulate; seeds 8–10.

Planted as an avenue tree, occasionally wild. It is fast growing tree. Fls.: Jan.–May.. Frs.: Throughout the year. One of the first trees to flower after the winter. Vern: Chitti-Kesara, Sunkesula. Eng.: White Gold Mohur.

The branches and leaves are used for leaf-manure. Wood is light and used in preparing cabinets and for churns, combs and matches. The leaves are said to be used in rheumatism and flatulence.

Ahobilam (KNL), *SSR* 15290; Ethipothala (GTR), VRK 3583; SKU Campus (ATP), *NY* 854; Peapaly (KNL), *RVR* 3194; Nagarjuna konda (NLG), *KTH* 9824 (CAL); Tirupati (CTR), *DR* 154 (SVU).

INDIA : Cultivated throughout.
WORLD : Wild in Ethiopia and Arabia.

**Delonix regia** (Bojer ex Hook.) Raf. Fl. Tell. 2: 92. 1837; Gamble 1: 396 (280). *Poinciana regia* Bojer. ex Hook. in Bot. Mag. t. 2884. 1829; FBI 2: 260; Brandis 249.

Deciduous tree, up to 20 m tall; crown spreading, widely branched; bark grey to pale-brown, smooth; wood white, soft. Leaves 10–30 cm long, bipinnate, pinnae 8–20 pairs, petiole to 7 cm; leaflets 12–30 pairs, coriaceous, oblong, 0.8–1 × 0.4–0.5 cm, glabrous, slightly puberulous beneath, apex obtuse, margin entire, base oblique or obtuse, sessile. Flowers bright scarlet, 10–12 cm across, terminal in branched racemes. Pod broadly linear, large, 30–80 × 3–6 cm, woody, dark brown, compressed, flat; seeds 20–30 in distinct cavities, greyish-white, hard boney, 1.5–2 cm in length.

It is widely culivated and planted in gardens and avenues, both in dry and moist regions. It is capable of growing in poor dry soil. Leaf fall: Mar.–Apr. New foliage: June–July. Fls.: Apr.–July. Frs.: Aug.–Oct. Pod remain on the tree for 1 year. Vern.: Pedda turayi, Turayi, Erra sunkesula, Kodi juttu, Seema sunkesula.

Wood is soft, flexible and of little value. The seed contains a gum which may be used in food and textile industries.

S.V. University (CTR), *SSR* 19684; Itikalapalli (ATP), *NY* 176; Gulemallabad RF (KNL), *RVR* 1438; Simhachalam hill (VSKP), *DCSR* 1849; Digwal (MDK), *CP* 11682; Sunkarimetta (VSKP), *PRS* 15473; Kalyani dam (CTR), *DR* 157 (SVU).

INDIA : A native of Madagascar.
**Note** : Its superficial spreading roots prevent grass and other plants growing in the vicinity and it is unsuitable as a shade tree in plantation.

## HARDWICKIA Roxburgh

**Hardwickia binata** Roxb. Pl. Cor. t. 209. 1811; FBI 2: 270; Gamble 1: 412 (292); Brandis 250; Beddome Fl. Sylv. S. India t. 26. 1869.

Deciduous tree, up to 30 m tall; trunk straight, cylindrical; bark dark grey, rough, longitudinally fissured; wood dark red, streaked with black or purple, irregularly inter-locked-grained and coarse textured, branchlets glaucous, slender, drooping. Leaves two-foliolate, petiole to 1 cm; leaflets thick-coriaceous, ovate-trapezoid, 2.7–4.5 × 1.5–2 cm, apex obtuse, margin entire, base oblique, ca. 5-nerved, sub-sessile. Flowers greenish-cream, small in 8 cm long axillary and terminal lax panicled racemes. Pod strap-shaped or oblong-lanceolate, to 7 × 1.8 cm, brownish, flat, with parallel longitudinal veins; seed 1, near the top, orbicular, to 2 × 0.8 cm, pointed at one end rounded at the other with a fairly hard testa.

Common, in moist deciduous forests to dry deciduous forests to 800 m. For its best development, a deep porous soil with well-fissured underlying rock is required. It is drought resistant and reproduces vegetatively by root suckers. Leaf fall: Jan. New foliage: Apr., Coppery, turning green. Fls.: Aug.–Dec. Fruits ripen from July–Sept. Vern.: Yepi, Nara yepi.

Wood extremely hard, very heavy and durable. Used for bridge and house posts, wells, cots, planking, furniture. Bark yields a red brown strong fibre extensively used for well ropes and other agricultural purposes. Gum exudate from the bark used in varnishes. The leaves are cut for manure and the branchlets for the fibre of their bark.

SKU Campus (ATP), *SSR* 13799; Dorasani bavi (KNL), *SSR* 15505; Bogada (KNL), *PRS & SSR* 15455; Macherla RF (GTR), *VRK* 3541; Kotpalli RF (ADB), *GO* 4518; Venkatagiri (NLR), *BSN* 3523. Sudekulagadu (PKSM), *RVK* 15713; Molachintapally (MBNR), *TDCK* 15301; Bhakarapet RF (CTR), *DR* 7335 (SVU).

INDIA : Peninsula, Central India, Delhi, Rajasthan, Uttar Pradesh—endemic.

## PARKINSONIA Linnaeus

**Parkinsonia aculeata** L. Sp. Pl. 375, 1752; FBI 2: 260; Gamble 1: 392 (281); Brandis 249; Beddome Fl. Sylv. S. India 91, Anal. Gen. 1.13. t. 2. 1867.

Small tree, up to 8 m tall; bark greyish green; branchlets puberulous, spines woody. Leaves bipinnate, 8–30 cm long, petiole short, thick, ending in a stout spine, often with pair of stipular thorns, pinnae 2–4 pairs, flattened; leaflets minute, numerous, chartaceous, elliptic-oblanceolate, margin entire, base acute or obtuse, petiolule sessile. Flowers yellow, in 20 cm long, lax-axillary racemes. Pod linear, 7.5–15 × 0.6–1 cm, moniliform, torulose, beaked, dehiscent, marked with parallel lines; seeds 1–6, linear-oblong, 5 mm, brown, smooth.

In plains, moist places or by arable lands to 300 m. It is recommended for plantations in black cotton soil and sandy waste lands. Fls.: Oct.–Mar. Frs.: Through the year.

Hyderabad (HYD), *SSR* 19341; Mulachintapalli forest (MBNR), *TDCK* 13641; way to Talakona (CTR), *DR* 291 (SVU).

INDIA : Cultivated as well as naturalised throughout arid belts.
WORLD : Native of tropical America, widely cultivated throughout tropics.

## PELTOPHORUM (Vogel) Bentham *nom. cons.*

**Peltophorum pterocarpum** (DC.) Baker ex Heyne Nutt. Pl. Med. Ind. 2: 755. 1927. *Inga pterocarpum* DC. Prodr. 2: 441. 1825. *Peltophorum ferrugineum* Benth. Fl. Austral. 2: 270. 1864; Brandis 247.

Evergreen tree, up to 15 m tall; foliage dense and dark, crown round, spreading; bark greyish, rough, branchlets rusty tomentose. Leaves 15–28 cm long, pinnae 10 pairs, petiole to 4.5 cm; leaflets 15–16 pairs, oblong, 1.3–2 × 0.6–0.9 cm, glabrous, apex emarginate, margin entire, base oblique, subcordate, lateral nerves 3–4 pairs forming loops, not prominent, petiolule 0. Flowers golden yellow on 50 cm long rusty tomentose panicles. Pod dark brown, oblong, 9 × 3 cm, compressed, flat, closely longitudinally veined, indehiscent; seeds 3–4, brown, oblong, 1 cm long.

Frequently planted as an avenue tree. It is very hard and fast growing. Fls.: May–Aug. Frs.: Apr.–Dec. Vern.: Kondachinta, Pacha Sunkesulu. Eng.: The copper-pod tree or Yellow Gold Mohur.

SKU Campus (ATP), *NY* 553; Kurnool, *AA* 2215, V.P. South (GTR), *VRK* 6832; Siddipet-Hyderabad road (MDK), *CP* 11380; Machareddy (NZB), *BR & KH* 9055; Jaipur RF (ADB), *GO & PVP* 4945.

INDIA : Wild in Andaman & Nicobar Islands, cultivated throughout.
WORLD : Indonesia, Cambodia, Thailand, Vietnam, Malaysia, Philippines to Australia. Native of Sri Lanka.

## SARACA Linnaeus

**Saraca asoca** (Roxb.) De Wilde in Blumea 15: 393. 1968. *Jonesia asoca* Roxb. in Asiat. Res. 4: 365. 1799. *Saraca indica auct. non* L. 1769; FBI 2: 271; Gamble 1: 409 (289); Brandis 250. Beddome Fl. Sylv. S. India t. 57. 1870.

Evergreen tree, up to 5 m tall; trunk short; bark dark brown to grey, warty; wood reddish-brown, soft. Leaves abruptly pari-pinnate, to 24 cm long, petiole to 5 mm, rachis glabrous, corky at the base, stipules intrapetiolar; leaflets 4–6 pairs, coriaceous, oblong or oblong-lanceolate, 9–22 × 2–5 cm, glabrous, apex acuminate, margin entire, base cuneate, rarely glandular just above the petiolule; secondary nerves 12–16 pairs, parallel. Flowers yellow-orange or red, scented (in night) in lateral dense corymbs. Pod oblong, 12.5–

25 × 4–5 cm, black, compressed, flat, leathery, woody with reticulations, tapering at both-ends, glabrous; seeds 4–8, oblong, 3.8 cm long, slightly compressed.

It is generally found wild, along streams or in the shades of evergreen forests. It is frequent in the hill forest of Visakhapatnam district. Frequently planted especially in gardens and near temples. It is one of sacred trees of the Hindus, Buddhists, the flowers being much used for religious ceremonies and temple decorations. Fls. & Frs.: All seasons. Vern.: Asoka Kankeli, Vanjulamu. Eng. Sorrowless tree, Asoka.

Wood soft, used for making ploughs and shafts. Leaves possess blood purifying properties and their juice mixed with cumin seeds is used for stomachache. Flowers and bark also used in some of ailments.

Kusimi forest (SKLM), *SSR & KSM* 15903; Palakonda hills (VSKP), *JLE* 25513 (MH).

INDIA : Western Ghats, North-eastern India, often cultivated throughtout India.
WORLD : Myanmar, Sri Lanka.

## **TAMARINDUS** Linnaeus

**Tamarindus indica** L. Sp. Pl. 34. 1753; FBI 2: 273; Gamble 1: 409 (290); Brandis 252; Beddome Fl. Sylv. S. India t. 184. 1872.

Large evergreen tree; up to 25 m tall; crown umbrella shaped; bark dark grey, rough (fissured); wood dark purplish brown; branches spreading, branchlets warty, tomentose. Leaves even-pinnate, 8 cm long, petiole to 8 mm; leaflets 12–16 pairs, chartaceous, oblong, 1.6 × 0.5 cm, glabrous, base and apex obtuse, margin entire, sessile. Flowers yellow with pink striations in 6–10 cm long racemes. Pod oblong, slightly curved, 8–12 × 1–2 cm; brown, indehiscent with crustaceous brittle epicarp, mesocarp pulpy, endocarp septate, leathery; seeds 3–7 or more, obovoid-orbicular, dark brown.

Common in all plains to 600 m, cultivated and self-sown. It grows best in deep alluvial soils and termite mounds but comes up in any type of soil. It is extensively introduced in enrichment planting in dry deciduous and thorn forests. Fls: Apr.–June and occasionally in Sept.–Oct. Frs.: Dec.–Feb. Vern.: Chinta.

Fruit pulp is made into preserves and also exported as a drug. Ripe fruit regarded as refrigerant, digestive, carminative, and laxative and useful in intoxication caused by alcohol or any poisonous substances. Wood is used for many household and agricultural purposes. Young leaves are eaten.

Eturnagaram (WGL), *SSR* 18548; Nirmal-Nizamabad road (ADB), *GO & DAM* 4987; Pennahobilam (ATP), *TP & NY* 375; North Dhone RF (KNL), *TP & RVR* 1312; Macherla RF (GTR), *VRK* 3832; Ramayampet (MDK), *BR*

*& CP* 11511; Tekrial (NZB), *TP & BR* 6423; On the way towards Punyagiri hills (VSKP), *GVS* 19401 (MH); 12th mile stone, Paderu (VSKP), *RPS* 15485; Kakavada (EG), *PRS* 15406; Peddachama (PKSM), *RVK* 15706, Kuppam (CTR), *DR* 1301 (SVU) .

INDIA : Cultivated throughout.
WORLD : Native of tropical Africa.

## MIMOSACEAE

1. Plants armed, flowers in spikes rarely globose heads.
   2. Inflorescence bi or tri-coloured—Flowers polygamous ........................ ***Dichrostachys***
   2. Inflorescence concolourous:
      3. Stamens definite ........................ ***Prosopis***
      3. Stamens indefinite:
         4. Pinnae 1-pair ........................ ***Pithecellobium***
         4. Pinnae 3 or more pairs ........................ ***Acacia*** *p.p*
1. Plants unarmed. Flowers in globose heads, rarely spikes (in *Acacia*)
   5. Pod dehiscent:
      6. Pods coiled. Phyllodes falcate—introduced ........................ ***Acacia*** *p.p*
      6. Pods not coiled. Phyllodes 0:
         7. Foliar glands 0 ........................ ***Calliandra***
         7. Foliar glands present
            8. Pinnae 1 pair ........................ ***Xylia***
            8. Pinnae more than 5 pairs ........................ ***Leucaena***
   5. Pod indehiscent:
      9. Stamens definite. Flower heads large, 4–6 × 3.5–5.5 cm ........................ ***Parkia***
      9. Stamens indefinite. Flower heads not large, 1–3 × 0.5–2.3 cm:
         10. Pods septate between the seeds. Flowers sub-sessile ........................ ***Samanea***
         10. Pods not septate between the seeds. Flowers pedicellate, rarely sessile ....... ***Albizia***

## **ACACIA** P. Miller

Charkrabarthy and Gangopadhyay (1996) have reviewed *Acacia* in India where they reported 20 indigenous species and 14 exotic species, now more or less naturalised in India. They reduced *Acacia chundra* and *A. torta* to synonymy.

1. Leaves are modified as phyllodes: — introduced and cultivated:
   2. Branchlets slightly angules, phyllodes falcate ........................ ***A. auriculiformis***
   2. Branchlets angled, phyllodes elliptic-oblong to obovate:
      3. Phyllodes thick, 4 cm broad; branchlets silvery white ........................ ***A. mangium***
      3. Phyllodes thin, 2 cm broad; branchlets dark brown ........................ ***A. holosericea***
1. Leaves are normal as leaf:
   4. Flowers in heads:
      5. Heads are in elongated, spreading panicles. Bark grey, smooth ....... ***A. leucophloea***
      5. Heads solitary or 2–3 together. Bark other than above:
         6. Flowers greenish-white ........................ ***A. tomentosa***
         6. Flowers yellow:
            7. Pod moniliform ........................ ***A. nilotica***

7. Pod not moniliform:
    8. Pod turgid:
        9. Leaf-rachis ending with a spine. Pod straight to slightly curved; seeds 20–30 .......... ***A. farnesiana***
        9. Leaf-rachis without spine. Pod coiled into a ring, seeds 5–8 .......... ***A. planifrons***
    8. Pod not turgid:
        10. Pinnae 2–5 pairs. Leaflets 5–12 pairs .......... ***A. eburnea***
        10. Pinnae 7–15 pairs. Leaflets 30–50 pairs .......... ***A. decurrens***

4. Flowers in spikes:
    11. Axillary, stipular spines to 6 cm long .......... ***A. horrida***
    11. Infra-axillary, stipular spines to 3 cm long:
        12. Pinnae 4–6 pairs. Pod narrowed, margin straight .......... ***A. ferruginea***
        12. Pinnae 8–18 pairs. Pod broad, slightly sinuate .......... ***A. catechu***

***Acacia auriculiformis*** A. Cunn. ex Benth. in Hook. J. Bot. 1: 377. 1842; Sen & Naskar in Bull. Bot. Surv. India 7: 31. 1968. *A. moniliformis* Griseb. in Goett. Abh. 19: 136. 1874.

Tree, up to 15 m tall; unarmed; branchlets glabrous, compressed, slightly angular. Phyllodes thick-coriaceous, oblong, elliptic, falcate, 11–21.5 × 4–3.5 cm, glabrous, glossy, apex obtuse, margin entire, base narrowed or attenuate, palminerved, with 3–6, subparallel, prominent main nerves; petiole 8–10 mm. Flowers bright yellow, small, fragrant; heads in 7–10 cm long, axillary, racemose spikes. Pod woody, brown, much twisted in irregular coils, 3–9 × 0.3–0.9 cm, glabrous, dehiscing along the edges; seeds 4–6, dark brown with yellow aril.

Commonly planted along road sides, gardens, and also as plantations in disturbed forests to 800 m. It has been scuccessfully introduced in almost all districts. It is quick growing tree and grows on a variety of soils and dunes. It is sensitive to water stress and light demander, cannot withstand competition. It produces root suckers and coppices well. Fls.: Aug.–Dec. Frs.: Jan.–Mar. Vern: Australian Tumma. Eng.: Australian wattle.

The wood is suitable for making wrapping-paper and bleeched pulps for making writing and printing paper and also be used for furniture-making. It is a good fire-wood.

Acchampet (GTR), *VRK* 5876; SKU Campus (ATP), *NY* 552; Medak, *CP* 11732; Bijjur Guest house (ADB), *GO & PVP* 4973; Vempali plantation sight (ADB), *GO* 4456; Appapur (MBNR), *TDCK* 16834.

INDIA : Planted throughout.
WORLD : Native of Australia, New Guinea and Queensland.

**Acacia catechu** (L.f.) Willd. Sp. Pl. 4: 1079. 1806; FBI 2: 295; Gamble 1: 427 (303); Brandis 267. *Mimosa catechu* L.f. Suppl. Pl. 439. 1781. *A. chundra* (Roxb. ex Rottler) Willd. Sp. Pl. 4: 1078. 1806; FBI 2: 295; Gamble 1: 428 (303) "*sundra*". *A. suma* (Roxb.) Buch.-Ham. ex Voigt Hort. Suburb. Calcutta 260, 1845; Gamble 1: 427 (302). *Acacia catechu*

Willd. var. *sundra* Prain in J. Asiat. Soc. Bengal 66: 508. 1897; Brandis 268. *Acacia sundra* (Roxb.) DC. Prodr. 2: 458. 1825; Beddome Fl. Sylv. S. India t. 50. 1870.

Deciduous tree, 3–15 m tall; bark blackish grey or dark brown, peeling off in thin vertical strips; wood red, close-grained; branchlets brown-blackish, terete, glabrous; infra axillary spines 3–8 mm long, paired. Leaves bipinnate; rachis 6–15 cm with small glands in between uppermost pinnae; petiole 1.5–4 cm, with one circular or cupular large gland (1.5 to 3 mm dia.); pinnae 8–18 pairs, 1–4 cm long; leaflets 10–40 pairs, chartaceous, linear to oblong, 3–6 × 0.5–1.3 mm, pilose to glabrous, apex obtuse or subacute, margin entire, base oblique, rounded to acute; mid vein subcentral, nerves obscure. Flowers pale yellow or creamy white, fragrant in solitary or 2–4 together, axillary, 5–10 cm long spikes. Pod dark brown to reddish brown, narrow-oblong, 5–12 × 1–6 cm, flat, sinuate along margins, thin-walled, beaked at apex, base narrowed, glossy; seeds 3–10, orbicular or ovate, 4–5 mm in dia., flattened.

Common in dry deciduous forests or in open forests to 500 m. Leaf fall: Feb. New foliage: Apr.–May. Fls.: Mar.–Aug. Frs.: Sept.–Dec. Vern.: Kasu, Khadiramu, Sandra etc. Eng.: Red catechu, Cutch tree.

The bark is bitter and acrid, used for cooling, astringent to the bowels, anthelmintic, antidysentric and antipyretic; it also cures itching, sore throat, bronchitis, indigestion, ulcers, boils, psoriasis, inflammation, leprosy, anaemia, leucoderma. Bark yields the important brown 'Khathi' dye. The 'Khari' or 'Kath' extracted from the heart wood is widely used with betel. The wood is also used for construction of houses and agricultural fields.

Kalasamudramu (ATP), *SSR* 13754 & *NY* 2850; Lower Ahobilam (KNL), *TP & RVR* 2583; Anupu (GTR), *VRK* 3523; Cheedipalem (EG), *GVS* 27437 (MH); Itchapuram (SKLM), *KCJB* 17101 (MH).

INDIA : Throughout.
WORLD : Sri Lanka, Pakistan to South-Eastern Asia and South China.

**Acacia decurrens** (Wendl.) Willd., Sp. Pl. 4:1072. 1806; Gamble 1:430. *Mimosa decurrens* Wendl., Bot. Beobach. 59. 1798.

Evergreen tree, 6–15m tall; bark olive-green to dark grey, branchlets greyish, angled, glabrous, leaves 3-pinnate, base decurrent, rachis 5–10 cm long, glabrous, petiole 1–2.5 cm long; pinnae 7–15 pairs, 1.5–7 cm long; glands circular in between each pair of pinnae, leaflets 30–50 pairs, linear, 4–10 × 0.5–1 mm, glabrous, apex and base rounded, margin entire, veins obscure. Flowers bright yellow, fragrant, sessile; heads 5–8 mm in dia., terminal or axillary in panicles. Pod narrow, 5–11 × 0.5–0.8 cm, flat, constricted, glabrous or hairy, apex shortly acuminate, stipe short; seeds 3–14, semicircular, 5–6 × 3.5–4 mm.

Introduced, naturalised in few areas of Visakhapatnam district. Fl.:Dec.–May. Fr. Nov.–June. Eng. Green wattle, common wattle.

It is cultivated for timber. The bark is used in tanning industry in north-east India. The wood is mostly used as fuel. The wood is a fairly good raw material for the paper pulp. Sunkarimetta (VSKP), *GVS* 22674.

INDIA : Introduced. Naturalized in South Indian hills (between 1400–2000 m) and North East India (to 1500 m).
WORLD : Native of Australia.

**Acacia eburnea** (L.f.) Willd. Sp. Pl. 4: 1081. 1806; FBI 2: 293; *p.p.*; Gamble 1: 426 (302). Brandis 264. *Mimosa eburnea* L.f. Suppl. Pl. 437. 1781. *A. campbellii* Arn. in Nov. Act. Phys. - Med. Acad. Caes. Leop.-Carol. Nat. Cur. 18–333. 1836.

Deciduous tree, up to 7 m tall; bark grey; branchlets brown, warty, terete; spines paired, 1–2.3 cm long, straight, whitish, unequal, often connate at very base. Leaves 3–5 in a cluster, rachis 1–4 cm long, densely ochraceous or tawny puberulous, petiole 3–10 mm; pinnae 2–5 pairs, 0.3–2 cm long, pubescent; glands circular in between uppermost pair of pinnae; leaflets 5–12 pairs, coriaceous, linear-oblong or oblong, 0.5–5 × 0.3–1 mm, glabrous, apex obtuse or sub acute, margin entire, base oblique; sessile. Flowers yellow, foetid, sessile, in axillary, solitary or 2-together 8–12 mm dia. heads. Pod linear, flat, curved, 6–12 × 0.3–0.8 cm, slightly constricted, glabrous, thin, apex acute, base shortly attenuate; seeds 4–10, oblong-elliptic.

Common, in scrubs or dry deciduous forests up to 600 m sandy or black scils. Fl.: Nov.–Mar. Fr.: Jan.–July. Vern.: Marmati.

Chandragiri (CTR), *CECF* 4264, Kolavenna (KSN), *P. Venkanna* 5959 (AU); Venkatayapalem (KSN), *JSG* 18640; Kodur (CDP), *JSG* 15010 (MH).

INDIA : Karnataka, Madhya Pradesh, Maharashtra, Rajasthan, Tamil Nadu, Uttar Pradesh.
WORLD : Sri Lanka, Pakistan, Afghanistan, Arabia.

**Acacia farnesina** (L.) Willd. Sp. Pl. 4:1083; 1806; FBI 2:292; Gamble 1:425(301); Brandis 263; Beddome Fl. Sylv. S. India t 52, 1870. *Mimosa farnesiana* L. Sp. Pl. 521. 1753. *Vachellia farnesiana* (L.) Wt. & Arn. Prodr. Fl. Ind. Orient. 272. 1834.

Small armed tree, up to 5 m tall; bark brownish; branchlets reddish or brownish lenticellate, glabrous. Stipular spines whitish, 1–3 cm long. Leaves 2–5 on tubercles of branchlets, 2.5–3 cm long, pinnae 2–7 pairs, rachis, pubescent, and ending with 1 mm long spine, petiole 1.5 cm long; leaflets overlapping, 13–22 pairs, thin, narrowly oblong, 3.5 × 1 mm, glabrous, apex obtuse, margin entire, base truncate, mid nerve middle to towards margin; petiolule 0. Flowers yellowish-orange, 2 mm across, very fragrant in globose,

1.2 cm across, 1–3, axillary heads, peduncle to 1.3 cm long. Pod brown, oblong, 4.5–6.2 × 1.2–1.7 cm, turgid, pulpy, margined, straight to slightly curved, apex hooked; seeds 20–30, 2-seriate, brown, 0.6 × 0.5 mm, irregularly globose, smooth, glabrous.

Occasional, in scrub forests to 600 m. Fls.: Feb. Frs.: Mar.–May. Vern.: Kasturi, Kampa tumma, Arimidamu, Murki Tumna. Eng.: Fragrant acacia, Cassie flower, Sponge tree, stinking acacia.

The bark is alexiteric, anthelmintic, antidysentric and cures stomatitis, caries of teeth, itching, bronchitis, leucoderma, ulcers, inflammation etc. The tree yields a gum suitable for confectionary. Wood is hard and close-grained and used for tent pegs and ship knee.

Gandhari (NZB), *TP & BR* 6352; Kalava near Hussenapuram (KNL), *RVR* 3130; Devanpally (NZB), *BR & KH* 9634; Pochammaralu (MDK), *BR & CP* 11500; Mokarlabad (RR), *MSM & KH* 10262; Domalapenta (MBNR), *TDCK* 16826.

INDIA : Introduced, but naturalized throughout.
WORLD : Native of tropical America, now Pantropical.

**Acacia ferruginea** DC. Prodr. 2: 458. 1875; FBI 2: 295; Gamble 1: 425 (303); Brandis 266; Beddome Fl. Sylv. S. India t. 51. 1870. *Mimosa ferruginea* Roxb. [Hort. Beng. 41. 1814, nomen] Fl. Ind. 2. 561. 1832.

Deciduous armed tree, up to 10 m tall; bark greyish-white, fissured, cracks oblong; wood straight grained and coarse textured; branchlets brownish or blackish, glabrous; stipular spines to 0.5 cm long, straight or slightly curved. Leaf rachis to 8.5 cm long, petiole to 3 cm; pinnae 4–6 pairs, 2.5–8 cm long, glands present only between upper most pinnae; leaflets 13–22 pairs, narrowly oblong, slightly falcate, 0.8–1.3 × 2–3.5 cm, glabrous, greyish when dry, apex obtuse-subacute, margin entire, base truncate, mid vein towards the upper margin; petiolule 0. Flowers yellow, 2.8 mm across in 6.5–7.5 cm long, paired spikes. Pod dark-brown, flat, strap-shaped, 10–17 × 2.5 cm, stipitate, glabrous, reticulately veined, apex obtuse with apicule, base narrowed, seeds 3–7.

Occasional, in dry deciduous forests and scrubs, up to 900 m. Fls.: Mar.–May. Frs.: Nov.–Jan. Vern.: Ansadra, Inupa Tumma, Val-sandra, Vanni sundra.

Gum obtained from the plant is similar to gum-arabic and it is demulcent, emolient and nutrient. The bark is bitter, acrid and anthelmintic. The extract of leaves and bark are used as astringent. Wood hard and very heavy used in cart wheels, posts, beams and agricultural implements.

Tirumala (CTR), *SSR* 15972; Donubai (SKLM), *PRS & SSR* 15467; Venkatagiri Kona (NLR), *BSN* 3883; Neelwai RF (ADB), *TP & GO* 5447.

INDIA : South India to Rajasthan.
WORLD : Sri Lanka, Myanmar.

**Acacia holosericea** A. Cunn. ex G. Don. Gen. Hist. 2: 402. 1832; Sen & Naskar in Bull. Bot. Surv. India 7: 31. 1965.

Tree, 4–6 m tall; branchlets dark brown, angled; glabrescent; young parts silky-pubescent. Phyllodes coriaceous, oblong-elliptic to obovate, 8–15 × 1–2 cm, thinly puberulous on mid rib, greenish-brown when dry, apex acute-mucronate, margin entire, base attenuate and decurrent into a petiole; 4-nerved, conspicuous; reticulations obscure. Flowers yellow, dense, in 5–6 cm long axillary spikes. Pod black, strap-shaped, twisted, 8–11 × 3–4 cm, curved, constricted, glabrous, stipe short; seeds black, with yellow aril.

Cultivated in Tirumala gardens. Fls. & Frs. Jan.–July.

INDIA : Cultivated throughout.
WORLD : Native of Australia.

**Acacia horrida** (L.) Willd. Sp. Pl. 4. 1082. 1806. *Mimosa horrida* L. Suppl. Pl. 438. 1781. *Acacia latronum* (L.f.) Willd. Sp. Pl. 4: 1077. 1806; FBI 2: 296; Gamble 1: 427 (303); Beddome Fl. Sylv. S. India, 95. 1871.

Small or low tree, 3–5 m tall; crown spreading; branchlets black or brownish warty, terete, glabrous; spines straight, divergent, 2–6 cm long, white, paired, larger one hollow and horn like. Leaves bipinnate, rachis to 3 cm long, sparsely yellowish puberulous, petiole 2–3 cm long with sessile, circular, solitary gland near the base; pinnae 2–5 pairs, 0.5–1.5 cm long; leaflets 8–12 pairs, firmly chartaceous, linear or oblong, to 4 × 1 mm, glabrous, apex obtuse or subacute, margin entire, base subequal or rounded, mid vein conspicuous, reticulations obscure. Flowers pale yellowish, fragrant in solitary or 2–3 together in glabrous, axillary spikes. Pod dark reddish brown, 2.5–4 × 1–1.2 cm, falcate-oblong, turgid, curved, apex rounded to obtuse, glabrous, prominently veined, shortly stipitate; seed 1–4, ovate or suborbicular, flattened, 4.5–5.5 mm dia.

Frequent in dry forests or scrub. Fls.: July-Mar. Frs.: Sept.–May. Vern.: Paki-tumma.

Guvvalacheruvu Road (CDP), *KMS* 6841 (CAL); Muchukota RF (ATP), *NY* 503; Macherla RF (GTR), *VRK* 3838; Nagarjunasagar (NLG), *KMS* 7722 (MH); Kalasamudram RF (ATP), *EC* 2376.

INDIA : Maharashtra, South India — Endemic.
WORLD : Africa.

**Acacia leucophloea** (Roxb.) Willd. Sp. Pl. 4: 1083. 1806; FBI 2: 294; Gamble 1: 427 (302); Brandis 265; Beddome Fl. Sylv. S. India t. 48. 1870. *Mimosa leucophloea* Roxb. Pl. Cor. t. 150. 1800.

Deciduous armed tree, up to 15 m tall; bark grey, smooth when young, exfoliating in irregular scales; wood reddish-brown with darker streaks;

branchlets pale, brownish, striate, puberulous; spines straight, strong, 0.5–2.5 cm, tomentose; petiole to 10 mm; pinnae 4–7 pairs, 2.5 cm long, glands circular in between the 4 upper most and lower most pinnae; leaflets 12–28 pairs, chartaceous, elliptic or linear-oblong, 2 × 0.5 cm, glabrous-sparsely puberulous beneath, apex acute or rounded, margin ciliate, base oblique, mid vein and nerves obscure, sessile. Flowers cream; globose heads in panicles. Pod linear, curved, 10–20 × 0.5–1 cm, flat, rusty tomentose, somewhat bullate along seed, thin-walled, apex rounded, tardily dehiscent; seeds 10–20, irregularly elliptical, 5–6 × 4–5 mm, dark brown, smooth, and shining.

Common, in dry deciduous forests or on waste lands and open places, to 500 m. Recommended for plantations in degraded soils and it can also be tried in alluvial soils.Fls.: July–Nov. Frs.: Apr.–June or sometimes Feb.–Mar. Vern.: Tella Tumma.

The tree yields a gum which is used in indigenous medicine; it is demulcent and used in emulcifying agent. Paste of fresh leaves are applied on local swellings. Young fruits are used as vegetables. Bark yields a fibre, used for preparing fishing nets and cordage; bark mainly used in distillaries for clearing and flavouring liquors. The timber hard, tough and is valuable, used for making agricultural implements.

Dorasani bavi (KNL), *PRS & SSR* 15449; Errakonda South RF-Hindupur (ATP), *NY* 1061; Pulikunta RF (ADB), *PVP* 9470; Ghanapur (NZB), *BR & KH* 7117; Ponnala fields (MDK), *CP* 11393; Macherla RF (GTR), *VRK* 3972; Palakonda hills (CDP), *CS* 8201; Sri Rangapur (MBNR), *TDCK* 13675; Tirupati (CTR), *DR* 732 (SVU).

INDIA : Delhi, Gujarat, Madhya Pradesh, Maharashtra, Orissa, Punjab, Rajasthan, Tamil Nadu, Uttar Pradesh, West Bengal.

WORLD : Pakistan, Sri Lanka, Myanmar, Indochina, Thailand, Indonesia.

**Acacia mangium** Willd. Sp. Pl. 4: 1053. 1806; Maheshw, Fl. Delhi 15. 1968.

Small to medium sized tree; branchlets angular, whitish pubescent. Phyllodes chartaceous, thick, elliptic-oblong, to 2.1 × 4 cm, greyish when dry, slightly falcate, apex sub-acute, margin entire, base obliquely narrowed, 3-nerved from base up to apex; petiole to 0.8 cm. Flowers bright yellow, dense in 1–2 axillary elongated spikes. Pod twisted in irregular coils, splitting along the edges; seeds purely black, aril dark yellow.

Introduced. This is planted as plantation crop in afforestation programme in almost all areas. Grows on eroded, rocky, thin mineral soils. Fls.: Dec.–Feb. Frs.: Apr.–June.

SKU Campus (ATP), *SSR* 19672.

**Acacia nilotica** (L.) Willd. ex Del. subsp. ***indica*** (Benth.) Brenan in Kew Bull. 12: 84. 1957. *Mimosa nilotica* L. Sp. Pl. 521. 1753. *A. arabica*

(Lam.) Willd. var *indica* Benth. in London J. Bot. 1: 500. 1842. *A. arabica auct. non* (Lam.) Willd. 1806; FBI 2: 293; Gamble 1: 425 (301); Brandis 264; Beddome Fl. Sylv. S. India t. 42. 1870.

Evergreen tree, up to 15 m tall; crown spreading; bark dark brown or black, logitudinally fissured, or deeply cracked; wood pale red when fresh cut, nearly colourless on exposure turning reddish-brown; branchlets blackish or greyish, glabrous, stipular spines straight, to 5 cm long, slender, white in pairs. Leaves bipinnate, rachis 4–7 cm long, greyish-puberulous; petiole to 2 cm; pinnae 4 pairs, 2–5 cm long, cup-shaped glands at the base of lowest pair and upper most pair of pinnae; leaflets 15–21 pairs, oblong, 5 × 1 mm, apex rotund, margin entire, base obtuse, nerves obscure; petiolule 0. Flowers yellow, fragrant, in axilllary, globose, 10–12 mm dia. heads. Pod moniliform, to 15 × 1.6 cm, whitish-tomentose, constricted, bullate over seeds, apex horned, base attenuate; seeds ca. 12, ovoid, dark brown, compressed, shining, 5 × 4 mm.

Common, in waste lands, hedges, tank bunds, almost all districts in dry localities to 300 m. Recommended for plantations in alluvial loam, tank silt, black-cotton-soils, alkaline soils. It is a strong light demander, poor coppicer and moderately drought resistent species. It is never leafless the New foliage comes out in Feb.–Apr. Fls.: June–Sept. Frs. ripen in Apr.–May. Vern.: Nalla thumma. Eng.: Black Babul.

It is used for several purposes. The pods make an excellent cattle fodder. Wood is good, durable and valuable for agricultural implements. Gum exudes largely from wounds in the bark. Mostly during Mar.–May. The bark and pod are used for tanning. A decoction of the bark used as substitute for soap.

Kumbum (PKSM), *PRS* 15458; Martur forest (EG), *PRS* 15430; Samathagram (ATP), *NY* 187; Kurnool-Anantapur Road (KNL), *AA* 1944; Penganga river (ADB), *GO & PVP* 4287; Katasintha (HYD), *KMS* 6005 (MH); Aklaspur (KNR), *GVS* 22519 (MH); Chandragiri (CTR), *DR* 2279 (SVU).

INDIA : Throughout plains and hills.
WORLD : Sri Lanka, Nepal, Pakistan, Arabia, Egypt, tropical Africa.

**Acacia planifrons** Wt. & Arn. Prodr. Fl. Ind. Orient. 276. 1836; FBI 2: 293–1878, *p.p.*; Gamble 1: 426 (302); Brandis 264; Beddome Fl. Sylv. S. India 95. 1871.

Deciduous armed tree, up to 10 m tall; crown flat, umbrella-like branches spreading; bark greyish-brown, rough, wood pale yellow; branchlets hairy, dark brown, terete. Spines of two kinds, one is 2–5 cm long, cylindric, ivory white, another one short, 1 cm long, slightly curved, brown. Leaves 3–5 in a cluster, bipinnate, rachis to 6 cm long, glabrous or partly or sparsely whitish-pilose, petiole to 1 cm long, pinnae 5–6 pairs; one gland near the base of the petiole and another gland is in between upper most pinnae; leaflets 10–16 pairs, coriaceous, linear-oblong, 3–6 × 1–2 mm, pubescent, apex acute, margin entire, base sub-cordate. Flowers yellow in globose heads, heads in axillary fascicles. Pod subcylindrical, coiled into ring, to 8 × 1.5 cm, slightly turgid, base rounded to acute, glabrous; seeds 5–8.

Occasional in waste lands. Fls.: Jan.–Mar. Frs.: June onwards. Vern.: Sale, Godugu tumma, Godugu-jali. Wood hard and heavy, used for the manufacture of agricultural implements and as fuel. Pods contain tannin, used as fodder. The tree has been recorded as a host for lac insect.

Borgaum-Manchippa Road (NZB), *BR & KH* 9041; Gooty west (ATP), *JSG* 11052 (DD); Karempudi Ghat (GTR), *CAB* 4753 (MH): Guttur Kona (ATP), NY 980.

INDIA : Western Peninsula
WORLD : Sri Lanka.

**Acacia tomentosa** Willd. Sp. Pl. 4: 1087. 1806; FBI 2: 294; Gamble 1: 426 (302); Brandis 265.

Small tree, up to 6 m tall; bark yellowish, with large lenticels; young branches tomentose. Spines two, 0.5–1.5 cm long. Leaves to 6.5 cm long, rachis tomentose, glandular; pinnae 6–12 pairs, 1.2–4 cm long, leaflets 15–20 pairs, oblong, to 0.25 × 0.1 cm, pubescent, apex obtuse, margin entire, base truncate, petiolule 0. Flowers greenish-white in axillary heads; heads 1–5, 1.2 cm across, on stout 2.5–3.8 cm long peduncles. Pod flat, thick, falcate-slightly curved, pale grey, tomentose, apex and base bluntly apiculate; seeds 6–13, ellipsoid, pale grey, 8 mm long, compressed.

Rare in waste lands and hedges. Fls.: May.–June. Frs.: Dec.–Feb.

Indervelli (ADB), *GO & PVP* 4542; Sone (ADB), *TP & PVP* 4034.

INDIA : Peninsula, Rajasthan, Uttar Pradesh.
WORLD : Sri Lanka, Myanmar, Thailand, Indo-China.

## ADENANTHERA Linnaeus

**Adenanthera pavonia** L. Sp. Pl. 384. 1753; FBI 2: 287; Gamble 1: 418 (296); Brandis 262; Beddome Fl. Sylv. S. India t. 46. 1870.

Deciduous tree, up to 12 m tall; bark grey, slightly rough; wood red, cross-grained; branchlets glabrous. Leaves bipinnate, 20–30 cm long, petiole 5–12 cm; pinnae 4–6 pairs; leaflets papery, 4–8 pairs, elliptic-oblong, inequilateral, 1.8–3.2 × 1–1.2 cm, puberulous, dark green above, glaucous beneath, apex obtuse, margin entire, base truncate to shortly cuneate; petiolule to 2 mm. Flowers pale yellow and white, fragrant, 6 mm across, in spiciform, 10–20 cm long racemes. Pod brown, linear-flat, falcately curved, 13–24 × 0.8–1.2 cm, spirally twisted after dehiscence; seed 8–14, bright scarlet (very showy), smooth, globose, 0.7 cm dia, lenticular, compressed.

Commonly cultivated in gardens, University campus areas, village sides and also as an avenue plant. It is fast growing. Fls.: Mar.–May. Frs.:Aug.–Oct. Vern.: Bandi gurivinda. Eng.: Red wood, Bead tree, Coral wood tree.

Seeds are used as weights and are also strung beads for using as ornaments. Seeds are a source of dye. Wood hard, durable and polishes well. It is used in general construction purposes and in furniture and cabinate making. Gum is also produced by the tree. The wood is powdered and used as dye. A decoction made from both the seeds and the wood is used in pulmonary disorders and as an internal application in chronic ophthalmia. The seeds roasted and sheel are sometimes eaten. They are used for boils and inflammation and for the treatment of cholera and general paralysis. It is an excellent fuel.

Tirumala gardens (CTR), *SSR* 15973.

INDIA : Eastern Himalaya, Peninsula, Assam, Meghalaya, Tripura, Uttar Pradesh, Andaman & Nicobar islands, often planted.

WORLD : Sri Lanka, Myanmar, Thailand, Malaysia, China.

## **ALBIZIA** Durazz

Chakrabarthy and Gangopadhyay (1996) have reviewed *Albizia* in India, where they excluded *A. sikharamensis.*

1. Pinnae 2–5 pairs:
    2. Leaflets 3.5–7 × 2.5–4 cm; petiolule to 3 mm ........ ***A. procera***
    2. Leaflets 0.3–4.8 × 0.5–2.2 cm; petiolule 0–1 mm:
        3. Flowers sessile; heads in panicles. Pod brownish. Leaflets 4–15 pairs ........ ***A. odoratissima***
        3. Flowers pedicellate; heads in umbellate clusters. Pod yellow. Leaflets 4–8 pairs ........ ***A. lebbeck***
1. Pinnae 5–13 pairs:
    4. Leaflets 1–2 × 0.5-0.7 cm. Pods reddish ........ ***A. thompsonii***
    4. Leaflets 3–9 × 1–2 mm. Pod yellow, or yellowish brown:
        5. Stipules foliaceous and conspicuous, > 2 cm long. Mid vein of leaflets strictly marginal ........ ***A. chinensis***
        5. Stipules not foliaceous and conspicuous, < 2 cm long. Mid vein of leaflets ascending slightly away from distal margin ........ ***A. amara***

**Albizia amara** (Roxb.) Boiv in Encycl. 19: 34. 1838; FBI 2: 301;Gamble 1: 432 (306); Brandis 272; Beddome Fl. Sylv. S. India t. 61. 1870. *Mimosa amara* Roxb. Pl. Cor. t. 122. 1799.

Deciduous tree, up to 15 m tall; crown spreading; bark yellow or grey, smooth with horizontal lines, scaly; wood purplish-brown, straight grained, medium textured; branchlets densely yellowish or grey pubescent. Leaves bipinnate; rachis to 12 cm long, densely yellowish puberulous, petiole 1–3 cm long, gland near middle and above it; pinnae 5–10 pairs, 1.5–6 cm long, glands in between 4 upper most pinnae; leaflets 15–25 pairs, linear-oblong, 3.5 × 1 mm, glabrous, apex subacute, margin entire, base unequal; mid vein nearly ascending, slightly away from distal margin; sessile. Flowers whitish-yellow; heads in axillary fascicles. Pod yellowish-brown, straight, strap-shaped, flat, 10–20 × 2–3.5 cm, thin, margins undulated; seeds 6–12, ovate to suborbicular, 7 × 5.5 mm.

Common, in dry deciduous forests to 500 m. It grows on very poor and shallow soils. Leaf fall: Dec.–Jan.. New foliage: Mar.–May. Fls.: Apr.–June. Frs.: Nov.–Jan. Vern.: Narlingi chikreni, Chigara, Nalla ragi, Konda chigara, Chikoni. Eng: Oil cake tree.

The leaves provide green manure used as an emergency cattle feed. The flowers are externally applied for skin treatment. Wood is considered as a first class wood for cabinet making; it is also an excellent fuel wood. The tree is used for shelter belts, wind breaks and for afforestation.

Tirumala (CTR), *SSR* 13735; Srisailam (KNL), *SSR* 15520; Pangidi (KNL), *SSR* 15992; Akkampet (MDK), *BR & CP* 11564; Palakonda hills (CDP), *CS* 9357; Machareddy RF (NZB), *BR & KH* 9050; Macherla RF (GTR), *VRK* 3818; Pennahobilam (ATP), *TP & NY* 330; Mohammadabad (RR), *MSM & KH* 10210; Anantagiri RF (VSKP), *KMS* 2285 (MH); Sangam (NLR), *MSR* 1133 (CAL); Aklaspur (KNR), *GVS* 25607 (MH); Papampet (EG), *PRS* 15418; Diguvametta (PKSM), *RVK* 15734.

INDIA : South India, Gujarat, cultivated elsewhere.
WORLD : Sri Lanka, East Africa, Abyssinia.

**Albizia chinensis** (Osb.) Merr. in Amer. J. Bot. 3: 575. 1916. *Mimosa chinensis* Osb. Dagbok. Ostind. Resa 233. 1757. *Albizia stipulata* (Roxb.) Boiv. in Encycl. 19: 33. 1838; FBI 2: 300; Brandis 272; Beddome Fl. Sylv. S. India t. 55. 1870. *A. marginata* (Lam.) Merr. in Philipp. J. Sci. 5. 23. 1910; Gamble 1: 433 (307).

Deciduous tree, up to 20 m tall; crown broad, spreading, flat-topped; bark pale grey or greenish-grey, smooth, horizontally furrowed; wood brown, branchlets yellowish pubescent; stipules foliaceous, unequally ovate-oblong to sub reniform or subcordate. Leaves bipinnate, rachis 11–18 cm long, tomentose; petiole to 4.5 cm, gland just below the middle; pinnae 10–13 pairs, 4–8 cm long, small glands in between all pairs of pinnae; leaflets 15–35 pairs, chartaceous, linear-oblong, falcate, 5–9 × 1.5–2 mm, pubescent, pale green above, glaucous beneath, apex acute and pointed, margin ciliate, base semi truncate or oblique; mid vein starting and ascending along the distal margin; sessile. Flowers pinkish-white, light yellow or reddish, fragrant in axillary or terminal, racemose panicles. Pod reddish or yellowish brown, glabrous, 9–11.5 × 1.5–1.7 cm, linear-oblong, flat, compressed, apex rounded to acute, base acute, shortly stipitate.

Occasional, in deciduous forests, particularly along river banks or streams to 700 m. Leaf fall: Jan.–Feb. New foliage: Mar. Fls.: Mar.–July. Frs.: Sept.–

Apr. Vern.: Nalla regi, Bandi chindugu, Konda-chiragu. Eng.: Stipulate siris. Leaves and twigs lopped for fodder. Wood is soft and of less value, used for making planks, packing boxes, dug-outs etc. An infusion of the bark is used as a lotion for cuts, scabies and other skin diseases.

Anantagiri (RR), *SSR* 18510; Sunkarimetta (VSKP), *PRS & SSR* 15475; Borra caves (VSKP), *PRS* 15477; Molachintapally (MBNR), *TDCK* 15303.

INDIA : Throughout.
WORLD : Nepal, Sri Lanka, Pakistan, Bhutan, China, Myanmar, Malaysia, Thailand, Cambodia, Laos, Vietnam, often cultivated.

**Albizia lebbeck** (L.) Benth. in Hook. J. Bot. 3: 87. 1844. *p.p*, '*lebbek*' et Trans. Linn. Soc. 30: 562. 1875; FBI 2: 298; Gamble 1: 432 (306); Brandis 271; Beddome Fl. Sylv. S. India t. 53. 1870. *Mimosa lebbeck* L. Sp. Pl. 516. 1753.

Deciduous tree, 10–20 m tall; crown spreading, bark brownish-grey, streaked; wood dark brown, shining, mottled; branchlets yellowish or brownish, glabrous, lenticellate. Leaves bipinnate; rachis to 18 cm long; petiole 5–8 cm long, gland below the middle; pinnae 2–4 pairs, 5–12 cm long, glands on the 2–5 upper pair of pinnae; leaflets 4–8 pairs, oblong, terminal pair of leaflets obovate, 2–4.8 × 0.8–2.2 cm, minute glands in between the leaflets, apex rounded or obtuse, margin entire, base obliquely obtuse, drying green; mid vein towards distal margin, secondary nerves prominent; petiolule 1 mm. Flowers greenish-yellow, fragrant; heads in axillary umbellate clusters. Pod pale yellow, linear-oblong, strap-shaped, 15–25 × 2–4.8 cm, smooth, glabrous, apex narrowed and rounded, base acute to attenuate, margin shortly sinuate, some what bullate over seeds, seeds 2–12, ovate-oblong, pale brown, flattened, 9–10 × 6–7 mm.

Common in moist forests, all plains, widely planted along roadsides and villages to 800 m. It is well-coppicer and grows well in deep loamy soils with good moisture, while stiff clay, gravelly soils restrict the growth. Leaf fall: Mar.–Apr. New foliage: May. Fls.: Mar.–June. Frs.: Jan. and remains on the tree till the following March and April. Vern.: Pedda dirisinam, Dirisana, Sirisamu.

Leaves and twigs are used for fodder. A reddish-brown gum exudes from bark and serves as substitute of the gum arabic. The flowers on steam distillation gives a colourless sweet-smelling oil. Wood hard and useful for building purposes.

SK Univeristy Campus (ATP), *SSR* 15548; Palakonda RF — Sanipaya Range (CDP), *KNS* 2387 (DD); Gani RF (KNL), *TP & RVR* 1799; Vijayapuri South (GTR), *VRK* 3950; Urlakonda (NLR), *MSR* 1316 (CAL); Jaggayyapet (SKLM), *GVS* 62396 (MH); Kundaram RF (ADB), *GO* 4493; Neela (NZB), *BR & GO* 9222; Diguvametta (PKSM), *RVK* 18306; Tirupati (CTR), *DR* 1330 (SVU).

INDIA : Throughtout.
WORLD : Tropical, subtropical Asia and Africa; often cultivated throughout tropics.

**Albizia odoratissima** (L.f.) Benth. in London J. Bot. 3: 88. 1844; FBI 2.299; Gamble 431 (306), Brandis 27·1; Beddome Fl. Sylv. S. India t. 54. 1870. *Mimosa odoratissima* L.f. Suppl. Pl. 437. 1781.

Large deciduous tree, 15–25 m tall; though never quite leafless; foliage drooping; bark dark-grey, marked with numerous, narrow, horizontal wrinkles; wood dark brown; branchlets blackish-brown, glabrescent. Leaves bipinnate, rachis 20 cm long, brown tomentose, petiole 7.2 cm long, gland near the middle or towards base; pinnae 2–5 pairs, 8–14 cm long, glands in between upper-most pair of pinnae; leaflets 4–15 pairs. chartaceous or thinly coriaceous, obliquely oblong, 0.3–1.8 × 0.5–1.4 cm, white pilose to glabrous on both surfaces, apex obtuse, margin entire, base obliquely rounded, usually drying grey; mid vein closer to upper margin, secondary nerves conspicuous beneath; petiolule 1 mm. Flowers pale yellowish-white, fragrant; heads in 25 cm long terminal panicles. Pod oblong, reddish-brown, flat, 1·5–30 × 1.8–3 cm, with darker blotches indicating the position of the seeds, glabrous, finely reticulate veined, apex rounded, base obtuse, shortly stipitate.

Common, in evergreen patches in deciduous forests to 1000 m. It coppices well and produces root suckers. Fls.: Apr.–June. Frs.: Oct.–Feb. Vern.: Konda chigara, Chinta yelagaku, Chindugu, Bandi sindiga, Karuvaghe, Ganera.

The twigs are lopped for fodder. The leaves boiled in ghee are used as cough remedy. The timber is widely employed for many purposes, particularly for construction of houses, agricultural implements, carts etc. The plant yields gum and dye (from bark).

Tirumala (CTR), *SSR* 13707; Srisailam (KNL), *SSR* 15521; Pinjeri Konda (EG), *PRS & SSR* 15414; Narsapur (MDK), *SSR & CP* 18503; Vijayapuri South (GTR), *VRK* 6815; Palakonda hills (CDP), *CS* 9353; Jalapur RF (NZB), *BR & KH* 9663; Kalmam Kalva (RR), *MSM & KH* 10304; Bheemram (ADB), *GO & DAM* 5186; Barmakonda (VSKP), *KCJB* 17144; Murremkonda (NLR), *MSR* 1282 (CAL); Farahabad (MBNR), *TDCK* 15364.

INDIA : Andaman Islands, Assam, Bihar, Delhi, Gujarat, Karnataka, Madhaya Pradesh, Meghalaya, Orissa, Rajasthan, Tamil Nadu, West Bengal.
WORLD : Sri Lanka, Pakistan, Nepal, Bhutan, tropical and sub-tropical South-East Africa; cultivated.

**Albizia procera** (Roxb.) Benth. in Hook. London J. Bot. 3: 89. 1844 and in Trans. Linn. Soc. London 30. 564. 1875; FBI 2: 299; Gamble 1: 432 (306); Brandis 271. *Mimosa procera* Roxb. Pl. Cor. t. 121. 1798.

Deciduous tree, up to 25 m tall; bark smooth, yellowish-white; wood brown, close-grained; young branches pale, lenticellate. Leaves bipinnate; rachis 6–10 cm long, glabrous; petiole 8–14 cm, large gland just little above the base; pinnae 2–3 pairs, 8–20 cm long, oval glands below the 2 uppermost pinnae on the rachis; leaflets 3–6 pairs, chartaceous to coriaceous, obliquely ovate to elliptic, glabrous above, adpressed whitish pilose beneath, apex obtuse, margin entire, base oblique; mid vein subcentral to slightly towards margin; secondary nerves 8–10 pairs; petiolule 2–3 mm. Flowers yellowish-white in heads on 10–15 cm long axillary and terminal panicles. Pod oblong, strap-shaped, 10–20 × 2–2.7 cm, flat, thin, orange-brown, glabrous, glossy, obscurely reticulate veined; seeds 4–10, ovate-elliptic, flattened, 7 × 4.5 mm.

Occasional, in deciduous forests, on river banks and along streams to 600 m. It is suitable for plantations in sandy and sandy loams with good moisture supply. Fls.:May–Aug. Frs.: Apr.–May of the following year. Vern.: Pasaraganni, Pachardu, Tella-chinduga, Tella chinta. Eng.: White siris.

All parts of the plant are reported to show anticancerous activity. A decoction of the bark is given in rheumatism and haemorrhage. It is used as fish poison. Wood hard and useful for many purposes, particularly for houses, furniture, carts and carriage, tea-boxes, dug-outs, rice pounders etc.

Rampur (ADB), *SSR* 19305; Gopalpet (NZB), *BR* 9505; Gazulapalli (KNL), *CAB* 8052 (CAL); Galikonda (VSKP), *CAB* 11768 (MH).

INDIA : Arunachal Pradesh, Bihar, Delhi, Gujarat, Jammu, Meghalaya, Nagaland, Rajasthan, Uttar Pradesh, Assam, Karnataka, Himachal Pradesh, Madhya Pradesh, Manipur, Meghalaya, Tamil Nadu.

WORLD : Tropical Asia, North-east Australia.

**Albizia thompsonii** Brandis in Ind. For. 25: 284. 1899; Gamble 1: 432 (306); Brandis 271.

Large deciduous tree; branches spreading; bark grey, smooth, slightly cracked below; branchlets brownish, terete, pubescent. Leaves bipinnate; rachis 10–15 cm long; petiole 3–8 cm long, one large gland near base; pinnae 3–7 pairs, 5–12 cm long, small glands in between the uppermost pinnae; leaflets 10–18 pairs, oblong, slightly falcate, 10–20 × 5–7 mm, apex obtuse or acute, margin entire, base semicordate, ashy grey when dry; mid vein central at base or slightly towards distal margin, secondary nerves obscure to faint. Flowers yellowish, fragrant, pedicellate in axillary fascicles, bearing few-flowered corymbs. Pod rich brown, 15–24 × 2.5–2.8 cm, moderately thick, beaked, glabrous, stipitate; seeds 7–10, elliptic, 5.2 × 4 mm.

Occasional, in deciduous forests, between 300–900 m altitudes. Leaf fall: Jan.–Feb. New foliage: Mar. Fls.: Feb.–Apr. Frs. July–Feb. Vern.: Kondala.

Mannanur RF (MBNR), *SSR* 15531; Appapur (MBNR), *SSR* 15545; Kalasamudram RF (ATP), EC 2978; Sunkarimetta (VSKP), *PRS* 15474.

INDIA : Madhya Pradesh, Orissa, Tamil Nadu. Endemic.

## CALLIANDRA Bentham

**Calliandra inermis** (L.) Druce, Rep. Bot. Exch. Club Brit. Isles 3: 415. 1914. *Gleditsia inermis* L. Sp. Pl. (ed. 2), 1509, 1763.

Small tree, up to 4 m tall. Leaves 2-pinnate; leaflets ca. 50 pairs, elliptic, 5–9 × 1–2 mm; membranous, margin sparsely ciliate. Flowers purplish in racemes. Pod oblong, 11 × 1.5 cm, flattened, thickned along margins, seeds ca. 7.

Occasional in gardens. Fls. & Frs.: July–Dec.

Vijayawada (KSN), *PV* 6037.

INDIA : Cultivated in gardens.
WORLD : Native of Mexico.

## DICHROSTACHYS (A.P. deCandolle) R. Wight & Arnott *nom. cons.*

**Dichrostachys cinerea** (L.) Wt. & Arn. Prodr. Fl. Ind. Orient. 271. 1834; FBI 2: 288; Gamble 1: 419 (297); Brandis 261; Beddome Fl. Sylv. S. India t. 185. 1872. ssp. *cinerea* var. *cinerea* Brenan & Brummit in Bot. Soc. Brot (Ser. 2) 39. 110. 1965; Kosterm. in Dassan. & Fosb.Rev. Handb. Fl. Ceylon. T. 466. 1980. *Mimosa cinerea* L. Sp. Pl. 520. 1753 non L. 517. 1753.

Armed, small tree, up to 7 m tall; bark grey, smooth; wood dark purple; branchlets spiny tipped. Leaves alternate or 2 to 3 in a cluster, bipinnate, to 6 cm long, petiole to 1 cm, rachis softly pubescent, with small glands between each pair of pinnae, pinnae 4–10 pairs; leaflets 10–25 pairs, linear-oblong, to 3 × 1 mm, close, glabrous above, pubescent beneath, apex subacute, margin entire, base oblique, sessile. Flowers numerous, dense in axillary spikes, spikes cylindric, the upper half yellow, the lower pink. Pod linear, flat, 5–7 cm long, irregularly twisted, dark brown, glabrous; seeds 6–10, ovoid, compressed.

Common, in scrub jungles and open forests, mainly in dry stony hills to 500 m. Fls.: May.–Aug. Frs.: Nov.–Dec. Vern.: Velthuru, Nalla veneturu, Nela jammi, Yertur. Wood hard and tough, used as a good fuel.

Penjera konda (EG), *PRS & SSR* 15413; Palakonda hills (CDP), *CS* 9301; Kalasamudaram RF (ATP), *NY* 680; Narsapur RF (MDK), *TP & CP* 14057; North Dhone RF (KNL), *TP & RVR* 1313; Macherla RF (GTR), VRK 3858; Gadirayala (RR), *MSM & KH* 10252, Tirupati (CTR), *DR* 120 (SVU).

INDIA : Karnataka, Madhya Pradesh, Maharashtra, Orissa, Rajasthan, Tamil Nadu, Uttar Pradesh. Endemic.

## **LEUCAENA** Bentham

**Leucaena latisiliqua** (L.) Gillis in Taxon 23: 190. 1974. *Mimosa latisiliqua* L. Sp. Pl. 519. 1753. *Leucaena leucocephala* (Lam.) de Wit in Taxon 10: 53. 1961. *Mimosa leucocephala* Lam. Encycl. 1:12, 1783. *Mimosa glauca* L. Sp. Pl. ed 2: 1504, 1763 non L. 1753. *Leucaena glauca auct. non* Benth. in J. Bot. (Hooker) 4: 416. 1845; FBI 2: 290; Gamble: 419 (297).

Deciduous tree, up to 12 m tall; bark smooth, branches glabrescent. Leaves bipinnate, to 30 cm long, pinnae 4–8 pairs, rachis pubescent, ending in a weak spine, petiole gland just below the first pair of pinnae. Leaflets 15–20 pairs, membranous, linear-oblong; 0.4–1.6 × 0.2–0.3 mm, glabrous above, sparsely or finely pubescent beneath, apex acute, margin entire, base oblique. Flowers white in solitary globose heads, or two in leaf axils. Pod flat, linear, oblong, compressed, 8–10 × 1–1.5 cm, reddish brown; seeds 15–30.

Commonly found near villages and naturalized. It is a native of Central America. It is recommended for plantation in a variety of soils although its best growth is obtained on deep and fertile soils. It is drought resistent but cannot tolerate prolonged drought. It is good coppicer. Fls.: May to July and Nov.–Feb. Frs.: July to Feb. It is cultivated for green manure, fodder, fuel. Vern: Kaniti, Subabul.

Thutigunta (WG), *PRS & SSR* 15420; Bank side of Godavari (EG), *PRS* 15420; Mamalapadu (KNL), *AA* 1970; Nalla Padu (GTR), *VRK* 6761; Anantapur town (ATP), *NY* 329; Chandragiri (CTR), *CAB* 10095 (MH); Araku towards Jolaput (VSKP), *GVS* 19691 (MH); Siddipet (MDK), *CP* 11375; Dharmaram (NZB), *BR & KH* 6460.

INDIA : Throughout.

## **PARKIA** R. Brown

**Parkia biglandulosa** Wt. & Arn. Prod. 279. 1834; FBI 2: 289. *Mimosa pedunculata* Roxb. Fl. Ind. 2: 551. 1832. non Poir. 1787. *Parkia pedunculata* (Roxb.) Macbr., Contrib. Gray Herb. n.s. 59: 19. 1919.

Unarmed tree, up to 30 m tall; branchlets pubescent. Leaves bipinnate, to 30 cm long, petiole to 3 cm, one pair of basal glands on the base, rachis pubescent, pinnae 20–35 pairs, glands in between the 6 uppermost pairs of pinnae, leaflets many (up to 80) pairs; linear-oblong, 3–5 × 0.2-0.4 mm;

glabrous above, puberulous below, apex sub acute-obtuse, margin entire, base truncate; sessile. Flowers small in large, globose, solitary heads. Flower heads yellow, looks like Batminton ball and pendulous. Pod linear-oblong-compressed, to 25 × 2 cm, dark brown,stipitate, to 3 cm, glabrous; horned; seeds 6–20, ellipsoid to ovoid, 1.1 × 0.8 cm, light brown.

Introduced, often planted. Fls.: Feb.-Mar. Frs.: May onwards.

Tirumala Ghat Road (CTR), *SSR* 16189; SVU Campus (CTR), *SSR* 15557; Araku (VSKP) *PRS* 15488.

INDIA : In most parts. Often cultivated.
WORLD : Throughout South-Eastern Asia.

## **PITHECELLOBIUM** C.F.P. Martius

**Pithecellobium dulce** (Roxb.) Benth. in London J. Bot. 3: 199. 1844; FBI 2: 285; Gamble 1: 416 (295); Brandis 273; Beddome Fl. Sylv. S. India t. 188. 1972. *Mimosa dulcis* Roxb. Pl. Cor. t. 99. 1798. *Inga dulcis* (Roxb.) Willd. Sp. Pl. 4: 1005. 1806.

Evergreen, armed tree, up to 18 m tall; bark grey, smooth, with yellowish white lenticels. Wood reddish-white, branchlets glabrous. Leaves bipinnate, petiole to 3.1 cm; pinnae 2 pairs, with a gland on the petiole and one below each pair of pinnae; leaflets paired, chartaceous, elliptic-oblong, 2–4.5 × 1–2.2 cm, glabrous, apex obtuse with apicule, margin entire, base obliquely obtuse, sessile. Flowers white in globose heads on 20 cm long, axillary and terminal panicles. Pod moniliform, torulose, 10–15 × 0.5–1.2 cm, pale green, reddish when ripe, spirally twisted; seeds 5–9, shining black with reddish-white aril.

Common, in dry plains and cultivated as a hedge plant to 800 m. It grows on variety of soils. It coppices well and can withstand any amount of pruning, lopping or browsing by animals. Fls.: Nov.–Feb. Frs.: Mar.–June Vern.: Simachinta. Eng.: Manila Tamarind.

Wood strong and useful to general construction as packing cases, fence posts, cart-building and agricultural implements; it is an excellent fuel wood. Pods (aril part) are edible. Leaves and pods are used as fodder for cattle and other livestock. The bark contains yellow dye.

Bhimudipakalu (EG), *PRS & SSR* 15410; Zaheerabad (MDK), *CP* 11598; Mugalapadu (KNL), *AA* 2040; Vijayapuri South (GTR), *VRK* 3939; Anantapur town (ATP), *NY* 1206; Mamidipally (NZB), *BR & KH* 9011; Way to Tirumala (CTR), *DR* 2069 (SVU).

INDIA : Cultivated and also naturalized throughout.
WORLD : Native of Central America.

## **PROSOPIS** Linnaeus

1. Leaflets > 1.8 cm long. .................................................................................. ***P. glandulosa***
1. Leaflets < 1.8 cm long:
    2. Pod cylindrical, prickles small. Leaflets 7–10 pairs, petiole to 1 cm. .......... ***P. cineraria***
    2. Pod compressed, thorns large. Leaflets 15–22 pairs, petiole to 3 cm. ................ ***P. juliflora***

**Prosopis cineraria** (L.) Druce. Rep. Bot. Soc. Exch. Club Brit. Isles 3: 422. 1914. *Mimosa cineraria* L. Sp. Pl. ed. 2: 1500. 1763 *excl. ref.* Br (*cinerea*). *Prosopis spicigera* L. Mant. 68. 1767; FBI 2: 288; Gamble 1: 421 (297); Brandis 260.

Evergreen tree, up to 10 m tall; tap root long; bark grey, rough, with deep fissures and cracks; wood dark brown, spines conical. Leaves bipinnate, 3–5 cm long, pinnae 2 pairs, leaflets 7–10 pairs, oblong, 0.5–1.3 × 0.3–0.5 cm, appressed hairy, apex rounded, margin entire, base oblique; petiole to 1 cm. Flowers small, yellow in 6–12 cm long slender spikes. Pods cylindric, pendent, straight or curved, glabrous, 12.5–22.5 cm long, coriaceous, filled with a dry sweetish pulp, seeds 10–15, oblong, compressed, moderately hard.

Common on dry stony lands, waste places in open forests, more rarely in interior forests to 400 m. It is light demander and coppices well, and also it is remarkable plant for the indication of locality of extreme temperature. Fls.: Oct.-Dec. Frs.: Mar.–June. Vern.: Jammi, Jambichani.

The pods form a valuable food for cattle and goats. Wood hard, suitable for interior construction work, such as windows, doors etc. and used as a good fuel. The bark (which has sweetish taste) as well as the galls formed on the leaves are used for tanning. Leaves are useful for fodder and green manuring. The tree exudes gum from the cut ends of branches.

Vanipenta (KNL), *PRS & SSR* 15447; Maredumilli (EG), *PRS & SSR* 15421; Reddipalli form (ATP), *NY* 12551; Vaddemanu (KNL), *RVR* 2540; Macherla RF (GTR), *VRK* 1292; Bhimavaram (WG), *CAB* 4300 (MH); Chandragiri (CTR), *KCJB* 10094 (MH); Sidhout (CDP), *JSG* 10869 (DD); Annavaram (EG), *JSG* 16103 (CAL); Duttalur (NLR), *CECF* 4732 (CAL).

INDIA : South India, Bihar, Delhi, Gujarat, Punjab, Rajasthan, Uttar Pradesh.
WORLD : Pakistan, Afghanistan, Iran, Arabia.

**Prosopis glandulosa** Torrey in Ann. Lyceum Nat. His. (Mist) New York. 2: 192. t. 2. 1827. var. *glandulosa* Johnston in Brittonia 14: 82. 1962; Burkart in J. Arnold Arb. 57: 564. 1976. *Prosopis julifloria* (Sw.) DC. var. *glandulosa* (Torrey) Cockerell. in Bull. Agri. Exp. Stat. New Mexico 15: 58. 1895, Parker, For. Fl. Punjab. 204. 1924.

Small tree, up to 4 m tall; bark grey; branchlets glabrous. Leaves bipinnate, petiole to 5 cm; pinnae 1 pair, to 15 cm long; leaflets 12–14 pairs, linear oblong, 2–3.5 × 0.3-0.5 cm, glabrous, apex sub-acute, margin entire, base cuneate; petiolule 1 mm. Flowers creamish-yellow in 7.5 cm long, axillary spikes. Pods straight, beaked, 14–18 × 0.8–1 cm; seeds more than 10.

Rare, in Owk forest of Kurnool district. Fls.: Jan.–Feb. Frs.: Mar.–Apr.
Owk RF (KNL), *RVR* 3049.

INDIA : Introduced and naturalised. Delhi, Punjab, Rajasthan.
WORLD : Native of USA.

**Prosopis juliflora** (Swartz) DC., Prod. 2: 447. 1825. *Mimosa juliflora* Swartz. Prodr. 85. 1788. *Prosopis chilensis* (Molina) Stuntz in USDA, Bur. Pl. Industr. Invent. Seeds 31: 85. 1914.

Small evergreen tree, up to 7 m tall; branches drooping; bark greyish brown, wood dark brown, irregular grained; branchlets glabrous, spinescent; Leaves alternate, 2-pinnate, to 6 cm; pinnae 1 or 2 pairs; leaflets 15–22 pairs, chartaceous, linear-oblong, 0.6–1.6 × 0.2–0.3 cm, apex slightly acute, margin entire, base obtuse; petiole to 3 cm. Flowers cream in 12 cm long spikes. Pod yellow, slightly curved, flat or cylindrical with transverse depressions between seeds, 10–18 × 0.5–0.8 cm, indehiscent, pendent; seeds 12–34, ovoid, flattened, 1.7 × 3 mm, hard, yellowish-brown, shiny.

Common, in waste places, rare in dry deciduous forests. It can grow up in rocky hills, sandy and saline patches, shifting sand dunes and coastal sands. It cannot tolerate water logged conditions. It is suitable for afforestation for arid and semi-arid lands. Fls.: Sept.–Oct. and Feb.–Mar. Frs.: Nov.–Dec. and Apr.–June. Vern.: Mulla thumma, Sarear thumma, Jalichettu.

It is an immediate source of fuel for local people and for agricultural implements. It yields smooth, light yellowish brown gum which can be used as emulcifying agent. Wood is very hard, strong, heavy and durable, which is used for fence posts.

Upper Ahobilam (KNL), *PRS & SSR* 1544; Jannaram (ADB), *TP & PVP* 4209; Itikalapalli (ATP), *NY* 168; Palakonda hills (CDP), *CS* 7761; Anupu (GTR), *VRK* 3595; Rangapuram RF (KNL), *RVR* 1425; Patancheruv (MDK), *CP* 11692; Indelwai (NZB), *TP & BR* 6410, Nerabylu (CTR), *DR* 317 (SVU).

INDIA : Introduced for afforestation, now naturalised throughout arid areas.
WORLD : Native of tropical America.

## SAMANEA (A. de Candolle) Merrill

**Samanea saman** (Jacq.) Merr.in J. Wash. Acad. Sci. 6: 46. 1916. *Mimosa saman* Jacq. Frag. Bot. 15, t, 9. 1800. *Inga saman* Willd. Sp. Pl. 4: 1024. 1806. *Enterlobium saman* (Jacq.) Prain. in J. Asiat. Soc. Beng. 66: 252, 1897.

Densely foliaceous, evergreen tree, up to 15 m tall; trunk short, crown broad, spreading; bark dark grey, rough; branchlets yellowish, tomentose. Leaves bipinnate to 12 cm; petiole to 1 cm, eglandular, pinnae 4–5 pairs, glands in between the all pairs of pinnae, leaflets 4–7 pairs, small glands are in between the all pairs of leaflets, upper ones obliquely obovate, to 1.6 × 0.3 cm, lower ones obliquely-elliptic, to 2.5 × 1 cm, glabrous above, yellow pubescent below, apex obtuse with minute apicule, margin entire, base truncate, sessile. Flowers numerous, rose coloured in terminal or axillary racemes. Pod linear-oblong, to 15 × 2 cm, dark brown, straight, thick margined, flattened, leathery, indehiscent; seeds up to 20, to 1.3 × 0.6 cm, brown, surrounded by brownish pulp.

Introduced. It is planted for shade along avenues and also in pasture and it is also fast growing tree. It thrives well in heavy to moderate rainfall areas, best in hot moist locality. It can be grown on dry and barren soils. Suitable for planting in saline and alkaline areas. Fls.: May–June. Frs.: Slowly and fully ripen in Mar.–Apr. of the following year. Vern.: Nidraganneru. Eng.: Rain tree.

The leaves and pods are esteemed as fodder for livestock. The pulp of the ripe pods is rich in fermentable sugars and may be employed for the production of alcohol. It is one of the recorded host plants of the lac insect in south India. The lac is of poor quality; it is reddish and is small brittle. The tree yields inferior quality of dark coloured gum. Wood is light and soft, suitable for furniture or even as fuel.

Sunnipenta (KNL), *PRS* 15457.

INDIA : Cultivated almost throughout.
WORLD : A native of Central America, West Indies.

## **XYLIA** Bentham

**Xylia xylocarpa** (Roxb.) Taub. in Bot. Central 1. 47. 395. 1891; Gamble 1: 417 (295). *Mimosa xylocarpa* Roxb. Pl. Cor. t. 100. 1798. *Xylia dolabriformis* Benth. in J. Bot. (Hooker) 4: 47. 1842; Brandis 262. Beddome Fl. Sylv. S. India t. 186. 1872.

Moderate to large deciduous tree, up to 15 m tall; bark thick, smooth, reddish-grey exfoliating in large irregular plates; wood reddish brown; young parts tomentose. Leaves bipinnate, 15–20 cm long, petiole to 4.9 cm with a gland at tip; pinnae 2, terminal on the petiole; leaflets 3–5 pairs, diminishing downwards, often a small unopposed leaflets situated below the lowest pair of leaflets, sub-coriaceous, elliptic-lanceolate to oblong, 2–14 × 2–4.9 cm, margin entire, apex acute, base rounded; petiole to 2 mm. Flowers sessile, yellowish white, sweet scented in long pedunculate globose heads. Pod falcate-oblong, flat, woody, 10–16 × 2.5–4.5 cm, light brown; glabrous; seeds 6–10, oblong, ellipsoid, brown, 1.3–1.6 × 0.5–1 cm, smooth, polished.

Common, in forests of North circars, Warangal and Adilabad districts. The tree coppices and pollards well and produces root suckers in abundance, particularly in burnt areas where the roots are exposed. Leaf fall: Feb. New foliage: Apr. Fls.: Mar.–Apr. Frs.: The pod commence ripening next December and the seed fall during Mar.–Apr. Vern.: Konda tangedu, Boja, Yerra Chennangi.

Wood very hard, durable, liable to split in seasoning used for house and bridge construction. Seeds are edible after roasting, its taste is like that of cashew nut.

Vantla mamidi (VSKP), *SSR* 15948; Pakhal (WGL), *SSR* 18523; Kakavada (EG), *PRS & SSR* 15401; Kusimi (SKLM), *PRS* 15472; Jaipur RF (ADB) GO & PVP 4947; Pakhal RF (WGL), *ANH* 15936 (MH); Sirivaka (EG), *CAB* 5258 (CAL).

INDIA : Eastern and Western Ghats, Bihar.
WORLD : Myanmar, Malaysia.

## ROSACEAE
### PRUNUS Linnaeus

1. Leaves distinctly serrate, secondary nerves 10–15 pairs .......... *P. jenkinsii*
1. Leaves entire or nearly so; secondary nerves 6 pairs .......... *P. ceylanica*

**Prunus ceylanica** (Wt.) Miq. Fl. Ind. Bat. 1: 366. 1855. *Polydontia ceylanica* Wt.. Ill 1:203. 1840 *'Polydonta'*. *Pygeum acuminatum* Coleb. in Trans. Linn. Soc. London 12: 360. t. 18. 1818. FBI 2: 318; Gamble 1: 438 (311). *Pygeum wightianum* Bl. ex C. Muell. in Walp. Ann. 4: 642. 1858; FBI 2: 319. 1878; Gamble 1: 439 (311); Brandis 282. *P. gardeneri* Hook. f. Fl. Brit. India 2: 321. 1878; Gamble 1: 439 (311). *P. sisparense* Gamble, in Kew. Bull. 1918: 238, 1918, Fl. Pres. Madras 1: 438 (311). 1919.

Trees up to 15 m tall; bark dark brown, smooth; wood yellow-pink, mottled and streaked; branchlets pubescent with white lenticels. Leaves coriaceous, ovate, elliptic or oblong, 5–18 × 3–8 cm, glabrescent, glossy, dark green above, paler beneath, apex acute or shortly acuminate, margin entire to dentate, base subacute or rounded, oblique, lateral nerves arched, spreading, prominent beneath, 2 large, oval glands immersed on either side of the midrib; petioles 1.5–2 cm long, stout, channeled above. Flowers yellowish-green, 6 mm in dia., in lax pubescent, solitary, axillary, 5–10 cm long racemes. Drupe 1.5 × 2 cm, transversely ellipsoid, bilobed, glabrous, endocarp thin, woody; seed solitary.

Occasional in shady hill valleys along streams in East Godavari and Viskhapatnam districts to 600 m. Fls.: Aug. Frs.: Nov.

Wood is used for making crude furniture, boxes, planks and booths and also it is a good fuel for burning bricks. Bark, leaves and fruits emits a smell of bitter almond when crushed. The kernel of the fruit is used as fish-poison.

Rampa hills (EG), *JSG* 15911 (CAL & DD); Gudem valley (VSKP), *VNS* 406 (CAL); Sapparla (VSKP), *GVS* 29706 (MH).

INDIA : Sikkim, Nagaland, Meghalaya, Assam, West Bengal, Bihar, Orissa, Uttar Pradesh, Maharashtra, Karnataka, Tamil Nadu, Kerala and Andaman & Nicobar islands.

WORLD : Nepal, Sri Lanka, Bangladesh, Myanmar, Thailand, Laos, South Vietnam.

**Prunus jenkinsii** Hook. f. Fl. Brit. Ind. 2: 317. 1878, Subba Rao & Kumari in Bull. Bot. Surv. India 13. 155. 1971; Brandis 231.

Medium sized evergreen tree; branches black, stout, glabrous. Leaves lanceolate or oblong-lanceolate, 12.5–17.5 cm long, glabrous, shining, red brown above and paler beneath when dry, apex caudate-acuminate, margin serrate, tipped with glands, base cuneate or acute; secondary nerves 10–15 pairs, obliquely arching, petiole 8 mm, eglandular. Flowers white, dense in 3.5–6.2 cm long, axillary racemes. Drupe ovoid, obovoid or sub-globose, ca 2.3 cm long, stone rugose with a thick ridge on one side; wall thick and hard.

Very rare, in Visakhapatnam district. Limmagudi hill near Araku (VSKP), *GVS* 32894 (MH).

INDIA : Assam.

## RHIZOPHORACEAE

1. Radicle minute, not protruded while the fruit is on the tree. Cymes long, many-flowered ... ........ ***Carallia***
1. Radicle large, protruded while the fruit is on the tree. Cymes short, few-flowered:
    2. Calyx lobes and petals 8–14, bracteoles absent; ovary 4-loculed ........ ***Bruguiera***
    2. Calyx lobes and petals 4 or 5–6; bracteoles present; ovary 2–3-loculed:
        3. Aerial or stilt roots concealing the trunk. Leaf tip with mucro. Calyx lobes and petals 4, entire; stamens 8. Hypocotyl long ........ ***Rhizophora***
        3. Aerial roots not prominent. Leaf tip without mucro. Calyx lobes and petals 5–6, petals longitudinally folded; stamens 10–12. Hypocotyl not very long ........ ***Ceriops***

### **BRUGUIERA** Lamarck

1. Flowers small, greenish-white in cymes. Leaves oblanceolate; Calyx lobes erect in fruit; hypocotyl clindrical, grooved ........ ***B. cylindrica***
1. Flowers large, orange-red, solitary. Leaves lanceolate. Calyx lobes recurved in fruit; hypocotyl terete, not grooved ........ ***B. gymnorhiza***

**Bruguiera cylindrica** (L.) Bl., Enum. Pl. Jav. 1. 93. 1827, quoad basionym; Gamble 1: 459 (325). *Rhizophora cylindrica* L. Sp. Pl. 44. 1753. *Bruguiera* *caryophylloides* (Burm.f.) Bl.. Enum. Pl. Jav. 1: 93. 1827; FBI 2: 438. Brandis 305. *Rhizophora caryophylloides* Burm. f. Fl. Ind. 109. 1768.

Small evergreen tree, up to 5 m tall; erect; branched from the base grey with few lenticels; bark smooth; wood close grained. Leaves scarcely coriaceous, oblanceolate, 5.5–10 × 2–5 cm, shining above, dull beneath, glabrous, light green, apex acute, margin entire, base cuneate; secondary nerves 5–6 pairs; petiole to 2.5 cm. Flowers greenish-white in axillary, 2–3-flowered cymes. Fruit oblong-obovoid; protruding radicle slender, cylindric, green; hypocotyle 7–15 × 0.5–1 cm, often curved, obscurely ribbed.

Common, in the tidal forests. Fls.: Jan.–Apr. Frs.: July–Oct. Vern.: Vurada. Wood useful as fuel.

Nizampatnam Port (GTR), *VRK* 6916; Vasavi canal (EG), *GVS* 68704 (MH); Delta Coringa (EG), *G.F.F. Foulkes* 100 (CAL).

INDIA : Throughout the Indian Coasts.
WORLD : S.E. Asia , Malaysia, N. Queensland.

**Bruguiera gymnorrhiza** (L.) Savigny in Lam. Encycl. 4. 696. 1798; FBI 2: 437. Brandis 305. *Rhizophora gymnorhiza* L. Sp. 443. 1753. *Bruguiera conjugata* Merr. in Philipp. J. Sci. 9: 118. 1914; Gamble 1: 458 (324). *B. rheedii* Bl. Enum. Pl. Javae. 1: 92. 1827; Beddome Fl. Sylv. S. India Anal: Gen. t. 1. 1872.

Evergreen tree, up to 5 m tall; bark grey, nearly smooth; wood red; young branches smooth. Leaves thickly coriaceous, elliptic or lanceolate, 7–15 × 3–6 cm; glabrous, bright green and paler benath, apex acute, margin entire, base cuneate; petiole to 3.5 cm. Flowers orange-red, large, solitary, drooping. Fruit oblong-obovoid, crowned by the slightly enlarged calyx lobes; radicle stout, rugose, grooved; hypocotyl elongate.

Common mangrove tree along the sea coast. Fls. & Frs. Oct.-May. Vern.: Thuddu panna.

Wood extremely hard used for building and fuel. The radicles form an article of diet for the local people.

Kothapalem (GTR), *VRK* 5764; Krishna delta (KSN), Coll. 17100 (MH); Coringa estuary (EG), *ARK Sastry* 7348 (CAL); Tumalapenta (NLR), *JSG* 21818 (DD); Nizampatnam port (GTR), *VRK* 5764.

INDIA : Throughout the coast of India.
WORLD : Tropical S. and E. Africa, Madagascar, Scychelles, Sri Lanka, S.E. Asia, Malaysia.

## CARALLIA Roxburgh

**Carallia brachiata** (Lour.) Merr. in Phillipp. J. Sci. 15. 249. 1919. *Diatoma brachiata* Lour. Fl. Cochinch. 296. 1790. *Carallia integerrima* DC. Prodr.

3. 33. 1828; FBI 2: 439; Gamble 1: 456 (325); Beddome. Fl. Sylv. S. India t. 193. 1972.

Evergreen tree, up to 15 m tall; trunk cylindric, crown dense; bark dark grey, corky; wood yellowish-red; branches opposite and glabrous. Leaves thin-coriaceous, broadly elliptic-obovate, 7–10.5 × 4–6 cm, glabrous, shining above and paler beneath, apex rounded or shortly acute, margin entire, base cuneate; secondary nerves 12 pairs, spreading and interarching freely; petiole very short and stout. Flowers white, small in axillary trichotomous cymes; buds exuding a sticky resionous substance. Fruit filiform, smooth, red; seed one with a thick orange testa.

Rare, in moist deciduous forests of Cuddapah and Chittoor districts, occasionally near streams at 600–1000 m. Fls.: Dec. Frs.: Mar.-May. Vern.: Kaaralli, Kaarvalli, Gijuru chettu.

Wood hard, moderately heavy, very handsome and used for panelling furniture, picture frames and similar purposes. It is also employed as fuel for steamers. The fruits are fleshy and edible.

Tumburatheertham (CTR), *SSR* 16166; Tirumala hills (CTR), *SSR* 15272; Cuddapah forests (CDP), Coll? 19843 (MH).

INDIA : Bengal, Assam, Sikkim, Peninsula.

WORLD : Sri Lanka, Myanmar, Malaysia, China, Australia, Bangladesh, Thailand, Vietnam, Laos, Cambodia, Indonesia, Australia, Madagascar.

**Note** : Gamble (lc) reports its occurrence from Cuddapah hills, but this was reported neither by Ellis (1987) nor by Madusudana Rao (1989) during their Floristic Studies on Cuddapah district. There is one old collection (MH, 19843) from Cuddapah hills without name, date and exact locality. The above finding indicates that the above collections might be from Seshachalam hills which combine both Cuddapah and Chittoor districts.

## **CERIOPS** Arnott

1. Leaf apex obtuse-retuse. Flowers greenish-white in peduncled heads; petals tipped with many ciliae ........ *C. decandra*
1. Leaf apex rounded or slightly emariginate. Flowers white, in clustered cymes; petals not tipped with ciliae ........ *C. tagal*

**Ceriops decandra** (Griff.) Ding Hou. in Steenis. Fl. Males (Ser. 1) 5: 471. 1958. *Bruguiera decandra* Griff. in Trans. Med. Soc. Calcutta 8. 10. 1836. *Ceriops roxburghiana* Arn. in Ann. Nat. Hist. 1: 364. 1838; FBI 2: 436; Gamble 1: 457 (321); Brandis 304.

Evergreen small tree, pneumatophores numerous. Leaves coriaceous, obovate-rounded, 4.5–9 × 2.5–4 cm, apex obtuse-retuse, margin entire, base cuneate; secondary nerves obscure; petiole 1–2.5 cm. Flowers greenish-white, sessile in short peduncled, dense heads. Fruit ovoid-conical, 1.2–1.4 cm long, enclosed by erect calyx lobes about one-third from the base; cotyledons red, protruding from the fruit; hypocotyl club shaped, sharply angular and sulcate, 10–12 cm long.

Common, in tidal forests of Godavari and Krishna deltas (Rolla S. Rao *et al.*, 1986).

INDIA : West Bengal, Bangladesh.
WORLD : Sri Lanka, Myanmar, Thailand, Cambodia, Vietnam, Malaysia, New Guinea.

**Ceriops tagal** (Perr.) Robins in Philipp. J. Sci. 3: 306, 1908. *Rhizophora tagal* Perr. in Mem. Soc. Linn. Paris 3: 138. 1824. *Ceriops candolleana* Arn. in Ann. Nat. Hist. 1: 364. 1837; *nom illeg*; FBI 2: 436. 1878; Gamble 1: 456 (324). Brandis 304.

Evergreen, small tree, up to 5 m tall; base buttressed with stilt roots; bark red, wood orange-red, peeling off in thin flakes; branchlets marked with scars of fallen leaves and stipules. Leaves coriaceous, obovate, 3–9 × 1–4 cm, glabrous, thick, fleshy, apex rounded or slightly emarginate, margin entire, base attenuate; petiole to 2 cm. Flowers white in axillary cymes. Fruit slightly conical, 1.2–1.4 cm long, enclosed by the calyx about 1/3 from the base; hypocotyl 15–20 long, club-shaped, pointed or acute towards radical end.

Common mangrove along sea coast. Fls.: Mar.–Sept. Frs.: Sept.–Dec. Vern.: Gatharu, Gedera.Bark and fruits used in tanning.

Nizampatnam port (GTR), *VRK* 6807; Vasavi canal (EG), *GVS* 68703 (MH); Delta Coringa reserve (EG), *G.F.F. Foulkees* 91 (CAL).

INDIA : West Coast, Andamans.
WORLD : Sea coast of tropical Asia, Africa and Austrlaia.

### RHIZOPHORA **Linnaeus**

1. Leaves elliptic, apex with long mucronate. Cymes 3–7-flowered; petals fleshy. Fruit 3–3.5 cm long, calyx lobes reflexed ........ ***R. mucronata***
1. Leaves oblong-lanceolate, apex short mucronate. Cymes 2-flowered; petals thin. Fruit 2.5 cm long, calyx lobes erect ........ ***R. apiculata***

**Rhizophora apiculata** Bl. Enum. Pl. Jan. 1: 91. 1827. *R. candelaria* DC. Prodr. 3: 32. 1829; Gamble 1: 456 (323). *R. conjugata auct. non* L. 1753; FBI 2: 436; Brandis 304.

Evergreen tree, up to 6 m tall; stiff roots branched; bark grey, smooth with vertical fissures. Leaves oblong-lanceolate, 5–14 × 3.5–5.3 cm, dark green above and paler beneath, apex acute with mucro, margin entire, base tapering; secondary nerves distinct; petiole to 2 cm. Flowers orange-yellow,

1–2 cm long, subsessile; cymes axillary, two flowered on stout peduncles. Fruit somewhat conical or obovate, 2.5 cm, rugose, surrounded at the base by the reflexed calyx-lobes; protruding radical cylindric, more than 20 cm long before falling from the tree.

Common mangrove tree, along the sea coast in the tidal forests. Fls.: Mar.–July. Frs.: July–Oct. Vern.: Uppu ponna, Kaaki ponna.

Bark used for tanning; wood is used for heavy construction.

Kothapalem (GTR), *VRK* 5762; Chinnagollapalem (KSN), *PV* 5770 (AU).

INDIA : Throughout Indian coast.
WORLD : Tropical South East Asia, Sri Lanka, Malaysia.

**Rhizophora mucronata** Poir. in Lam. Tabl. Encycl.(text) 2: 517. 1794, t. 396. f. 2. 1797 & Encycl. 6: 189. 1804; FBI 2: 435; Gamble 1: 456 (323); Beddome Fl. Sylv. S. India Anal. Gen. t. 13. f 4. 1872.

Evergreen small tree, often butressed, with aerial, stilt roots and pneumatophores; bark brown with horizontal fissures; wood dark red; branchlets significantly marked with leaf and stipular scars. Leaves coriaceous, elliptic, 10–15 × 5–8 cm, brown or black dotted beneath and also pale-punctulate, apex with long mucro, margin entire, base narrowed. Flowers yellowish, subsessile, 3–7 in long pedunculate axillary cymes. Fruit brown, obpyriform, 3.5–4.5 cm long, crowned with persistent, reflexed sepals; hypocotyl cylindric, ca. 30 cm long, rugose.

Common, in the outer fringes of the tidal forests. It is strong light demander and early seed-bearer. Fls. & Frs.: Mar.–Oct. Vern.: Uppo Ponna, Kandia, Adaviponna. Eng.: True mangrove, salt ponna, Demon candel, White candel mangrove.

Bark is a powerful, astringent and also gives a valuable tanning material. Wood very hard and heavy used for temporary bridges and heavy construction, also an excellent fuel. Fruit is sweet and edible; the juice is made into a light wine.

Vasavi canal (EG), *GVS* 68701 (MH); Tehmalapenta (NLR), *JSG* 21818 (DD); Nizampatnam (GTR), *VRK* 6952.

INDIA : Throughout Indian coast.
WORLD : Pakistan, Sri Lanka, Bangaladesh, Myanmar, Malaysia, Africa and Austrila.

## COMBRETACEAE

Gangopadhyay and Chakrabarthy (1997) have reviewed the family Combretaceae in India, where they reported 18 indigenous species and 3 exotic species. They kept *Terminalia arjuna* under *T. cuneata*, while *T.*

*coriacea* and *T. alata* were reduced to synonymy with *T. elliptica*. However we followed earlier regional Floras in keeping the above as separate taxa.

1. Petals present (4 or 5). Mangrove plants ........................................................ ***Lumnitzera***
1. Petals absent. Terrestrial plants:
    2. Flowers in globose heads. Leaves eglandular. Fruit small, 2 winged ............ ***Anogeissus***
    2. Flowers in elongate simple or branched spikes. Leaves glandular rarely eglandular (only 2 sps) on the base of petiole or at the base of the leaf lamina. Fruit entire-ridged or 3–5 winged .............................................................................................. ***Terminalia***

## ANOGEISSUS (A.P. de Candolle) Guillemin & Perrottet

1. Bark grey, smooth; branches straight. Leaves elliptic or oblong-elliptic, apex obtuse, rotund; secondary nerves 8–10 pairs. Flower heads in cymes. Fruit wings entire ............ ***A. latifolia***
1. Bark dark grey, rough; branches drooping. Leaves elliptic-lanceolate, apex acute-obtuse; secondary nerves 6-pairs. Flower heads solitary. Fruit wings dentate ............ ***A. acuminata***

**Anogeissus acuminata** (Roxb. ex DC.) Guill. & Perr. Fl. Seneg. Tent 7: 280. 1832; FBI 2: 230. Gamble 1: 466 (330), Beddome Fl. Sylv. S. India t. 16. 1969; *Conocarpus acuminata* Roxb. ex DC. Prodr. 3: 16. 1828. *Anogeissus acuminata* (Roxb. ex DC.) Guill. & Perr. var. *lanceolaa* Wall. ex C.B. Clarke in Hook. f. Fl. Brit. India 2: 451. 1878.

Large deciduous tree, up to 25 m tall; bark dark grey, rough, granulated when old; wood grey, shining. Leaves elliptic or elliptic-lanceolate, 1.7–4.5 × 1.5–2.4 cm, glabrous above, appressed pubescent beneath, apex acute-obtuse with mucronate tip, margin entire, base obtuse; secondary nerves 6 pairs, slender and distinct; petiole to 3 mm. Flowers greenish; heads solitary, axillary. Fruits brown, small, 6.8 mm in dia., broader than long, with upper most pubescent beak much shorter, wings dentate, often lobed.

Usually on river banks from Srikakulam to Godavari. Leaf fall: Dec.–Jan. New foliage: Apr.–May. Fls.: Jan.–Mar. Frs.: Apr.–May. Vern.: Pasichetty, Pachi manu, Pedda manu, Bo-chakram.

Wood is similar to that of *Terminalia chebula* in appearance, not so strong and good as that of *Anogeissus latifolia;* warps and cracks in seasoning and is not durable especially when exposed to wet; it is much stronger than teak under all stresses, a first class timber for any work where strength and elasticity are required. One of the best Indian woods for such articles as hammer and tool handles, axe-helves, golf class etc.

Galikonda (VSKP), *SSR* 19643; Bheemram (ADB), *GO & DAM* 5172; Rampachodavaram (EG), *GVS* 27231 (MH); Telganda (Godavary), *JSG* 15989 (MH); Sambavi Rangi (EG), s.n. (DD).

**Anogeissus latifolia** (Roxb. ex DC.) Wall. ex Guill. & Perr. Fl. Seneg. Tent. 7: 780. 1832. FBI 2: 450; Gamble 1: 466 (33); Brandis 315; Beddome Fl. Sylv. S. India t. 15. 1869. *Conocarpus latifolia* Roxb. ex DC. Prodr. 3: 16. 1828.

Large deciduous tree, up to 30 m tall; bark grey, smooth; wood small, purplish; branchlets grey-pubescent. Leaves chartaceous, elliptic or oblong-elliptic, 5.5–11 × 4–8 cm, glabrous when fully grown, pale, dull glaucous-green, apex obtuse-rotund, margin entire, base rounded; secondary nerves 8–10 pairs, conspicuous beneath; petiole to 1 cm. Flowers small, numerous in pedunculate heads. Drupe small in globose heads, nearly orbicular, 0.8 cm across, yellowish brown, compressed, narrowly 2-winged with entire margins, crowned by persistent calyx tubes which are modified into stiff beak; seed solitary and minute.

Very common, in mixed deciduous forests at 600–1200 m. Tree is useful in clothing dry hill slopes. Leaf fall: Feb., Apr.–May. Fls.: June–Sept. Frs.: Dec.–Mar. Vern.: Tirman. Chiru manu, Velama. Eng.: Axle-wood tree.

Wood very tough, strong, useful for cart axils, shoulder poles and rafters, boat building and agricultural implements and other purposes. Timber makes an excellent charcoal and fuel. Leaves are rich in tannin and are collected for canning and yields a gum much used in calico-printing. Foliage is used for fodder.

Talakona (CTR), *SSR* 13731; Avagondapalem (ATP), *SSR* 18578; Penchala kona (NLR), *SSR* 19334; Khadam river dam (ADB), *TP & PVP* 4138; Vinukonda RF (GTR), *VRK* 6858; Sardhana road (MDK), *BR & CP* 11621; Mosra (NZB), *TP & BR* 6105; Palakonda hills (CDP), *CS & DAM* 7711; Kothapalli (RR), *MSM* 10447; Pedamantanala (PKSM), *RVK* 15838; lower Ahobilam (KNL), *TP & RVR* 2586.

INDIA : Bihar, Chota Nagapur, Uttar Pradesh, Madhya Pradesh, Maharashtra, Karnataka, Orissa.

WORLD : Sri Lanka, Pakistan.

## LUMNITZERA Willdenow

**Lumnitzera racemosa** Willd. in Ges. Naturf. Freunde Berlin Neue Schriften 4: 187. 1803, FBI 2: 452, Gamble 468 (331), Brandis 314; Beddome Fl. Sylv. S. India Anal. Gen. t. 21. f. 2. 1869.

Evergreen tree of shrubby growth, up to 4 m tall; bark thick, brown, rough; wood greyish-brown; branchlets glabrous. Leaves thick-coriaceous, elliptic-oblanceolate rarely obovate, 4–7.8 × 1.4–2.2 cm, bright green, glossy, glabrous, apex obtuse-shallowly emarginate, margin crenate, base attenuate; mid nerve prominent beneath, lateral nerves 3-pairs, inconspicuous, reticulations obscure. Petiole 0. Flowers white, sessile, 1 cm in terminal or axillary, 4–6.5 cm spikes. Fruit elliptic-oblong, woody, to 1 cm long, longitudinally striate when dry, clasped at the base by two oblong-spathulate bracteoles, crowned by the persistent calyx-tube.

Common in coastal-mangroves. Fls.: Apr.–Aug. Frs.: Aug. onwards. Vern.: Kadavi.

Wood hard and used for posts and as fuel.

Nizampatnam (GTR), *VRK* 68108; Kothapalem (GTR), *VRK* 5767.

INDIA : Coastal area.
WORLD : Eastern tropical Africa, Madagascar, tropical Asia, North Australia and Polynesia, throughout Malaysia.

## TERMINALIA Linnaeus

### KEY 1 (Vegetative and Flowering)

1. Leaf lamina glabrous or glabrescent. Glands present or absent:
   2. Leaves alternate or clustered at the ends of branchlets:
      3. Glands 0; petiole long, to 8 cm; secondary nerves 6–8 pairs, to 2 cm apart. Spikes solitary or paired in leaf axils ............ ***T. bellirica***
      3. Sessile glands present at the base of leaf lamina, petiole short, to 1.5 cm; secondary nerves 10–12 pairs, to 1 cm apart , spikes paniculate ............ ***T. paniculata***
   2. Leaves opposite or sub-opposite:
      4. Glands absent. Leaves broadly oblong or elliptic-oblong, apex obtuse or emarginate, thick, glaucous ............ ***T. pallida***
      4. Glands present:
         5. Glands at the base of leaf lamina. Spikes branched ............ ***T. arjuna***
         5. Glands on the petiole close to the base of leaf lamina. Spikes simple ... ***T. chebula***
1. Leaf lamina pubescent or tomentose, glands present:
   6. Leaves obovate ............ ***T. catappa***
   6. Leaves other than obovate:
      7. Spikes simple. Leaves villous beneath; glands sessile at the base of lamina ............ ***T. gella***
      7. Spikes branched. Leaves tomentose beneath; glands subsessile on the mid vein of the base of the lamina:
         8. Leaves yellowish tomentose beneath ............ ***T. coriacea***
         8. Leaves brownish tomentose beneath ............ ***T. alata***

### KEY 2 (Vegetative & Flowering)

1. Foliar glands absent. Spikes simple:
   2. Branchlets and petiole whitish pubescent, leaves obovate or broadly elliptic, apex obtuse; secondary nerves regular, straight and conspicuous, petiole to 8 cm ............ ***T. bellirica***
   2. Branchlets and petiole glabrous, leaves elliptic-oblong, apex emarginate; secondary nerves irregular, not straight and conspicuous; petiole to 1 cm ............ ***T. pallida***
1. Foliar glands present. Spikes simple or branched:
   3. Glands stalked or subsessile, turbinate on the mid vein of lamina, margin sub-entire to crenulate:
      4. Under surface of leaf brownish tomentose ............ ***T. alata***
      4. Under surface of leaf yellowish pubescent ............ ***T. coriacea***
   3. Glands sessile on the petiole of the base of lamina, margin entire:
      5. Glands at the base of the lamina. Spikes solitary or paired:
         6. Leaf base sub-cordate ............ ***T. catappa***
         6. Leaf base rounded:

7. Leaves villous to pubescent beneath ........ ***T. gella***
7. Leaves glabrous on both surfaces ........ ***T. chebula***

5. Glands close to base of the lamina. Inflorescence branched spikes or in panicles:
8. Leaves elliptic-lanceolate, 6.3–13.5 × 2.8–9 cm, apex acuminate, margin entire........ **T. paniculata**
8. Leaves oblong or oblanceolate, 7–11 × 3–4.5 cm, apex obtuse, margin crenate-serrate ........ ***T. arjuna***

## KEY 3 (Fruiting)

1. Drupe winged;
2. Wings 3, unequal, 1 long (front ridge of ovary growing out into a long wing) to 2 cm long cm, 2 very short, 2–3 mm long ........ ***T. paniculata***
2. Wings 5, equal, 3–6 cm long:
3. Drupe notched at the apex; wings thick, narrow, irregularly marked with ascending lines, dark brown at maturity ........ ***T. arjuna***
3. Drupe not conspicuously notched; wings thin, broad, coriaceous, regularly marked with numerous horizontal, straight lines running from axis to edges, pale brown when maturity.
4. Fruit glabrous ........ ***T. alata***
4. Fruit velvety yellowish tomentose ........ ***T. coriacea***

1. Drupe not winged:
5. Drupe with minutely brown or grey tomentose:
6. Drupe ovoid or ellipsoid, ± 5, ridged when dry ........ ***T. gella***
6. Drupe globular, not ridged when dry, often abruptly narrowed to short stalk........ ***T. bellirica***
5. Drupe glabrous:
7. Drupe ellipsoid, yellowish-orange or reddish, 5.5–7 × 2.5–3.1 cm, conspicuously keeled along compressed side ........ ***T. catappa***
7. Drupe ovoid or obovoid, greenish or yellowish-green, 2.5–4 × 1.3–1.7 cm, not keeled along margins:
8. Fruit yellowish green, faintly angled ........ ***T. chebula***
8. Fruit pale green, faintly 5-ridged when dry ........ ***T. pallida***

**Terminalia alata** Heyne ex Roth, Nov. Pl. Spec. 379. 1821. *Terminalia tomentose* (Roxb. ex DC.) Wt. & Arn. Prodr. Fl. Ind. Orient. 314. 1834; FBI 2: 447; Gamble 1: 465 (329); Brandis 310; Beddome Fl. Sylv. S. India t. 17. 1869. *Pentaptera tomentose* Roxb. ex DC. Prodr. 3: 14. 1828.

Deciduous tree, up to 25 m tall; crown heavy; branches spreading; bark black, with deep vertical fissures and transverse cracks; wood dark brown; branchlets villous. Leaves thick-coriaceous, oblong or elliptic, 10–26 × 4–10 cm, glabrous above, tomentose beneath, apex obtuse, margin crenulate, base often cordate; secondary nerves 18–21 pairs, atriculate; petiole to 2 cm; glands turbinate on the mid nerve at the base of leaf blade. Flowers light yellow in axillary spikes or in dense panicles. Drupe brown, 4 cm in dia., glabrous; wings 5, equal, lines numerous, straight, running horizontally from the axis.

Common, in moist and dry deciduous forests of high hill above 700 m. The tree attains largest dimensions on deep, rich, alluvial soil; it favours stiff clay soil. It is a light demander and rapidly suppressed under shade. It coppices well and fairly tolerant to damage by fire. It is one of the worst browsed species in the forest by deer and pigs. Leaf fall: Jan.-Feb. New foliage: May–June. Fls.: May–July Frs.: Oct.–Feb. Vern.: Nalla Maddi, Inu Maddi, Maddi.

Timber extensively used for house building, railway wagons, mine props, bed steads, furniture, carts etc. The bark yields tannin and exudes a gel like fluid, used as purgative and as an adhesive.

Bodhuluru (EG), *SSR* 15238; Mothugudem (KMM), *SSR* 15246; Narsapur RF (MDK), *SSR* 16199; Bheemaram RF (ADB), *GO & PVP* 4494; Sirnapalli RF (NZB), *BR* 9086; Talakona (GTR), *DR* 735 (SVU); Addatigala (EG), *GVS* 68571 (MH).

INDIA : Throughout.
WORLD : Myanmar, Indo-China, Sri Lanka, Thailand.

**Terminalia arjuna** (Roxb. ex DC.) Wt. & Arn. Prodr. Fl. Ind. Orient. 314: 1834; FBI 2: 447; Gamble 1: 465 (329); Brandis 311., Beddome Fl. Sylv. S. India t. 28. 1869. *Pentaptera arjuna* Roxb. ex. DC. Prodr. 3: 14. 1828.

Evergreen, handsome tree, up to 30 m tall; often buttressed; crown spreading, outer bark stiff, flaking-off in pieces, inner bark white and smooth outside, flesh coloured when cut, exuding red resin; wood brown, variegated with darker streaks. Leaves thick-coriaceous, oblong or oblanceolate, 7–11 × 3.4–5 cm, glabrous, apex obtuse, margin crenate-serrate, base obtuse-subcordate; secondary nerves 14–16 pairs, parallel; petiole to 1.5 cm; 2 glands on the petiole, close to the base of lamina. Flowers dull yellow; spikes axillary in panicles. Drupe dark brown, 2–4 cm long, obovoid-oblong, dark, woody, fibrous, apex notched; wings 5, equal, projecting, striated with numerous curved lines.

Frequent, along streams, river banks, ravines and dry water-courses; reaching a large size on fertile alluvial loam up to 1300 m. The tree is a moderae shade-bearer and does not tolerate dense overhead shade. It has more or less superficial root system which radially along the banks of streams, forming its favourable habitats. Fls.: Apr.–July. Frs.: Dec. Persistent. Vern.: Tella maddi, Tittu maddi, Vag-maddi.

Wood hard, used for agricultural implements and boat building. Bark is acrid and stiptic, tonic, febrifugal, antidysenteric, hypotensive, diuretic and for dyeing and tanning and leaves are fed to tasar silk worms.

Bogada (PKSM), *SSR* 1619; Talakona (CTR), *SSR* 16138; Avagondapalem RF (ATP), *NY* 1030; Somayajulapalli RF (KNL), *RVR* 2354; Kunavaram (KMM), *DFO* 70475; Kundaram RF (ADB), *TP & GO* 5438; Ibrahiamapet block (NZB), *TP & BR* 6122; Lothuvagu (RR), *MSM* 10294; Kudichintala bailu (MBNR), *TDCK* 15386.

**Fig. 34.** *Terminalia arjuna* (DC.) Wight & Arn.
A. Twig, B & C. Bracteoles, D. Flower, E. Calyx split open, F & G. Stamens, H. Pistil.

INDIA : Widely distributed
WORLD : Sri Lanka.

**Terminalia bellirica** (Gaertn.) Roxb. Pl. Cor. t. 198. 1805; "*bellerica*" FBI 2: 445; Gamble 1: 463 (328); Brandis 307. Beddome Fl. Sylv. S. India t. 19. 1869. *Myrobalanus bellirica* Gaertn. Fruct. 2: 90. t. 97. 1791.

Deciduous tree, up to 30 m tall; crown extensive; bark bluish-grey with vertical cracks; wood yellowish-grey; branchlets warty, whitish-pubescent with persistent leaf scars. Leaves crowded at the ends of branches, coriaceous,

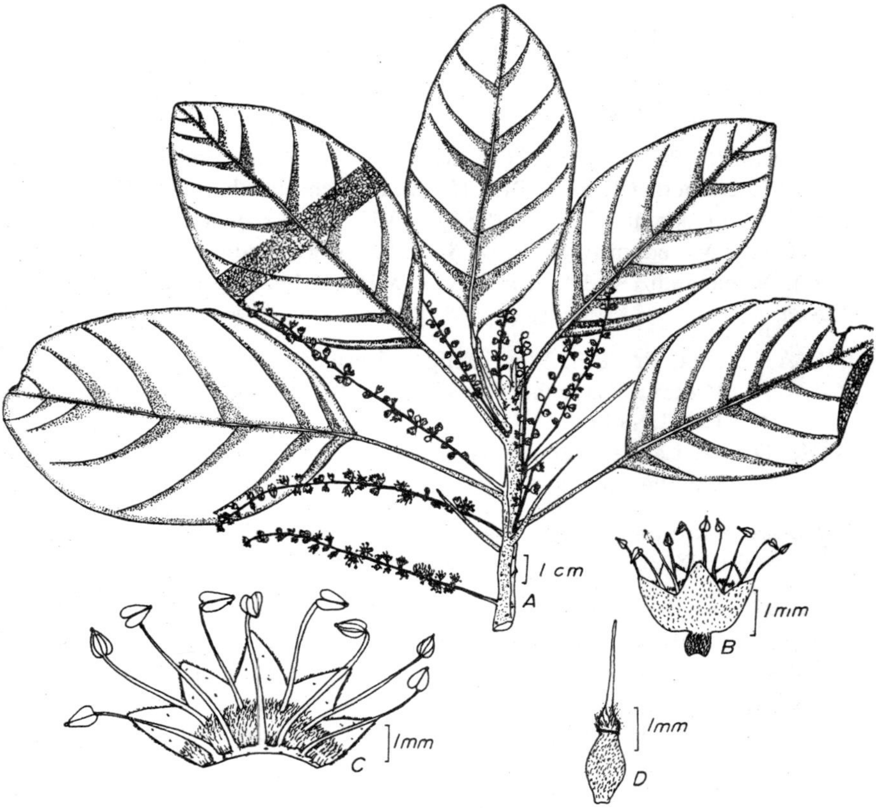


**Fig. 35.** *Terminalia bellirica* (Gaertn.) Roxb.
A. Twig, B. Flower, C. Calyx with stamens, D. Pistil.

broadly elliptic or obovate, 6–19 × 3.2–13.5 cm, glabrous, minutely puberulous when young, finely pellucid dotted above, greyish-green, apex rotund or obtuse, margin entire-subcrenulate, base cuneate, oblique; secondary nerves 6–8 pairs; petiole to 8 cm. Flowers cream coloured, scented in axillary, solitary or paired spikes. Drupe globular, 3 × 2 cm, minutely grey-tomentose, obscurely 5-angled when dry, horned.

Common in hilly deciduous forests, especially near forest villages, 500–1400 m. It is a light demander. Leaf fall: Nov. New foliage: Apr. Fls.: Apr.–May. Frs.: Nov.–Feb. Vern.: Tadi, Thandra, Bhutavasamu, Tani, Vibhitakamu. Commercial name: Belleric myrobalan.

The fruit is one of the important constituents of the ayurvedic drug, "Triphala churna" which is used against diseases like asthma, bronchites, and all other similar diseases. Belleric myrobalan used in tanning, but of less value than that of *T. chebula*. Fruits are used for dyeing and also used to increase the potency of spirits. Kernal yields an oil which is reported to have antibiotic activity against a wide variety of micro organisms. Wood hard, not durable and little used, but useful if carefully seasoned.

Donubai (SKLM), *SSR* 15591; Kalasamudram (ATP), *SSR* 13745; Pakhal (WGL), *SSR* 18522; Nirmal (ADB), *SSR* 19333; Loyapalli (GTR), *VRK* 3978; Diguvametta (PKSM), *RVK* 15750; Kudichintala bailu (MBNR), *TDCK* 13606; Kambakkam (CTR), *DR* 864 (SVU); Sunkarimetta (VSKP), *NPBK* 720 (CAL); Maredumilli (EG), *GVS* 27252 (MH); Kodur (CDP), *JLE* 14363 (MH); Mahanandi (KNL), *JLE* 25503 (MH); Urlakonda (NLR), *MSR* 1308 (CAL).

INDIA : Uttar Pradesh, West Bengal, Karnataka, Assam, Chota Nagpur, Madhya Pradesh, Maharashtra, Nagaland, Orissa, Punjab and Tamil Nadu.

WORLD : Sri Lanka, Nepal, Myanmar, Thailand, Indo-China, Malaysia, Bangladesh.

**Terminalia catappa** L. Syst. Nat. ed. 12.2: 674. 1767; FBI 2: 444; Gamble 1: 463 (328); Brandis 307; Beddome Fl. Sylv. S. India t. 18. 1869.

Deciduous tree, up to 20 m tall; crown widely spreading, trunk base somewhat buttressed, fluted slightly above; bark greyish, furrowed; wood red; branchlets whorled. Leaves clustered at the ends of branchlets, thin-coriaceous, obovate, 15–28 × 9–15 cm, glabrous above, tomentose beneath, apex obtuse-subacute, margin entire, base sub-cordate; secondary nerves 8–12 pairs, 2–3 cm apart, prominent; petiole to 1.7 cm, brownish pubescent; glands 2, sessile at the base of leaf blade. Flowers creamish-white, pedicellate in 13 cm long spikes. Drupe reddish, ellipsoid, 5.5–7 × 2.5–3.1 cm, glabrous, compressed, smooth, conspicuously keeled; pericarp fibrous and fleshy, endocarp hard; seed one.

Commonly planted in almost all the places like households, gardens and avenues. It grows well in coastal regions. The soil on which it thrives is permeable siliceous sandstone, consisting typically of the beach detritus brought down by streams and of sand and shingles banked up by wind and waves. It can be grown on sandy soil. Leaf fall: Jan.–Feb. turn red when before fallen. New foliage: Mar. Fls.: Mar.–Nov. Frs.: May–Apr. Vern.: Badamchettu, Bedam, Ingudu, Natubadam. Eng.: Indian Almond-wood.

Seeds are edible, the oil which extracted from dried kernal is used as a substitute for almond oil. The shell comprising the fibrous covering, husk and hard endocarp is a good source for making furfurel. The bark contains a small amount of colouring matter. The Tasar silk worm is fed on these leaves. The leaf juice is applied externally for scabies, leprosy, skin diseases and headache. The fruit is bitter, sweet and acrid.

Anantapur town (ATP), *NY* 584.

INDIA : Commonly planted throughout.
WORLD : Tropical Asia, North Australia, beach forests of Andamans and Malay Peninsula.

**Terminalia chebula** Retz. Obs. Bot. 5: 31. 1789. incl. forma *tomentosa*; FBI 2: 446. *p.p.*; Gamble 1: 464. (328); Brandis 308; Beddome Fl. Sylv. S. India t. 27. 1869. *Myrobalanus chebula* (Retz.) Gaertn. Fruct. 2.91. t. 97. 1791.

A moderate sized deciduous tree, up to 20 m tall; branchlets glabrescent; bark dark brown, shallowly or deeply fissured; wood brownish grey. Leaves alternate, subopposite, thin-coriaceous, ovate, elliptic-obovate, 5–18 × 3–8.5 cm, glabrous, apex obtuse and usually apiculate, margin entire, base rounded; nerves regular, parallel; petiole to 1.5 cm; glands 2, just below the leaf blade. Flowers dull white in simple spikes. Drupe yellowish-green, 2.5–3.8 × 1.3–1.5, ovoid or obovoid, pendulous, faintly 5-angled, glabrous, stone oblong, very thick.

Occasional, in deciduous forests at 700–1400 m. Leaf fall: Nov. New foliage: Mar.–May. Fls.: Aug.–Nov.. Frs.: Dec. onwards. Vern.: Karakkai chettu, Nalla karaka.

Fruits are an important source of tannin material and used externally for chronic ulcers, wounds etc. Dye is also extracted from fruits. Wood very hard used in building and for agricultural work. Bark is used for tanning and dyeing. 'Triphala' the traditional Indian medicine of Ayurveda consisting of the fruits of *T. chebula, T. bellirica* and *Phyllanthus emblica* is widely used for treatment of any sort of digestive or intestinal disorders.

Kalasamudram (ATP), *SSR* 13750; Tirumala hills (CTR), *SSR* 15276; Galikonda (VSKP), *SSR* 15922; Avagondapalem (ATP), *SSR & KSM* 15568; Jaipur RF (ADB), *GO* 4967; Siddulaiah kona (NLR), *SSR* 19379; Balapalli (CDP), *JSG* 16547 (CAL); Papikonda RF (WG), *DCSR* 50 (CAL); Rampachodavaram (EG), *GVS* 24472 (MH); Aklaspur (KNR), *GVS* 2022 (MH); Tummala bailu (PKSM), *RVK* 15889; Farhabad (MBNR), *TDCK* 15381.

INDIA : Himachal Pradesh, Karnataka, Kerala, Madhya Pradesh, Maharashtra, Sikkim, Tamil Nadu, Uttar Pradesh, West Bengal.
WORLD : Nepal, Sri Lanka, Myanmar, Bangladesh.

**Terminalia coriacea** (Roxb.) Wt. & Arn. Prodr. Fl. Ind. Orient. 315. 1834; Gamble 1: 465 (329). *Pentaptera coriacea* Roxb. Fl. Ind. 2: 438. 1832. *Terminalia tomentosa* (Roxb. ex DC.) Wt. & Arn. var. *coriacea* (Roxb.) C.B. Clarke in Fl. Brit. India 2: 448. 1878.

Large deciduous tree, up to 18 m tall; branchlets velvety pubescent with persistent leaf scars; bark deeply cracked, grey; wood hard, dark brown,

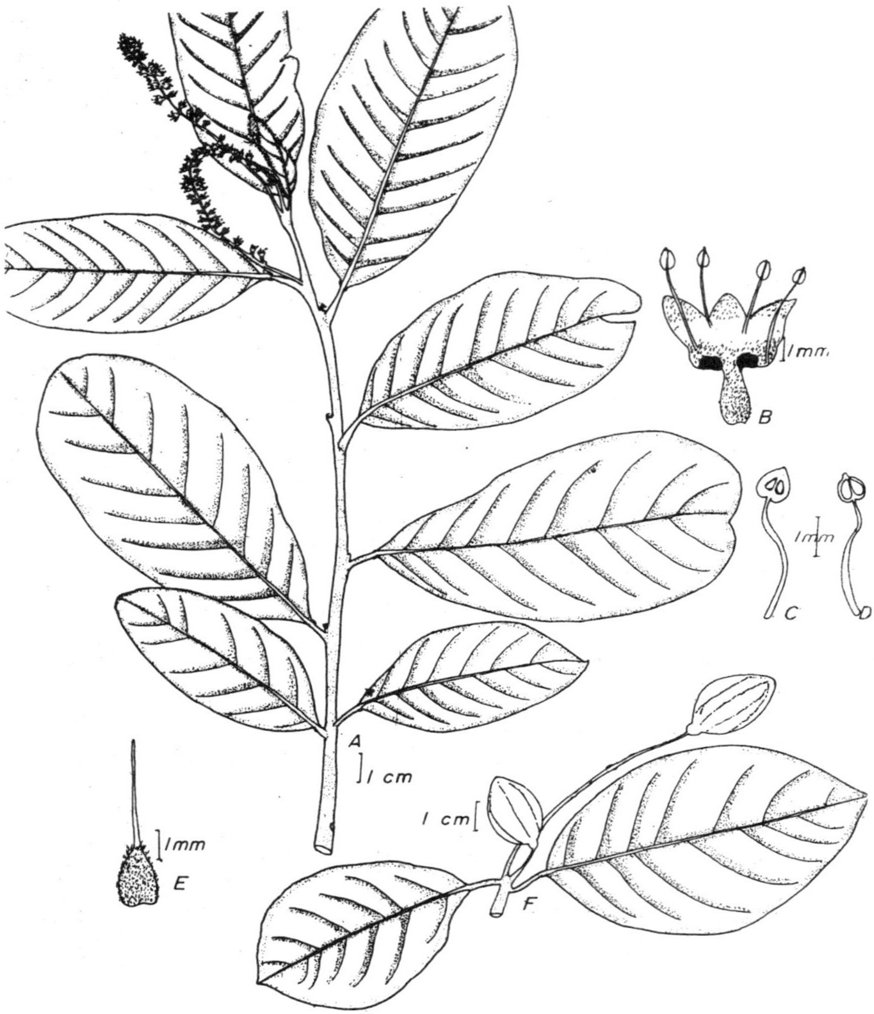


**Fig. 36.** *Terminalia chebula* Retz.
A. Twig, B . Part of Calyx with stamens, C & D. Stamens, E. Pistil, F. Fruiting Branch.

noticeable for the yellowish velvety down. Leaves alternate or sub-opposite, coriaceous, elliptic-ovate, oblong, 10–20 × 4.7–9.5 cm, puberulous above, densely tomentose beneath, apex obtuse, margin entire-slightly crenate, base unequally cordate; secondary nerves 13–20 pairs; petiole to 2 cm; glands 2, sessile at the base of midrib beneath. Flowers dull yellow in axillary or terminal panicles. Drupe yellowish-white, 4.5 × 3 cm, wings 5, equal, yellowish velvety, thin, lines horizontal, apex rounded.

Common, on dry hills in deciduous forests, at 400–1500 m. Fls.: June–July. Frs.: Aug. onwards. Vern.: Jani, Nalla maddi. Wood is very useful as timber.

Tirumala hills (CTR), *SSR* 13703; Horsely hills (CTR), *SSR* 13792; Anantagiri (RR), *SSR & KSM* 18518; Avagondapalem (ATP), *SSR & KSM* 18564; Chelama (KNL), *JLE* 16763 (MH); Maredumilli (EG), *GVS* 2422 (MH); Kodimial (KNR), *GVS* 20143 (MH); Kodur (CDP), *JLE* 14335 (CAL); Pagari Kutta-Naraspur (MDK), *KMS* 6648 (CAL); Veligonda (NLR), *MSR* 1388 (CAL).

INDIA : Peninsula

**Terminalia gella** Dalz. in J. Bot. (Hooker) 3: 27. 1840. 1841; Gamble 1: 464 (328).

Medium sized tree, up to 15 m tall; bark greyish. Leaves coriaceous, ovate or ovate-oblong, 3–15 × 2–8.4 cm, glabrous above, tawny villous beneath, apex obtuse, margin entire, base rounded; secondary nerves 5–10 pairs, prominent beneath; petiole to 2 cm, villous; glands sessile at the base of the lamina. Flowers pale yellow; spikes axillary, pubescent, to 5 cm long. Drupe ovoid or ellipsoid, not winged, faintly 5-ridged when dry, minutely brown tomentose.

Rare, in moist forests of Visakhapatnam and Srikakulam districts at 800–1000 m.

Anantagiri (VSKP), *SSR* 19604; Sapparla (VSKP), *SSR* 19653; Chintapalli (VSKP), *GVS* 28141 (MH); Ichapur near Mondemkhal (SKLM), *GVS* 62490 (MH).

**Terminalia pallida** Brandis Indian Trees 308. 1921; Gamble 1: 464 (328).

Evergreen tree, up to 10 m tall; bark greyish; branchlets glabrous. Leaves thick-coriaceous, broadly oblong or elliptic-oblong, glabrous, glaucous, apex obtuse or emarginate, margin entire, base rounded; secondary nerves 7–9 pairs; petiole to 1 cm, orange-reddish. Flowers pale yellow in simple terminal spikes. Drupe obovoid, 2.5–4 × 1.3–1.7 cm, very faintly 5-ridged when dry, glabrous.

Occasional, chiefly on rocky hills of dry deciduous forests on high hills of Cuddapah and Chittoor districts and it occurs occasionally in Kurnool district at 700–800 m. Fls.: Mar.–Apr. Frs.: Aug.–Sept. Vern.: Tella karaka, Velama karaka.

Fruit is used in tanning and dyeing, also used as a substitute for *T. chebula*.

Tirumala hills (CTR), *SSR* 13744; Pulivendala RF (CDP), *SSR* 16149; Towards Talakona (CTR), *GVS* 46004 (MH).

**Note** : This is endemic to the Eastern Ghats with limited distribution. Tirumala hills and its surrounding areas contain the largest number of trees compared with that of other areas of Chittoor and Cuddapah districts.

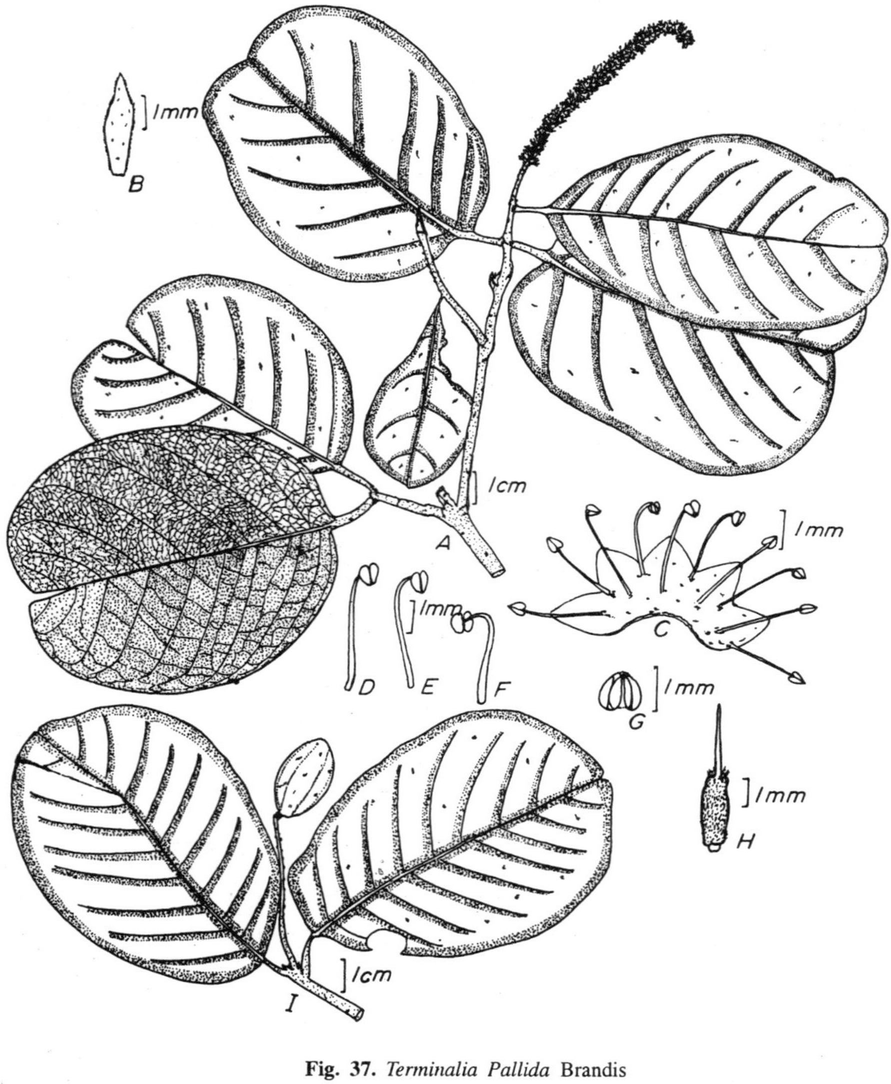


**Fig. 37.** *Terminalia Pallida* Brandis
A. Twig, B. Bract, C. Calyx with stamens, D, E & F. Stamens, G. Anther, H. Pistil, I. Fruiting twig.

**Terminalia paniculata** Roth, Nov. Pl. Sp. 383. 1821; FBI 2: 448; Gamble 1: 465 (329); Brandis 311; Beddome Fl. Sylv. S. India t. 20. 1869.

Deciduous tree, up to 20 m tall; bark dark brown, cracked; wood pale brown; branchlets rusty pubescent. Leaves thin-coriaceous, elliptic-oblong, or elliptic-lanceolate, 6.3–13.5 × 2.8–9 cm, glabrescent, apex acute or shortly acuminate, margin entire, base rounded; secondary nerves 10–12 pairs, parallel; petiole to 1.5 cm; glands two, sessile, close to the base of the lamina. Flowers yellowish in 15 cm long spicate panicles. Drupe rusty pubescent, wings 3, unequal, 1 long (front ridge of ovary growing out into long wing), 1.5–2 × 1–1.3 cm, 2 very short.

Rare, only in the deciduous forests of Cuddapah and Chittoor districts at 700–850 m. Fls. July–Oct. Frs.: Mar. Vern.: Neemari, Nerali, Chirumanu.

Wood very hard, used for constructional work, furniture, ship-building, agricultural implements and plywood for tea-chests. Bark and fruits yield tannin.

Tirumala (CTR), *SSR* 15268 & 13775; Balapalli (CDP), *JLE* 15760 (MH); Talakona (CTR), *DR* 385 (SVU).

INDIA : Western Peninsula.

## MYRTACEAE

1. Fruit berry:
    2. Berry large in size, seeds numerous—cultivated ................................................ ***Psidium***
    2. Berry small; seeds few (1 or 2):
        3. Flowers 1–2, long pedicellate, calyx tube not produced above the ovary .... ***Eugenia***
        3. Flowers many, short pedicellate, calyx tube produced above the ovary ..... ***Syzygium***
1. Fruit capsule:
    4. Flowers bright red or crimson in elongate spikes; petals distinct, free ......... ***Callistemon***
    4. Flowers cream coloured in capitate umbels; petals fused and form a calyptra as a lid ... .......................................................................................................... ***Eucalyptus***

## **CALLISTEMON** R. Brown

**Callistemon citrinus** (Curtis) Skeels U.S.D.A. Bur. Pl. Industr. Bull. 282: 49. 1913. *Metrosideros citrina* Curtis. Bot. Mag. t. 260. 1794.

Evergreen tree, 3–6 m tall; branches spreading, silky-hairy when young. Leaves linear-lanceolate, 3–8 × 0.4–0.9 cm, rigid, apex-acute, margin entire, base narrowed; midrib and secondary nerves prominent; glands lateral. Flowers bright red or crimson in 5–10 cm long, terminal spikes. Capsule woody, depressed-globose, 4 mm long, glabrous, contracted at summit.

Cultivated in gardens. Fls. & Frs.: Throughout the year.

SVU Campus (CTR), *SSR* 19395.

INDIA : Widely cultivated.
WORLD : Native in Eastern Australia.

## EUCALYPTUS L'Heritier de Brutelle

1. Bud single in the axil, warty.—Flowers large ........................................ *E. globulus*
1. Buds 3 or more per peduncle, smooth:
   2. Umbels paniculate ........................................ *E. citriodora*
   2. Umbels not paniculate:
      3. Fruits 4–6 mm dia m........................................ *E. camaldulensis*
      3. Fruits 6–9 mm dia m ........................................ *E. tereticornis*

**Eucalyptus camaldulensis** Dehnh. Cat. Pl. Hort. Comald. ed. 2. 20. 1832. *E. rostrata* Schl. Linnaea 20: 365 (381). 1922.

Tall tree, branches drooping; bark smooth, white or ashy, decorticating in rounded plates. Leaves lanceolate, falcate, 10–20 × 1–2 cm. Flowers whitish in axillary, 3–12-flowered umbels. Fruit 4–6 mm in dia., rim convex, valves exerted.

Planted (Rao. Rolla *et al.*, 1986). Fls. & Frs.: Nov.–Jan.

The wood is strong and durable. It is used for engineering construction, railway sleepers, felloes, piles, posts and pavings. It is the principal source of commercial Eucalyptus gum, also called Eucalyptus kino or Red gum. The gum is occasionally used in diarrhoea, relaxed throats and astringent in bleeding gums, cuts, etc.

**Eucalyptus citriodora** Hook. in Mitch. Exped. Trop. Aust. 235. 1848.

Tall tree; bark smooth, grey, decorticating in small plates. Leaves lanceolate (peltate in young) often falcate, 10–20 × 1–3.5 cm, strongly lemon scented, petiolate. Flowers cream; umbels 3–5-flowered, arranged in panicles. Fruits ovoid-urceolate, 1–1.4 × 1 cm, valves enclosed.

Commonly planted. Fls.: May–June. Eng.: Lemon scented Eucalypt. Essential oil from the leaves is of commercial importance, oil is employed in soap perfumery and as a source of citronella used in the manufacture of citronellol, hydroxy citronellal and methanol. The wood is strong, tough, elastic and durable, used for general construction purposes, carriage building, ship building, mine props, tool handles, wood paving and sleepers. It has a pleasant odour.

INDIA : Indigenous in Queensland; now cultivated in many parts of the world.

**Eucalyptus globulus** Labill. Voy. Rech. Perouse 1: 153. t. 13. 1799; Maiden, Crit. Revis. Eucalyptus 2: 249. 1913; Gamble 1: 486 (343); Hall *et al.*, Forest trees Australia (ed. 3) 128. 1970; Brandis 327.

Large tree; bark smooth, sheds in long strips or sheet; young branches glaucous-grey. Leaves elliptic-ovate, lanceolate or falcate, 7–15 × 3–8 cm, glaucous, apex acute, base obtuse or rounded; petiole 2.5–5 cm. Flowers whitish, 4 cm across, solitary or rarely 3, axillary, peduncle reduced. Fruit sessile, globular to hemispherical, 1–1.5 × 1.5–3 cm, 4-ribbed, warty.

Planted. A cool, moist, equitable climate and deep fertile soil, which is not calcareous or saline are favourable for the growth of this tree. Timber is fire-resistant and used for the construction of godowns, sheds, stations and platforms. It is used mainly as good fuel. Tree yields an oil from the leaves and terminal branchlets, largely used as a mosquito and Vermin repellent and as an ingredient of germicidal and disinfecting preparation. Dried leaves are used in the form of tincture in asthma and chronic bronchitis. The leaf decoction is used as insect and vermin repellant.

INDIA : Cultivated (Thammanna *et al.*, 1994).
WORLD : Native of Australia.

**Eucalyptus tereticornis** Sm. Spec. Bot. New. Holl. 41. 1795; Brandis 327.

Trees tall; bark smooth, grey, decorticating in long plates. Leaves lanceolate, often falcate, 10–20 × 1.2–2.5 cm, glabrous, apex acute, margin entire, base obtuse or truncate; lateral nerves and intra marginal vein slender but distinct. Flowers cream, 5–12, in axillary umbels. Fruit 6–9 × 8–10 mm, semi-spherical to turbinate, valves exerted.

Planted (Thammanna *et al.*, 1994). Fls. & Frs.: Oct.–Apr.

INDIA : Widely planted in many parts of India.
WORLD : Indigenous in the southern coast of Australia and New Guinea.

## EUGENIA Linnaeus

**Eugenia bracteata** (Willd.) Roxb. ex DC. Prodr. 3:264, 1828; FBI 2:50. 1879; incl var. ***roxburghii;*** Gamble 1: 484 (342): Brandis 325. *Myrtus bracteata* Willd. Sp. Pl. 2:969. 1799.

Tree, up to 6 m tall; bark yellowish-grey, smooth; wood grey, close-grained; young shoots rusty pubescent. Leaves sub-coriaceous, gland-dotted, opposite -decussate, elliptic, 3–6.5 × 1.7–3.3 cm, apex obtusely acute, margin entire, base cuneate; secondary nerves not prominent, petiole to 5 cm. Flowers white, showy, axillary. Fruit globose, orange coloured, 8 × 7 mm diam, crowned with calyx lobes; seeds 1 or 2.

Common in coastal districts on sandy ground at 1000–1250 m. Often cultivated. Fls.: Mar.–June. Frs.: July–Nov. Ripe fruits edible. Vern. : Arivita

Simhachalam (VSKP), *SSR & KSM* 15571; Tettipet-Sriharikota (NLR), *YYR* 577; Mandasa Rest house (SKLM), *GVS* 62323 (MH); Kambakkam hills (CTR), *DR* 821 (SVU).

INDIA : Peninsula.
WORLD : Sri Lanka.

## **PSIDIUM** Linnaeus

**Psidium guajava** L. Sp. Pl. 470. 1753; FBI 2: 468; *"guyava"* Gamble 1: 472 (331); Brandis 326.

Evergreen tree, up to 8 m tall; bark smooth, whitish-grey; wood smooth; branchlets appressed tomentose. Leaves coriaceous, oblong or ovate-lanceolate, 5–13 × 3–5.5 cm, glabrous above, glabrescent beneath, apex acute with apicule, margin entire, base acute; secondary nerves 15–20 pairs, conspicuous beneath. Flowers white, axillary, solitary or in pairs. Berry green, yellow when ripe, smooth, large; seeds numerous, woody.

Cultivated, planted near households for fruits and also run wild. Fls. & Frs.: Throughout the year. Vern.: Jama, Ettajama, Goyya, Tella jama. Eng.: Common Guava.

The fruit is sweet, juicy and highly flavoured; eaten mostly as fresh fruit. It is used in the preparation of jellies. It is one of the richest natural source of vitamin 'C' and contains 4–10 times more of this vitamin than in citrus fruits. Fruit is tonic, cooling and laxative and it is good in colic and bleeding gums. The fruit and its conserve are astringent and used in diarrhoea and dysentry. Wood is used for implements.

S.K. University (ATP), *NY* 539; Mungala padu (KNL), *AA* 2047; Ghani RF (KNL), *RVR* 3132; Chodavaram (EG), *MSR* 1491 (CAL); Ramapatnam (NLR), *JSG* 12244 (DD); Appapur (MBNR), *TDCK* 16812; Tirumala (CTR), *DR* 1649 (SVU).

## **SYZYGIUM** J. Gaertner *nom. cons.*

### KEY 1

1. Leaves alternate ........ ***S. alternifolium***
1. Leaves opposite:
    2. Petiole < 5 mm long:
        3. Intramarginal nerve double, secondary nerves 11 pairs, distant to 1.5 cm apart ........ ***S. samarangense***
        3. Intramarginal nerve single, secondary nerves close ........ ***S. calophyllifolium***
    2. Petiole > 8 mm long:
        4. Lateral nerves not producing a significant intramarginal vein ........ ***S. operculatum***
        4. Lateral nerves producing a distinct intramarginal vein:
            5. Leaf apex obtuse, secondary nerves 4–7 pairs, intra-marginal vein inconspicuous ........ ***S. malabaricum***
            5. Leaf apex acuminate, secondary nerves numerous, intramarginal vein conspicuous:
                6. Petiole reddish, leaf base slightly narrowed ........ ***S. cumini***
                6. Petiole green, leaf base cuneate ........ ***S. heyneanum***

## KEY 2 (Flowering & Fruiting)

1. Corymbiform racemes terminal .......... ***S. calophyllifolium***
1. Cymes lateral or axillary:
   2. Petals not calyptrate (free); cymes short:
      3. Flowers greenish-white, small; calyx without thickened staminal disc. Berry spherical or globose .......... ***S. malabaricum***
      3. Flowers white, large; calyx with thickened staminal disc. Berry turbinate-flat above .......... ***S. samarangense***
   2. Petals calyptrate (united); cymes somewhat longer:
      4. Flowers greenish .......... ***E. operculatum***
      4. Flowers whitish or yellowish:
         5. Flowers yellowish-white, cymes condensed .......... ***S. alternifolium***
         5. Flowers white or creamish white, cymes laxly branched:
            6. Berry globose or subglobose, dark purple, crowned with the truncate calyx-limb .......... ***S. cumini***
            6. Berry obliquely oblong-ellipsoid, pale purple, crowned with cup shaped calyx-limb .......... ***S. heyneanum***

**Syzygium alternifolium** (Wt.) Walp. Rep. 2: 19. 1843; Gamble 1: 481 (340). *Eugenia alternifolia* Wt. Ic. t. 537. 1842, FBI 2: 497; Brandis 322. Beddome Fl. Sylv. S. India t. 198. 1872.

Deciduous tree, up to 12 m tall; bark greyish, slightly fissured; wood dark red, close and irregulary grained, fine textured; branchlets pale, glabrous. Leaves thick-coriaceous, ovate-elliptic to suborbicular, 10–12.3 × 7–9 cm, glabrous, dark green and shining above, paler beneath, apex obtuse, margin entire, base unequally rounded; main lateral nerves parallel, numerous; petiole up to 3 cm, reddish. Flowers cream or yellowish white, sweet scented in 8 cm long, axillary cymes with divaricating branches. Berry globose, 0.5–1.5 × 0.5–0.8 cm, dark purple, crowned at top with persistent calyx tube.

Common, in the hills of open dry deciduous forests of Nallamalais and Seshachalam hill ranges at 600–850 m. Leaf fall: Dec.–Jan. New foliage: Mar. Fls.: Mar.–Apr. Frs.: Apr.–June. Vern.: Mogi, Movi, Manchi moyadi.

Fruits are edible. Wood hard to very hard, heavy and suitable for making rafters, scantlings, beams and posts.

Tirumala hills (CTR), *SSR* 15971; Chintala (PKSM), *SSR* 16122; Diguvametta (PKSM), *RVK* 15753; Cuddapah forests (CDP), *CBDM* 20717 (MH).

The distribution of this tree is restricted to southern part of the Eastern Ghats and usually associated with *T. pallida* and *Pterocarpus santalinus*.

INDIA : Duthie in FBI stated its occurrence from Nagari hills of Assam.

**Note** : This plant shows considerable variations in size, shape, colour of the fruits in same locality.

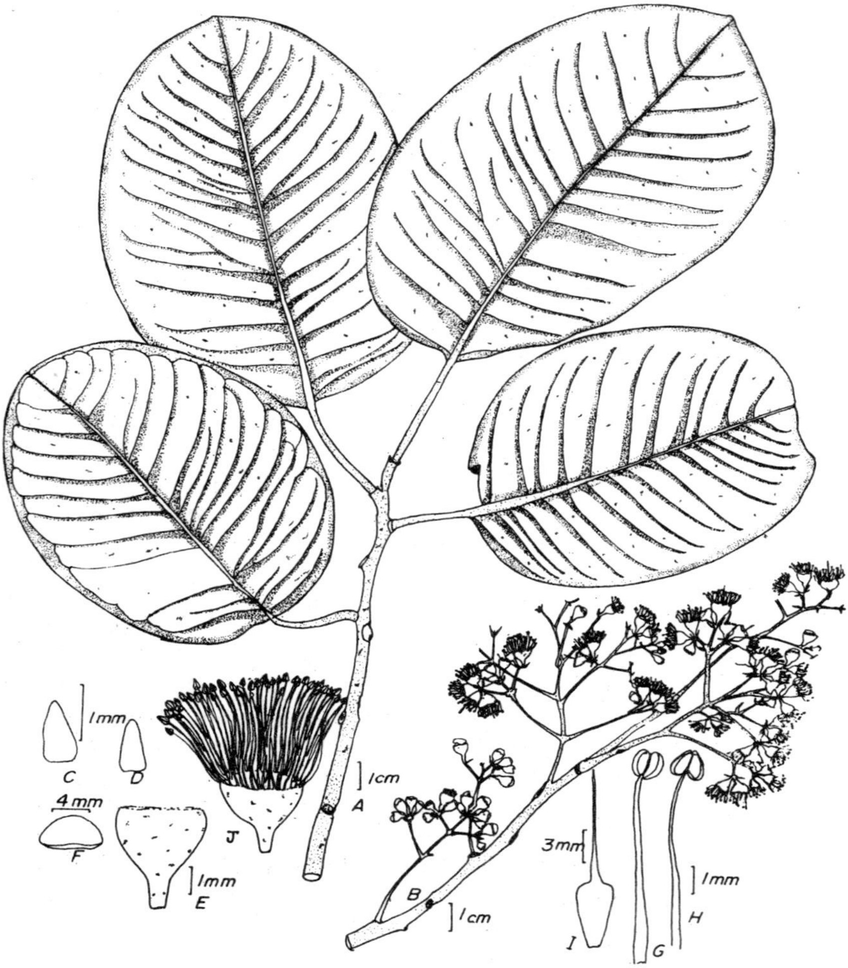


**Fig. 38.** *Syzygium alternifolium* (Wight) Walp.
A. Twig, B. Flowering branch, C. Bract, D. Bracteole, E. Calyx, F. Calyptra, G & H. Stamens, I. Pistil, J. Flower

**Syzygium calophyllifolium** Walp. Rep. 2: 180. 1843; Gamble 1: 480 (339). *Eugenia calophyllifolia* (Walp.) Wt. Ic. t. 1000. 1845; FBI 2: 494; Brandis 324.

Small or medium-sized tree; branches spreading, young branches tetragonous, pale. Leaves coriaceous, elliptic or obovate, 1.8–3 × 1.5–2 cm, glabrous, brownish when dry, apex obtuse, rounded, margin entire, base acute; secondary nerves numerous, close; petiole 1 mm. Flowers white, sessile, in 2 cm long, terminal in trichotomous, corymbiform racemes. Berry dark purple, oblong or obovoid.

It is included here on the authority of Beddome's collection from Cuddapah forest.

Fruits edible.

Cuddapah, Beddome 20711 (MH).

**Syzygium cumini** (L.) Skeels in U.S.D.A Bur. Pl. Industr. Bull. 248.2. 1912. *Myrtus cumini* L. Sp. Pl. 471. 1753. *Eugenia jambolana* Lam. Encycl. 3: 198. 1789; FBI 2: 499; Brandis 523. Beddome Fl. Sylv. S. India t. 197. 1872. *Syzygium jambolanum* (Lam.) DC. Prodr. 3: 259. 1828; Gamble 1: 481 (340).

Evergreen, densely foliaceous tree, up to 20 m tall; branchlets often pendulous; bark smooth, light grey with larger patches of darker colour with shallow depressions; wood reddish brown to reddish grey. Leaves opposite - decussate, coriaceous, variable, lanceolate, elliptic-oblong or broadly ovate-elliptic, 8–16 × 4–6.5 cm, glabrous, shining, pellucid-dotted, apex acuminate, margin entire, base slightly narrowed; secondary nerves numerous, prominent on the lower surface, petiole to 1.8 cm, reddish. Flowers pale greenish-white, sweet scented in 6–15 cm long trichotomous cymes on old-wood below the leaves. Berries globose or sub globose, 1–2.5 × 1–1.5 cm, purplish red and black when ripe, glabrous; seed solitary with fleshy cotyledons.

Common, in river banks and forests both in the plains and hills up to 1500 m. Planted along roads and in gardens. It is a very good avenue tree except in very dry situations. Fls.: Feb.–May. Frs.: June–Aug. It renews leaves about the time of flowering through the inflorescence. Vern.: Neredu, Jambu, Jinna.

Fruits are edible. Wood is moderately hard, strong, used for building and agricultural purposes. It is a very good heating fuel. The ripe fruits are widely eaten and a preparation is given in stomachache, as a remedy for enlarged-spleen and in chronic diarrhoea. The bark and seed also used for diabetes. Bark is used in dyeing and tanning works and for colouring fishing nets.

Palakonda (VSKP), *SSR* 15924; Jambugumbala RF (ATP), *NY* 580; Madhavaram RF (KNL), *TP & RVR* 1350; Vijayapuri south (GTR), *VRK* 3949; Between Gummada & Donubai (VSKP), *GVS* 6242 (MH & CAL); Janakpur (ADB), *PVP* 9492; Akkampet (MDK), *BR & CP* 11563; Mamidypalli (NZB), *BR & KH* 9006; Appapur (MBNR), *TDCK* 15327; Horsely hills (CTR), *DR* 1423 (SVU).

INDIA : Throughout.
WORLD : Sri Lanka, Malaysia, Australia.

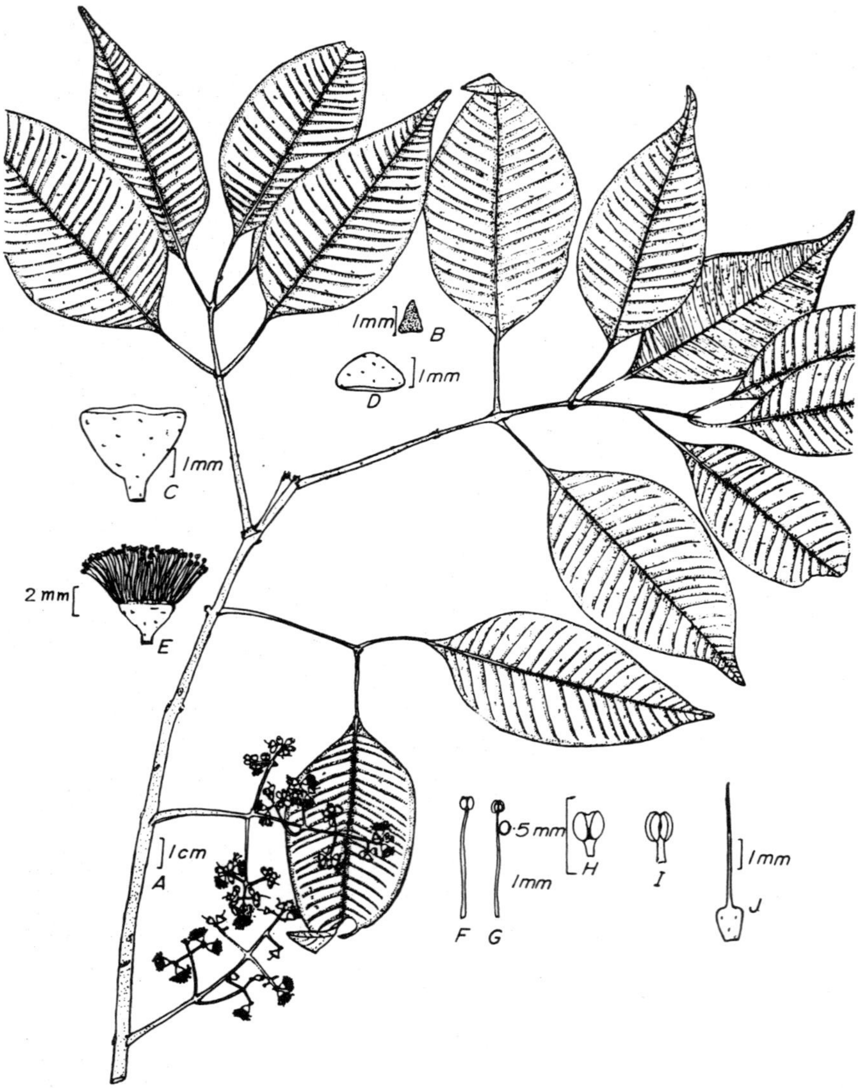


**Fig. 39.** *Syzygium cumini* (L.) Skeels.
A. Twig, B. Bract, C. Calyx, D. Calyptra, E. Calyx with stamens, F & G. Stamens, H & I. Anthers, J. Pistil.

**Note** : Galls on branchlets (oblong) and leaves (globose). This plant shows considerable variation in the size of fruits depending on the habitat. In wild state in valleys fruits are smaller up to 1 cm long and sweeter than in cultivation where the size measures up to 2.5 cm across.

**Syzygium heyneanum** (Duthie) Wall. ex Gamble Fl. Pres. Madras 1:482 (341). *Eugenia heyneana* Duthie in FBI 2: 500; Brandis 323.

Evergreen tree, up to 6 m tall; bark smooth, greyish; branchlets greenish white, glabrous. Leaves opposite, elliptic-oblong or lanceolate, 8–12 × 1.5–5 cm, glabrous, pellucid-dotted, shining and brown above, paler beneath, apex acute-acuminate, margin entire, base cuneate; secondary nerves (main lateral nerves) numerous, slender, with reticulate intermediate nerves, petiole to 1 cm. Flowers white, small in 10–15 cm long laxly branched cymes from the scars of fallen leaves. Berries oblong, to 1.2 × 0.8 cm, crowned with cup-like calyx limb, pale purple, glabrous.

Rare, in most of the districts along streams in dry deciduous forests and in river beds up to 600 m. Fls. & Frs.: Mar.–May. Vern.: Chinna neredu.

Jalapur RF (NZB), *BR* 9659; Krishna district. *J.N. Hearsey* s.n. (DD); Savathulagundam (ADB), *NRR & TRS* 8655 (MH).

INDIA : Bihar and Orissa; Western Peninsula.

**Syzygium malabaricum** (Beddome) Gamble Fl. Pres. Madras 1: 481 (340). 1919. FBI 2: 500. *Eugenia malabarica* Beddome Fl. Sylv. S. India t. 199. 1872; FBI 2: 497; Brandis 322.

Middle sized tree; branchlets yellowish brown. Leaves lanceolate or obovate, 4–9.2 × 2.2–6 cm, glabrous, minutely dotted beneath, apex obtuse, margin entire, base cuneate; secondary nerves 4–7 pairs, intramarginal venation inconspicuous; petiole to 1 cm, reddish. Flowers very small, greenish-white, sessile in 2 cm long, compound cymes. Berry spherical or globose, 0.5 cm in dia.

Rare, in Visakhapatnam district.

Korai Road (VSKP), *GVS* 22633 (MH).

INDIA : Wainad.

**Syzygium operculatum** (Roxb.) Nicolezu in Pfamilien. Il. 7. 85. 1893; Gamble 1: 481 (340). *Eugenia operculata* Roxb. Fl. Ind. 2: 486. 1832; FBI 2: 498. *p.p.* Brandis 322.

Middle sized tree; bark grey or light brown, smooth; wood reddish-grey; branchlets subterete or tetragonous, pale. Leaves thinly coriaceous, broadly ovate or elliptic, rarely obovate, glabrous, apex rounded or obtusely acuminate, margin entire, base narrowed; secondary nerves 8–12 pairs. Flowers greenish, small, sessile, odorous, in broad, compound, trichotomous cymes in leafless axils. Berry dark purple, globose or ovoid, 0.6–1.2 cm in dia., juicy, crowned by calyx-rim; seed 1.

Northern Circars, at higher elevations (1200–1500 m) Fls.: Apr.–May.

Fruits edible. The wood is used for building and agricultural implements.

Between Gummada and Donubavi (VSKP), *GVS* 62423 (MH).

**Syzygium samarangense** (Bl.) Merr. & Perr. in Journ. Arnold. Arbor. 19: 115. 1938; & in Mem. Amer. Acad. Arts. & Sci. 18: 167 1939. *Myrtus samarangensis* Bl. Bijdr. 1084. 1820. *Eugenia javanica* Lam. Encycl. 3: 200. 1789; FBI 2: 474.

Small tree; trunk short; crown wide; branchlets terete or terete-compressed. Leaves ovate-oblong or elliptic-oblong, 12–22.5 × 6.5–11.6 cm, glabrous, slightly glaucous above, apex obtusely acute-acuminate, margin entire, base semi-cordate or rounded; secondary nerves 11 pairs, 1.5 cm apart, prominent beneath, intra-marginal vein double; petiole very short. Flowers white, 3 × 1.2 cm, many in terminal and lateral trichotomously branched, long peduncled, lax cymes. Berry pinkish-white, turbinate, flattened at the apex, crowned by the fleshy inflexed calyx lobes; seeds 0.5–0.8 cm in dia.

Rare, introduced and cultivated. Fls.: Mar.–Apr. Frs.: May–June. Vern.: Gulabi jammi chettu. Eng.: Samarang Apple.

Fruits juicy and slightly acidic, flavourless, but edible. Wood hard.

Rakthachandanam thota (SKLM), *GVS* 62362 (MH).

INDIA : Cultivated; Andaman & Nicobar.
WORLD : Native of Malacca.

## LECYTHIDACEAE

### KEY 1 (Vegetative)

1. Leaves puberulous beneath; secondary nerves up to 7 pairs, irregular, apex obtuse or sub-acute ........ ***Barringtonia***
1. Leaves glabrous on both surfaces; secondary nerves regular, apex shortly acuminate-long acuminate, acute or obtuse:
    2. Leaves thin; secondary nerves curved upwards, angle of divergence narrow; margin entire; base cuneate or acute. Cultivated ........ ***Couroupita***
    2. Leaves thick; secondary nerves curved towards margins, angle of divergence broad, margin crenulate-denticulate; base attenuate. Wild. ........ ***Careya***

### KEY 2 (Flowering)

1. Flowers on the trunk or old wood; stamens fused into 2 unequal bundles and forming a hood over ovary cultivated ........ ***Couroupita***
1. Flowers in the axils of leaves, stamens free:
    2. Flowers yellowish-white, 6 cm across on short spikes, all staminal filaments antheriferous ........ ***Careya***
    2. Flowers pale pink, 1.3 cm across in axillary pendulous racemes few staminal filaments antheriferous ........ ***Barringtonia***

### KEY 3 (Fruiting)

1. Berry small, ovoid, angular; seed one ........ ***Barringtonia***
1. Berry large, globose, rounded; seeds many embedded in pulp:

2. Fruit not foetid, crowned by persistent calyx lobes .............................................. ***Careya***
2. Fruit foetid, not crowned by calyx lobes .................................................... ***Couroupita***

## BARRINGTONIA J.R. Forster & J.G.A. Forster

**Barringtonia acutangula** (L.) Gaertn. Fruct. Sem. Pl. 2: 97. t. 101. 1790; sub. sp *acutangula* Bedd. Fl. Sylv. S. India t. 204. 1872; FBI 2: 508; Gamble 1: 487 (344); Brandis 330. *Eugenia acutangula* L. Sp. Pl. 471. 1753.

Evergreen tree, up to 10 m tall; trunk short; bark thick, dark grey, distinctly furrowed on old trees; wood pale or reddish brown; branchlets densely pubescent. Leaves thin-coriaceous, oblanceolate to obovate, 3–15 × 2–9 cm, glabrous above, yellowish puberulous on nerves beneath, apex obtuse or subacute, margin denticulate-crenate, base narrowly attenuate; secondary nerves 8–15 pairs, prominent beneath; petiole to 8 mm. Flowers pale pink or red with very caducous petals in 45 cm long, axillary, pendulous, spicate racemes. Berry ovoid, quadrangular, narrowed at both ends, 3 × 1 cm, glabrous, fibrous, crowned with persistent calyx tube; seed solitary, ovoid; embryo thick.

Common, in plains along streams or water swamp forests. It generally occupies low lying areas along the edges of the streams, rivers and around the edges of fresh water swamps. It prefers heavy and clayey soil but does not grow under saline conditions. Fls.: Apr.–June. Frs.: Dec.–Jan. Vern.: Kurpa, Kadmi, Kanapa, Tarepu, Kudapa, Kalapa, Nir-kanti, Nir-kanapa. Eng.: Indian-Oak.

Wood soft, used for furniture. Bark is used for fish-poisoning, kernals are used in diarrhoea.

Seethampet (SKLM), *SSR* 15577; Pakhal (WGL), *SSR & KSM* 18519; Racherla RF (KNL), *RVR* 1373; Bheemaram (ADB), *GO & DAM* 5166; Jodevagu-Bheemaram RF (ADB), *TP & GO* 5453; Vempalli (KNR), *GVS* 20184 (MH); Gokavaram to Chodavaram (EG), *MSR* 1953 (CAL); Sriharikota (NLR), *J. Prasad* 103330 & 103339 (DD); Kambakkam hills (CTR), *MCB* 4516 (MH).

INDIA : Throughout.
WORLD : Widespread from Afghanistan to Northern Australia.

## CAREYA Roxburgh

**Careya arborea** Roxb. Pl. Cor. t. 218. 1811 & Fl. Ind. 2: 638. 1832; FBI 2: 511; Gamble 1: 488 (345); Brandis 332; Beddome Fl. Sylv. S. Inda. t. 205. 1872.

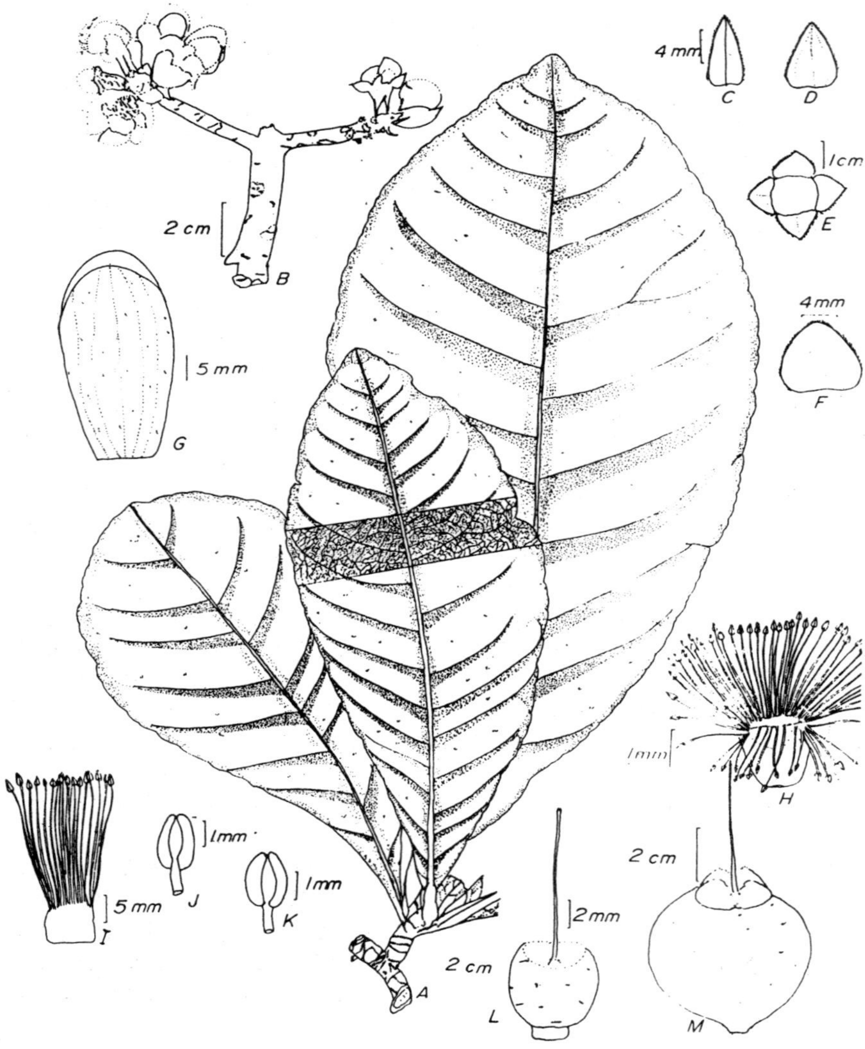


**Fig. 40.** *Careya arborea* Roxb.
A. Twig, B. Inflorescence, C & D. Bracts, E. Calyx, F. Sepal, G. Petal, H. Stamens with calyx, I. Stamens, J & K. Anthers, L. Pistil, M. Fruit.

Deciduous tree, 15–20 m tall; bark thick, dark brown, rather smooth with exfoliating scales; wood red, straight-grained, medium coarse textured; branchlets pubescent with persistent leaf scars. Leaves apically clustered, coriaceous, obovate, 18–32 × 6–14.4 cm, glabrous, apex shortly acuminate, margin crenulate, base attenuate; secondary nerves 16–18 pairs; petiole to 1 cm. Flowers yellowish-white on woody peduncles in spikes. Berry globular,

5–7 cm dia., yellowish, rind thick, crowned with the limb of calyx; seeds many, embedded in fleshy pulp.

Common, in all deciduous forests. It is a fire resistant and good coppicer. Leaf fall: Feb., old leaves often red. New foliage: Apr.–May. Fls.: Mar–Apr. Frs.:June–July. Vern.: Kumbhi, Budda budwa, Guda-Dhermi, Dudippa, Gadava, Budadhermi.

Bark gives a good fibre and can be used for making brown paper and coarse cordage, bark extract used in body swellings. Wood heavy, hard, strong, and of good quality but only used for fuel and agricultural implements. Leaves are used for wrapping bidis. The fruit is edible and aromatic. It contains astringent gum. A decoction of the fruit is given to promote digestion. The seeds are reported to be poisonous.

Upper Ahobilam (KNL), *SSR* 15297; Farhabad (MBNR), *SSR* 15537; Bheemaram (ADB), *GO & DAM* 5168; Maredumilli (EG), *GVS* 24415 (MH); Gummada (SKLM), *GVS* 62450 (MH); Narsipatnam (VSKP), *GVS* 30090 (MH); Tirumala (CTR), *DR* 2072 (SVU).

INDIA : Throughout.
WORLD : Myanmar, Sri Lanka, Malayan Peninsula, Siam.

## **COUROUPITA** Aublet

**Couroupita guianensis** Aublet Hist. Pl. Guiane 708. t. 282. 1775; Curtis. Bot. Mag. tt. 3158 & 3159. 1832.

Deciduous tree, up to 20 m tall; bark brownish-grey, smooth; branchlets glabrous. Leaves clustered at the ends of branchlets, thin coriaceous, oblanceolate, or oblong-obovate, 14–40 × 8–17 cm, glabrous, pubescent on the nerves beneath, apex acute or obtuse, margin entire or obscurely serrulate, base acute or cuneate, often oblique; secondary nerves 15–20 pairs, prominent beneath; petiole to 1.8 cm. Flowers reddish or tinged yellow on the outside, 10 cm across, cauliflorus; stamens fused into 2 unequal bundles and forming a hood over ovary. Fruit pale reddish-brown, woody, pendulous, 15 cm in dia., inner pulp ill-smelling.

Introduced. Cultivated in gardens. Leaf fall: Dec.–Jan. New foliage: Feb.-Mar. Fls.: Throughout the year. Vern.: Nagalingam. Eng.: Cannon-ball tree.

Tirumala gardens (CTR), *SSR* 18597; Kailasa Kona (CTR), *SSR* 19684.

INDIA : Cultivated throughout.
WORLD : French Guyana, Java.

## MELASTOMATACEAE
## **MEMECYLON** Linnaeus

1. Leaves to 10 cm long ........ ***M. edule***
1. Leaves to 6.5 cm long:
   2. Flowers shortly pedicellate or sessile ........ ***M. lushingtonii***
   2. Flowers clearly pedicellate :
      3. Leaves narrowly ovate-oblong; branchlets quadrangular. Cymes few-flowered ........ ***M. molestum***
      3. Leaves ovate or elliptic-ovate; branchlets terete. Cymes many-flowered ........ ***M. umbellaum***

**Memecylon edule** Roxb. Pl. Cor. t. 82. 1798; FBI 2: 563. *p.p.*; Gamble 1: 504 (355); Brandis 335.

Densely foliaceous, evergreen tree, up to 10 m tall; bark deep-brown, closely and regularly fluted; wood light brown. Leaves coriaceous, ovate, 5–10 × 3–3.5 cm, glabrous, thick, black dotted beneath, green above, brown beneath, apex obtusely acute or narrowly acute, margin entire, base rounded; secondary nerves 8–10 pairs, intramarginal vein and nerves inconspicuous; petiole to 4 mm. Flowers blue, very numerous in clustered, compound, 2 cm long, umbellate cymes. Berry globose, blue or dark purple, 2 mm in dia., glabrous; seed to 5 mm.

Common, in dry evergreen forests along rocky ravines at 800–1300 m. Fls.: Apr.–June. Frs.: June–July. Vern.: Alli. Fruits edible.

Near Battigudem (KNR), *NRR, TRS* 86590 (MH); Diguvametta (KNL), *JLE* 42184 (MH); Kavali (NLR), *BSN* 4928 (MH); Dharakonda (VSKP), *CAB* 17076 (MH); Rampa hills (EG), *MSR* 1551 (CAL).

INDIA : Tamil Nadu.
WORLD : Sri Lanka, Myanmar, Thailand, Malaysia.

**Memecylon lusingtonii** Gamble in Kew Bull. 1919 : 226. 1919 & Fl. Pres. Madras 1 : 503 (355) 1919.

Small tree branchlets slender, pale, tuberculate. Leaves elliptic or elliptic lanceolate, upto 5 × 2.5 cm, glabrous, green when dry, apex obtusely acute, margin entire, base rounded to acute, nerves obscure; petiole short. Flowers pale blue in umbellules. Berry globose, 1-seeded.

Rare, in Chittoor district at above 500m. Fls. & Frs : Feb.–June.

Talakona (CTR), *SSR* 19680; Horsely hills (CTR), *BSN* 3739, *CECF* 4746 (CAL); *JSG* 15110 (CAL)

INDIA : Western ghats.

**Memecylon molestum** (C.B. Clarke) Cogn. in DC. Monog. Phan. 7: 1169. 1891; Gamble 1: 505 (356). *M. edule* Roxb. var. *molesta* C.B. Clarke in Hook. f. Fl. Brit. India 2: 564. 1879.

Small tree, up to 4 m tall; branchlets distinctly quadrangular. Leaves narrowly ovate-oblong, elliptic or ovate, to 5 × 2 cm, glabrous, yellow when dry, apex obtuse or emarginate, margin entire, base acute; secondary nerves 6–8 pairs, not prominent, intramarginal nerve visible; petiole short. Flowers bright blue in few-flowered, lateral, 1.2 cm in dia., cymes. Drupe globose or slightly ovoid.

Occasional, in Cuddapah and Chittoor districts.

Balapalle (CDP); *JLE* 15706 (MH); *JSG* 17118 (CAL).

**Memecylon umbellatum** Burm. f. Fl. Ind. 37. 1768; Gamble 1: 504 (353).
*M. edule* Roxb. Pl. Cor. t. 82. 1795; Clarke in FBI 2: 563; Gamble 1: 504 (556); Brandis 335; Beddome Fl. Sylv. S. India t. 206.

Trees, up to 4 m tall; branches terete, glabrous, glaucous, bark corky, fluted. Leaves coriaceous, ovate, elliptic-ovate, 6.3 × 3 cm; glabrous, dull above, yellowish beneath when dry; apex shortly and bluntly acuminate, margin slightly revolute, base attenuate, midrib prominent; secondary nerves 6–9, obscure, petiole to 5 mm. Flowers bluish purple or blue, many in 3 cm long umbellate cymes. Berry globose, 1 × 1 cm, reddish tinged, ripening black; seed one.

Along rocky ravines and on the tops of high rocky hills at 600–1000 m. Fls.: Apr.–June. Frs.: June–Aug. Vern.: Alli, Peddalli, Kukka Alli.

Owk RF (KNL), *RVR* 3156; Kavali (NLR), *BSN* 4829; Srisailam (KNL), *JLE* 1686 (MH); Kambakkam (CTR), *DR* 1358 (SVU).

INDIA : Eastern Peninsula.
WORLD : Sri Lanka, Myanmar, Thailand, Malaysia.

## LYTHRACEAE

### LAGERSTROEMIA Linnaeus

1. Leaves elliptic or elliptic-oblong, 4–9.5 × 1.8–3.5 cm, apex acute with apicule. Flowers white in panicles; petals 1 cm long, claw slender. Calyx smooth in capsule ......... ***L. parviflora***
1. Leaves oblong-lanceolate or elliptic, 10.5–17 × 5.5–9 cm, apex shortly acuminate. Flowers pinkish in racemes, petals 2.5–4.5 cm long, claw not slender. Calyx ribbed in capsule ........ ***L. reginae***

**Lagerstroemia parviflora** Roxb. Pl. Cor. t. 66. 1796; FBI 2: 575; Gamble 1: 513 (362); Brandis 335; Beddome Fl. Sylv. S. India. t. 31. 1867.

Deciduous tree, up to 20 m tall; bark light grey to reddish, smooth, exfoliating in narrow longitudinal flakes; wood light brown; branchlets pubescent with very short simple hairs. Leaves thin-coriaceous, elliptic or elliptic-oblong, 4–9.5 × 1.8–3.5 cm, green, glabrous and shining above, pale or greyish, some times hoary-tomentose beneath, apex apiculate, margin entire, base sub-acute; secondary nerves 6–8 pairs, arcuate, prominent beneath; petiole to 4 mm. Flowers white, fragrant, in 13 cm long, terminal

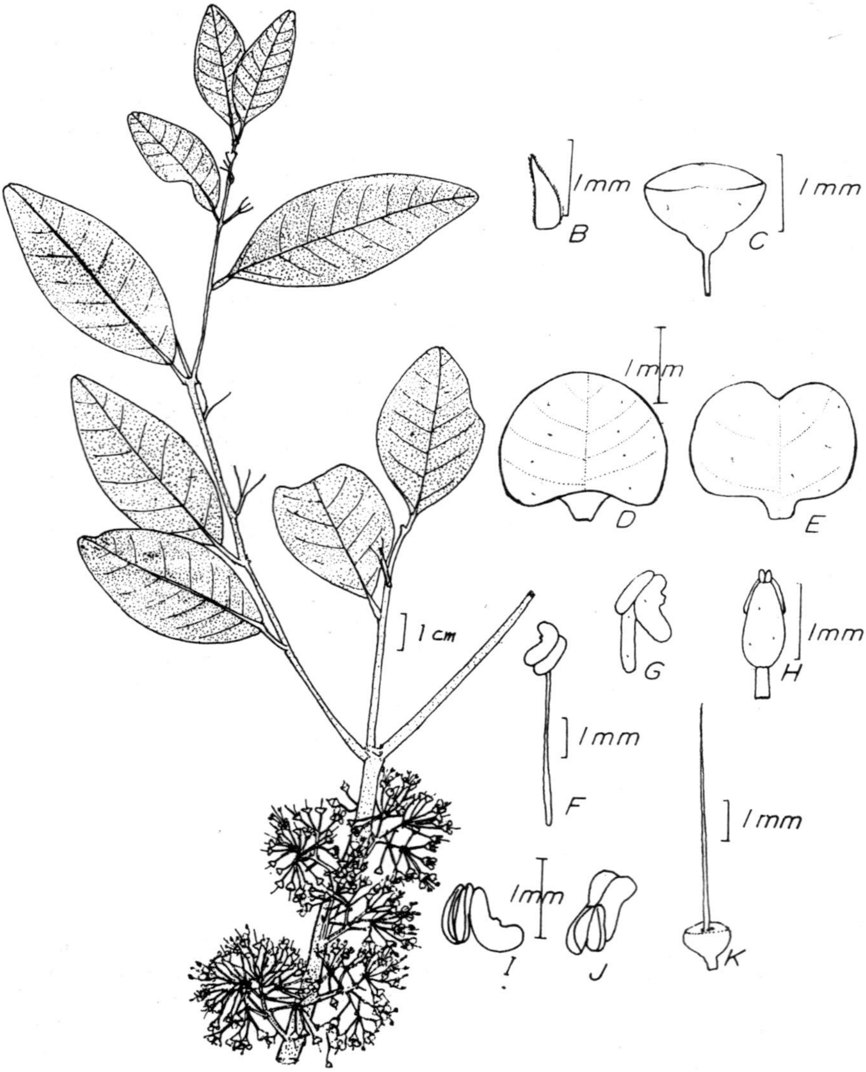


**Fig. 41.** *Memecylon umbellatum* Burm f.
A. Twig, B. Bract, C. Calyx, D & E. Petals, F. Stamen, G–J. Anthers, K. Pistil.

or axillary lax panicles. Capsule ellipsoid, brown, 3 × 2 cm, woody, adnate to persistent calyx, 3–4-valved, dehiscent; seeds numerous, 1.5 × 0.5 cm, flattened, winged, which are longer than seed.

Common, in moist and dry deciduous forests at 500–1300 m. It is light demander, drought resistant, good coppicer, non-browsable and fire-resistant. It thrives best on deep porous loamy soils. Leaf fall: Mar.–Apr. New foliage:

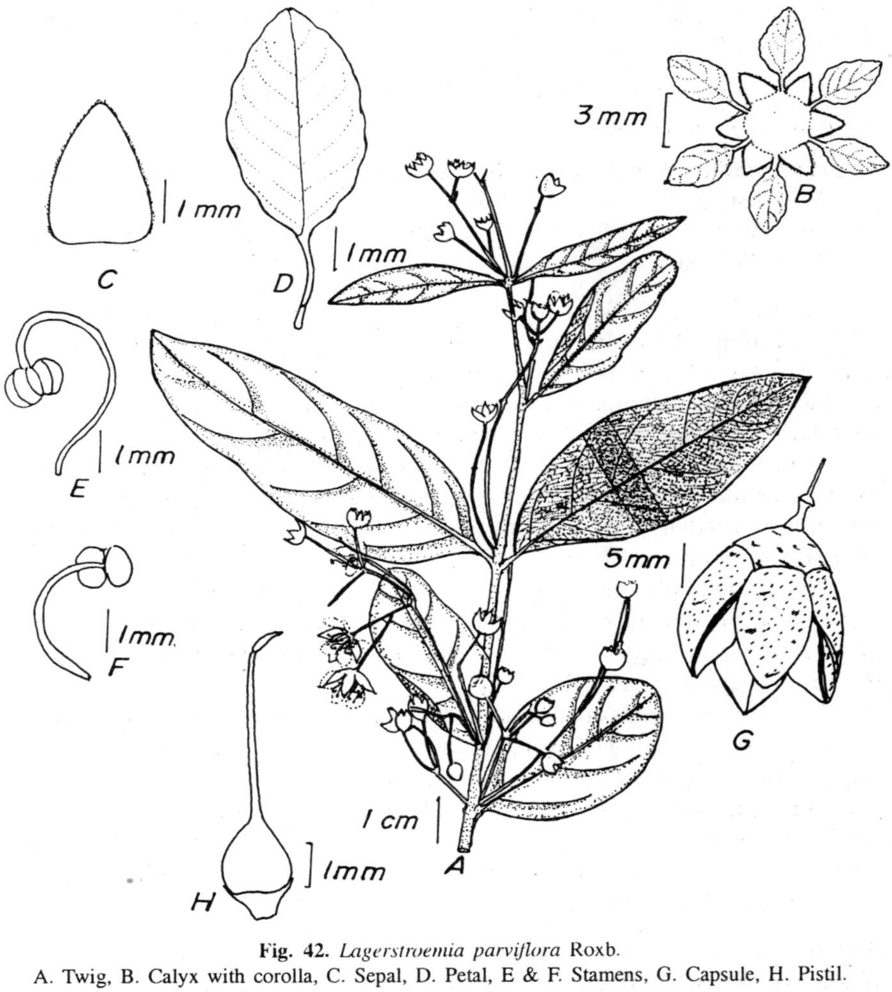


**Fig. 42.** *Lagerstroemia parviflora* Roxb.
A. Twig, B. Calyx with corolla, C. Sepal, D. Petal, E & F. Stamens, G. Capsule, H. Pistil.

May. Fls.: May, short lived. Frs.: Ripen in Dec.–Feb. Vern.: Chennangi, Gullakaraka, Chinnagi.

The wood is valued for timber, it is tough, elastic and durable, used for building constructions, fencing, bridges, boats, oars and agricultural implements, also for carts, wood pipes and water tanks. It is suitable for making boxes, tool handles, tent pegs, bent-wood furniture, picker arms, telegraph poles etc. The tree is lopped for fodder and it is a host for tasar silk worm. It is reported to yield a sweet edible gum.

Pangidi (KNL), *SSR* 15217; Mothugudem (KMM), *SSR*15243; G. Madugula (VSKP), *SSR* 15952; Narsapur (MDK), *SSR & KSM* 18506;

Penchalakona (NLR); *SSR* 19347; Satruguda (ADB), *GO* 4898; Mamandur (CTR), *DR* 1273 (SVU).

INDIA : Assam, Central India and Western Peninsula.
WORLD : Myanmar, Java.

**Lagerstroemia reginae** Roxb. Pl. Cor. t. 65. 1795. *L. flos-reginae* Retz. Obs. Bot. 5: 25. 1788. *p.p.* FBI 2: 577; Gamble 1: 513 (362); Brandis 339; Beddome Fl. Sylv. S. Inda t. 29. 1869. *L. speciosa* (L.) Pers. Syn. Pl. 2: 72. 1807.

Deciduous tree, 10–15 m tall; crown rounded; bark light grey or pale brown, flaking-off in irregular flakes. Leaves thick, oblong-lanceolate or elliptic, 10.5–17 × 5.5–9 cm, glabrous, green above, paler beneath, apex abruptly-shortly acuminate, margin entire, base acute or rounded; secondary nerves 10–15 pairs, conspicuous, curving upwards and forming loops; petiole to 5 mm. Flowers pinkish, 2–3 cm across in 12 cm long racemes; pedicel stout. Capsule ellipsoid, to 2.5 × 1.5 cm in dia., 4-valved, apiculate, half of the fruit enclosed by the persistent calyx; seeds many, compressed, ascending, pale brown, winged, 1.6 × 0.5 cm, glabrous.

It is widely cultivated in gardens and avenues for showy flowers. It is moderate light demander and coppices well. Leaf fall: Jan.–Feb. New foliage: May. Fls.: July–Aug. Frs.: Nov.–Jan. Vern.: Varagogu. Eng.: Queen of flowers, Queen crape Myrtle.

The bark and seeds are considered as stimulant, febrifuge and astringent. The inner bark yields coarse fibre, and tree exudes a resin. Wood is hard, strong, elastic and moderately heavy, durable and is used for building purposes, planking, ship building, piles, bridges, water tanks, railway carriages, motor lorry bodies, floor boards, rice pounders, etc. It is suitable for telegraph poles, leather cutting blocks, match boxes and splints.

Tirumala (CTR), *SSR* 18595.

INDIA : From Assam to Malaca; hills of Deccan Peninsula.
WORLD : Pakistan, Myanmar, New Guinea, China, Indonesia.

## SONNERATIACEAE

### **SONNERATIA** Linnaeus. f. *nom. cons.*

1. Small tree. Leaves broadly obovate. Flowers dark red, 4.5 cm across; petals 6; stigma capitate. Capsule 2.5–5 cm dia. ........ ***S. caseolaris***
1. Medium sized tree. Leaves narrow elliptic or oblong lanceolate. Flowers whitish, 2.5 cm across; petals 0; stigma mushroom shaped. Capsule 1.5–2 cm dia. ........ ***S. apetala***

**Sonneratia apetala** Buch-Ham. in Symes Embassy Ava. 3.477, 1800; FBI 2: 579; Gamble 1: 515 (363); Brandis 341.

Evergreen tree, up to 12 m tall; pneumatophores numerous, conical; bark black, smooth, rough below on old tree; wood reddish brown; branches slender, drooping; foliage light glaucous-green. Leaves coriaceous, narrow elliptic or oblong-lanceolate, 3.5–10.5 × 2.2–5 cm, glaucous, glabrous, thick, apex obtuse or acute, margin entire, base attenuate; secondary nerves 10–12 pairs, obscure, slightly prominent above; petiole 0.5–1.2 cm. Flowers whitish, 2.5 cm across, solitary or 3–5 in terminal and lateral corymbs. Capsule depressed-globose, smooth, 1.5–2 cm dia., tipped with the style, convex at the top, supported by persistent calyx; seeds numerous, small, curved, angular.

Common, in tidal forests. Fls.: Mar.–Apr. Frs.: June. Vern.: Kalinga, Kyalanki. Wood hard and used as good timber.

Coringa river estuary (EG), *T. Ananda Rao* 7292 (CAL), *A.R.K. Sastry* 7357 (CAL); Coringa Delta (EG), *G.F.F. Foulkes* 95 (CAL); Coringa canal Bank (EG), *GVS* 68688 (MH).

INDIA : West Bengal, Deccan Peninsula.
WORLD : Bangladesh, Myanmar, Sri Lanka.

**Sonneratia caseolaris** (L.) Engler & Prantl Pflazenf. 267. 1897. *Rhizophora caseolaris* L. in Stickman, Herb. Amb. 31. 1754. *Sonneratia acida* L.f. Suppl. 253. 1751; FBI 2: 579; Brandis 340.

Small, evergreen tree, 5–6 m tall; bark white; wood grey, even-grained. Leaves sub-fleshy, broadly obovate, elliptic-oblong, 7.5–10 × 2.5–4.5 cm, apex obtuse, margin entire, base much attenuated; secondary nerves ca. 10 pairs, conspicuous above; petiole very short. Flowers dark red, 4–5 cm across, solitary, terminal at the ends of branchlets. Capsule depressed-globose, 2.5–5 cm in dia., supported by the persistent calyx and tipped by persistent tapering stylar base.

Rare, in Coringa estuary in East Godavari (Venkateswarlu, 1944). Fls.: Mar.–June. Frs.: Aug. Vern.: Kandia. Fruits edible.

INDIA : West Bengal; Deccan Peninsula.
WORLD : Tropical S.E. Asia and Sri Lanka to North Australia, Solomon Island and New Hebrides.

## PUNICACEAE

### **PUNICA** Linnaeus

**Punica granatum** L. Sp. Pl. 472. 1753; FBI 2. 581; Brandis 341.

Deciduous, small tree, up to 6 m tall; bark smooth, dark grey; branchlets often spine-tipped. Leaves clustered in condensed branchlets, coriaceous, oblanceolate or oblong, 3.5–8 × 1.4–2 cm, apex obtuse or subacute, margin entire, base acute or attenuate; petiole to 0.8 cm. Flowers bright red, solitary or 2–4 together, 3.7–5 cm long, terminal. Berry yellowish, globose, indehiscent, crowned by persistent calyx-lobes; interior septate with membranous wall; pulp pellucid, pinkish, formed from the outer layer of the testa of the seed; rind thick and coriaceous; seeds numerous, angular.

Planted for the fruits throughout Andhra Pradesh, in gardens and backyards of houses. Fls. & Frs.: Throughout the year. Vern.: Danimma. Eng.: Pomegranate.

Tirumala (CTR), *SSR* 15266.

Seed pulp is sweet and edible. The fruit is a good source of sugar, vitamin C, and a fair source of iron. The rind of the fruit is also the source of a dye, which gives yellowish-brown shades and has been used for dyeing wool and silk. The flower yields a light red dye used for dyeing cloth. The powdered flower buds are used in bronchitis. Fruits are to be good in inflammation of stomach and heart pain. The fresh pomegranate juice is used as an ingredient of cooling and refrigerent mixtures.

## CARICACEAE

### **CARICA** Linnaeus

**Carica papaya** L. Sp. Pl. 1036. 1753; FBI 2: 599; Gamble 1: 525 (371); Brandis 345.

Tree, 5–8 m tall; dioecious, crown round; trunk usually unbranched; bark pale-greenish, smooth; wood soft with milky juice. Leaves 5.5 cm across, palmately 7-lobed; lobes pinnately lobed, orbicular, glabrous, apex acuminate, margin serrate, base rounded, petiole hollow, long. Flowers pale yellowish, fragrant, 2–5 cm long; male flowers slender, long tubed in drooping, 45 cm long panicles; female flowers broader, solitary or few together on short peduncles. Fruit greenish-yellow to orange, oblong, fleshy with central cavity, to 50 cm long, indehiscent; seeds numerous, black, 0.3 cm dia.

Planted in gardens, house compounds for its edible fruits. Fls. & Frs.: Throughout the year. Vern.: Boppasi, Boppayi. Eng.: Papaya, Pawpaw, Melon tree. Fruit pulp yellow, sweet, and edible and it is used as abortifacient.

INDIA : Cultivated throughout.
WORLD : Native of West Indies. Cultivated in other parts.

## ALANGIACEAE

### **ALANGIUM** Lam. *nom. cons.*

1. Leaves orbicular to broadly oblong, margin angular. Flowers in cymes; stamens as many as petals. Berry ovoid, obscurely ribbed, glabrous .......... ***A. chinense***
1. Leaves ovate-oblong or elliptic, lanceolate, margin entire. Flowers in fascicles; stamens double the number of petals or more. Berry ellipsoid, not ribbed, pubescent. ... ***A.salvifolium***

**Alangium chinense** (Lour.) Harms. in Ber. Deutsch. Bot. Ges. 15: 24.1897. *Stylidium chinense* Lour. Fl. Cochinch. 1: 221. 1790. *Marlea begonifolia* Roxb. Pl. Cor. 3: 80. t. 283. 1820; FBI 2: 743. *Alangium begonifolium* Baillon Hist. Pl. 6: 270. 1876; Gamble 1: 572 (404).

Deciduous tree, 15–20 m tall; branches horizontal, internodes clearly angular; bark silvery grey, smooth, wood white, even grained. Leaves orbicular to broadly oblong, 9–10 × 6–7 cm, glabrous with tufts of hairs in the nerve-axils beneath, apex acuminate, margin entire or angular, lobed, base rounded or sub-cordate; petiole 1.5–2.3 cm. Flowers whitish, fragrant, 6–8 merous, 8–12, in 6 cm long, axillary, dichotomous cymes; filaments hairy. Berry ovoid, 8 × 6 mm, obscurely ribbed, glabrous.

Rare, in Visakhapatnam and Vijayanagaram districts, occurs generaly in moist ravines. Fls. & Frs.: Mar.–June.

Leaves are used as fodder. Wood soft, sometimes used for making furniture and axe handles but it is not much important.

Punyagiri (VSKP), *GVS* 23309 (MH).

INDIA : Sikkim, Himalayan tracts and Kumaun, extending eastwards to Bihar, North Bengal, the hills of Assam, Maghalaya, Manipur, Arunachal Pradesh.

**Alangium salvifolium** (L.f.) Wangerin in Engl. Pflanzen. IV. 220 b.41: 9.1910. ssp *salvifolium* Turill & Milne. Redh. Fl. Trop. E. Africa. *Alangiaceae* 3. 1958; Gamble 1: 572 (404). *Grewia salvifolia* L.f. Suppl. Pl. 409. 1781. *Alangium lamarckii* Thw. Enum. Pl. Zeyl. 133. 1859; FBI 2: 741; Brandis 354; Beddome Fl. Sylv. S. India t. 215. 1872.

Deciduous bushy tree, up to 12 m tall; trunk often with numerous holes; bark grey, orange-yellow when young, deeply fissured; wood olive-brown, close grained; branchlets appressed tomentose. Leaves chartaceous, ovate-oblong or elliptic-lanceolate, 6–15 × 1.8–4.3 cm, glabrous above, glabrescent or puberulous below, apex subacute to acuminate, margin entire, base oblique, 3-nerved from base; secondary nerves 6 pairs, irregular; petiole to 1 cm, tomentose. Flowers white, scented in irregular axillary fascicles. Berry ellipsoid, 2 × 1 cm, red or blackish purple, pubescent, succulent with bony endocarp, crowned by calyx lobes; seed solitary, ovoid, 0.8 × 0.4 cm.

Common, in dry regions, in the plains and low hills and on road sides, especially in forest outskirts, up to 800 m. Its growth moderate. Leaf fall: Feb. New foliage: May–June. Fls.: Feb.–Apr. Frs.: July. Vern.: Uduga, Ooduga, Ankolamu, Urgu, Nalla Uduga.

Fruits are astringent, acidic and edible. They are anti-phlegmatic, laxative, tonic and refrigerent. Wood hard, useful for ornamental work and also as a good fuel. Root bark used as a purgative and for skin diseases.

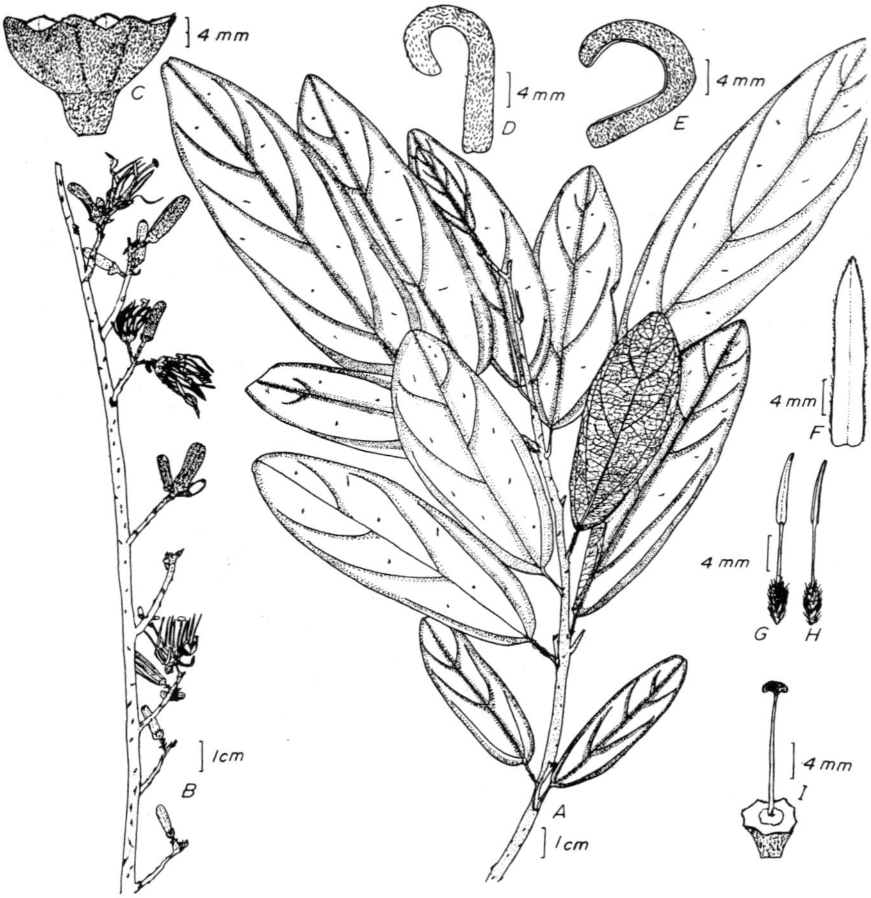


**Fig. 43.** *Alangium salvifolium* (L.f) Wang.
A. Twig, B. Inflorescence branch, C. Calyx, D, E & F. Petals, G & H. Stamens, I. Pistil.

Lower Ahobilam (KNL), *SSR* 15295; Simhachalam (VSKP), *SSR* 5563; Rampur (ADB), *SSR* 19311; Durgam forest (SKLM), *GVS* 62516 (MH); Servivaram (VSKP), *CAB* 17058 (MH); Diguvametta (PRSM), *RVK* 15723; Vatvarlapally (MBNR), *TDCK* 13602; Kailasakona(CTR), *DR* 1435 (SVU).

INDIA : Throughout.

WORLD : Tropics from East Africa, Sri Lanka, Myanmar, Malaysia and Phillippines.

# RUBIACEAE

1. Inflorescence a capitate head:
   2. Calyx tubes coharent; corolla spreading, >1.5 cm across; ovule solitary in each carpel ... **Morinda**
   2. Calyx tubes free, corolla not spreading, < 1 cm across; ovules numerous in each carpel:
      3. Stipules foliaceous. Corolla lobes imbricate:
         4. Bracteoles linear. Peduncle axillary, 1–3 together. Calyx lobed; stigma clavate, rarely capitate. Capsule cuneate, not ribbed. Leaves cordate at base ... **Haldinia**
         4. Bracteoles spathulate. Peduncle terminal, solitary. Calyx truncate, not lobed; stigma mitriform. Capsule oblong, faintly 10-ribbed. Leaves truncate or acute at base ... **Mitragyna**
      3. Stipules lanceolate or entire, flat. Corolla lobes valvate:
         5. Seeds winged. Stigma globose ... **Nauclea**
         5. Seeds not winged. Stigma fusiform ... **Anthocephalus**
1. Inflorescence not capitate:
   6. Trees armed:
      7. Ovary 1-locular ... **Ceriscoides**
      7. Ovary 2–5-locular:
         8. Flowers many in cymes; ovule solitary in each locule. Fruit drupe ... **Meyna**
         8. Flowers solitary; ovules 2–numerous in each locule. Fruit berry:
            9. Berry ovoid or ellipsoid, 6 × 4 cm. Flowers white ... **Tamilnadia**
            9. Berry globose, 2.5 × 2 cm. Flowers yellowish-white ... **Catunaregum**
   6. Trees unarmed :
      10. Inflorescence spicate; bracts large, foliaceous ... **Hymenodictyon**
      10. Inflorescence not spicate; bracts small, not foliaceous:
          11. Fruit capsular. Flowers in panicles ... **Wendlandia**.
          11. Fruit not capsular. Flowers in cymes, corymbs or solitary:
              12. Ovule solitary; flowers 4 merous:
                  13. Corolla lobes valvate ... **Canthium**
                  13. Corolla lobes twisted:
                      14. Style twice as long as tube, stigma 2-fid ... **Pavetta**
                      14. Style less than twice as long as tube, stigma fusiform ... **Ixora**
              12. Ovules 2 or numerous; flowers 5-merous.
                  15. Ovules 2, 1 erect, 1 pendulous ... **Scyphiphora**
                  15. Ovules numerous:
                      16. Corolla tube villous within:
                          17. Cymes terminal ... **Chomelia**
                          17. Cymes axillary or leaf-opposed ... **Deccania**
                      16. Corolla tube glabrous within:
                          18. Corolla lobes twisted; calyx limb persistent. Fruit berry ... **Gardenia**
                          18. Corolla lobes imbricate; calyx, limb deciduous. Frit drupaceous ... **Guettardia**

## ANTHOCEPHALUS A. Rich

**Anthocephalus chinensis** (Lam.) A. Rich. ex Walp. Repert. 2: 491. 1843. *Cephalanthus chinensis* Lam. Encycl. 1: 678. 1785. *Anthocephalus cadamba* (Roxb.) Miq. Fl. Ind. Bat. 2: 135. 1856; FBI 3: 23; Brandis 367. *Anthocephalus indicus* A. Rich. in Mem. Soc. Hist. Nat. Paris 5: 238. 1834; Gamble 2: 583 (411). *Nauclea cadamba* (Roxb.) Beddome Fl. Sylv. S. India t. 35. 1869.

Large, deciduous tree; branches spreading, sub whorled; bark grey, fissured; wood white or yellowish, even grained. Leaves coriaceous, elliptic-oblong or ovate, 12–28 × 10–15 cm, glabrescent beneath, apex acute or shortly acuminate, margin entire, base rounded or cuneate; secondary nerves 8–15, strong; petiole 2.5–6 cm., stipules 1.2–1.5 cm. Flowers greenish-yellow or orange, small in globular heads. Pseudocarp orange, fleshy, 5–6.2 cm dia.; seeds muricate.

Occasional, along river banks and hill valleys and is also cultivated for ornament and as a shade tree in plantation. It prefers deep well-drained moist alluviam. It is a light demander and coppices vigorously. Fls.: May–July. Frs.: Jan.–Aug. Vern.: Kadamba.

The ripe fruit is acidic and pleasantly flavoured and is eaten both raw and cooked. The fruit is cooling and it is said to destroy the phlegm and impurities of blood when eaten. The juice is given in fever with persistent thirst. The flowers are used as vegetable. Leaves are used as fodder for cattle. Wood moderately hard, chiefly employed for cheap boarding, plywood and for packing cases and tea boxes; it is also used as beams, rafters and light constructional work.

Kilagada (VSKP), *GVS* 44935 (MH); Sujanakotapeta (VSKP), *GVS* 32900 (MH); Godavari district *JSG* 16092 (MH).

INDIA : Throughout; often cultivated.
WORLD : Sri Lanka, Sumatra.

## **CANTHIUM** Lamarck

**Canthium dicoccum** (Gaertn.) Teijsm. & Binn. Cat. Hort. Bog. 113. 1866. var. *umbellatum* (Wt.) Sant. & Merch. in Bull. Bot. Surv. India 3: 107. 1961. *C. umbellatum* Wt. Ic. t. 1034. 1845; FBI 3: 132. *Plectronia didyma* (Gaertn.) Kurz. var. *umbellatum* (Wt.). Santapau & Merch. Bull. Bot. Surv. India 3: 107. 1962; Gamble 2: 624 (440). *Canthium didymum auct. non* Gaertn; Beddome Fl. Sylv. S. India t. 221. 1872.

Densely foliaceous, evergreen tree, up to 10 m tall; wood pale-yellow, straight grained; young branches four-angled, glabrous, stipules lanceolate. Leaves coriaceous, elliptic-obovate, 7–15 × 3–5 cm, glabrous, apex acuminate, margin entire, base acute-attenuate; petiole to 1 cm. Flowers whitish in axillary, pedunculate umbels. Drupe obovoid, 0.6 × 0.7 cm.

Common, in deciduous forests up to 1200 m. Fls.: Dec.–Feb. Frs.: Apr.–May. Vern.: Alli, Balusu, Nalla Balusu, Nakkina, Nakkeri, Niralli.

Wood is used for making combs and for agricultural implements; sometimes used as posts; it is also used for fuel and that it burns very readily without much smoke.

Tirumala (CTR), *SSR* 13709 & 15265; Galikonda (VSKP), *SSR* 15931; Anantagiri (VSKP), *SSR* 18522; Siddulaiah Kona (NLR), *SSR* 19374; Kalasamudram RF (ATP), *NY* 1197; Gupalapatnam (ADB), *GO* 4407; Maredumilli (EG), *GVS* 2424 (MH).

INDIA : South India, Maharashtra, Bihar, Orissa.

## CATUNAREGUM Wolf

**Catunaregum tomentosa** (Bl. ex DC.) Tirveng. in Taxon 27: 515. 1978. *Gardenia tomentosa* Bl. ex DC. Prodr. 4: 379. 1830. *Randia tomentosa* (Bl. ex DC.) Hook. f. Fl. Brit. India 3: 110. 1880 non Wt. & Arn.

Armed, deciduous tree, 4–6 m tall; bark grey, rough; branchlets greyish, glabrous; spines to 3 cm long. Leaves obovate-spathulate, 5–9 × 2–3.5 cm, glabrous to slightly puberulous above, slightly pubescent beneath, apex obtuse to acute, margin entire, base attenuate; secondary nerves 5–6 pairs; petiole to 1 cm. Flowers yellowish-white, solitary, axillary on spines, fragrant. Berry globose, glabrous, 2.5 × 2 cm.

In deciduous forests. Fls.: Apr.–May. Frs.: Aug.–Oct. Vern.: Kotuvagochh. Crushed fruits used as fish-poison.

G. Madugula (VSKP), *SSR* 15957; Bogada (PKSM), *SSR* 16107; on the way to Nulakamaddi (EG), *GVS* 24540 (CAL).

## CERISCOIDES Tirvengadum

**Ceriscoides turgida** (Roxb.) Tirveng. in Bull. Mus. Hist. Nat. Pairs 35: 15. 1978. *Gardenia turgida* Roxb. Fl. Ind. 2: 557. 1824; FBI 3: 118; Gamble 2: 619 (437); Brandis 379. *G. montana* Roxb. Fl. Ind. 1: 709. 1832.

Armed, deciduous tree, up to 10 m tall; bark smooth, bluish grey, thick; wood white and close grained; branchlets tomentose with strong spines. Leaves thick, chartaceous, variable, dimorphic, 1.5–2.5 × 0.8–1.5 cm, hispid above, hirsute below, apex obtuse, margin entire, base attenuate; secondary nerves 8–10 pairs; subsessile. Flowers white, dimorphic; male flowers in fascicles; female flowers solitary. Berry globose, 5 × 4 cm, grey, turgid; seeds numerous, embedded in pulp.

Occasional, in the interior forest areas near streams at 300–700 m. Leaf fall: Feb. New foliage: Mar.–Apr. Fls.: Mar. Frs.: June. Vern.: Konda Manga. Karinguva, Erri Bikki, Botakari, Pedda Bikki. Tender fruits used as vegetable by local people.

Mothugudem (KMM), *SSR* 15244; Bairluti (KNL), *SSR* 15519; Tiryani RF (ADB), *PVP* 9801; Jalalpur RF (NZB), *BR* 9660.

INDIA : Base of Himalaya, Bihar, Western Peninsula, Central and South India.
WORLD : Myanmar, Sri Lanka, Thailand, Indo-China.

## CHOMELIA Linnaeus *nom. cons.*

**Chomelia asiatica** (L.) O. Kuntze. Rev. Gen. Pl. 1: 278. 1891; Gamble 2: 613 (432). *Rondeletia asiatica* L. Sp. Pl. 172. 1753. *Webera corymbosa* Willd. Sp. Pl. 1: 1224. 1798; FBI 3: 102; Brandis 377. *Tarenna asiatica* (L.) O. Kuntze ex Schum. in Bot. Tiddskr. 24: 332. 1902. *Webera asiatica* (L.) Beddome Fl. Sylv. S. India Anal. Gen. 16. f. 2. 1869.

Evergreen shrub to small tree (in high elevations); bark dark grey, fairly rough; wood light brown, hard, close-grained; branchlets glabrous. Leaves coriaceous, oblong or elliptic-oblong, 6–13 × 2–5 cm, glabrous, black when dry, apex acute-acuminate, margin entire, base acute or rounded; secondary nerves 10–12 pairs, turns upwards and close to the margins; petiole to 1.8 cm; stipules with broad persistent triangular base, cuspidate blade. Flowers white, fragrant, to 1.5 cm across, in terminal, corymbose cymes. Berry black, globose, 6 mm dia., with 2 mm long crown of calyx lobes; seeds 6–8.

Common, in scrub forests and also in evergreen forests up to 1200 m. Fls.: Apr.–July. Frs.: Oct.–Dec. Vern.: Komari, Papidi, Konda papidi, Komi.

Talakona (CTR), *SSR* 13787; Simhachalam hills (VSKP), *MSG* 5608; Penchalakona (NLR), *MSG & DAM* 5660; Loyapalli (GTR), *VRK* 3986; Kalasamudram RF (ATP), *EC* 2916; Devipatnam (EG), *GVS* 24437 (CAL); Palakonda RF (CDP), *KNS* 2375 (DD).

INDIA : Western Peninsula
WORLD : Sri Lanka, Malaysia.

## DECCANIA Tirvengadum

**Deccania pubescens** (Roth) Tirveng. in Nordic J. Bot. 3: 456. 1983. var. ***candolleana*** (Wt. & Arn.) Tirveng. l.c. *Randia candolleana* Wt. & Arn. Prodr. Fl. Ind. Orient. 39. 1834; FBI 3: 113; Gamble 2 : 616 (425); Brandis 383. *Gardenia pubescens* Roth, Nov. Pl. Sp. Ind. Or. 151. 1821.

Tree, up to 6 m tall; bark greyish, rough; wood light brown, close grained; branchlets with elevated leaf scars. Leaves chartaceous, orbicular or ovate to obovate, 6–16.5 × 3.4–12.4 cm, lower surface with densely silky tomentose than upper surface when young, apex obtuse, margin entire, base attenuate; secondary nerves 6–8 pairs, prominent; petiole to 4 cm. Flowers white in short axillary cymes. Berries clustered, globose, 1 × 0.5 cm, rind thin, leathery with lines on the surface; seeds smooth.

Common, in open lands of deciduous forests up to 1000 m. Fls.: May–July. Frs.: Sept.–Oct. Vern.: Konda manga, Pedda manga. Jidi manga.

Kalasamudramu (ATP), *SSR* 13757; Appapur (MBNR), *SSR* 15544; Talakona (CTR), *SSR* 15258; Diguvametta (PKSM), *SSR* 16102, Avagondapalem RF (ATP), *SSR* 18558; Javuku (GTR), *VRK* 3874; North Dhone (KNL), *RVR* 2518; Penchalakona (NLR), *MSG & DAM* 5647.

## GARDENIA J. Ellis *nom. cons.*

1. Leaf nerve axils glandular-cultivated. .......... ***G. jasminoides***
1. Leaf nerve axils eglandular:
   2. Leaves petiolate; secondary nerves slender, parallel, very prominent beneath .......... ***G. resinifera***
   2. Leaves sessile or subsessile; secondary nerves other than above:
      3. Leaves 7–30 × 2.5–18.5 cm. Flowers pedicellate, calyx-lobes linear-lanceolate; stamens included. Fruit large .......... ***G. latifolia***
      3. Leaves 3–8.5 × 1.5–6 cm. Flowers sessile; calyx lobes triangular; stamens exserted. Fruit small. .......... ***G. gummifera.***

**Gardenia gummifera** L.f. Suppl. Pl. 164. 1781; FBI 3: 116; Gamble 2: 618 (436).

Deciduous tree, up to 7 m tall; bark greyish brown, smooth; wood yellowish-white, close-grained; branches rough and twisted. Leaves obovate-oblong, 3–8.5 × 1.5–6 cm, apex obtuse or bluntly acute, margin entire, base rounded to acute; secondary nerves 13–20 pairs, nerve axils with villous glandular pits, transverse nerves irregular; petiole 0. Flowers white change into yellow, 1–3 in axils. Berry ellipsoid or oblong with numerous longitudinal elevated lines, beak stout, smooth, 4 × 3 cm; seeds many, rugose.

Common, in deciduous forests above 600 m on the degraded slopes. Leaf fall: Dec.–Jan. New foliage: Apr. Fls.: Jan.–Apr. Frs.: May. Vern.: Bikki, Manchi Bikki, Chitta mitta, Chit-mit.

Leaf buds and the young shoots yield a yellow resinous exudation, which is transparant with a sharp pungent taste and is used in medicine. Wood is used locally for combs. Fruits edible.

Kalasamudram (ATP), *SSR*, 13744; Tirumala (CTR), *SSR* 13786; Bogada (PKSM), *SSR* 16108; Penchalakona (NLR), *SSR* 19350; Palakonda hills (CDP), *CS* 1338; Manchippa (NZB), *BR* 9027; Narsapur (MDK), *TP & CP* 11774; Upper Ahobilam (KNL), *RVR* 2763; Gopalapatnam (ADB), *GO* 4403.

INDIA : Peninsula.

**Gardenia jasminoides** Ellis. Phil. Trans. 5: 51. 1761. *G. floria* L. Sp. Pl. ed 2. 305. 1762.

Small tree. Leaves opposite or ternately whorled, obovate or oblanceolate, 7–15 × 3–6.5 cm, glabrous or puberulous beneath, apex very shortly, bluntly acuminate or shortly cuspidate, margin entire, base attenuate or narrowed; secondary nerves 13–20 pairs, strong, glands in the nerve axils; petiole 0 or short. Flowers white, fragrant, 6 cm long, often double, solitary, terminal in the forks of the branches or becoming lateral by displacement on single shoots.

Cultivated in gardens. Fls.: Sept.–Apr.

Nellore town (NLR), *SSR* 19382.

WORLD : Native of China.

**Gardenia latifolia** Aiton Hort. Kew. 1: 294. 1789; FBI 3: 116;Gamble 2: 619 (436); Brandis 379.

Densely foliaceous, deciduous tree, up to 10 m tall; bark glaucous-grey, exfoliating and leaving round depressions; wood yellowish-brown, close grained, smooth. Leaves broadly elliptic, 6–30 × 2.5–18.5 cm, glabrous, apex obtuse to subacute, margin entire, base rounded to narrowed into short broad petiole; secondary nerves 13 pairs, small, hairy glands in their axils, nervules regular, parallel. Berry globose, 3–5.5–5 5 cm; seeds many, rugose, flattened.

Common, in moist deciduous forests at 500–900 m. Leaf fall: Jan. New foliage: May. Fls.: Feb.–Apr. Frs.: May–Nov. Vern.: Pedda bikki, Adavi bikki, Tharpi chettu, Karin-guva. Pedda Manga. Eng.: Indian Box wood.

The growing tip exude a yellow resin. Wood hard,strong, and heavy, used for making combs and turnery articles, light furniture, tobacco jars, engraving, shuttles, mallet heads, toys, mathematical instruments, egg cup etc. Fruits edible.

Bairluti (KNL), *SSR* 15291; Diguvametta (PKSM), *SSR* 16105; Penchalakona (NLR), *MSG* 5648; Mahaboob Ghat (ADB), *GO & PVP* 4264; Narsapur (MDK), *TP & CP* 11978; Vinukonda (GTR), *VRK* 6886; Gundumala (ATP), *NY* 11232; Guvvalacheruvu (CDP), *RVR* 8408.

INDIA : Throughout.
WORLD : Sri Lanka.

**Gardenia resinifera** Roth, Nov. Pl. Spec. 150. 1821. *G. lucida* Roxb. Fl. Ind. 2: 553. 1824; FBI 3: 115; Gamble 2: 618 (436); Brandis 379.

Tree, up to 10 m tall; bark greenish-grey, flaking-off; wood yellowish white, and close-grained. Leaves thin-coriaceous, elliptic-oblong, 5.5–16 × 2.3–7 cm, bullate in the nerve axils, apex obtusely acute, margin entire, base rounded-attenuate; secondary nerves 14–25 pairs, nervules regular, parallel; petiole to 0.7 cm. Flowers yellow, fragrant, solitary, axillary. Berry ellipsoid, 2 × 1.5 cm, with a crown of lanceolate calyx lobes; seeds rugose.

Occasional, in the interior hilly areas to 800 m. Leaf fall: Jan. New foliage: Feb. Fls. Mar.–June. Frs.: July–Nov. Vern.: Karinga, Chinna karinga, Verri bikki.

**Fig. 44.** *Gardenia Latifolia* Ait.
A. Twig, B. Flowering twig, C. Corolla split open, D. Calyx, E & F. Stamens, G. Pistil.

The growing tips exude a yellow resin, used in cutaneous diseases and to keep off flies and worms. Wood useful for small turned articles.

Appapur (MBNR), *SSR* 15542; Diguvametta (PKSM), *SSR* 15999; Penchalakona (NLR), *SSR* 19357; Upper Ahobilam (KNL), *RVR* 2758; Kalasamudram (ATP), *NY* 675; Gupalapatnam (ADB), *GO* 4402.

INDIA : Peninsula.
WORLD : Bangladesh, Myanmar.

## GUETTARDA Linnaeus

**Guettarda speciosa** L. Sp. Pl. 991. 1753; FBI 3: 126. 1886; Beddome Fl. Sylv. S. India, Anal. Gen. t. 17. f. 2. 1869.

Evergreen tree, up to 10 m tall; branchlets stout. Leaves clustered, subcoriaceous, ovate-obovate, 8–14 × 5–8 cm, puberulous along nerves above, pubescent beneath, apex obtuse-shortly acute, margin entire, base obtuse or cordate; secondary nerves 6–9 pairs; petiole 2.5–3.7 cm; stipules foliaceous, ovate, pubescent, caducous. Flowers white, pubescent, polygamous, 3.7 cm long in axillary, long peduncled, spreading, dichotomously branched cymes. Drupe globose, woody, obscurely lobed.

Often cultivated along East Coast.

Venkatagiri hills (NLR), *BSN* 4112 (VV).

INDIA : Western and Eastern Peninsula, Andaman Islands.
WORLD : Sri Lanka, Malay Peninsula; Tropical shores of the Old and New World.

## HALDINIA Ridsdale

**Haldinia cordifolia** (Roxb.) Ridsd. in Blumea 24: 361. 1978.*Nauclea cordifolia* Roxb. Pl. Cor. t. 53. 1796, Fl. Ind. 1: 514. 1832; Beddome Fl. Sylv. S. India t. 33. 1869. *Adina cordifolia* (Roxb.) Hook. f. ex Brandis Forest Fl. N.W. India t. 33. 1874; FBI 3: 24; Gamble 2:584 (412); Brandis 368.

Large and beautiful deciduous tree up to 18 m tall; crown large, spreading; trunk erect with horizontal branches; base often buttressed and fluted; bark brownish-grey, thick, furrowed; wood yellow, even-grained, fine and even textured; young branches pubescent. Leaves thick-chartaceous, orbicular-ovate, 7–22.5 × 7.5–25.6 cm, dark green, puberulous along nerves above, densely pubescent and pale beneath, apex shortly acuminate, margin entire, base cordate; secondary nerves 5–7 pairs; petiole to 5.7 cm. Flowers yellowish in axillary heads. Capsule cuneate, densely downy, 5 mm, breaking into 2 follicular cocci; seeds 6, elongate, tailed above or winged, 0.30 × 0.15 mm.

Common, in dry deciduous forests to 1200 m. It is a light demander and grows well on well drained ground along the lower slopes of hills. Leaf fall: Feb.–May. New foliage: May–June. Fls.: June–Sept. Frs.: Feb.–May of the following year. Vern.: Pasupu ganapa, Pasupu Kadamba, Bandari, Rudra ganapa, Bandaru, Pedda Kamba.

The roots are used as an astringent in dysentry. The bark is acrid and bitter used in biliousness; it is also regarded as febrifgue and antiseptic. Leaves are used as fodder. Wood moderately hard, durable under water for panelling, railway carriages and for bobbins, also variety of other purposes such as frame work, door and window leaves, ceiling, railway sleepers, furniture, tea-chests, canoes, dugouts, packing cases, matches, cigar boxes, combs, yokes etc.

Eturnagaram (WGL), *SSR & KSM* 18544; Rampur (ADB), *SSR* 19313; Vinukonda (GTR), *VRK* 6847; Anantagiri (MDK), *MSM & NV* 10413; Balugram (KNL), *TP & RVR* 2742; Penchalakona (NLR), *MSG & KH* 5661; Jalapur RF (NZB), *BR & KH* 9655.

INDIA : Throughout.
WORLD : Sri Lanka, Myanmar, Indonesia, Sikkim, Bhutan, China, Vietnam, Thailand.

### HYMENODICTYON Wallich *nom. cons.*

1. Leaves ovate-elliptic. Flowers pale green, pedicellate. Capsule reflexed. ......... ***H. orixense***
1. Leaves obovate. Flowers pale yellow, sessile. Capsule erect. .......................... ***H. obovatum***

**Hymenodictyon obovatum** Wall. in Roxb. Fl. Ind. 2: 153. 1824; FBI 3: 36; Gamble 2: 589 (416); Beddome Fl. Sylv. S. India t. 219. 1872.

Deciduous tree, 6–15 m tall; bark exfoliating in irregular scales; wood brownish-grey. Leaves obovate, elliptic or orbicular, 7–15 × 5–9 cm, glabrous or slightly pubescent along the nerves beneath, apex acuminate, margin entire, base acute to cuneate; secondary nerves 6–10 pairs, 1 cm apart; petiole to 7 cm. Flowers pale yellow, sessile in axillary and terminal spikes. Capsule oblong, erect, 1 × 0.5 cm, loculicidal; seeds narrowly winged, papery, 6 mm long.

Rare, in deciduous forests of Nallamalais. Fls. & Frs.: Aug.–Dec. Vern.: Dudippa.

Wood soft and light to moderately heavy. It is useful as a good timber yielding tree.

Modugamamisala (KNL), *JLE* 42198 (MH).

INDIA : Maharashtra to Tamil Nadu, Orissa, Assam, Meghalaya in slopes near streams in hilly forest at 150–400 m (Peninsula).

**Hymenodictyon orixense** (Roxb.) Mabb. in Taxon 31: 66. 1982. *Cinchona orixense* Roxb. Bot. Descr. Swietenia 21. 1793 and Med. Facts. Obs. 6: 152. 1795. *Hymenodictyon excelsum* (Roxb.) Wall in Roxb. Fl. Ind. 2: 149. 1824; FBI 3: 35; Gamble 2: 589 (416); Brandis 371.

Deciduous tree, up to 25 m tall; crown round; trunk cylindrical; bark dark grey, thick, soft, corky and furrowed in old; wood white when fresh cut turning darker; young branches brown puberulous. Leaves membranous, ovate-elliptic, 18 × 12 cm, puberulous along nerves above, pubescent beneath, apex abruptly acuminate, margin entire, base acute; secondary nerves 8–10 pairs; petiole to 7 cm; stipules broadly oblong, recurved. Bracts foliaceous; flowers pale green, fragrant in clusters along the rachis of dense, cylindrical, pubescent, spiciform racemes in 22 cm long, terminal panicles. Capsule ellipsoid, or oblong brown, 1.6–2 × 0.3–0.8 cm, loculicidal, puberulous when young, warty, lenticelled, wrinkled when maturity on stout, decurved pedicel; seeds ca. 20, 11–12 × 5–6 mm, broadly ovate with flattened wings and irregularly denticulate margin; testa reticulate, brown.

Common, in dry, mixed forest, particularly in dry rocky hills at 600–900 m. It is a good coppicer and strong light demander. Leaf fall: Nov.–Feb. New foliage: June. Fls.: June–Aug. Frs.: Jan.–Mar. Vern.: Dudippa, Burja, Burija, Chedippa, Udippa.

Leaves are used as cattle fodder and for dye. Bark and root are used as febrifgue; bark also used for tanning. Wood is not durable, used for box planking, scabbards, grain measures, rollers, bobbins, jute milling, picture and slate frames, mathematical instruments, pencils, drums and toys; suitable for cheap furniture.

Basapuram (KNL), *TP & SSR* 15210; Vanipenta (KNL), *SSR* 15501; Balapalli hills (CDP), *JSG* 10801 (CAL); Kullumidi extension RF (EG), *GVS* 68562 (MH); Krishnapuram (VSKP), *JSG* 1986 (MH).

INDIA : Throughout, excluding Jammu & Kashmir.
WORLD : Sri Lanka, Myanmar, Malaysia, Nepal, Bhutan, Bangladesh.

## **IXORA** Linnaeus

1. Calyx lobes longer than tube; flowers in terminal branched cymes. Leaf base rounded or sub-cordate ........ ***I. pavetta***
1. Calyx lobes reduced to teeth and shorter than tube; flowers in terminal corymbs. Leaf base acute or cuneate ........ ***I. finlaysoniana***

**Ixora finlaysoniana** Wall. ex G. Don Gen. Syst. 3: 572. 1834. Gamble 2. 629 (444).

Small tree, up to 8 m tall; branchlets glabrous, brownish-white. Leaves coriaceous, oblanceolate, elliptic or obovate, 5.5–15.5 × 2.8–4.5 cm, glabrous, apex obtuse to subacute with mucro, margin entire, base acute cuneate; secondary nerves 8–12 pairs, not distinct, irregular; petiole to 6 mm. Flowers white, fragrant in terminal, peduncled, 4 cm long, corymbose cymes. Berry globose.

Occasional, in forests, and also cultivated for its flowers. Fls.: Jan.–Mar. Frs.: May–June.

K.V.R. College (KNL), *MSG* 5668; V.P. South (GTR), *VRK* 6726; Yanam (EG), *GVS* 68708 (MH); Vijayawada town (KSN), *PVK* 5292 (MH).

Distribution : East India.

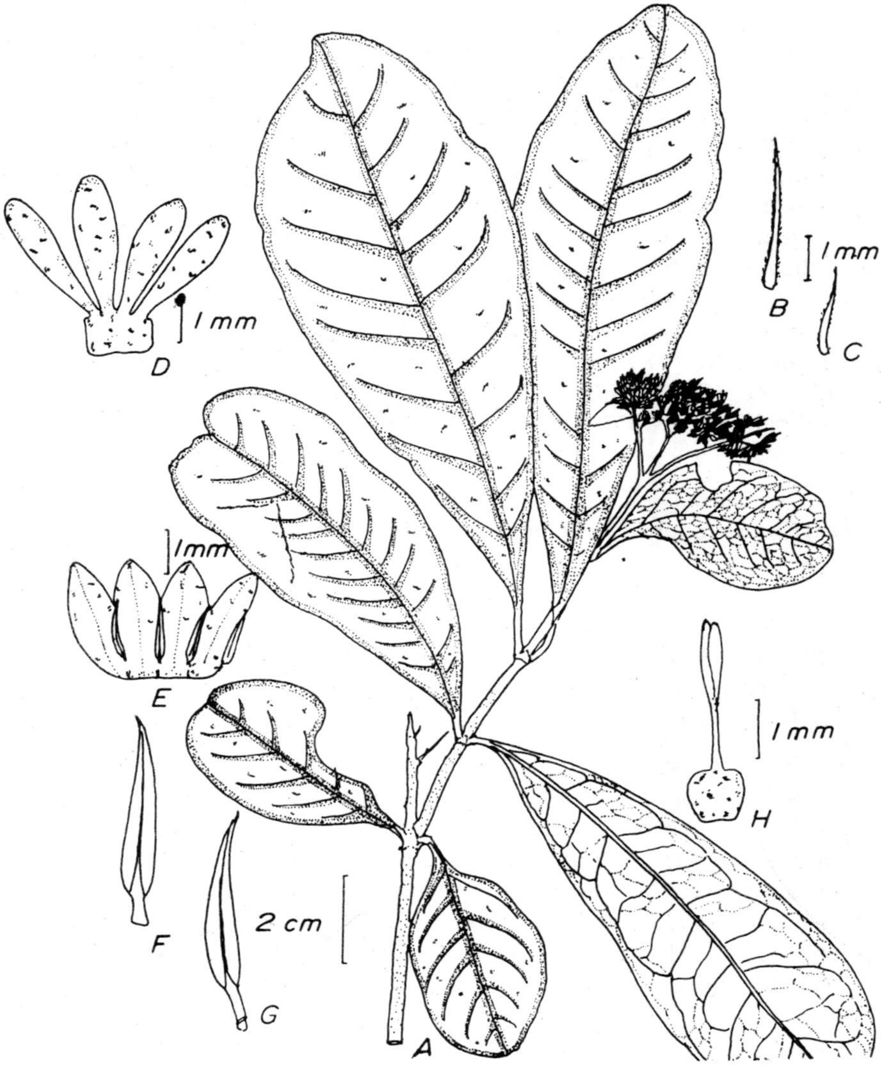


**Fig. 45.** *Ixora finalysoniana* Wall. ex G. Don.
A. Twig, B. Bract, C. Bracteole, D. Calyx, E. Corolla spread out, F & G. Stamens, H. Pistil.

**Ixora pavetta** Andrews Bot. Rep. t. 78. 1799. *I. parviflora* Vahl Symb. Bot.3: 11. t. 52. 1754; FBI 3: 142; Gamble 2:631 (445); Brandis 388; Beddome Fl. Sylv. S. India t. 222. 1872. *Ixora arborea* Roxb. ex Smith. in Rees Cyclop. 19: 5. 1811.

Small, evergreen tree, up to 7 m tall; crown columnar; bark thick, pale-grey, nearly smooth; wood light reddish brown, close-grained; branchlets glabrous. Leaves coriaceous, hard and shining, oblong, elliptic or obovate, 5–15.6 × 1.7–6 cm; glabrous, apex subacute, margin entire, base rounded or sub-cordate or sometimes cuneate; secondary nerves 8–12 pairs, faint: petiole to 5 mm, stout. Flowers white, odorous, small and very numerous in terminal and branched cymes. Drupe depressed-globose, 0.8–1 cm dia., shining, black.

Common, in deciduous forests and dry ever green forests particularly in shady habitats at 300–1000 m. Fls.: Jan.–Apr. Frs.: May–June. Vern.: Kollangi, Korivi chettu, Putta pala, Gorivi.

The branches are used as torches.

Donubai RF (SKLM), *SSR* 15585; UpperAhobilam (KNL), *SSR* 15287; Chitvel RF (CDP), *SSR* 16157; Devapur (ADB), *GO & PVP* 4242; Mamidipally RF (NZB), *BR & KH* 9020; Narasihmulu Konda (NLR), *MSG* 5667.

INDIA : Throughout.
WORLD : Sri Lanka, Myanmar, Bangladesh and dry localities.

## MEYNA Roxburgh ex Link

**Meyna spinosa** Roxb. ex Link Jahrb. Gewachrk 3: 32. 1820. var. *pubescens* Robyns in Bull. Jard. Brux. 11: 230. 1928. *Vangueria spinosa* Roxb. Fl. Ind. 2: 172. 1924; FBI 3: 136; Gamble 2: 626 (442); Brandis 386.

Armed, small tree; bark dark-coloured, nearly smooth; spines straight, sharp, 1.2–3.5 cm long. Leaves membranous, ovate or elliptic-oblong, 4–10 × 2–6.5 cm, slightly pubescent above, densely pubescent beneath, apex bluntly or obtusely acuminate, margin entire, base rounded or acute; secondary nerves 7–9 pairs, tertiary nerves prominent below; petiole to 1 cm, densely hairy; stipules short, broad, very obtuse, caducous. Flowers pale green, very faintly fragrant on short pedunculate axillary cymes. Drupe yellowish, subglobose, smooth with persistent calyx lobes; pyrenes 4–5, woody.

Rare, in dry deciduous forests at high elevations. Fls.: Apr.–May. Frs.: Aug.–Dec. Vern.: Cegagadda, Veliki, Viskilamu.

Cherukonda (VSKP), *GVS* 28156 (MH); Elvinpeta block (SKLM), *GVS* 62469 (MH).

INDIA : Northern Bengal to Canara, Tamil Nadu, Karnataka, Kerala.
WORLD : Myanmar, Java.

## MITRAGYNA Korthals *nom. cons.*

**Mitragyna parvifolia** (Roxb.) Korth. Observ. Naud. India. 19. 1839; Gamble 2: 585 (413). *Nauclea parvifolia* Roxb. Pl. Cor. t. 52. 1796 & Fl. Ind. 1: 513. 1832; Beddome Fl. Sylv. S. India t. 34.1869. *Stephegyne parvifolia* Korth. Verh. Nat. Gesch. Ned. Bezitt. Bot. 161. 1840. FBI 3: 25; Brandis 369.

Deciduous tree, up to 20 m tall; crown round, trunk short, buttressed; bark light grey, smooth, dark patches of exfoliating scales when old; wood light pinkish-brown, even-grained. Leaves sub-coriaceous, very variable in shape and size, elliptic, oblong, sub-orbicular or ovate, 3.5–17.5 × 3–11.5 cm, glabrous, apex obtuse-acute, margin entire, base truncate or acute rarely cordate; secondary nerves 6–8 pairs; petiole to 5.4 cm. Flowers pale yellow or cream, fragrant in globose heads. Capsules aggregated in globular head, obovoid, faintly angled, 5 × 2 mm, dark blue when ripe; seeds flattened.

Occasional, in hilly tracts, of deciduous forests, more in the plains to 800 m. It reaches its best development on well drained ground with deep soil. It is fairly drought hardy and coppices well. Leaf fall: Feb.–Mar. New foliage: May. Vern.: Rudraksh-kamba, Battaganapa, Botruga, Pacha pararari, Nir-ka.

The wood is strong, hard and moderately heavy, but not as good as that of *Haldina cordifolia*, and used for furniture, agricultural implements, utensils, combs, cups, spoons, carved articles. The bark and roots are used in fever and colic. The bark yields a cordage fibre. The branches are often lopped for fodder.

Pothuraju Penta (KNL), *SSR* 15984; Talakona (CTR), *SSR* 13729; Narsapur (MDK), *SSR & CP* 16197; Rampur (ADB), *SSR* 19317; Mamidipally RF (NZB), *BR & KH* 9019; Kalasamudramu RF (ATP), *NY* 1195; Loyapalle (GTR), *VRK* 3815; Palakonda hills (CDP), *CS* 7740.

INDIA : Throughout.
WORLD : Sri Lanka, Myanmar.

## MORINDA Linnaeus

1. Leaves glabrous above. Flowers solitary, leaf-opposed. Drupes free .............. ***M. angustifolia***
1. Leaves pubescent. Flowers in axillary globose heads. Drupes cohering ........ ***M. pubescens***

**Morinda angustifolia** Roxb. Pl. Cor. t. 287. 815; FBI 3: 156; Gamble 2: 651 (459).

Small tree, up to 4 m tall. Leaves coriaceous, oblong-lanceolate or obovate, 15–23 × 5.5–9.3 cm, glabrous above, slightly puberulous beneath, apex acute to shortly acuminate, margin entire, base narrowed or tappered;

secondary nerves 10–12 pairs, prominent below; petiole to 3.5 cm; stipules acute. Flowers white, solitary, leaf-opposed, rarely 2–3-nate. Fruit black, turbinate, 2 cm.

Rare, in North Circars and Nallamalais. Fls.: & Frs.: May–Dec.

The roots yield a yellow dye, which is used for dyeing cotton, yarn and cloth.

Ragupaliam (VSKP), *CAB* 1587 (MH); Diguvametta (KNL), *CAB*, 8024 (MH).

INDIA : Assam, Bihar, Orissa, Eastern Himalays.

**Morinda pubescense** J.E. Smith in Rees. Cyclop. 24: n. 3. 1813 var. ***pubescens*** Verdc. in Kew Bull. 37: 543. 1983. *M. tinctoria* Roxb. var. *tomentosa* (Heyne ex Roth.) Hook. f. Fl. Brit. India 3: 156. 1880; Gamble 2:652 (459); Brandis 892, *M. tinctoria* Heyne ex Roth, Nov. Pl. Sp. 147. 1821.

Deciduous tree, up to 12 m tall; bark corky, deeply cracked, reddish-brown; wood red, streaked with yellow, straight-grained, even and medium textured; branches 4-angled. Stipules bifid. Leaves chartaceous, elliptic or lanceolate-ovate, 6.5–19 × 2.4–11 cm, pubescent on both surfaces, apex acuminate, margin slightly undulate, base suddenly narrowed into a petiole; secondary nerves 8–9 pairs; petiole to 2 cm. Flowers white in globose heads, solitary or 2-nate, leaf-opposed or terminal. Syncarpium of false berries due to coalescence of succulent calyces, whitish green with the boundaries of the separate fruits marked by lines; pyrenes stony; seeds oblong or ovoid.

Common, in dry forests and waste lands to 500 m. Leaf fall: Nov.–Dec. New foliage: May. Fls.: May–Sept. Frs.: Oct. onwards. Vern.: Mogi, Togar-Mogili, Chogarmaddi, Tzogur, Togaru, Maddi, Banda maddi.

Wood moderately hard, used for wooden plates and dishes, medium class furniture, toys, cotton roals, slate frames etc. The root gives a red or yellow dye.

Avagondapalem (ATP), *SSR* 18576; Muchukota RF (ATP), *TP & NY* 1141; Jaipur RF (ADB), *GO & PVP* 4824; Narsapur (MDK), *CP* 11578; Guvvalacheruvu (CDP), *R.V. Reddy* 8401; Manchippa RF (NZB), *BR & KH* 9032; Mohammedaba (RR), *MSM & KH* 10240; Simhachalam hills (VSKP), *MSG & KH* 5610.

INDIA : Peninsula, Bengal, Assam, Central Province, Rajputana, Gujarat, Deccan, Karnataka, Bihar.

WORLD : Sri Lanka, Myanmar, Malaysia.

## NEONAUCLEA Merril

**Neonauclea purpurea** (Roxb.) Merr. Inter. Rumph. 483. 1917; Gamble 2: 584 (412), *Nauclea purpurea* Roxb. Pl. Cor. t. 41. 1795; FBI 3: 26; Brandis 368.

Tree, up to 15 m tall; wood light red, even-grained, soft. Leaves coriaceous, elliptic-oblong or lanceolate, 8–15 × 4–8 cm, glabrous, apex acute to shortly acuminate, margin entire, base tapering; secondary nerves 10–12 pairs, 1 cm apart, conspicuous; petiole to 3 cm, drying black; stipules flat, membranous. Flowers purple; heads solitary, globose, small, on 1–2 cm long peduncle. Fruit of 2-dehiscent many-seeded cocci, crowned with persistent base of calyx-lobes.

In moist valleys of evergreen forests at 600–700 m. Fls.: Feb. Frs.: July–Sept. Vern.: Bagada.

Rampa hills (EG), *VNS* 1547 (CAL); Near Ammavaridhara, Dharalammagudi (VSKP), *GVS* 47404 (CAL), Chintapalli (VSKP), *GVS* 28232 (CAL).

INDIA : Tamil Nadu, Maharashtra.

## PAVETTA Linnaeus

1. Calyx teeth short. Leaves black when dry ........ ***P. hispidula***
1. Calyx teeth minute:
   2. Inflorescence and under surface of leaves tomentose ........ ***P. tomentosa***
   2. Inflorescence and under surface of leaves glabrous ........ ***P. indica***

**Pavetta hispidula** Wt. & Arn. Prodr. Fl. Ind. Orient. 431. 1834; FBI 3: 151;Gamble 2: 633 (446).

Small tree, up to 3 m tall. Leaves chartaceous, obovate-oblong or oblanceolate, 7–15 × 2.5–4.5 cm, pubescent above, hispid below, apex acuminate-emarginate, margin entire, base rounded; petiole to 2.5 cm, stipules connate. Flowers white, in terrminal or axillary pedunculate cymes. Berry green, globose, 6 mm dia.

Rare, in forests of East Godavari district. Fls. & Frs.: Oct.–May.

Bison hills (EG), *CAB* 51669 (MH).

**Pavetta indica** L. Sp. Pl. 110. 1753; FBI 3: 150; Gamble 2: 633 (446); Brandis 387.

Small tree, up to 6 m tall. Leaves chartaceous, oblong-obovate to lanceolate, 7–15 × 2.5–4.5 cm, glabrous, apex subacute to shortly acuminate, margin entire, base acute-attenuate; petiole 1.5 cm; stipules connate. Flowers white in terminal or axillary, trichotomously branched, corymbose cymes. Berry black, globose, fleshy, 6 × 4 mm, smooth.

Common, in deciduous forests. Fls.: May–Aug. Frs.: July onwards. Vern.: Papidi, Papatokammi. Fruits edible.

Penchalakona (NLR), *MSG* 5652; Balugram RF (KNL), *TP & RVR* 2737; Macherla RF (GTR), *VRK* 3836; Rampachodavaram (EG), *GVS* 24456 (MH), Guvvalacheruvu (CDP), *JSG* 15021 (MH); Thandipandal (CTR), *MCB* 45089 (MH); Kodimial (KRN), *GVS* 20122 (MH).

INDIA : Throughout.
WORLD : Malay Archipelago, South China, North Australia, Sri Lanka.

**Pavetta tomentosa** Roxb. ex Smith in Rees Cyclop. 26: n.2. 1819; Gamble 3: 1879 (1301); *P. indica* L. var. *tomentosa* (Smith) Hook. f. Fl. Brit. India 3: 150. 1880; Gamble 2: 683 (446); Brandis 387.

Shrub to small tree, up to 7 m tall; young branches softly pubescent. Leaves chartaceous, elliptic or oblanceolate, 15–28 × 5.7–10 cm, apex acute or shortly, bluntly acuminate, margin entire, base acute; secondary nerves 10–13 pairs, slightly curved upwards and ending near to the margin; petiole to 3 cm. Flowers white, 1.5 cm long, in much branched, long pedunculate, densely tomentose, corymbose cymes. Fruit black, orbicular-globose, 7 mm dia., not distinctly didymous, hairy; seeds 2.

Common, in deciduous and semi-evergreen forests. Fls.: May–July. Frs.: Oct. Vern.: Tella papidi.

Tirumala (CTR), *SSR* 13720; Pangidi (KNL), *SSR* 15221.

INDIA : Throughout.
WORLD : Indo-China, Malay Archipelago.

## **SCYPHIPHORA** Gaertner

**Scyphiphora hydrophyllacea** Gaertn. Fruct. 3: 91. t. 196. 1805; FBI 3: 125; Gamble 2: 620 (438); Brandis 384.

Small, evergreen tree; trunk short; crown rounded. Leaves coriaceous, ovate-oblong, obovate, 2–5.5 × 2.4–3.7 cm, glabrous, shining, brown when dry, apex obtuse, margin entire, base acute; secondary nerves 4–5 pairs, not obscure; petiole to 1.3 cm; stipules short, connate into a sheath around the stem. Flowers white in umbellate, short pedunculate clusters. Drupe oblong, 2 × 6 mm, grooved.

Rare, East coast in mangrove forest. Fls. & Frs.: Mar.–Sept.

Sacremento light house area, Kanaikuppa (EG), *J.S.S.N. Raju* s.n. (MH).

INDIA : Mangrove swamps on the Carnatic coast, Malayan Peninsula, the Andaman Islands.
WORLD : Sri Lanka.

## **TAMILNADIA** Tirvengadum

**Tamilnadia uliginosa** (Retz.) Tirveng. & Sastre in Mauritius Inst. Bull. 8: 85. 1979. *Gardenia uliginosa* Retz. Obs. Bot. 2: 14. 1781. *Randia uliginosa* (Retz.) DC. Prodr. 4: 386. 1830; FBI 3: 110; Gamble 2:615 (434); Brandis 381.

Deciduous, armed, rigid tree, up to 7 m tall; dark reddish-brown, scaly; wood greyish-white, close grained; branches numerous, thick, quadrangular, horizontal, many of them short and terminating in 1–2 pairs of strong, sharp thorns or 1–2 pairs at the ends of arrested branchlets. Leaves coriaceous, fascicled on suppressed branchlets, thin, obovate to oblanceolate, 4–17 × 1.5–5.6 cm, glabrous and shining above, often pubescent beneath, apex obtuse, margin entire, base narrowed; secondary nerves 6–8 pairs; petiole to 1.5 cm; stipules interpetiolar, acute from a broad base. Flowers solitary, white, fragrant at the ends of arrested branchlets. Fruit ovoid, ellipsoid to 6 × 4 cm, yellowish brown, smooth, calyx-limb persistent; seeds numerous, compressed, closely packed in pulp, dark-brown, hard, shiny, obscurely angular.

In open dry forests in wet places and on black-cotton soils. It produces root-suckers freely and hardy against frost and drought. Leaf fall: Dec.–Jan. New foliage: May–June. Fls.: May–July. Frs.: Dec.–Feb. Vern.: Nalaika, Boliki, Kukka-elka, Goanke, Adavi-manga.

Fruits are edible and makes a good vegetable when cooked; unripe fruit is astringent. The leaves are boiled and eaten; they are used as fodder for deer and cattle. The roots are considered to possess cooling, diuretic and tonic properties; they are employed in biliousness and boils in children and in diarrhoea and dysentry. Wood hard and fairly heavy, not much used and it is suitable for turning and small articles.

Devipatnam (EG), *SSR* 15251; Talakona (CTR), *SSR* 13785; Eturnagaram (WGL), *SSR* 18535; way to Kunthala water falls (ADB), *GO & DAM* 5113.

INDIA : Throughout.
WORLD : Sri Lanka, Myanmar in dry localities.

## **WENDLANDIA** Bartling ex A. de Candolle *nom. cons.*

1. Calyx tube glabrous:
   2. Leaves whorled ........ ***W. angustifolia***
   2. Leaves opposite:
      3. Ultimate twigs and rachis of inflorescence puberulous to pubescent; flowers 7 mm long:
         4. Flowers 5–6.2 mm long, glabrous within. Stipules truncate or abruptly cuspidate ........ ***W. glabrata***
         4. Flowers 2–3 m long, villous within. Stipules bifid ........ ***W. bicuspidata***
      3. Ultimate twigs and rachis of inflorescence glabrous; flowers 7–11 mm long ........ ***W. gamblei***

1. Calyx tube pubescent:
   5. Leaves hoary or silvery pubescent; stipules recurved ........ ***W. heynei***
   5. Leaves not hoary or silvery pubescent; stipules erect:
      6. Flowers in dense thyrsoid cymes ........ ***W. thyrsoidea***
      6. Flowers in panicled cymes ........ ***W. tinctoria***

**Wendlandia angustifolia** Wt. ex Hook. f. Fl. Brit. India 3: 40. 1880; Gamble 2: 588 (415).

Small tree, up to 4 m tall. Leaves coriaceous, whorled, linear-lanceolate or elliptic-lanceolate, glabrous, shining, apex acute-shortly acuminate, margin entire, base narrowed; secondary nerves 8-pairs, slender, midrib strong; stipules triangular-ovate, persistent. Flowers white in slender pyramidal panicles. Capsule globose.

Rare, forests of Cuddapah district. Fls. & Frs.: Mar.–Nov.

Cuddapah forests, *RHB* 1880 (MH).

INDIA : Tamil Nadu.

**Wendlandia bicuspidata** Wt. & Arn. Prodr. Fl. Ind. Orient. 1:403. 1834; Gamble 2:588 (415). *W. notoniana* var.. *bicuspidata* (Wt. & Arn.) Hook. f. FBI 3: 38. 1880. *W. notoniana* var. *zeylanica* Hook. f. Flora of British India 3: 40. 1880.

Small trees; branchlets pubescent. Leaves coriaceous, elliptic-obovate, 6–13 × 2.5–7 cm, glabrous, brown when dry, apex obtusely, shortly acuminate, margin entire, base acute; secondary nerves 6–8 pairs; petiole to 1 cm; stipules interpetiolar, oblong, bifid at apex. Flowers whitish or yellowish cream, fragrant, villous within, in 6–10 cm long, terminal panicled cymes. Capsule brown, globose, 2-valved.

Rare, a few plants found on the rock-crevices near a waterfall on Venkatagiri durgam. Fls. & Frs.: May.

Venkatagiri durgam (NLR), *ASR* 8674 (VV).

INDIA : Peninsula.
WORLD : Sri Lanka.

**Wendlandia gamblei** Cowan in Notes Royal Bot. Gard. Edinb. 16:271. 1932.

Tree, up to 8 m tall; bark greyish, rough; branchlets glabrous. Leaves coriaceous, elliptic-oblong or elliptic-lanceolate, 7–11 × 2.8–3.5 cm, glabrous, apex acute or shortly acuminate, margin entire, base cuneate or attenuate; secondary nerves 8–9 pairs; petiole to 2 cm. Flowers yellowish, subsessile in 8 cm long, axillary or terminal panicles. Capsule globose, 2 mm in dia., brown, 2-valved, crowned by dentate lobes; seeds 2.

In forests of Visakhapatnam and East Godavari districts. Fls.: Feb.–Mar. Frs.: Apr.

Sunkarimetta (VSKP), *SSR* 15912; Anantagiri (VSKP), *SSR* 79618; Minumuluru (VSKP), *GVS* 29606 (MH).

INDIA : Orissa.

**Wendlandia glabrata** DC. Prodr. 4: 411. 1830; FBI 3: 39; Gamble 2: 587 (415); Brandis 374.

Small evergreen tree; branchlets glabrous. Leaves elliptic-lanceolate, glabrous, apex acuminate, margin entire; petiole to 2.5 cm; stipules short, truncate or abruptly cuspidate. Flowers white, rather large (5–6.2 mm long). Capsule globose, 1.5 mm across.

North Circars, hills of Visakhapatnam and Godavari up to 1,500 m. Fls. & Frs.: Feb.–Sept.

Way to Tirumala (CTR), *KS* 7851 (MH); Rampa hills (EG), *JSG* 18727 (DD); Mantralakanama (KNL), *JSG* 16014 (DD).

INDIA : Orissa, Tamil Nadu, Manipur, Karnataka.
WORLD : Myanmar.

**Wendlandia heynei** (Roem. & Schult.) Santapau & Merchant in Bull. Bot. Surv. India 3: 110. 1961. *Rondeletia heynei* Roem. & Schult. Syst. Veg. 5: 234. 1819. *Wendlandia exserta* (Roxb.) DC. Prodr. 4: 411. 1830; FBI 3: 37. Gamble 2: 587 (415); Brandis 373.

Deciduous tree, up to 8 m tall; greyish or silky tomentose; bark dark brown, rough; wood close grained, reddish-brown; branchlets tomentose. Leaves coriaceous, lanceolate, ovate-lanceolate, 4–11 × 0.6–4 cm, slightly pubescent and dull above, densely greyish pubescent beneath, apex acuminate, margin entire, base acute, often inequilateral; secondary nerves 10–17 pairs; petiole to 2.5 cm; stipules interpetiolar, ovate, acute, the upper part recurved. Flowers white, sweet scented, small, subsessile in 10–18 cm long tomentose panicles. Capsule globose, 2 mm in dia., hoary pubescent, crowned by the erect lanceolate sepals; seeds small or minute, black.

Rare, in open forests. It comes up gregariously on newly exposed ground, particularly on land slips and abandoned cultivation. It is light demander and useful for reclothing bare hill slopes and clearings. Fls.: Mar.–Apr. Frs.: Apr.–June.

Leaves are used as fodder. Wood hard, heavy and durable, useful for building, house posts and agricultural implements and yields good fuel.

Kokkarampalli (VSKP), *SSR & KSM* 15959.

INDIA : Orissa, Central India, Western Peninsula, Deccan Konkan.

It is new record for Andhra Pradesh from Visakhapatnam.

**Wendlandia thyrsoidea** (Roem. & Schult.) Steud. Nom. Bot. (ed. 2) 2: 786. 1841. *Canthium thyrsoideum* Roem. & Schult. in L. Syst. Veg. 6: 207. 1820. *Wendlandia notoniana* Wall. ex Wt. & Arn. Prodr. Fl. Ind. Orient. 403. 1834; FBI 3: 40; Gamble 2: 588 (415); Brandis 374; Beddome Fl. Sylv. S. India t. 224. 1872.

**Fig. 46.** *Wendlandia heynei* (Roem. & Schult.) Santapau & Merchent.
A. Twig, B. Inflorescence part, C. & D. Bracteole dorsal and ventral view, E. Calyx, F. Calyx split open, G. Corolla spread out with stamens, H. Stamens, I. Calyx with pistil, J. Pistil, K. Fruit.

Shrub to tree; bark orange-red, fibrous; wood reddish-brown; young branches pubescent, often reddish. Leaves subcoriaceous, usually ternately whorled, rarely opposite, elliptic-lanceolate or elliptic-obovate, 6–12 × 3–6 cm, tomentose beneath, brown when dry, apex acute-shortly acuminate; margin entire, base much tapered; secondary nerves 8–10 pairs, prominent, pink in young leaves; petiole to 1 cm; stipules variable, up to 0.5 cm long. Flowers pale yellow, fragrant in dense pyramidal, hirsute thyrsoid cymes. Capsule globose, to 2 mm in dia., pubescent, crowned with the calyx tube.

Rare, hill of Cuddapah forests (Gamble) above 1000 m.

INDIA : Maharashtra, Tamil Nadu.

**Wendlandia tinctoria** (Roxb.) DC. Prodr. 4: 411. 1830; FBI 3: 38; Gamble 2: 588 (415); Brandis 374. *Rondeletia tinctoria* Roxb. Fl. Ind 1: 522. 1824.

Small, evergreen tree; bark brown, stringy; wood rose-white; branchlets puberulous. Leaves coriaceous, elliptic or elliptic-oblong, 7–17.5 × 3–8.7 cm, apex acute to shortly acuminate, margin entire, base acute or narrowed; secondary nerves 6–8 pairs, slightly puberulous on the nerves beneath; petiole 2 cm; stipules ovate-truncate, puberulous. Flowers white, in terminal, dense, 15 cm long panicles. Capsule globose, 3 × 2 mm, loculicidal, 2-valved.

Occasional, in deciduous and semi-evergreen forests. Fls.: Jan.–Mar. Frs. Apr.

The bark is employed in Assam and West Bengal as a mordant in dyeing. An extract prepared from the bark is used to relieve cramps in patients suffering from cholera.

Way to Pahrabad (MBNR), *SSR* 15534; Talakona (CTR), *SSR* 1553; Nallamalais (KNL), *JLE* 23848 (MH).

INDIA : Orissa, Assam.
WORLD : Myanmar.

## MYRSINACEAE

1. Fruit follicle, cylindric; seeds exalbuminous ........ ***Aegiceras***
1. Fruit drupe or berry, globose; seeds albuminous:
   2. Corolla lobes nearly or quite polypetalous, imbricate, flowers greenish ........ ***Embelia***
   2. Corolla gamopetalous, lobes twisted; flowers pinkish ........ ***Ardisia***

### **AEGICERAS** Gaertner

**Aegiceras corniculatus** (L.) Blanco. Fl. Filip. 1: 79. 1837; Gamble 2: 757 (532). *Rhizophora corniculata* L. Amoen. Acad. 4: 123. 1729. *Aegiceras majus* Fruct. Semi Pl. 1: 216 t. 46. f. 1. 1788; FBI 3: 533; Brandis 421; Beddome Fl. Sylv. S. India, Anal. Gen. t. 19. f. 3. 1869.

Small, evergreen tree, up to 5 m tall; bark brownish-grey, smooth; wood reddish-brown, often streaked with yellow, even-grained; branchlets glabrous. Leaves coriaceous, obovate, 3.5–5.5 × 2–3 cm, glabrous, pale green above, slightly glaucous beneath, apex emarginate, margin entire, base narrowed; secondary nerves numerous, faint, midrib strong; petiole to 1.5 cm. Flowers white, fragrant in axillary sessile or subsessile umbels. Fruit brown, cylindric, curved, 2.5–5 cm long, tip acute, dehiscing longitudinally; seed 1, exalbuminous.

Frequent in tidal forests. It coppices well. Fls.: Feb.–Apr. Frs.: June–Sept. Vern.: Dudumara, Guggilam.

Nizampatnam port (GTR), *VRK* 6812; Kothapalam (GTR), *VRK* 5766; Coringa Canal bank (EG), *GVS* 68697; Thummalapenta (NLR), *JSG* 12596 (MH).

INDIA : Sea coast of all the states.
WORLD : Tropical shores from Pakistan to Australia.

### ARDISIA Swartz *nom. cons.*

**Ardisia solanacea** Roxb. Pl. Cor. t. 27. 1795 & Fl. Ind 1: 580. 1832; Gamble, 2: 756 (531). *A. humilis auct. non* Vahl: Wt.. Icon. Pl. Ind. Orient. t. 1212. 1849 & Spicil. Neilgh. t. 138. 1851; FBI 3: 529 in part. Brandis 418.

Evergreen, small tree, up to 8 m tall; bark brown, smooth; wood grey with broad medullary rays; branchlets stout, glabrous. Leaves coriaceous, somewhat succulent when fresh, lanceolate, 9–23 × 3.2–7.2 cm, gland dotted, glabrous, apex sub-acuminate, margin entire, base cuneate; secondary nerves numerous: petiole to 1.5 cm. Flowers pinkish, in axillary corymbs. Drupe globose, 0.8 cm, purple-black; seeds globose.

Almost all the districts, usually in moist places and along streams, up to 1500 m. Fls.: Mar.–May. Frs.: Aug.-Sept. Vern.: Konda Mayur, Chilu Karekka, Kunti, Konda Bhogada. Adavi Mayuri.

The flowers are eaten as pot herb. The juice of the berries is said to yield a yellow dye.

Donubai RF (SKLM), *SSR* 15594; Chintapalli (VSKP), *SSR & KSM* 16140; Mulyaguda-Borra caves (VSKP), *TP & EC* 7410; Pamuleru (EG), *KCSR & TP* 12531; Anantagiri (VSKP), *DAM* 6506.

INDIA : Throughout (except in the drier portion).
WORLD : Sri Lanka, Myanmar.

### EMBELIA N. Burman *nom. cons.*

1. Leaves glabrous beneath. Racemes glabrous or rustry tomentose .............. *E. tsjeriam-cottam*
1. Leaves villous beneath. Racemes villous .............. *E. villosa*

**Embelia tsjeriam-cottam** (Roem. & Schult.) DC. Trans. Linn. Soc. Lond. 17: 131. 1834; Gamble 2: 753 (531). *Ardisia tsjeriam-cottam* Roem. & Schult. Syst. 4: 578. 1819. *Embelia robusta auct. non* Roxb; FBI 3: 315. *p.p.*

Small tree; bark brown, horizontally cracked; wood reddish, resinous with broad medullary rays; branchlets light green, lenticellate, glabrous. Leaves elliptic or obovate, 3–15 × 2–7 cm, glabrous, paler, above, often reddish beneath. apex abruptly acuminate, margin entire, base subacute or obtuse;

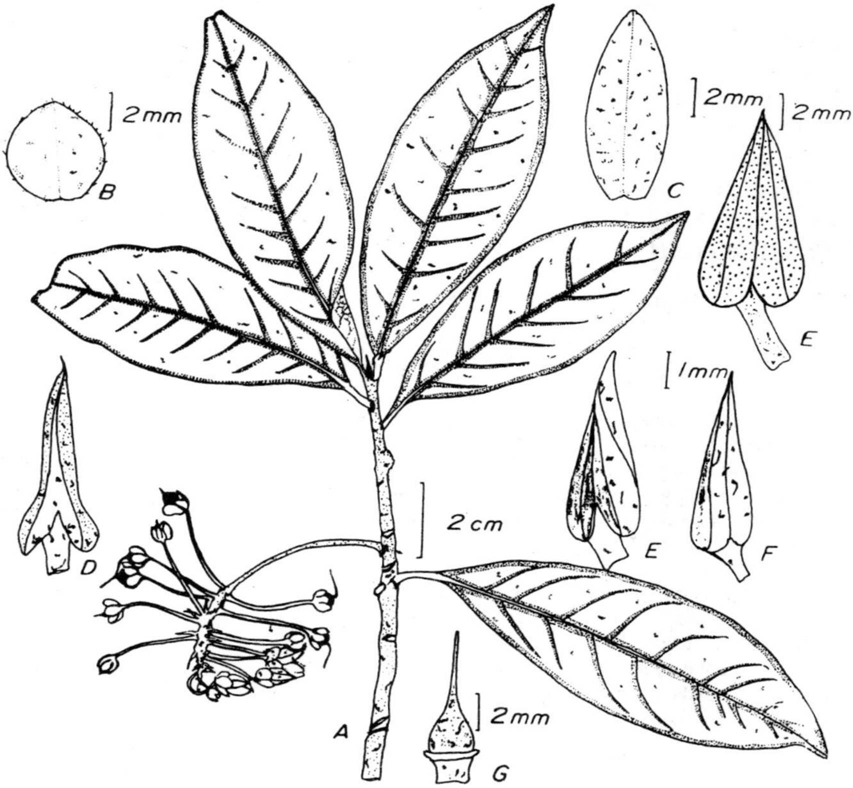


**Fig. 47.** *Ardisia solanacea* Roxb.
A. Twig, B. Sepal, C. Petal, D. E & F. Anthers, G. Pistil.

petiole 7.5–12 mm. Flowers greenish-white, small, dioecious in axillary and extra-axillary racemes. Fruit red, sub-globose, 3–4.2 mm dia., tipped by the style, epicarp crustaceous; endocarp fleshy; seed 1.

Occasional in the Circars and Deccan. Fls.: May–July. Frs.: Oct.–Jan.

Fruit is used as antispasmodic and carminative. The dried bark of the root is used as toothache.

Ramannapenta (KNL), *JLE* 32636 (MH).

INDIA : Throughout greater part.
WORLD : Sri Lanka.

**Embelia villosa** Wall. in Roxb. Fl. Ind. 2: 289. 1824; Gamble 2: 753 (529); *Embelia robusta auct. non* Roxb. C.B. Clarke in Hook. f. Fl. Brit. India 3: 515. 1882. *p.p.* Brandis 416.

Small tree; branches long spreading, scrambling; bark brown; branchlets rusty pubescent. Leaves membranous, elliptic-obovate, to 10 × 6.2 cm, villous or stellate-pubescent beneath, glabrous above, apex abruptly acuminate, margin entire, base cuneate; petiole 0.6–1.2 cm. Flowers glandular in terminal or axillary racemes. Berry globose, small, 1-seeded.

Occasional, in semi-evergreen forests. Fls.: & Frs.: July–Oct.

Gundlabrahmeswaram. (KNL), *JLE* 16938 (MH); Galikonda (VSKP), *GVS* 44374 (MH); Maredumilli (EG), *GVS* 27291 (MH); Araku (VSKP), *NPBK* 10743 (MH).

INDIA : Deccan, hills of Karnataka, West Bengal; upper Gangetic plain.

## SAPOTACEAE

1. Calyx lobes 1-seriate
   2. Staminodes absent; pedicel to 4 mm. Leaves oblong-lanceolate, glabrous beneath; secondary nerves numerous, parallel. Seeds 2–6 .................. ***Chrysophyllum***
   2. Staminodes present, pedicel 6–10 mm. Leaves elliptic or obovate, tomentose beneath; secondary nerves 8–13 pairs, not distinctly parallel. Seed usually 1 .................. ***Xantolis***
1. Calyx-lobes 2-seriate;
   3. Calyx lobes 4:
      4. Stamens 8. Petiole 3 mm long .................. ***Isonandra***
      4. Stamens 12 or more. Petiole 2–3.5 cm long .................. ***Madhuca***
   3. Calyx lobes 6–8:
      5. Secondary nerves few, distant .................. ***Mimusops***
      5. Secondary nerves numerous, close .................. ***Manilkara***

## CHRYSOPHYLLUM Linnaeus

**Chrysophyllum roxburghii** Don. Gen. Hist. 4: 33. 1838; FBI 3: 535; Gamble 2: 758 (533); Brandis 423; Beddome Fl. Sylv. S. India t. 236. 1872.

Large evergreen tree; bark dark grey, smooth; wood greyish-white, cross-grained; branches horizontal; branchlets with rusty pubescence. Leaves coriaceous, oblong-lanceolate, 7–15 × 2.5–5 cm, glabrous, shining, apex caudate-acuminate or obtusely acuminate, margin entire, base obliquely acute or attenuate; secondary nerves numerous, close, parallel, conspicuous; petiole to 7 mm. Flowers small, greenish-white, pedicellate in axilllary umbels. Berry yellowish-green, globose, obscurely 5–6-angled, 2.5–3 cm in dia., tawny tomentose when young; seeds 2–6, brownish-yellow, ellipsoid, 1.2–1.8 cm long, polish, hard.

Northern Circars, Rampa hills of Godavari (Gamble). Fls.: Jan.–Mar. Frs.: May–June.

Fruits edible, but with sticky juice. Wood very hard, not durable, but used for buildings and shingles, it is said to be suitable for match splints, boxes and for sugarcane crushing mills.

INDIA : Peninsula, North-Eastern Indian region.
WORLD : Myanmar, Sri Lanka, Thailand, Vietnam, Laos, Cambodia, Cochin-China, Hainan, Malay Peninsula, Indonesia, Philippines, New Guinea.

## ISONANDRA Wight

**Isonandra villosa** Wt. Ic. t. 360. 1843. FBI 3: 538; Gamble 2: 761 (535). *I. lanceolata* Wt. Ic. P. 367. 1843. Brandis 424.

Tree; branchlets thick, obscurely hairy. Leaves coriaceous, broadly obovate, up to 7 × 6.2 cm, glabrous, apex rounded or emarginate, margin entire, base rounded; secondary nerves 12–15 pairs, regular, parallel, 5 mm apart; petiole to 3 mm. Flowers sessile in clusters on very short thick lateral spurs. Berry 1.2 × 0.5 cm, ellipsoid or oblong with a fleshy pericarp; seed 1, flattened.

Deccan, Veligonda hills of Nellore (Ramaswami, Cf. Gamble, & Srinivasa Rao 1991).

## MADHUCA Hamilton ex J.E. Gmelin

1. Leaves elliptic. Anther connectives apiculate. Pedicel stout. Berry ovoid, 1–4-seeded ........ ***M. indica***
1. Leaves oblanceolate. Anther connectives toothed. Pedicel slender. Berry ellipsoid, 1-seeded. ........ ***M. longifolia***

**Madhuca indica** J. Gmelin Syst. Nat. 2: 799. 1791. *M. longifolia* (J. Koenig) Macbr. var. *latifolia* A. Chev. Rev. Int. Bot. Appl. Agric. Trop. 23: 149. 1943. *Bassia latifolia* Roxb. Pl. Cor. t. 19. 1795 & Fl. Ind. 2: 526. 1832; FBI 3: 544. Gamble 2: 763 (536); Beddome Fl. Sylv. S. India t. 41. 1870.

Deciduous tree, up to 18 m tall; trunk short; bark thick, dark with vertical cracks, the inner bark red, milky; wood reddish-brown, straight or shallowly inter-locked grained, medium to fine textured; branchlets glabrous, spreading. Leaves clustered near the ends of branches, chartaceous, thick, elliptic or elliptic-oblong, 8–20 × 4–10 cm, glabrescent, apex acute or shortly acuminate, margin entire, base rounded or acute; secondary nerves ca. 12 pairs, 1.6 cm apart; petiole to 3.5 cm. Flowers cream, dense in fascicles on drooping densely tomentose pedicels. Berry ovoid-oblong, 4 × 2 cm, fleshy, greenish; seeds 1–4, brown, 1–3 cm long, polished, slightly broad.

Common, in mixed deciduous forests. It prefers a sandy soil but also grows on shallow dry soils. It is a drought hardy species, strong light demander and coppices well. Leaf fall: Jan.–Feb. New foliage: May. Fls.: Feb.–Apr. Frs.: June–July. Vern.: Ippa, Ippi. Eng.: Indian butter tree.

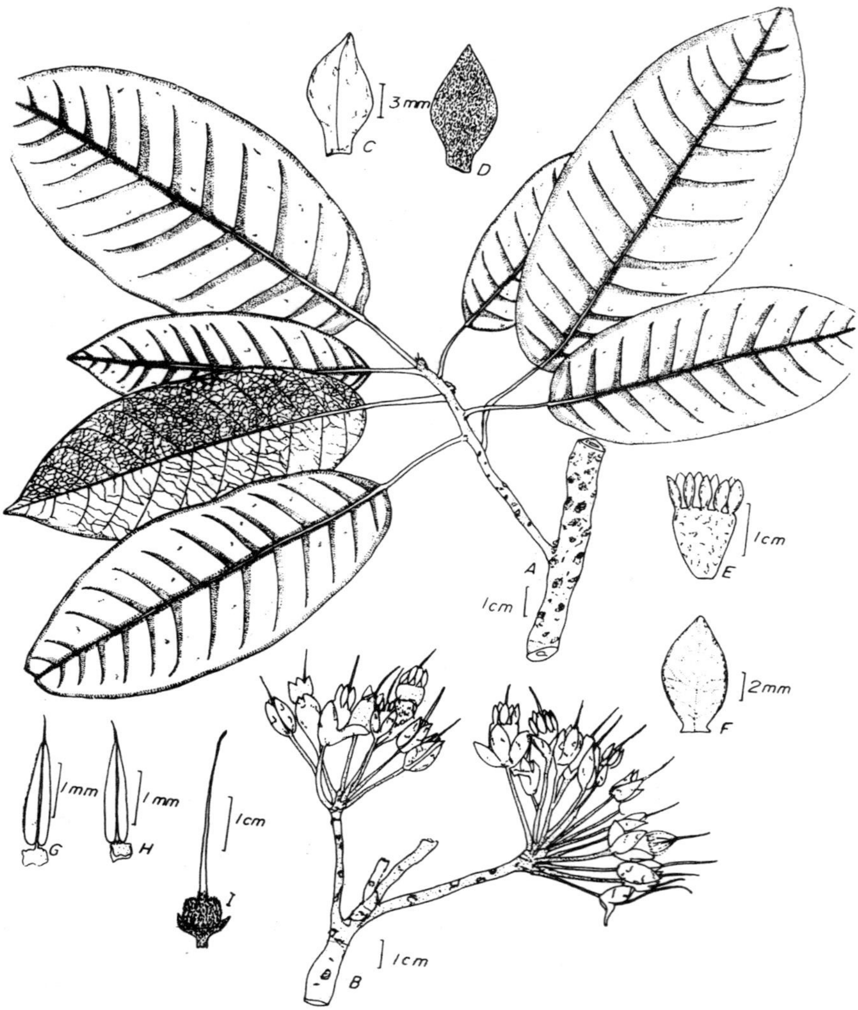


**Fig. 48.** *Madhuca indica* Gmel.
A. Twig, B. Inflorescence branch, C & D. Sepals, E. Corolla, F. Corolla lobe, G & H. Stamens, I. Pistil.

Corolla of flowers contain much sugar and are an important article of food and for the distillation of spirit. The fruits have large fleshy cotyledons which give valuable edible oil. Wood very hard, heavy to very heavy, used for all kinds of builidng purposes as beams, door and window frames, it is used as bridge piles, oil and sugar presses, frames and foot-boards of carts, axiles etc.

Rollapenta (KNL), *SSR* 15509; Seethampet forest (SKLM), *SSR* 15573; Sundergiri (ADB), *GO & PVP* 4236; Isukagundem to Devarakonda (CDP), *R.V. Reddy* 8174; Narsapur RF (MDK), *CP* 11587; Laxmapur RF (NZB), *TP & BR* 6251; Mohammedabad (RR), *MSM & KH* 10224.

INDIA : Throughout.
WORLD : Myanmar.
**Note** : Fresh seed has a high percentage of fertility, but the seed quickly loses its viability if kept and is much subject to insect and fungus attack. The tree suffers greatly from the attacks of *Loranthus*.

**Madhuca longifolia** (J. Koenig) Macbr. Contr. Gray Herb. (n.s) (53). 17. 1918. *Bassia longifolia* J. Koenig ex L. Mant. Pl. 563. 1771; FBI 3: 544; Gamble 2: 763 (537); Brandis 427; Beddome Fl. Sylv. S. India t. 42. 1870.

Deciduous tree, 15–20 m tall; bark dark yellowish-grey to brown, scaly; wood red, close grained, coarse and even textured; branchlets brownish-yellow, glabrous. Leaves chartaceous, oblanceolate, 5.5–8 × 2–3 cm, glabrous, green above, paler beneath, apex acute, margin entire, base shortly narrowed; secondary nerves 10–12 pairs, raised above; petiole to 2 cm. Flowers creamish-white, long pedicellate umbels. Berry ellipsoid, glabrous, 4 × 2 cm; seed 1, compressed, light brown, smooth, shining with a fairly thick and hard testa. The fertility of fresh seed is high, but the seeds do not retain their viability long.

Occasional, in open forests. Leaf fall: Dec.–Jan. New foliage: Mar. Fls.: Mar. Frs.: May–June. Vern.: Pedda Ippa.

Berries are eaten raw or cooked. Flowers are regarded as cooling, tonic and demulcent. They are used in coughs, colds and bronchitis. The corolla are a rich source of sugars and contain appreciable amounts of vitamins and calcium. Seeds yield an oil used mainly in the manufacture of soaps (particularly laundry soaps), oil has emollient properties and is used in skin diseases, rheumatism and headache. The bark contains tanin, used in dyeing and tanning. The wood is heavy to very heavy, used for ship building, in bridge construction, beams in house building, and also for carts and agricultural implements.

Gundumala RF (ATP), *NY* 1242.

INDIA : Western Ghats, Konkan, South India.
WORLD : Sri Lanka, Myanmar.

## MANILKARA Adanson

1. Leaf apex acute or retuse. Flowers solitary; staminodes absent. Berry 3–7-seeded .. ***M. zapota***
1. Leaf apex obtuse or emarginate. Flowers in fascicles; staminodes present. Berry 1-seeded:

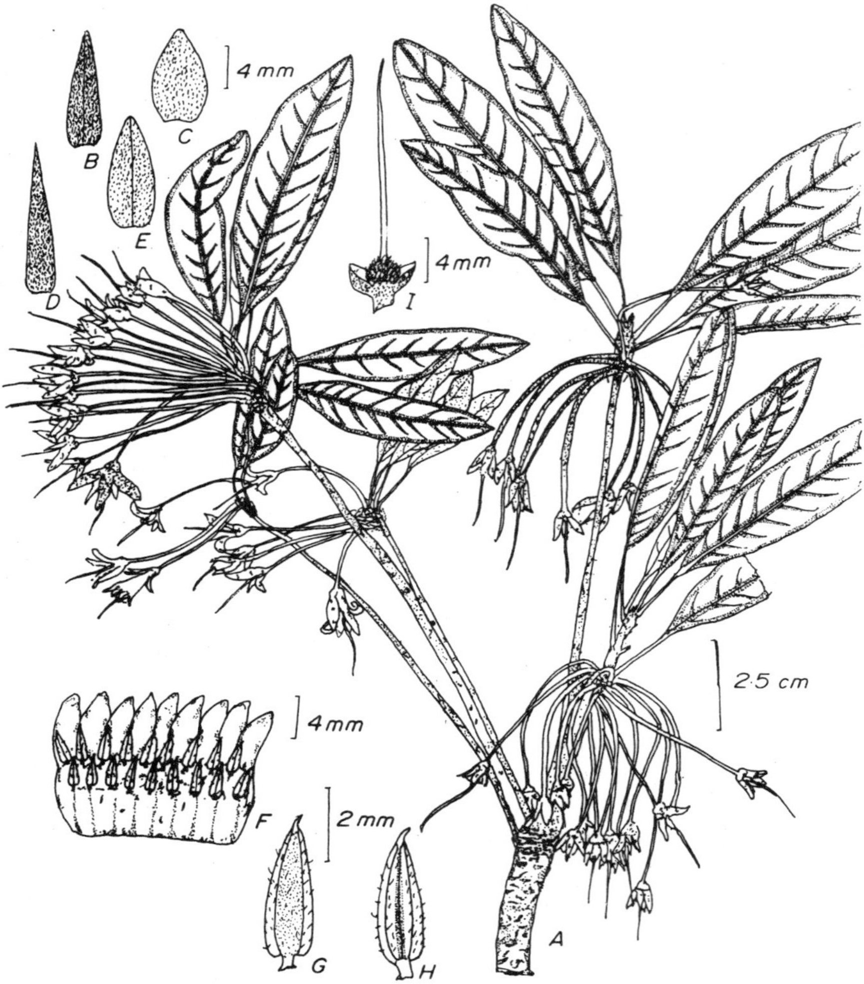


**Fig. 49.** *Madhuca longifolia* (Koen.) Macbr.
A. Twig, B–E. Sepals, F. Corolla spread out, G & H. Stamens, I. Pistil.

2. Leaves broadly elliptic; secondary nerves to 14 pairs; petiole to 3.5 cm. Flower pedicel to 4.5 cm long; sepals 6 mm long, oblong. Berry globose, depressed. ................ ............................................................................................ ***M. roxburghiana***
2. Leaves broadly oblong; secondary nerves numerous; petiole to 2 cm. Flower pedicel to 0.8 cm long; sepals 3 mm long, ovate. Berry ellipsoid. .......................... ***M. hexandra***

**Manilkara hexandra** (Roxb.) Dubard. in Ann. Inst. Bot. Geol Colon. Marseille (Ser. 3) 23. 9. f. 2. 1915; *Mimusops hexandra* Roxb. Pl. Cor. t. 15. 1795 & Fl. Ind. 2: 238; Gamble 2: 766 (538); FBI 3: 549; Brandis 425. *M. indica* A. DC. Prodr. 8: 205. 1844.

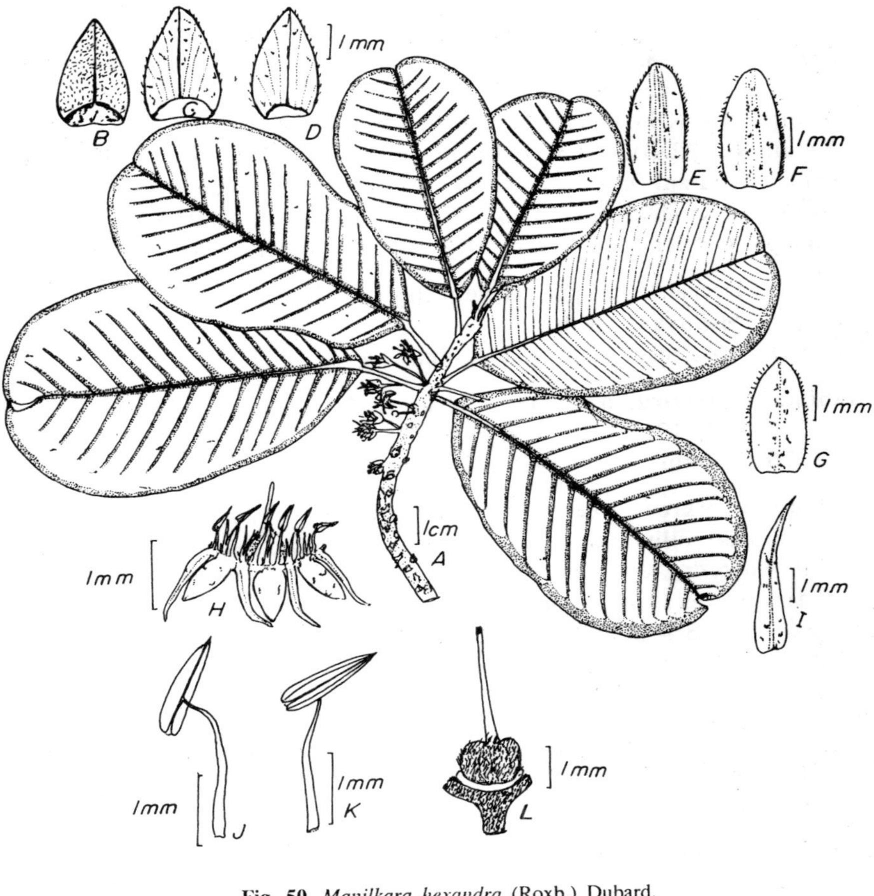


**Fig. 50.** *Manilkara hexandra* (Roxb.) Dubard.
A. Twig, B, C & D. Sepals, E, F & G. Petals, H. Part of flower, I. Petal, J & K. Stamens, L. Pistil.

Evergreen tree, up to 15 m tall with a shady crown; trunk erect; bark blackish-grey, crimson inside, deeply furrowed; wood red and close grained. Leaves clustered at the ends of branches, coriaceous, obovate or oblong, 4–11.3 × 4–6.2 cm, glabrous, dark green above, and paler beneath, apex emarginate, margin entire, base obtuse or acute; secondary nerves 20–25 pairs, parallel, impressed above, raised below; petiole to 2 cm. Flowers whitish, 3–6 in leaf axils. Berry reddish-yellow, ellipsoid, slightly curved, oblong 0.5–1.2 × 0.5–0.8 cm, smooth; seed 1, rarely 2, ovoid, reddish brown, 1–1.5 cm long, and shining.

Common, in dry evergreen forest to 500 m. It perfers sandy loam and loamy soils; in unfavourable conditions (dry situations) it becomes stunted and even shrubby like. It is a light demander. Fls.: Nov.–Jan. Frs.: Apr.–July. Vern.: Puttapala, Pala Nemmi, Manchipala.

Wood very hard, heavy and tough, used for house-posts, turnery, oil presses etc. Fruits are largely collected for its edible pulp. Branches used in religious ceremonies.

Chintala (PKSM), *SSR* 16110; Srisailam (KNL), *SSR & KJ* 12582; Inchapalli RF (ADB), *GO & DAM* 5151; Polavaram (WG), *CAB* 4795 (MH); Veerabhadrapuram (EG), *GVS* 68667 (MH); Loya Kondapalli (KSN), *PVK* 6066 (MH); Krishna river bank (NLG), *KSN* 9802 (MH); Kankunuru to Mechdupur (KNR), *NRR* 83186 (MH); Karaka (VSKP), *CAB* 1688 (MH).

INDIA : Deccan Peninsula, extending north to Gujarat, North Eastern region, Madhya Pradesh.
WORLD : Sri Lanka, Siam, Indo-China.

**Manilkara roxburghiana** (Wt.) Dubard in Ann. Mus. Col. Mars. (Ser. 3) 23: 10. f. 2. 1915. *Mimusops roxburghiana* Wt. Ic.4: t. 1588. 1850; FBI 3: 766; Gamble 2: 766 (538); Brandis 425.

Large evergreen tree; wood reddish-brown; branchlets puberulous. Leaves broadly-elliptic, 4–9 × 2.8–5.8 cm, glabrous, brownish above, paler beneath when dry, apex emarginate or obtuse, margin entire, base obtuse; secondary nerves 14 pairs, not very close, inconspicuous; petiole to 3.5 cm. Flowers white, 3–4 in axillary fascicles. Berry globose, 1 cm long.

Rare, in dry hills. Fls. & Frs.: Mar.–Dec. Vern.: Renga, Magula.

Wood hard and used in building construction.

Way to Pathalaganga (KNL), *JLE* 23750; Yerramalais (KNL), *JSG* 10912 (CAL).

INDIA : Tamil Nadu.

**Manilkara zapota** (L.) P. Royen in Blumea 7: 410. 1953. *Achras zapota* L. Sp. Pl. 1190. 1753. *Achras sapota* L. Sp. Pl. (ed. 2) 470. 1762; FBI 3: 534; Gamble 2: 759 (538); Brandis 424.

Evergreen tree, up to 12 m tall; bark greyish, smooth, scaly; branchlets puberulous. Leaves crowded at the ends of branchlets, elliptic-oblong, 7.5–11 × 2.3–3.9 cm, glabrous, apex acute or retuse, margin entire, base acute; secondary nerves 15–20 pairs, parallel; petiole 2–5 cm. Flowers white, solitary, axillary, pedicellate; calyx rusty tomentose. Berry brown, globose, 3–5 cm dia., fleshy, rough; seeds 3–7, black, glossy, oblong, compressed, 1–1.2 cm long.

Commonly cultivated for its fruits in plains and lower slopes. Fls.: Feb.–Apr. Frs.: May. Vern.: Sapota, Seema ippa. Eng.: Sapota. Fruits edible.

Tirumala (CTR), *SSR* 19681; SKU Campus (ATP), *NY* 544; Gani RF (KNL), *RVR* 3141.

INDIA : Widely cultivated.
WORLD : Native of South America, cultivated in tropics.

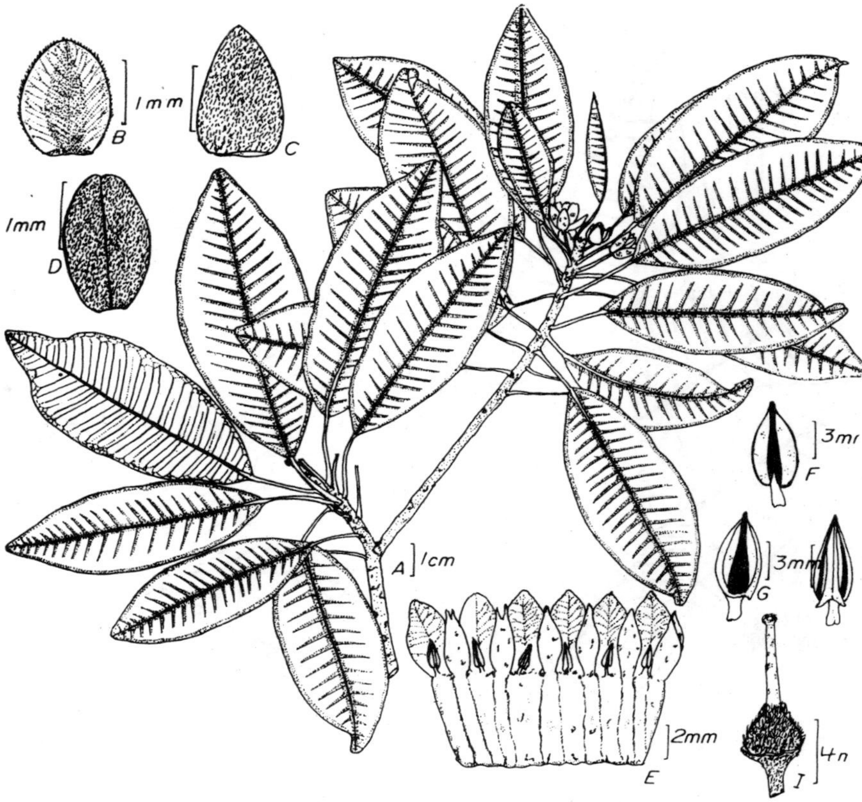


**Fig. 51.** *Manilkara zapota* (L.) P. Royen.
A. Twig, B, C & D. Sepals. E. Corolla spread out, F, G & H. Stamens, I. Pistil.

## MIMUSOPS Linnaeus

**Mimusops elengi** L. Sp. Pl. 349. 1753; FBI 3: 548; Gamble 2 : 765 (538); Brandis 425; Beddome Fl. Sylv. S. India t. 40. 1870.

Evergreen handsome tree, up to 10 m tall; crown hemispherical; trunk cylindrical; bark dark grey, deeply fissured; wood dark red, close-grained, shallowly inter-locked, even and very fine textured; branchlets glabrous. Leaves subcoriaceous, elliptic-ovate or elliptic-oblong, 4.5–15 × 4–7.5 cm, glabrous, apex acute or acuminate, margin entire, base acute, petiole to 2.5 cm. Flowers white, fragrant, 2–6 in fascicles. Berry orange-yellow, ovoid, fleshy, 1.5 cm across; seed solitary, ovoid, compressed, brown, shining, 1.5–2.3 × 1–1.3 cm, smooth with hard testa and soft whitish albumen.

It is often cultivated, wild in forests except in small patches in moist-evergreen forests of Eastern ghats especially in Northern Circars above 400 m. It grows on a variety of soils including laterite. It is a shade bearer and one

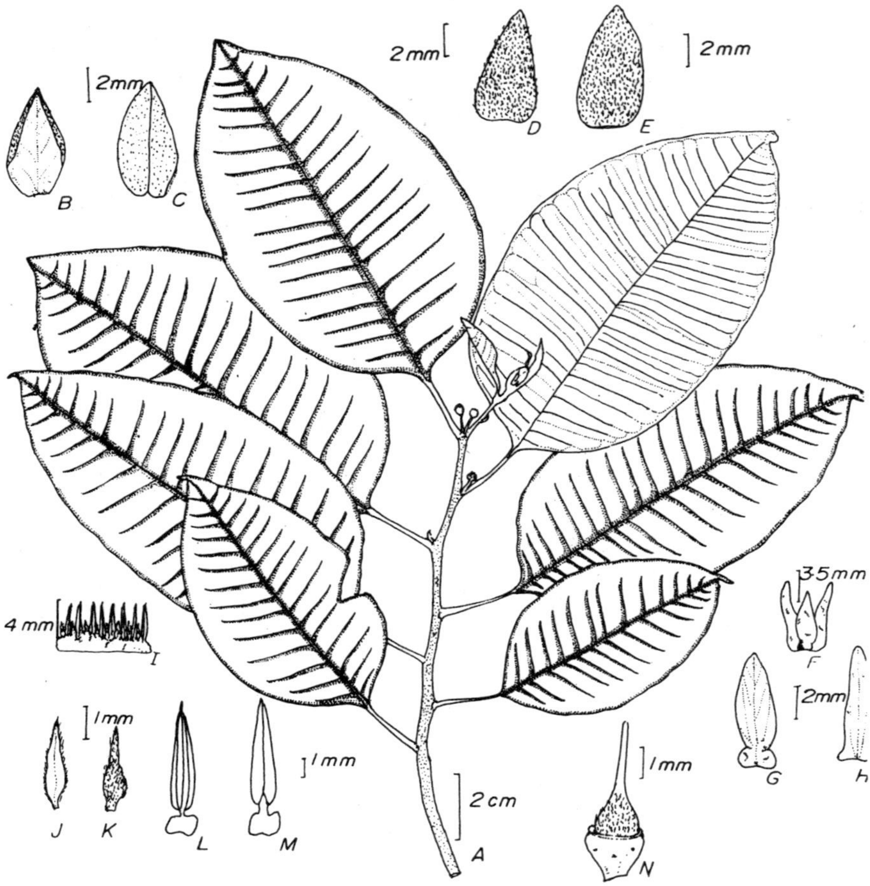


**Fig. 52.** *Mimusopos elengi* L.
A. Twig, B–E. Sepals, F. Part of corolla, G & H. Corolla lobes, I. Androecium, J & K. Staminodes, L & M. Stamens, N. Pistil.

of the best plants recommended for avenue plantations because of its scented flowers, edible fruits and dense crown. Under favourable conditions the tree reaches large dimensions. Fls.: Feb.–Apr. Frs.: Following Feb.–July. Vern.: Pogada, Bogada, Pagada, Vakulamu, Kesara. Eng.: Bullet wood.

Young twigs are used for cleaning teeth. The fragrant flowers are used for making garlands and a perfume is distilled from them. Fruits edible, sometimes used for making preservatives and pickles. Seeds yield an oil, used for cooking and lighting. The tree yields gum and lopped for fodder. Wood used for house building, carts, cabinet work, rice pounders.

Tirumala (CTR), *SSR* 13770 and 18594; V.P. South (GTR), *VRK* 3928; Pinjarakonda (EG), *KSR & TP* 12900; Bogada (VSKP), *PVK* 5498 (MH); Rakthachandanam thota (SKLM), *GVS* 62365 (MH).

INDIA : Indian Peninsula along Western ghats from Mumbai Southwards and on the east from northern circars, southwards, Andamans.
WORLD : Myanmar, Sri Lanka.

## **XANTOLIS** Rafinesque

**Xantolis tomentosa** (Roxb.) Raf. Sylv. Tell. 36. 1838. *Sideroxylon tomentosum* Roxb. Pl. Cor. 1. 28. t. 28. 1795; FBI 3; 538; Gamble 2: 759 (534); Brandis 422.

Evergreen tree, 8–12 m tall; usually spinescent; bark grey or light brown, cracked; wood yellowish-brown turning to light brown with age, bitter, straight-grained, very fine textured; branchlets with brown tomentum. Leaves sub-coriaceous, elliptic, elliptic-oblong or oblanceolate, 7–16 × 3–7 cm, glabrescent and dark green above, tomentose, and paler beneath, apex obtuse or abruptly acute, margin obscurely undulate, base acute, narrowed, cuneate; secondary nerves 8–13 pairs, straight, parallel, inter-secondary nerves present; petiole to 8 mm, tomentose. Flowers whitish, pedicellate, drooping, 5 mm long in fascicles from old leaf scars. Berry yellowish, globose, 2.5 cm dia., tomentose when young, seated on the calyx, style persistent; seed usually 1, black or deep brown.

Occasional, in valleys of moist evergreen forests up to 1500 m.Fls.: Apr.–June. Frs.: Nov.–Feb.

The wood is heavy to moderately heavy, used for planks and beams, suitable for all sorts of tool handles, mallet heads, rice pounders, etc. and also as fuel.

Galikonda (VSKP), *SSR* 15920; Anantagiri (VSKP), *SSR* 19628; Duggeru forest (SKLM), *GVS* 62515 (MH); Maredumilli (EG), *GVS* 24412 (MH).

INDIA : Western Peninsula, Orissa, Bihar, North Kanara.
WORLD : Sri Lanka, Myanmar, Indo-China.

## EBENACEAE

1. Flowers 3-merous; ovary 3-loculed. Fruiting calyx not accrescent ............................ ***Maba***
1. Flowers 4–5 merous; ovary 4–10-loculed. Fruiting-calyx accrescent (except *D. chloroxylon*) ............................ ***Diospyros***

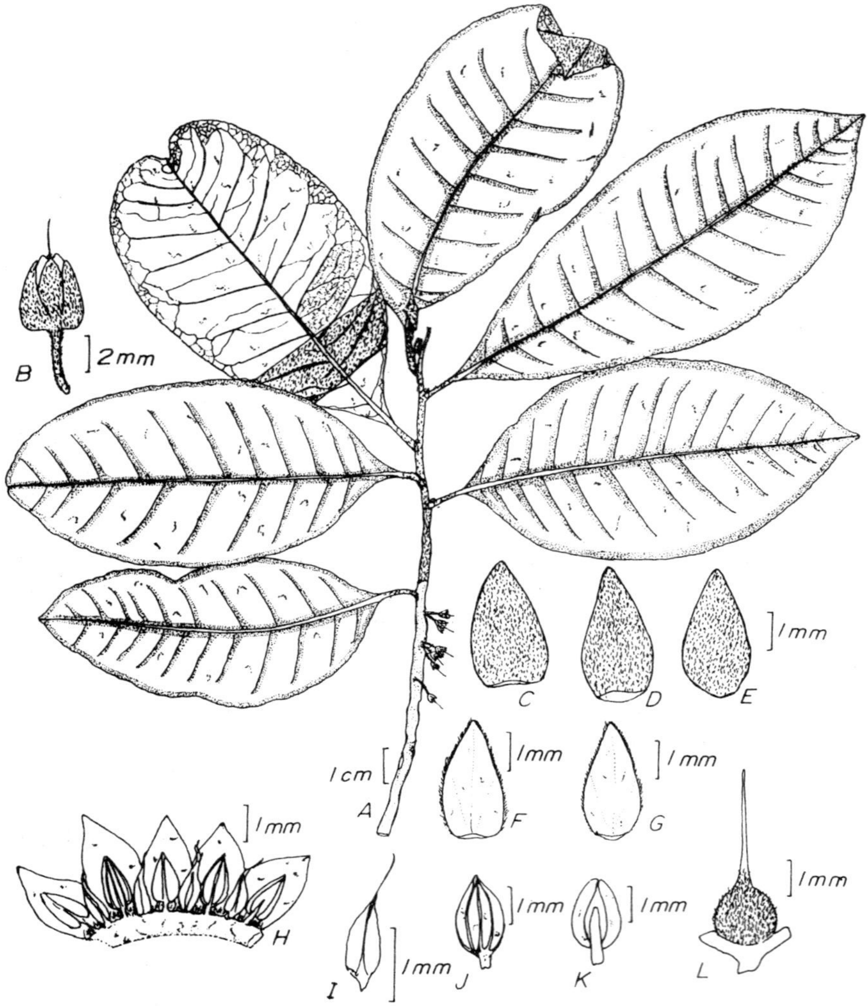


**Fig. 53.** *Xantolis tomentosa* (Roxb.) Raf.
A. Twig, B. Calyx, C–G. Calyx lobes, H. Corolla spread out, I–K. Stamens, L. Pistil.

## DIOSPYROS Linnaeus

1. Flowers 2–2.5 cm across, yellowish; stamens 2–4. Berry 6 cm across .............. ***D. malabarica***
1. Flowers 0.5–1.5 cm across, other than yellowish; stamens 6–16. Berry to 4.5 cm across:
    2. Leaves tomentose or pubescent in lower or both surfaces:
        3. Leaves softly pubescent on both surfaces .............................................. ***D. cordifolia***
        3. Leaves pubescent or tomentose on lower surface:
            4. Petiole to 5 mm. Female flowers sessile; calyx lobes 5, hardly enlarged in fruit ... ................................................................................................ ***D. chloroxylon***

4. Petiole to 1.5 cm long. Female flowers pedicellate; calyx lobes 4–6, enlarged in fruit:
   5. Leaves elliptic. Pedicel stout; corolla tubular; calyx thick to 8 mm, stamens 10-12. Pericarp crustaceous .................................................. ***D. melanoxylon***
   5. Leaves broadly ovate. Pedicel not stout; corolla not tubular; calyx thin to 3 mm; stamens 16. Pericarp leathery ............................................ ***D. exsculpta***

2. Leaves glabrous on both surfaces.
   6. Female flowers tawny tomentose or pubescent:
      7. Male flowers densely fascicled, corolla yellow tomentose without; stamens 10 ... ........................................................................................ ***D. candolleana***
      7. Male flowers in short peduncled cymes, corolla glabrous without; stamens 24–32........................................................................................ ***D. assamilis***
   6. Female flowers glabrous or little hairy:
      8. Male flowers in cymes.—Leaves black when dry ............................ ***D. sylvatica***
      8. Male flowers in umbels or fascicles:
         9. Berries in clusters .................................................................. ***D. ovalifolia***
         9. Berry solitary:
            10. Stamens 6–12 in uneven groups. Leaves oblong- elliptic, black when dry ... ..................................................................................... ***D. ebenum***
            10. Stamens 16 in pairs. Leaves elliptic-obovate, greyish-green when dry. ... ..................................................................................... ***D. montana***

**Diospyros assamilis** Beddome in Madr. For. Rep. 20 t. 1. 1866–67; FBI 3: 558; Gamble 2: 775 (545).

Large evergreen tree; bark black, thin, much cracked; wood black, streaked with grey and brown. Leaves membranous to chartaceous, elliptic-oblong, to 15 × 5.5 cm, glabrous, drying black, apex acuminate, margin entire, base somewhat rounded; secondary nerves usually opposite and from the base. Flowers pale yellow in short peduncled cymes or fascicles. Berry globose.

Northern circars, Palakonda hills of Visakhapatnam (cf. Gamble). Eng.: Malabar Ebony.

INDIA : Evergreen forests of Western Ghats of Southern Maharashtra, Karnataka and Kerala.

**Diospyros candolleana** Wt. Ic. t. 1221. 1848; FBI 3: 566; Gamble 2: 773 (543); Brandis 434.

Trees, up to 10 m tall; bark greenish-black, smooth; wood light pink with grey streaks; young branches hairy. Leaves coriaceous, elliptic-oblong, 5–15 × 3.3–6 cm, glabrous, apex shortly, obtusely acuminate, margin entire, base rounded or shortly cuneate, midrib prominent and raised beneath, depressed above; secondary nerves numerous, irregular, conspicuous above, petiole 0.8–1.5 cm. Flowers yellowish, tawny pubescent, 2–10 in axillary fascicles or in short cymes. Fruit pinkish, globose or ellipsoid, 2.5 cm in dia., glabrous; fruiting calyx segments triangular with reflexed margins; seeds 3–4, brown, 1.2 cm long, oblong, wrinkled; endosperm ruminate.

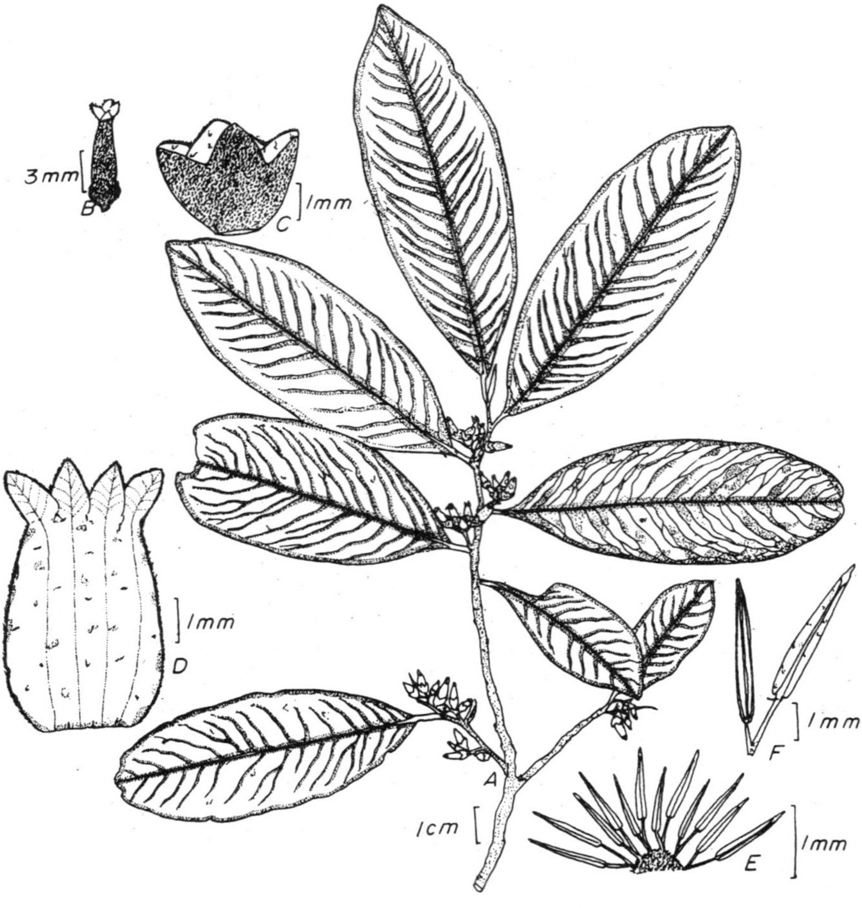


**Fig. 54.** *Diosypros candolleana* Wight.
Male: A. Twig, B. Flower, C. Calyx, D. Corolla spread out, E. Androecium, F. Pair of stamens.

Rare, in evergreen forests to 1500 m. Fls.: Feb.–June. Frs.: Mar.

Wood hard and heavy. A decoction of the root bark is given in cases of swellings and rheumatism.

Galikonda (VSKP), *SSR* 15923; Minumuluru (VSKP), *GVS* 29593 (MH).

INDIA : Tamil Nadu, Karnataka, Orissa, Kerala (Western Peninsula).

**Diospyros chloroxylon** Roxb. Pl. Cor. t. 4: 1795 & Fl. Ind. 2: 538. 1832; FBI 3: 560; Gamble 2: 776 (546); Brandis 430. *D. capitulata* Wt.. Icon. Pl. Ind. Orient. t. 1224. 1848 & t. 1588. 1850.

Evergreen tree, up to 10 m tall; sometimes spinescent; bark dark brown, deeply cracked, peeling off in rectangular small scales; wood yellowish-

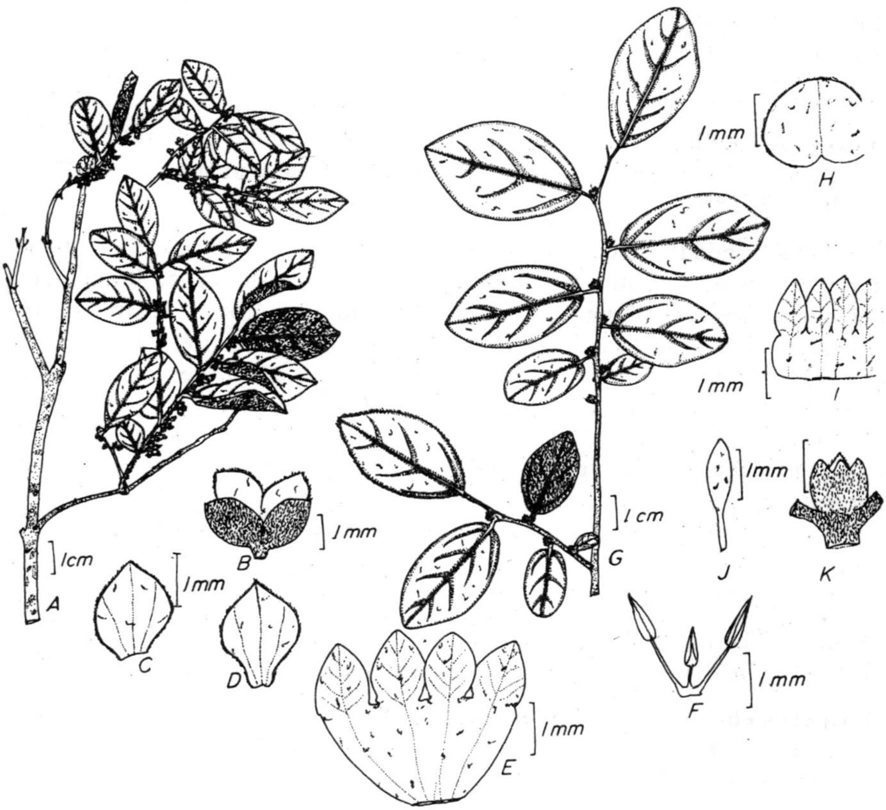


**Fig. 55.** *Diospyros chloroxylon* Roxb.
Male: A. Twig, B. Calyx, C & D. Calyx lobes, E. Corolla, F. Staminal part, G. Female twig, H. Calyx lobe, I. Corolla spread out, J. Staminode, K. Pistil.

grey; branchlets rusty tomentose. Leaves thinly-coriaceous, elliptic or elliptic-oblong, 1.6–5.6 × 0.8–3 cm, slightly pubescent above, fulvous hairy beneath, apex and base acute or obtuse; nerves inconspicuous; petiole to 5 mm. Male flowers small in axillary clusters; female ones solitary. Berry brown, globose, 1 cm across, smooth, shining; fruiting calyx enlarged, submembranous, flat; seeds 2–3.

Common, in dry deciduous forests. It produces root suckers, prefers laterite and sand stone hills. Fls.: Mar.–June. Frs.: Jan.–Apr. Vern.: Nela ulimiri, Ullinda, Ellinda, Thorika, Illinta.

Fruits are edible. Wood moderately hard and used as a good fuel. It is lopped for fodder.

Tirumala (CTR), *SSR* 13714; Pinjarakonda (EG), *SSR* 15231; Kalasamudram RF (ATP), *SSR* 13753; Eturnagaram (WGL), *SSR* 18538; Guvvalacheruvu (CDP), *RVR* 8430; Mohammedabad (RR), *MSM* 10247;

Burugidda (NZB), *BR* 9076; Jayanthipuram (GTR), *VRK* 6967; Vempalli (ADB), *GO* 4793; Kondapalli fort (KSN), *PVK* 5757 (MH).

INDIA : Central and Southern region, Orissa.

**Diospyros cordifolia** Roxb. Pl. Cor. t. 49. 1795; Gamble 2: 776 (545). *D. montana* Roxb. var. *cordifolia* (Roxb.) C.B. Clarke in Hook. f. Fl. Brit. India 3: 555. 1882.

Small or middle-sized, deciduous tree; often thorny with branching; bark greyish-brown, smooth; wood reddish or yellowish-white; branchlets pubescent. Leaves oblong, or oblong-lanceolate; 2.5–7 × 1.3–2 cm, softly pubescent on both surfaces, apex obtuse or acute, margin entire, base rounded or cordate; petiole 3–8 mm, hairy or tomentose. Male flowers white in 3-flowered, axillary cymes; female flowers white, turning brown, solitary, pedunculate. Berry orange-yellow, turning black, globose or ovoid, 2.4–3.7 cm; fruiting calyx thickened, glabrous; seeds brown, compressed; the percentage of fertility is high.

Rare, in dry deciduous and evergreen forests. Fls.: Mar.–June. Frs.: Following Dec.–Feb.

Regupalem (VSKP), *CAB* 1590; Godavari, *JSG* 18539 (MH).

INDIA : Throughout tropical region.
WORLD : Sri Lanka, Myanmar and North Australia.

**Diospyros ebenum** J. Koeng. ex Retz. Physiogr. Salsk. Hand 1: 176. 1780; FBI 3: 558; Gamble 2: 775(545); Brandis 434; Beddome Fl. Sylv. S. India t. 65. 1870.

Densely foliaceous, evergreen tree, up to 15 m tall; crown compact; bark dark grey, deeply fissured; wood orange-brown, very hard, close-grained; branches rigid, pendulous; branchlets glabrous. Leaves sub-coriaceous, 5–10 × 3–4 cm, oblong-elliptic, bright green and shining above, paler beneath, apex obtuse or bluntly subacute, rarely emarginate, margin entire, base acute, 3–5-nerved; secondary nerves 4–8 pairs, reticulate, conspicuous on both surfaces; petiole to 6 mm. Flowers greenish-yellow; male flowers few, sessile in short umbellate, axillary clusters; female flowers solitary, axillary, subsessile or with short pedicel. Berry globose, to 2 cm across; fruiting calyx forming a hemispheric, woody cup, reflexed with a short apical beak; seeds 3–8, dull-black.

Rare, in dry evergreen forests at 700–900 m. It prefers well-drained, rocky soil and grows well on sandy loam. Fls.: Mar.–Apr. Frs.: June. Vern.: Nallauti, Tuki. Eng.: Ebony.

Wood very durable and used for turnery, carving, piano-keys, veeners, musical instruments, sports goods, mathematical instruments and fine ornamental work of various kinds. Value of wood depends on the intensity of its blackness. Fruits edible. Tree yields gum and is used in medicines to clear obstruction in vision.

Tirumala (CTR), *SSR* 13764; Mamandur (CTR), *SSR* 16180.

INDIA : Orissa.

**Diospyros exsculpta** Buch.-Ham. in Trans. Linn. Soc. London 15: 110. 1827 excl. Syn.; non Dalz. 1861; Sant. & Wagh in Bull. Bot. Surv. India 5: 106. 1963. *D. tomentosa* Roxb. Fl. Ind. 2: 532. 1832; FBI 3: 564; Gamble 2: 772 (543); Brandis 433.

Medium-sized tree; bark greyish black, exfoliating in rectangular scales; wood reddish-brown, fairly straight or curly grained, fine and even textured; young branchlets densely rusty tomentose. Leaves broadly ovate, 7.5–15 × 5–8 cm, apex obtuse or slightly acute, margin entire, base rounded; nerves and reticulations conspicuous above; petiole short. Flowers white 4–5-merous; male flowers in short cymes; female flowers solitary. Berry globose, 2.5–3.7 cm dia, glabrous; seeds 1.2 cm long, shortly oblong, slightly compressed.

Northern circars down to the Godavari, in deciduous forests. Fls.: Apr.–May. Frs.: Oct.–Dec.

Fruits sweetish, edible. Wood hard and is used for small articles of ornament; carved walking sticks, tables, picture frames, cabinet work. Leaves are used for wrapping local cigarettes known as bidies.

Godavari *JSG* 12653 (MH); *RHB* s.n (MH).

INDIA : From Bengal, Bihar to Central India and southwards as far back as the Godavari, Siwalik and Sub-Himalayan forests of the Punjab, Westwards to the Ravi.

**Diospyros malabarica** (Desr.) Kostel. Allg. Med. Pharma. Fl. 3: 1099. 1834. *Garcinia malabarica* Desr. in Lamk. Ency. 3: 701. 15–2. *Diospyros embryopteris* Pers. Syn. Pl. 2: 624. 1807; FBI 3: 556. 1882; Beddome Fl. Sylv. S. India t. 69. 1870. *D. peregrina* Gurke in Eng. Nat. Pflanzenfam 4: 164. 1897.

Evergreen tree, up to 12 m tall; trunk fluted, branches spreading, almost to the ground; young foliage reddish; bark dark grey or black, smooth, flaking off in large pieces; wood grey with darker streaks. Leaves thick-coriaceous, oblong or oblong-lanceolate, 15–30 × 4–5.8 cm, glabrous, shining, glaucous beneath, apex acute, margin entire, base obtuse; secondary nerves 12–16 pairs, scarcely raised, very reticulate between, oblique; petiole to 1.3 cm. Flowers yellowish, fragrant, axillary; male flowers in clusters; female flowers solitary, pedicellate. Berry yellow, covered with a red velvety tomentum, subglobose, 6 cm in dia., fruiting calyx woody, reflexed; seeds ca. 8, large, compressed with a fairly thick testa, embedded in glutinous pulp.

**Fig. 56.** *Diospyros malabarica* (Desr.) Kostel.
A. Twig with immature fruit.

Occasional, chiefly along the streams. It is shade bearer. Fls.: Apr.–May. Frs.: Ripens following Apr. Vern.: Nitta Tumiki, Tumiki, Racha-Tumiki, tinduki, Gabu.

The pulp of the fruit is largely eaten by monkeys and fruit bats, sometimes eaten by human beings. Unripe fruits are rich in tannin and are employed for tanning and dyeing cloth. Pulped fruit used as a preservative for fishing nets and as glue for book binding. Rind and bark are used medicinally for the cure of diarrhoea. Infusion of the fruit used as gargle in aphthae and sore throat. Wood moderately hard, employed for building construction and boat making. The gum obtained from tree is used for paving the bottom of boats and seams of fishing nets to make them durable and also it is used in book binding since it preserves books from insects.

Tirumala (CTR), *SSR* 13780; Talakona (CTR), *SSR* 16134; Rajahmundry (EG), *GVS* 68665 (MH); Punyagiri (VSKP), *GVS* 23307 (MH); Towards Sunki from Salur (SKLM), *GVS* 62501; Kederipuram (VZN), *NRR* 83641.

INDIA : Throughout.
WORLD : Sri Lanka, Sikkam, Indo-China, Western Malaysia.

**Diospyros melanoxylon** Roxb. Pl. Cor. t. 46. 1795 & Fl. Ind. 2: 530. 1832; FBI 3: 564; Gamble 2: 772 (543); Brandis 432. *D. exsculpta* Beddome Fl. Sylv. S. India t. 66. 1870.

Deciduous tree, up to 18 m tall; bark dark grey or black, exfoliating in rectangular scales; wood reddish-brown with an irregular black heart wood; young branches tawny tomentose. Leaves coriaceous, broadly elliptic, 6.5–30 × 3.2–14 cm, glabrescent above, pubescent beneath, apex subacute or obtuse, margin entire, base acute or rounded; secondary nerves 8–10 pairs, prominent on both sides; petiole to 1 cm. Male flowers 3–12 in 2.5 cm long, tawny-tomentose cymes. Female flowers solitary. Berry globose, 3–4.5 cm across, yellow when ripe, smooth, supported by the thick calyx with wavy reflexed margins; seeds 4–6, compressed, oblong, brown, 5 mm long, shining with a wrinkled testa.

Common, in dry, mixed deciduous forests at 600–1100 m. It readily grows on poor denuded soil, hot dry hill slopes in moist valleys. It prefers loamy soil with dominant clay fraction and loose porous soil. It is a good coppicer and it also reproduces by root suckers. Leaf fall: Dec.–Jan. New foliage: May. Vern.: Tuniki, Tumiki, Beedi Aku, Tellagada, Tuki, Manchi Tumiki. Eng.: Ebony, Tendu.

The leaves are extensively used as a wrapper for local cigarettes known as bidies, it is a minor forest produce for Bidi industry. The leaves are diuretic, carminative, laxative and styptic. The bark is astringent and its decoction is used in diarrhoea and dysepsia. Fruits are edible, astringent, not unpleasant to taste. Wood hard, moderately heavy to heavy, used for posts, rafters, golf sticks, railway keys etc., it is good fuel wood.

Anantagiri RF (RR), *SSR* 18508; Eturnagaram (WGL), *SSR* 18530; Donubai RF (SKLM), *SSR* 15588 & 15593; Penchalakona (NLR), *SSR* 19352; Narsapur RF (MDK), *BR & CP* 14112; Manchippa RF (NZB), *BR & KH* 9025; Yemugorideva-Kottakota RF (ATP), *TP & NY* 2115; Neelvai RF (ADB), *GO & DAM* 5164; Near Mukunuru (KNR), *NRR & TRS* 86575 (MH); Madgole to Paderu (VSKP), *GVS* 28090 (MH); Triruvur (KSN), *PVK* 5713.

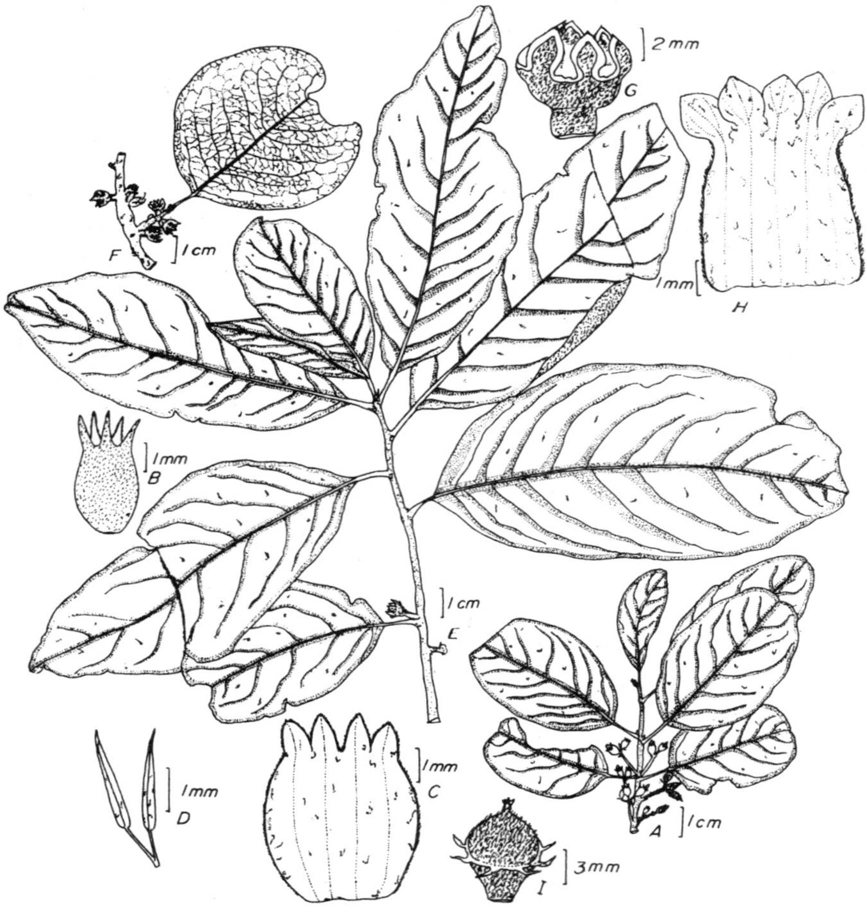


**Fig. 57.** *Diospyros melanoxylon* Roxb.
Male: A. Twig, B. Calyx, C. Corolla spread out, D. Staminal pair.
Female: E. Twig, F. Flowering twig, G. Calyx, H. Corolla spread out, I. Pistil.

INDIA : Punjab, Rajasthan, Madhya Pradesh, Orissa, Uttar Pradesh, Karnataka, Bihar, Tamil Nadu.

WORLD : Sri Lanka.

**Diospyros montana** Roxb. Pl. Cor. t. 45. 1795 & Fl. Ind. 2. 538. 1832; FBI 3: 555; Gamble 2: 776 (545); Brandis 431.

Semi-evergreen tree, up to 8 m tall, often spiny; bark greyish-black, peeling off in thin flakes; wood grey, tinged with yellow or brown with darker patches, and fine grained. Leaves thick-coriaceous, obovate or elliptic, 5–12 × 2–6 cm, glabrous, apex acute, margin entire, base obtuse; secondary nerves 10 pairs, faint; petiole 7 mm. Flowers white; male flowers few in short cymes; female flowers solitary, axillary, drooping on short pedicels. Berry yellow turning black, globose, 2.5–3 cm in dia., glaucous, pendulous; fruiting calyx thick, reflexed; seeds 5, brown compressed.

Common, in moist deciduous forests up to 1100 m. It coppices well. Leaf fall: Jan.–Feb. New foliage: Mar. Fls.: Mar.–May. Frs.: Dec.–Feb. Vern.: Erra-gatha, Erra Goda, Jagadagondi, Muchi Tunki.

Fruit is bitter with an unpleasant odour; applied externally for boils and also used for stupefying fish. Wood moderately hard and is used for making small articles of furniture also for carts, agricultural implements and indoor work, suitable for a rafters, matches and carvings; good fuel wood.

Kutravada (EG), *SSR* 15235; Donubai (SKLM), *SSR* 15583; Rampur (ADB), *SSR* 19318; Chedmal RF (NZB), *BR & CPR* 9254; Mohammedabad (RR), *MSM & KH* 10225; Brahmagundam (CTR), *GVS* 46940 (MH); Nagiri hills (CDP), *JSG* 24458 (MH); Tadikonda (VZN), *NRR* 83630 (MH); Pakhal RF (WGL), *KMS* 13170; Kilagada (VSKP), *GVS* 22588; Rachepalli (KRN), *MRRP* 205 (MH).

INDIA : Tamil Nadu, Uttar Pradesh, Maharashtra, Andaman and Nicobar Islands, Madhya Pradesh, Orissa and Kerala.
WORLD : Sri Lanka, Nepal, Malaysia, tropical Asia.

**Diospyros ovalifolia** Wt., Icon. Pl. Ind. Orient. t. 1227. 1848; FBI 3: 557; Gamble 2: 777 (546); Brandis 435.

Evergreen tree, up to 10 m tall; bark greyish, rough; streaked with black; branchlets lenticellate, glabrous. Leaves coriaceous, elliptic-oblong or rarely oblanceolate, 5.5–11 × 1.4–3.6 cm, glabrous, pale green above, pale yellowish beneath when dry, apex obtuse, slightly acute, margin entire, base acute or shortly cuneate; secondary nerves 7–10 pairs, irregular, prominent beneath and forming loops, midrib depressed above, raised beneath; petiole 5–7 mm long. Flowers sessile, yellowish-white in axillary fascicles; male flowers 3–3.5 mm across, 4–12 in each axil., female flowers 1–2 or rarely to 6 together in the axils of fallen leaves. Berry light yellowish-orange, globose, 1–1.8 cm dia., glabrous; fruiting calyx slightly enlarged, persistent, reflexed, tawny pubescent; seed 1.

Occasional, in moist evergreen forests at 700–1100 m, frequently found near streams. Fls.: Nov.–Dec. Frs.: Apr.–May. New foliage by June. Vern. Kukka-Tumiki.

Wood suitable for building construction. Fruits edible.

Talakona (CTR), *SSR* 18591; Penchalakona (NLR), *SSR* 19345; Dornal (KNL), *JSG* 18676 (MH); Near Borra caves (VSKP), *GVS* 44272.

INDIA : Southern Peninsular region.
WORLD : Sri Lanka.

**Diospyros sylvatica** Roxb. Pl. Cor. t. 47. 1795. Fl. Ind. 2: 537. 1832; FBI 3: 559; Gamble 2: 773 (543); Brandis 431; Beddome Fl. Sylv. S. India t. 143. 1872.

Evergreen tree, up to 10 m tall; bark black, smooth, thin; wood grey with black streaks and irregular patches; branchlets glabrous. Leaves coriaceous or membranous, elliptic, 3.5–9 × 2.5–3.5 cm, glabrous, shining above and paler beneath, black when dry, apex acuminate, margin entire, base acute or obtuse; petiole to 1 cm. Flowers white; male flowers few in 1.2 cm long, axillary cymes; Female flowers solitary, axillary. Berry globose, 1.3–1.6 cm in dia., yellowish, smooth; fruiting calyx enlarged, spreading, flat, foliaceous; seeds 2–8, oblong, 1 cm long, rugose, shining, brown.

Occasional, near streams and hill slopes in the moist deciduous forests at 700–1500 m. New foliage by March. Fls.: Apr.–May; Frs: Jan.–Mar. Vern.: Pasara Gatha, Tella gatha, Gatha.

Wood moderately hard and heavy used for fancy work. Fruits edible.

Anantagiri (VSKP), *SSR* 15936; Borra caves (VSKP), *SSR* 15939; Minumuluru (VSKP), *SSR* 15942; Galikonda (VSKP), *SSR* 19634; G. Konduru (KSN), *PVK* 6087; Near Peddarukla stream (PKSM), *NRR* 83762; Maredumilli (EG), *GVS* 24229; Near Mukunuru (KNR), *NRR & TRS* 86581 (MH).

INDIA : Karnataka, Tamil Nadu, Kerala, Orissa.
WORLD : Sri Lanka.

## MABA J.R. Forster & J.G.A. Forster

1. Leaves obovate-spathulate, apex obtuse-emarginate ........ ***M. buxifolia***
1. Leaves elliptic-lanceolate, apex acute or acuminate ........ ***M. neilgherrensis***

**Maba buxifolia** (Rottb.) A.L. Juss. Ann. Mus. Natl. Hist. Nat. 5: 418. 1804; FBI 3: 551; Gamble 2: 768 (539). *Pisonia buxifolia* Rottb. Nyesaml. Kongel Donskeviden S.K. Selsk. Skr. 2: 536. t. 4. f. 2. 1783.

Shrub to tree, up to 8 m tall; bark grey to black, rough; wood grey with darker streaks, close-grained; branchlets glabrous. Leaves coriaceous, obovate-spathulate, 1.5–4 × 0.6–1.2 cm, glabrous, apex obtuse or emarginate, margin entire, base acute or attenuate; secondary nerves not distinct; petiole to 2 mm. Flowers white to 4.5 mm long, 3-merous, bisexual in fascicles. Berry purplish black, oblong, 5 mm across, crowned by calyx; seeds one.

Common, in plains and also in hilly regions up to 800 m. Fls.: Apr.–May. Frs.: June–July. Vern.: Alli, Tella Alli, Chinna-ullingi, Uti, Yerruti, Pisinki.

Wood hard, heavy and durable, used for boat anchors, handles and sheaths of weapons and rafters. The fruits are pulpy and edible.

**Fig. 58.** *Diospyros sylvatica* Roxb.
Male: A. Twig, B. Calyx, C. Corolla, D. Staminal part, E. Calyx lobe.
Female: F. Twig, G. Calyx lobe, H. Corolla, I. Pistil with sepals, J. Fruit.

Simhachalam (VSKP), *SSR* 15568; Kambakkam (CTR), *SSR* 16165; Siddluliah kona (NLR), *SSR* 19375; Sattenapalli beat (ADB), *TP & PVP* 4070; Penchalakona (NLR), *KH* 5530; Owk RF (KNL), *RVR* 3030.

INDIA : Peninsula.
WORLD : Sri Lanka.

**Maba neilgherrensis** Wt., Ill. Ind. Bot. t. 148. 1850. *Maba ebenum* Wt., Icon. Pl. Ind. Orient. t.t.. 1228 & 1229. 1848. *Maba buxifolia* sensu Clarke in Hook. f. Fl. Brit. India 3. 555. 1882; Gamble 2: 768 (539).

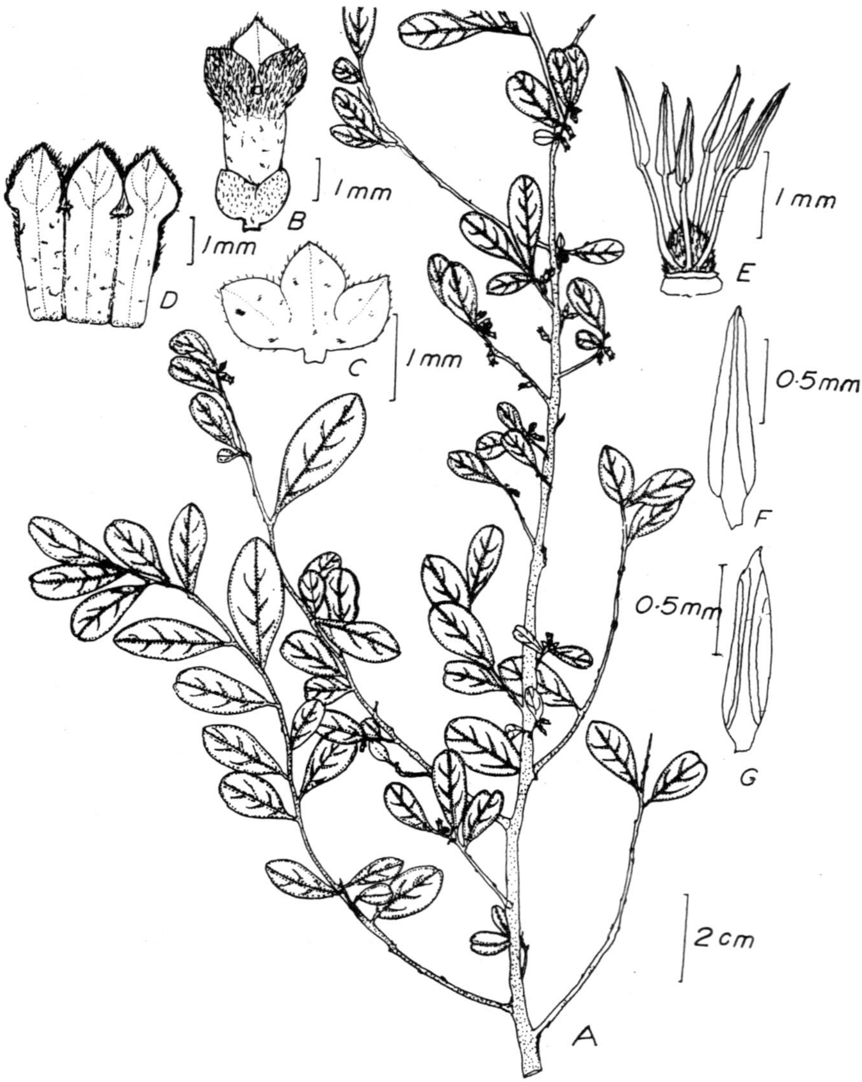


**Fig. 59.** *Maba buxifolia* (Rottb.) A.L. Juss.
A. Twig, B. Flower, C. Calyx, D. Corolla spread out, E. Androecium with pistillode, F & G. Anthers.

Evergreen tree, up to 15 m tall; bark black; wood brownish-yellow. Leaves chartaceous, elliptic-lanceolate, 2.5–7.5 × 2–3 cm, glabrous, apex acute or acuminate, margin entire, base rounded. Flowers yellowish-white, bisexual, 3-merous; male flowers in cymes; female flowers solitary. Berry to 1.2 cm dia., seeds 1–3.

Rare, in forests of Chittoor district. Fls.: Apr.–June. Frs.: Aug. Vern.: Karuti. Wood hard.

Near Akasadhara (CTR), *GVS* 45911 (MH).

INDIA : Peninsula.

## SYMPLOCACEAE

### **SYMPLOCOS** N.J. Jacquin

1. Leaves elliptic-obovate. Flowers in spikes .................. ***S. cochinchinensis*** sub sp.***laurina***
1. Leaves elliptic. Flowers in racemes .................................................................. ***S. theaefolia***

**Symplocos cochinchinensis** (Lour.) S. Moore in J. Bot. 52: 148. 1914 sub sp. **laurina** (Retz.) Nooteb. Rev. Symplocae. 156. 1976 & in Steenis Fl. Males. 1. 8: 248. 1977. *Dicalyx cochinchinensis* Lour. Fl. Cochinch. 1: 663. 1790. *Myrtus laurina* Retz. Obs. Bot. 4: 26. 1810. *Symplocos spicata* Roxb. Fl. Ind. 2: 542. 1832; FBI 3: 573; Gamble 2: 782 (549); Brandis 441; Beddome Fl. Sylv. S. India t. 149. 1872.

Evergreen, small tree; bark light grey, smooth; wood white. Leaves elliptic-obovate, 5–20 × 3–7 cm, glabrous, apex acute to acuminate, margin crenate-serrate, base cuneate or narrowed; secondary nerves 8–10 pairs, prominent; petiole to 1.2 cm. Flowers white, fragrant, to 7 mm across in simple or branched, axillary spikes. Drupe bluish-black, globose, 5 mm dia., glabrous, ribbed, crowned by small calyx; seed curved.

Very rare, in semi-evergreen forests at 1100–1350 m. Fls.: Jan.–Mar. Frs.: Apr.–June. Wood is used for building, posts and furniture and also for fuel. Dried leaves used as a dye-stuff.

Vangasara RF (VSKP), *SSR* 19668; Sunkarimetta (VSKP), SKW 20545 (BLAT).

INDIA : Throughout the greater part of India in the hills.
WORLD : Nepal, Bangladesh, Myanmar, Indonesia, Malay Peninsula, China, Japan, Pacific Islands, Sri Lanka.

**Symplocos theaefolia** D. Don, Prodr. Fl. Nepal. 145. 1825; FBI 3: 575; Gamble 2: 784 (551); Brandis 441.

Evergreen small tree; wood white; branchlets angular, glabrous. Leaves coriaceous, elliptic, to 12.5 × 5 cm, glabrous, shining, apex acuminate, margin entire, base cuneate. Flowers in 1.2 cm long racemes. Drupe ovoid, glabrous, to 1 cm long.

Hills of Visakhapatnam at Rangalu and Gomangi at 900–1350 m. (A.W. Lushington cf. Gamble).

Wood soft used for making rough house-posts and as fuel. The leaves serve the purpose of an auxillary in dyeing.

## OLEACEAE

1. Corolla > 1 cm long, lobes imbricate:
   2. Capsule orbicular, compressed, thin, erect; seeds not winged. Leaves rough above. Flowers white-orange ........ ***Nyctanthes***
   2. Capsule obovoid, woody, pendulous; seeds winged. Leaves smooth. Flowers yellowish-brown ........ ***Schrebera***
1. Corolla < 1 cm long, lobes valvate:
   3. Inflorescence terminal, rachis pubescent-cultivated ........ ***Ligustrum***
   3. Inflorescence axillary, rachis glabrous:
      4. Stamens exerted ........ ***Olea***
      4. Stamens inserted ........ ***Chionanthus***

## CHIONANTHUS Linnaeus

### KEY 1 (Vegetative)

1. Leaves thin, up to 20 × 6.5 cm; secondary nerves conspicuous.
   2. Leaves shining; secondary nerves 10–14 pairs ........ ***C. intermedia***
   2. Leaves rather dull; secondary nerves 8 pairs ........ ***C. macrophylla***
1. Leaves thick, up to 13.5 × 5.3 cm; secondary nerves obscure:
   3. Leaf apex acute or shortly and bluntly acuminate, margins recurved; secondary nerves numerous ........ ***C. mala-elengi***
   3. Leaf apex abruptly acute or emarginate, margins not recurved; secondary nerves 5–7 pairs ........ ***C. zeylanica***

### KEY 2 (Flowering & Fruiting)

1. Flowers in cymose fascicles. Seeds abluminous. Corolla lobes 6–7 mm long ........ ***C. mala-elengi***
1. Flowers in panicles. Seeds ex-albuminous. Corolla lobes 2–4 mm long:
   2. Panicles to 2 cm long. Drupe 5 × 2 mm ........ ***C. zeylanica***
   2. Panicles 5–10 cm long. Drupe 0.7–1.5 × 0.5–0.6 cm:
      3. Drupe ellipsoid, 0.7–1.2 cm long. Petals 2.5–3.7 mm long. ........ ***C. intermedia***
      3. Drupe oblong, 1.5–1.7 cm long. Petals 2.5 mm long. ........ ***C. macrophylla***

**Chionanthus intermedia** Wt. Icon. t. 1245. 1848; FBI 3. 609; Gamble 2. 794 (558); Beddome Fl. Sylv. S. India t. 239. 1872.

Densely foliaceous, evergreen tree, up to 15 m tall; bark light yellowish-brown; wood white, close grained; branches pendulous. Leaves thinly coriaceous, elliptic-oblong or elliptic-lanceolate, 6–18 × 2.6–6.5 cm, glabrous, shining, acuminate at both ends, margin entire; secondary nerves 10–14 pairs; petiole to 2.5 cm. Flowers small, cream in 5 cm long, axillary panicles. Drupe ellipsoid, 0.7–1 × 0.2–0.6 cm, greyish brown.

Rare, along the streams in dry-deciduous forests at 500–1000 m. Fls.: Feb. Frs.: Mar.–May. Vern.: Satapala, Verri Pogada. Wood moderately hard. Leaves are useful as precious fodder.

Gundlabrahmeswaram (KNL), *SSR* 15282; Bodhulur (EG), *SSR* 15239; Tirumala (CTR), *SSR* 15282; Galikonda (VSKP), *SSR* 15932; Rampa hills (EG), *JSG* 16009 (MH).

INDIA : Eastern and Southern Indian region.
WORLD : Taiwan, Malaysia, Australia, Myanmar, S. China, tropical Himalaya.

**Chionanthus macrophylla** Bl. Mus. Bot. Lugduno-Batavum 1: 317. 1851. *Linociera macrophylla* (Bl.) Wall. ex Clarke in Hook. f. Fl. Brit. India 3: 610. 1892; Gamble 2: 794 (558). *Chionanthus ramiflora* Roxb. Fl. Ind. 1: 106. 1820.

Small tree. Leaves oblong or elliptic-oblong, or slightly oblanceolate, to 20 × 6 cm, glabrous, apex shortly acute, margin entire, base acuminate; secondary nerves 8-pairs; petiole to 2.5 cm. Flowers white, in 8–10 cm long panicles. Drupe purple, oblong, 1.2–1.5 cm.

Rare, along moist and shady areas of deciduous forests of Visakhapatnam and East Godavari districts. Fls.: July–Oct. and Feb.–Mar. Frs.: Round the year. Vern.: Kurma Gochh.

Rampa hill (EG), *JSG* 16009 (MH).

INDIA : Eastern and Southern region.
WORLD : Myanmar, Bangladesh.

**Chionanthus mala-elengi** (Dennst.) Green. in Bull. Bot. Surv.India 26: 124. 1984. *Forsythia mala-elengi* Dennst. Schlues. Hort. Ind. Malab. 12–21 & 31. 1818. *Linociera malabarica* Wall. ex G. Don, Syst. 4: 53. 1838; FBI 3: 607; Gamble 2: 794 (558); Brandis 448. *Chionanthus malabarica* Beddome Fl. Sylv. S. India t. 239 a. 1872.

Small to moderate-sized evergreen tree; crown compact; bark light grey, smooth or lenticellate. Leaves coriaceous, lanceolate to obovate or elliptic-oblong, 5–13.3 × 1.8–5.3 cm, glabrous, shining, apex acute or shortly and bluntly acuminate, margin entire, base acute; secondary nerves numerous, obscure, united within the margin and forms an intramarginal vein; petiole to 5 mm. Flowers white, fragrant, sessile or subsessile in cymose fascicles; peduncle to 1.8 cm. Drupe bluish, obliquely oblong, 6 × 4 mm; endocarp thin; seeds 1 or 2.

Rare, in semi-evergreen forests near streams. Fls.: Nov.–Jan. Frs.: Following Mar.–Apr. Vern.: Punna Ganu, Chedda Naredi, Porumbolu.

Sapparla (VSKP), *GVS* 29718 (MH); Dharakonda gedda (VSKP), *GVS* 47424 (MH).

INDIA : Peninsula, Andaman & Nicobar Islands.

**Chionanthus zeylanica** L. Sp. Pl. 8. 1753; FBI 3: 608. *Linociera zeylanca* (L.) Gamble in Fl. Pres. Madras 2: 794 (558). 1922.

Tree, up to 12 m tall; bark grey; wood white or yellowish-white; branchlets glabrous. Leaves thick-chartaceous, obovate or ovate, 3.5–7.5 × 2.3 cm, glabrous, apex obtusely acute or emarginate, margin entire, base rounded; secondary nerves 5–7 pairs, faint; petiole to 5 mm. Flowers cream in 2 cm long, axillary panicles. Drupe obovoid, 5 × 2 mm, apex mucronate.

Common, in deciduous forests at 400–800 m. Fls.: Dec.–Feb. Frs.: Throughout the year. Vern.: Neeli Blabba, Punisi, Punagamu.

Tirumala (CTR), *SSR* 15279; Anantagiri (VSKP), *DAM* 6505; Chitvel RF (CDP), *SSR* 16142; Way to Japalitheertham (CTR), *KS* 7853 (MH).

INDIA : Peninsula.
WORLD : Sri Lanka.

## LIGUSTRUM Linnaeus

**Ligustrum lucidum** Ait. f. Hort. Kew. (ed. 2) 1: 19. 1789 (1810).

Small tree; bark yellowish-grey, smooth; branchlets lenticellate, glabrous. Leaves elliptic, 5.2–8.5 × 2.6–3 cm, glabrous, apex acuminate, margin entire, base acute; secondary nerves 5 pairs, sub impressed; petiole to 1 cm. Flowers white, shortly pedicellate in 12 cm long, terminal panicles.

Cultivated. Fls.: Jan.–Feb. Frs. Apr.–May.

Tirumala gardens (CTR), *SSR* 13708 & 15275.

Cultivated in the plains and in the hills up to 1500 m. Indigenous to China and Japan.

## NYCTANTHES Linnaeus

**Nyctanthes arbor-tristis** L. Sp. Pl. 6. 1753; FBI 3: 603; Gamble 2: 792 (556); Brandis 442; Beddome Fl. Sylv. S. India t. 240. 1872.

Small tree, 4–6 m tall; bark light brown, rough; wood pale red or yellowish-brown, close-grained; branchlets 4-angular, often pendent. Leaves coriaceous, ovate or elliptic, 7–14 × 3–7.2 cm, scabrous with short bulbous hairs above, pubescent beneath, apex short to long acuminate, margin entire, base obtuse or rounded; secondary nerves 3–5 pairs, prominent beneath; petiole to 7 mm. Flowers white-orange, very fragrant to 2.5 cm across in 3-chotomous, axillary and terminal cymes. Capsule brown, elliptic-oblong or orbicular, compressed, 2-valved, 0.7–1.6 × 0.8–1.8 cm, glabrous; epicarp thick; seeds 2.

Common, wild in deciduous forests and often cultivated in gardens. It coppices well. Fls. & Frs.: Throughout the year. Vern.: Parijatham, Krishti. Wood is mainly useful for fuel.

Penchalakona (NLR), *SSR* 19353; Sunnipenta (KNL), *RVR & NVP* 2331; Kothapalli (RR), *MSM & KH* 10446; on the way to Kunthala water falls (ADB), *GO & DAM* 5127; S.K. University (ATP), *VRK & EC* 3219; Vinukonda RF (GTR) *VRK* 6849; Narsapur RF (MDK), *TP & CP* 11973; Ibrahimpet RF (NZB), *TP & BR* 6135.

INDIA : Sub-tropical Himalaya, West Bengal, Bihar, Orissa, Assam, Central India, Tamil Nadu.

WORLD : Nepal, Myanmar.

**Notes** : *Nyctanthes arbor-tristis* was assigned to family Oleaceae till Airy Shaw (in Kew Bull. 272, 1952) transferred it to the family **Verbenaceae** under new sub-family **Nyctanthoideae** Airy Shaw. Stant (in Kew Bull. 273-276, 1952) pointed out several morphological and anatomical features different from those in Oleaceae, that prove Verbenaceous affinity of this plant, but nothing conclusive can be said at present, since the finding of other authors give weight to the arguments on either sides. Crete (1963) on embryological evidence and Das *et al.*,(1966) on chemotaxonomic evidences establish a close relationship between *Nyctanthes* and Oleaceae; Kapil and Vani (1966) support de Jussieu's (1789) treatment of assigning *Nyctanthes* to Oleaceae, Kundu (1966), however, on anatomical and palynological evidences suggests that *Nyclanthes* may belong neither to Oleaceae nor to Verbenaceae.

## **OLEA** Linnaeus

1. Flowers bisexual; corolla deeply lobed. Leaves chartaceous; secondary nerves 4–6 pairs, glandular at the nerve axils ........ ***O. paniculata***
1. Flowers polygamo-dioecious; corolla not deeply lobed. Leaves coriaceous; secondary nerves 7–12 pairs, not glandular at the nerve axils:
   2. Leaves up to 12.5 × 5 cm, not abruptly acuminate at apex........ ***O. dioica***
   2. Leaves up to 6.5 × 3.5 cm, abruptly acuminate at apex ........ ***O. polygama***

**Olea dioica** Roxb. Fl. Ind. 1: 105. 1820; FBI 3: 612; Gamble 2: 296 (559); Beddome Fl. Sylv. S. India 3: 153. 1872.

Medium to large evergreen tree; bark greyish-brown, rough; wood pale brown streaked and mottled, straight and close-grained; branchlets glabrous. Leaves elliptic-oblong or obovate, to 12.5 × 5 cm, glabrous, apex shortly acuminate, margin dentate, base round; secondary nerves 8–12 pairs; petiole to 1.5 cm. Flowers pedicellate in panicles at the axils of fallen leaves. Drupe deep purple, ob-ovoid or ovoid, 1.3–1.7 cm long; seed white, ovoid.

Rare, in semi-evergreen forests of Visakhapatnam district and also planted for shade at 800–1000 m. Fls.: Jan.–Mar. Frs.: Apr.–June.

Wood hard, strong, elastic and heavy; it is suitable for carving and cabinet work. The fruit is bitter to taste. Bark used as febrifuge.

Sunkarimetta (VSKP), *HS* 20566 (BLAT); Gudem (VSKP), *SKW* 2095; Anantagiri (VSKP), *SKW* 20615.

INDIA : Eastern Himalayas, Assam, Deccan Peninsula, chiefly in Western Ghats.
WORLD : Nepal.

**Olea paniculata** R. Br. Prodr. 523. 1810; Kiew, in Blumea 25: 312. 1979. *O. glandulifera* Wall. ex G. Don. Gen. Syst. 4: 49. 1838; FBI 3: 162. Gamble 2: 796 (559); Brandis 446; Beddome Fl. Sylv. S. India t. 238. 1872.

Densely foliaceous, evergreen tree, 12–15 m tall; trunk short; crown broad and umbrageous; bark grey, exfoliating in brittle scales; wood reddish-grey, close-grained; branchlets glabrous. Leaves chartaceous, rhomboid-lanceolate, elliptic or obovate, 6.2–16 × 3–7.5 cm, glabrous, dark green above, paler beneath; apex narrowly acuminate, margin slightly undulate, base acute or shortly attenuate; secondary nerves 4–6 pairs, irregular, prominent and glandular at nerve axils beneath; petiole to 2 cm. Flowers white in axillary or terminal to 8 cm long trichotomous panicles. Drupe purplish-black, ovoid, pendulous; epicarp rugose.

Rare, along banks of rivers and in shady ravines of moist forests at 900–1100 m. Fls.: Mar.–Apr. Frs.: June–Aug.

Wood hard and durable, takes a good polish, used for house construction, carpentary, agricultural implements and turnery.

Talakona (CTR), *SSR* 15550; Vanadurga temple, Minumuluru (VSKP), *GVS* 47355 (MH).

INDIA : Peninsula, Himalayas.
WORLD : Nepal, Java, New Guinea to New Caledonia and Australia.

**Olea polygama** Wt.. Icon t. 1239 & 1240. 1848; FBI 3: 613; Gamble 2; 796 (559).

Small tree. Leaves coriaceous, lanceolate to broadly elliptic, 2.5–6.5 × 1.5–3.5 cm, glabrous, drying reddish-brown, apex abruptly, shortly acuminate, margin entire, distantly minutely crenate, base acute to attenuate; secondary nerves 7–8 pairs, sub-conspicuous; petiole to 4 mm. Flowers white, polygamo-dioecious in condensed racemes.

Very rare. Fls. & Frs.: Nov.–May.

On way to Brahmagundam (CTR), *GVS* 46946 (MH to CAL).

INDIA : Southern region.
WORLD : Sri Lanka.

### SCHREBERA Roxburgh *nom. cons.*

**Schrebera swietenoides** Roxb. Pl. Cor. t. 101. 1799 & Fl. Ind. 1: 109. 1832; FBI 3: 604; Gamble 2: 792 (557); Brandis 444; Beddome Fl. Sylv. S. India t. 248. 1872.

Deciduous tree, up to 20 m tall; crown globose; bark pale grey, nearly smooth exfoliating in thin scales; wood brownish-grey, hard, close grained; branchlets lenticellate. Leaves imparipinnate, to 25 cm long, petiole to 7 cm, 7–9 foliolate; leaflets coriaceous, ovate to ovate-lanceolate, 2.5–8.5 × 1–3.5 cm, glabrous above, villous beneath, apex acute-acuminate, margin entire, base often unequal-sided; secondary nerves 5–9 pairs; petiolule to 3 mm. Flowers yellowish-brown, fragrant at night, numerous in 8 cm long, trichotomous cymes. Capsule woody, obovoid or pear shaped, to 6 × 2.5 cm, pendulous, light brown, very hard and woody, dotted with scabrous specks, loculicidal, 2-valved; seeds 20-30, flat, winged, 2–3.7 cm.

Common, in the hills of dry-deciduous forests at 600–1000 m. Leaffall: Dec.–Feb. New foliage: Apr.–May. Fls.: Apr.–May. Frs.: Throughout the year dehiscing by March. Vern.: Mokkem, Makkam, Ishuga Rashi, Magalingam, Mokkalapa.

Wood is used for house-posts, turning etc. The tree produces root-suckers.

Srisailam (KNL), *SSR* 15522; Kokkarampalli (VSKP), *SSR* 15960; Peccheruvu (KNL), *RVR* 2152; Rachepalli (KNR), *GVS* 22247 (MH): Rampachodavaram (EG), *GVS* 24470 (MH); Cherukonda (VSKP), *GVS* 28164 (MH).

INDIA : Tropical Himalaya; Central and South Indian region, Indian Peninsula.
WORLD : Myanmar.

## SALVADORACEAE

### SALVADORA Linnaeus

**Salvadora persica** L. Sp. Pl. 122. 1753; FBI 3: 619; Gamble 2: 799 (562); var. *wightiana* (Thw.) Verdc. in Kew Bull. 19. 150. 1964. *S. wightiana* Planchon ex Thw. Enum. Pl. Zeyl. 190. 1860; Beddome Fl. Sylv. S. India t. 247. 1872.

Evergreen tree, 8–12 m tall; crown compact, trunk short; branches drooping; bark grey, deeply cracked; wood white, soft; branchlets glabrous, pale yellowish. Leaves thick-chartaceous, ovate-lanceolate, or elliptic-oblong, 3.2–6.2 × 1.7–2.8 cm, glabrous, apex obtuse or subacute, margin entire, base rounded or obtuse; secondary nerves 3–5 pairs, slightly raised beneath; petiole 0.7–2 cm. Flowers yellowish-green, 3 mm across in axillary and terminal copious panicles. Drupe red, globose, 3–4 mm dia., smooth, juicy; seed 1, globose, 2 mm across.

Common, in open places, especially in saline lands and black cotton soils. Fls. & Frs.: Jan.–May. Vern.: Ghunia, Gonni, Gogu, Chekkara-chettu, Pedda Gogu, Wara Gogu. Eng.: Tooth-brush tree, Mustard tree. Twigs are used for tooth cleaning.

Owk RF (KNL), *RVR* 3026; Krishnapatnam basin (NLR), *KH* 5533; Rudroor (NZB), *BR & KH* 9636; Pennahobilam (ATP), *TP* 1272; Palakonda hills (CDP), *CS* 8292.

INDIA : Peninsula.
WORLD : Baluchistan, Sind, Sri Lanka.

## APOÇYNACEAE

1. Leaves alternate. Flowers large:
    2. Leaves thick, and large. Corolla hypocrateriform ........................................ ***Plumeria***
    2. Leaves not thick and large. Corolla infundibuliform:
        3. Calyx glandular within; flowers yellow. Leaves linear .............................. ***Cascabela***
        3. Calyx not glandular within; flowers white. Leaves oblanceolate ................. ***Cerbera***
1. Leaves not alternate. Flowers not large:
    4. Leaves whorled .................................................................................... ***Alstonia***
    4. Leaves opposite:
        5. Follicles few seeded—cultivated. ......................................................... ***Eravatamia***
        5. Follicles many seeded:
            6. Corolla tube shorter than lobes; calyx lobes ovate; stamens exerted .... ***Wrightia***
            6. Corolla tube longer than lobes; calyx lobes linear-ovate; stamens inserted at the base of swollen corolla tube ............................................................. ***Holarrhena***

## ALSTONIA R. Brown

1. Large tree. Flowers in umbellate cymes; corolla tube long; follicles to 50 cm long. Leaf apex obtuse ........................................................................................ ***A. scholaris***
1. Small tree. Flowers in corymbose cymes; corolla tube short; follicles 8.3 cm long. Leaf apex acuminate ........................................................................................ ***A. venenata***

**Alstonia scholaris** (L.) R. Br. Mem.Wern. Nat. Hist. Soc. 1: 76. 1811; Gamble 2. 810 (569); Brandis 459; Beddome Fl. Sylv. S. India t. 242. 1872. *Echites scholaris* L. Mant. Pl. 53. 1767; FBI 3: 642.

Large, evergreen tree; base buttressed; bark dark grey, nearly smooth, lenticellate, yellow inside, exuding milky juice when fresh cut; wood white, even-grained; branches spreading in tiers of whorls; branchlets copiously lenticellate. Leaves 4–7 in a whorl, coriaceous, elliptic-oblong, obovate or oblanceolate, 10–25 × 3.8–8 cm, glabrous, green above, paler and smooth beneath, apex obtusely acute or emarginate, margin entire, base acute or obtuse; secondary nerves numerous, conspicuous, parallel, connecting near margin and form a intramarginal vein; petiole to 5 mm. Flowers greenish-white, to 1.5

cm across, sub-capitate at the ends of the panicle branches. Follicles pendulous, cylindric, narrow, 15–50 × 0.3 cm; seeds numerous, linear-oblong, 6 mm long, flattened, hairs longer at both ends.

Rare, in deciduous forests, preferably in moist situations. Sometimes planted in gardens for ornament. Fls.: Dec.–Mar. Frs.: May–Sept. Vern.: Edakulapala, Palagaruda. Eng.: Shaitan wood, Devils tree, Dita-bark tree.

Wood soft and not durable, only used for boxes, furniture, scabboards and temporary purposes. Stem bark is used for Malaria; it is regarded as a bitter tonic and a mild febrifuge and possesses astringent, anthelminitic and galactogogue properties.

Donubai (SKLM), *NRR* 83617.

INDIA : Widely distributed, often planted.
WORLD : Sri Lanka, S.E. Asia from Myanmar and S. China through Malaysia to New Guinea and Queensland, also in Africa.

**Alstonia venenata** R. Br. Mem. Wern. Nat. Hist. Soc. 1: 77. 1811; FBI 3: 642; Gamble 2: 810 (569); Brandis 459.

Shrub to small tree; branches usually 3-nately whorled; bark thin, light brown; wood white, even-grained; branchlets glabrous. Leaves membranous in whorls of 3–6, oblong, oblong-lanceolate or narrowly oblong, 9–14.5 × 1.9–3.3 cm, glabrous, sometimes hairy on nerves above, pubescent or softly hairy beneath, apex narrowly acuminate, margin entire, base cuneate; secondary nerves numerous, parallel, slender, united and form an intramarginal vein; petiole to 1.3 cm. Flowers white, faintly smelled in pubescent cymose panicles. Follicle fusiform, 8.3 × 0.5 cm, tapering at both ends; seeds compressed, linear-oblong, to 1.2 cm long, densely hairy at both ends.

Occasional, in the hills. Fls.: Apr.–July. Frs.: Nov.–Feb. Vern.: Edakula-pala.

The plant is a rich source of indole alkaloids. The ripe fruits are stated to possess tonic and anthelmintic properties and are reported to be used as a remedy for impure blood, syphilis, insanity and epilepsy.

Minumuluru towards Ontally (VSKP), *GVS* 30015; Near Dora Jammu (SKLM), *GVS* 62475.

INDIA : Deccan peninsula, Western Ghats.

## CASCABELA Rafinsque

**Cascabela thevetia** (L.) Lipp. in Feddes Repert. 97. 52. 1980.*Cerbera thevetia* L. Sp. Pl. 209. 1753. *Thevetia nerifolia* Juss. ex. Steud. Nom. Bot. (ed. 2) 2: 680. 1841 ('*nereifolia*'); Gamble 2: 821 (577). *T. peruviana* (Pers.) Merr. in Philipp. J. Sci. 9: 130. 1914.

Small to mėdium sized tree; bark greyish. Leaves coriaceous, narrowly linear, to 12 × 0.8 cm, apex obtuse, margin entire, base acute, glabrous, green above, paler beneath; midrib prominent below; petiole to 1 cm; Flowers yellow or white, làrge, axillary, solitary, or in a few-flowered cymes. Drupe yellow; broadly turbinate, four-angled, fleshy, glabrous.

Cultivated in gardens for its showy flowers, rarely as an escape. Fls. & Frs.: Throughout the year. Vern.: Pachha ganneru.

The milky juice and the seeds are deadly poisonous and used for committing suicide. Plant gives an alkaloid 'Thevetin' which is used in heart troubles.

Madhavaram RF (KNL), *RVR* 1870; S.K. University (ATP), *NY* 391; Sangam (NLR), *MSR* 1137 (CAL); Ethipothala falls (GTR), *VRK* 3300.

WORLD : Native of America; widely naturalized elsewhere in the tropics.

## **CERBERA** Linnaeus

**Cerbera odollana** Gaertn. Fruct. 2: 193. t. 124. 1791; FBI 3: 638; *C. manglias* L. Sp. Pl. 208. 1753.

Small to moderate-sized tree; bark whitish-grey, papery; wood grey, spongy; branchlets whorled, stout, marked with leaf scars. Leaves lanceolate, obovate-oblong or oblanceolate, 10–28 × 3–7 cm, glabrous, apex abruptly acuminate, margin entire, base tapered; secondary nerves numerous, slender, horizontal, connected by an intra-marginal nerve; petiole to 3 cm. Flowers white with yellow throat, turning purple, 3.5–4.5 cm long, odorous in terminal, peduncled, paniculate cymes. Fruit green, subglobose, 5–10 cm long, smooth; seeds ovoid, 2–2.5 cm across, smooth, glabrous.

East-coast in salt swamps and backyards, cultivated as an ornamental. Fls.: Mar.–Apr. Frs.: Aug.

Anugam (NLR), *JSG* 20418 (MH).

INDIA : Coasts of Deccan Peninsula.
WORLD : Sri Lanka, Malaysia.

## **ERVATAMIA** (A. de Candolle) Stapf

**Eravatamia divaricata** (L.) Burkill in Rec. Bot. Surv. Ind. 10: 320. 1925, *Nerium divaricatum* L. Sp. Pl. 209. 1753. *Tabernaemontana coronaria* (Jacq.) Willd. Enum. Hort. Berol. 275. 1809; FBI 3: 646. *T. divaricata* (L.) R.Br. ex Roem. & Schult. Syst. 4: 427. 1819. *Eravatamia coronaria* (Jacq.) Stapf in Oliver, Fl. Trop. Africa 4: 127. 1902; Gamble 2: 813 (511).

Small tree. Leaves oblong-lanceolate or elliptic-oblong, unequal, 3–5 cm, glabrous, shining, apex acuminate or caudate, margin entire, base acute; nerve axils glandular; petiole 6.2–10 mm. Flowers white, 3.7–4 cm across; Follicles 3.7–5 cm long, turgidly oblong, spreading and recurved, 3–7-ribbed; seeds with a fleshy red aril.

Very commonly cultivated. Fls.: Throughout the year. Vern.: Nandivardanamu.

INDIA : Cultivated throughout.
WORLD : Throughout tropics.

## HOLARRHENA R. Brown

**Holarrhena pubescens** (Buch.-Ham.) Wall. ex G. Don in Gen. Hist. 4: 78. 1837. *Echites pubescens* Buch.-Ham in Trans. Linn. Soc. London 13: 521. 1821. *Holarrhena antidysenterica* Wall. ex DC. in Prodr. 8: 413. 1844; FBI 3: 644; Gamble 2: 811 (570); Brandis 459; Beddome Fl. Sylv. S. India Anal. Gen. 20. f. 6. 1869.

Deciduous small tree, up to 8 m tall; bark thick, brown, peeling off in irregular flakes; wood white, even-grained. Leaves chartaceous, elliptic-oblong, 8.5–14 × 5.5–9 cm, apex shortly acuminate, margin entire, base rounded; secondary nerves ca. 14 pairs; petiole to 4 mm. Flowers white to yellowish in terminal and axillary, paniculate or corymbose cymes. Follicles distinct, smooth, terete, slender, curved, to 20 cm long; seeds oblong-linear, coma apical, pale brown, testa thin, somewhat brittle, closely adhering.

Common, in deciduous forests and open lands above 250–500 m. The tree is useful as accessory species in clothing the ground and acting as a nurse to more valuable species; it is important for reclothing the waste lands. It coppices well and produces root suckers in abundance, sensitive to frost and drought hardy. Fls.: Apr.–July. Frs.: Throughout the year. Vern.: Kolamukki, Palavareni, Pala chettu, Tedlapala, Palankudu, Kodaga, Palakodise, Istari-pala, Kodisipala.

Stem and root bark are medicinal and used to cure dysentry. Wood is soft, and fairly durable, used for carving into picture- frames, domestic utensils and fancy articles of various kinds; also suitable for pen holders and Mathematical instruments, brush backs etc. Floss from the seeds is used for stuffing pillows.

Tirumala (CTR), *SSR* 13710; Donubai RF (SKLM), *SSR* 15595; Narsapur (MDK), *SSR & CP* 16198; Eturnagaram (WGL), *SSR* 18549; Avagondapalem (ATP), *SSR & KSM* 18567.

INDIA : Throughout.
WORLD : Myanmar.

## **PLUMERIA** Linnaeus

**Plumeria rubra** L. Sp. Pl. 209. 1753. *P. acuminata* R.Br. Ait. Hort. Kew 2. 2: 1789. *P. acutifolia* Poir. in Lam. Encycl. Suppl. 2: 667. 1812; FBI 3: 641; Gamble 2: 821 (577); Brandis 458.

Tree, up to 7 m tall; bark greyish, smooth. Leaves obovate, obovate-oblong, oblanceolate, 20–25 × 7–8 cm, glabrous to pubescent beneath, apex acute, obtuse or shortly acuminate, margin entire, base tapering; secondary nerves 10–15, 1.5 cm apart, conspicuous with marginal vein; petiole to 7 cm. Flowers red or white with yellow notch, fragrant, large in short, umbellately divided cymes with long peduncle. Follicles paired, to 30 × 3.5 cm; seeds flat, winged.

Cultivated as an ornamental in gardens, temples. Fls.: Most part of the year. Frs.: Jan.–Feb. Vern.: Devaganneru, Devakanchan, Nuruvarhaalu, Vaada ganneru. Eng.: Temple or Pagoda tree.

S.K. University (ATP), *NY* 513.

INDIA : Cultivated throughout.
WORLD : Native of tropical America; widely naturalised in the tropics.

## **WRIGHTIA** Robert Brown

1. Flowers greenish or cream coloured; calyx and corolla pubescent; corona scales oblong. Follicles connate throughout, rough with white tubercles. Leaves elliptic-obovate, tomentose, apex caudate-auminate .......... ***W. arborea***
1. Flowers white; calyx and corolla glabrous; corona scales fimbriate. Follicles jointed at apex only, smooth. Leaves elliptic-oblong or oblong-lanceolate, glabrous, apex abruptly acuminate .......... ***W. tinctoria***

**Wrightia arborea** (Dennst.) Mabb. in Taxon 26: 533. 1977. *Periploca arborea* Dennst. Schulssel Hortus Malab. 13. 23 & 25. 1818 & in Forts. Allg. Teutsch. Gaerten. Mag. 3. 22. 41 & 83. 1818. *W. tomentosa* Roem. & Schult. in Syst. Veg. 4: 414. 1819; FBI 3: 653; Gamble 2. 816 (573); Brandis 461.

Small, deciduous tree, up to 10 m tall; bark light grey, corky, with light coloured; wood white, even-grained; branchlets softly white tomentose. Leaves elliptic-obovate, 6.5–15 × 3.8–7.5 cm, slightly pubescent along the nerves above, tomentose beneath, drying dark green, apex acuminate, margin entire, base acute; main lateral nerves 12–15 pairs; petiole to 5 mm. Flowers greenish or cream in short, dense, erect, terminal, corymbose cymes. Fruit cylindrical, pendent; follicles connate throughout, rough with white tubercles, 11.5–27 cm, apically beaked; seeds slender, attenuated at the apex.

Occasional, in deciduous forests at 700–1350 m. It is a moderate light demander, fairly hardy; coppices well. Leaf fall: Feb.–Mar. New foliage:

Apr.–May. Fls.: Apr.–June. Frs.: Dec.–Feb. Vern.: Tella Pala, Tedla Pala, Pala adavi venkudu, Pala bariki.

The leaves are eaten as pot herb; the tender fruit is also edible. The fruit contain floss which is used for stuffing. Wood very light to light and moderately hard, used for turning, carving, domestic utensils, combs, toys, packing cases, pen holders, picture frames etc. The fresh seeds have a high percentage of fertility.

Upper Ahobilam (KNL), *SSR* 14366 & 15201; Kakavada (EG), *SSR* 15225; Appapur (MBNR), *SSR* 15539; Minumuluru (VSKP), *SSR* 15956; Rampur (ADB), *SSR* 19308; Balapalli (CDP), *JSG* 11359 (CAL); Narsapur RF (MDK), *KMS* 7953 (MH); Dorajammu (SKLM), *GVS* 62437 (CAL); Bhupatipur (KMN), *GVS* 20176 (MH).

INDIA : Throughout.
WORLD : Sri Lanka, Myanmar, Thailand, Indo-China, South-China, Malay Peninsula.

**Wrightia tinctoria** (Roxb.) R. Br. Mem. Wern. Nat. Hist. Soc. 1: 774. 1811; FBI 3: 653; Gamble 2: 815 (573); Brandis 461; Beddome Fl. Sylv. S. India t. 241. 1872. *Nerium tinctorium* Roxb. Orient. Repert. 1: 39. cum tab. 1791.

Densely foliaceous, deciduous tree, to 12 m tall; bark pale, scaly, smooth; wood white to lemon-yellow, even-grained; branchlets glabrous. Leaves membranous, elliptic-oblong or oblong lanceolate, 4–12 × 2.4–4.4 cm, glabrous, sparsely pubescent beneath, apex bluntly acuminate, margin entire, base truncate; secondary nerves 8–9 pairs; petiole to 2 mm. Flowers white, fragrant in terminal, lax, trichotomous cymes. Follicles 2, distinct, apically joined, cylindric, curved, 30 × 0.5 cm, green; seeds pointed at the apex with a deciduous coma often more than 2–3.5 cm long at the base.

Common, in open dry deciduous forests and on dry sandy soils, plains to 1500 m. It is moderate light demander and produces root suckers. Leaf fall: Dec.–Jan. New foliage: Mar.–June. Fls.: Apr.–May (hills); July–Aug. (plains). Frs.: Throughout the year. Vern.: Akupala, Doddipala, Teddapala, Ankudu, Palakordsha, Reppala, Chitti Ankudu, Jeddepaala.

The flowers are occasionally used as a vegetable. The leaves are the source of blue dye indigo called Mysore pala Indigo which is used as a substitute for indigo. Leaves are also used as wrapper for bidis. Wood commonly used for carving as frames, spoons, with carved handles, combs, screens.

Kalasamudram (ATP), *SSR* 13748; Simhachalam (VSKP), *SSR* 15564; Eswarmalai RF (ATP), *TP & NY* 828; Kadam RF (ADB), *TP & PVP* 4074; North Dhone RF (KNL), *RVR & TP* 1320; Air strip-Macherla RF (GTR), *VRK* 3533; Balapalle (CDP), *JLE* 14348 (CAL); Kadirinavani palli (NLR),

*MSR* 1217 (CAL); Konda-palle (KSN), *CAB* 8131 (MH); Rachepalle (KMN), *GVS* 22252 (MH); Near Addatigala (EG), *GVS* 68564 (MH).

INDIA : Throughout.
WORLD : Myanmar, Timor.

## LOGANIACEAE
### **STRYCHNOS** Linnaeus

1. Bark grey, smooth. Petiole to 1.2 cm. Flowers 8–12 mm long. Berry orange, 2.5–5 cm dia., seeds numerous .......... ***S. nux-vomica***
1. Bark blackish, cracked. Petiole to 5 mm. Flowers 4–5 mm long. Berry black, 1.3 cm dia., seeds 1 or 2. .......... ***S. potatorum***

**Strychnos nux-vomica** L. Sp. Pl. 189. 1753; FBI 4: 90; Gamble 2: 868 (610); Brandis 473; Beddome Fl. Sylv. S. India t. 243. 1872.

Deciduous tree, up to 15 m tall; bark thin, yellowish grey to blackish grey, smooth, covered with tubercles and containing chlorophyll tissue inside; wood white, turning yellowish grey, close grained, medium- and uneven textured. Leaves coriaceous, elliptic, ovate, orbicular, 3.5–7.8 × 3–6 cm, glabrous, dark green and shining above, apex obtuse or acute, margin entire, base rounded, 5-nerved, transverse nerves irregular, inconspicuous; petiole to 1.2 cm. Flowers greenish-white, fragrant, in terminal compound cymes. Fruit globose, 2.5–5 cm dia., thick-shelled, slightly rough, but shining, orange-red when ripe; seed numerous, discoid, 4 × 2.16 cm, much compressed, ash-grey, with silky hair radiating from the centre, concave on one side and convex on the other.

Common, in deciduous forests to 850 m. Recommended for plantations on laterite rock with clayey loam and alluvial soils. Leaf fall: Dec. New foliage: Feb. Fls.: Mar.–Apr. Frs.: Feb.–Aug. Vern.: Mushti, Mushini, Visha Mushti. Eng.: Snake-wood tree.

Strychnine and brucine obtained from the seeds of which are extremely bitter and most poisonous; it is also useful as an insecticide to kill vermin infection. Leaves are lopped for green manure. Wood is largely employed for axe-handles, plough, shares, cart wheels, cots etc.

Kalasamudram RF (ATP), *SSR* 13756; Tirumala (CTR), *SSR* 15270; Talakona (CTR), *SSR* 15551; Chintala forest (PKSM), *SSR* 16132; Chitvel RF (CDP), *SSR* 16160; Penchalakona (NLR), *KH* 5525; Loyalpalli (GTR), *VRK* 3811; Chinnoor (ADB), *GO* 4498; Bandamacheral (KSN), *AWL* 1201 (DD); Balapalle (CDP), *JLE* 1424 (MH, CAL); Pakhal West (WG), *ANH* 15963 (MH); Chelama (KNL), *JLE* 23709 (MH).

INDIA : West Bengal, Bihar, Maharashtra, Orissa, Central and South India.
WORLD : Sri Lanka, Indio-China, Malaysia, Myanmar, Siam.

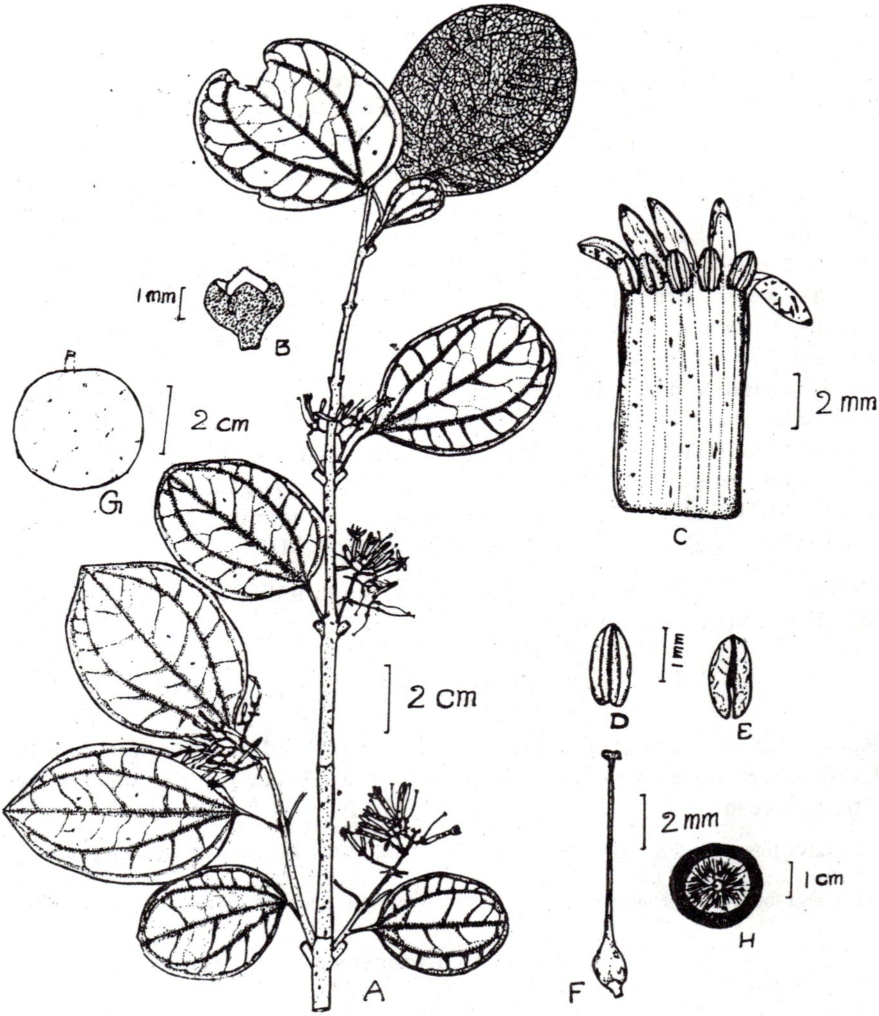


**Fig. 60.** *Strychnos nux-vomica* L.
A. Twig, B. Calyx, C. Corolla spread out, D & E. Anthers, F. Pistil, G. Fruit, H. Seed.

**Strychnos potatorum** L.f. Suppl. Pl. 148. 1781; FBI 4. 90; Gamble 2: 868 (610); Brandis 474.

Deciduous tree, to 15 m tall; bark dark, black or brownish, deeply verticle cracked; wood white, turning yellowish-grey, close-grained, medium-fine and uneven textured, branches fluted. Leaves coriaceous, ovate or elliptic, 4–7 × 2.5–4 cm, glabrous, shining, bright green, apex acute-subacuminate, margin entire, base rounded; 3–5-nerved, transverse nerves parallel, conspicuous; petiole to 5 mm. Flowers whitish in axillary cymes. Berry globose, 1.3 cm dia, black when ripe; seed 1, rarely 2, in whitish pulp, bluntly lenticular, shining with short adpressed silky hairs.

Common, in dry deciduous forests to 800 m. It is a shade bearer and drought resistent, coppices well and produces root suckers. Fls.: Feb.–May. Frs.: Oct.–Mar. Vern.: Induga, Indiba chettu, Chilaka Mushthi, Chilla Ginja, Katakamu, Chilla. Eng.: Cleaning nut tree.

The seed is non-poisonous and after being ground-up is used for cleaning muddy water. Fruit pulp is edible. Wood hard, heavy and durable; commerically not important, but is occasionally used for building huts, carts and agricultural implements.

Kalasamudram (ATP), *SSR* 13756; Diguvametta (PKSM), *SSR* 16000; Penchalakona (NLR), *KH* 5526; North Dhone RF (KNL), *RVR* 2520; Ethipothala (GTR), *VRK* 3906; Satharmat (ADB), *TP & PVP* 4105; Balaparallagutta (CTR), *GVS* 46793 (CAL & MH); Nigadi hill (CDP), *JSG* 15219 (MH); Salur (SKLM), *NPBK* 1066 (CAL); Kadimikonda (EG), *Bourne* 3541 (CAL); Lakshmipuram (VSKP), *CAB* 1729 (MH).

INDIA : West Bengal, Bihar, Central and South India.
WORLD : Sri Lanka, Myanmar.

## CORDIACEAE

Recent taxonomists place *Cordia* and *Ehretia* under a separate family Cordiaceae while in previous Floras, they are placed under the family Boraginaceae.

1. Calyx lobes divided to below the middle, deciduous when fruit setting; style once forked, arms 2. ........ ***Ehretia***
1. Calyx lobes connate more than half, accrescent in fruit; style twice forked, arms 4. .... ***Cordia***

## CORDIA **Linnaeus**

1. Flowers large, orange-scarlet; corolla tube longer than calyx. Cultivated. ........ ***C. sebestena***
1. Flowers small, whitish or yellow; corolla tube not longer than calyx:
    2. Fruiting calyx not ribbed:
        3. Leaves fulvous tomentose beneath, margin entire to slight undulate ...... ***C. wallichii***
        3. Leaves glabrous, sometimes hairy in the nerve axil; margin entire to remotely dentate or sinuate ........ ***C. dichotoma***
    2. Fruiting calyx ribbed:
        4. Leaves sub-opposite ........ ***C. gharaf***
        4. Leaves alternate:
            5. Leaves 10–24 × 10–20 cm ........ ***C. macleodii***
            5. Leaves 2–8 × 1–6.5 cm:
                6. Filaments glabrous. Leaves rough, glabrous or thinly pubescent .... ***C. monoica***
                6. Filaments hairy at the base. Leaves pubescent or tomentose beneath:

7. Flowers in corymbs, yellowish-white. Petiole to 2 cm long. ... ***C. domestica***
7. Flowers in cymose panicles, white. Petiole 2.5–3.2 cm long. ... ***C. evolutior***

**Cordia dichotoma** Forst. f. Prodr. 18. 1786. *C. obliqua* Willd. Phytogr. Lit. 4. 1794; FBI 4: 137. *p.p.*; Gamble 2: 887 (623). *Cordia myxa* auct. plur. (non. Linn. 1753); Cooke 2: 199. 1904.

Deciduous tree, up to 15 m tall; branches drooping; bark thick, greyish brown, with shallow furrows and wrinkles; wood greyish-brown; branchlets glabrous. Leaves elliptic-oblong or orbicular, 4–11 × 0.3–8 cm, glabrous, smooth, sometimes hairy in the nerve axils, apex obtuse or acute, margin entire or remotely dentate or sinuate, base acute or cuneate; 3-nerved, secondary nerves 3–4 pairs from above basal ones; petiole to 3 cm. Flowers white, fragrant in corymbose, glabrous panicles. Drupe pinkish, ovoid or oblong, tipped by short spine like structure, to 2 cm long, shiny, pulp scanty and very viscid, calyx accrescent, flat; seed one, large.

Very common, in open lands and mixed forests. Fls.: Feb.–Mar. Frs.: May–July. Vern.: Bankira, Banka Nakkara, Chinna-Nakkaru, Botgiri, Irid, Pedda irid, Botuku.

Fruits are edible, mucilage is demulcent and for coughs; unripe fruits are pickled by local people. The bark is astringent. Wood is used for agricultural implements and also as fuel. The young tender leaves are used as vegetable.

Vantlamamidi (VSKP), *SSR* 15954; Chintala forest (PKSM), *SSR* 16131; Narsapur (MDK), *SSR* 16200; Akuthotapalli (ATP), *NY* 1260; Satnella river (ADB), *GO & PVP* 4223; Sirnapalli RF (NZB), *BR & GO* 9095; Owk RF (KNL), *RVR* 1474; Kudichintalabailu (MBNR), *TDCK* 15387.

INDIA : Throughout.
WORLD : Pakistan, Sri Lanka, N. Africa.
**Note** : The plant is known under different names in our floras, more commonly as *Cordia myxa* Linn. or *C. obliqua* Willd. C.B. Clarke in Hook. f. FBI (loc. cit.) has treated the two as different species. Cooke (loc. cit.) opines that Willdenow plant *(C. obliqua)* cannot be correctly ascertained as its type specimen is not traceable but he has pointed that the description by Willdenow and the figure accompanying it agrees well with that of *C. myxa*, therefore, the two can be presumed to the same. However, the Indian plant is different from *C. myxa* of Linn. Its earliest valid name is *C. dichotoma* Forst. f.

**Cordia domestica** Roth, Nov. Pl.Sp. 123. 1821; Gamble 2: 888 (624). *C. myxa* var. *domestica* (Roth.) C.B. Clarke in Hook. f. Flora of British India 4: 137. 1883.

Small tree, up to 8 m tall; bark greyish brown, smooth or longitudinally wrinkled; wood greyish-brown, marked with darker bands of porous tissue. Leaves broadly ovate, 2–5 × 1–2.5 cm, fulvous pubescent above, densely beneath, apex obtuse to subacute, margin entire to sparsely dentate, base rounded; secondary nerves 3–5 pairs; petiole to 2 cm. Flowers yellowish white, few in 6 cm long, terminal corymbs. Drupe ellipsoid, yellowish or pinkish yellow, 5 mm long; fruiting calyx ribbed.

Rare, in scrub jungles, 800–1100 m. Fls.: Jan.–Mar. Frs.: Apr.–July. Vern.: Irki, Narkkeri, Urunakkera.

The fruits are eaten by monkeys and birds. The bast fibre is used for cordage. Wood not durable, used for boat building, well-curbs, gun stocks, and also a good fuel.

Guvvalacheruvu RF (CDP), *RVReddy* 8173.

INDIA : Throughout.
WORLD : Myanmar, Sri Lanka.

**Cordia evolutior** (C.B. Clarke) Gamble Fl. Pres. Madras 2: 888 (624). 1923.
*C. fulvosa* Wt. var. *evolutior* C.B. Clarke in Hook. f. FBI 4: 140. 1883.

Medium sized tree; crown low spreading, branches terete; young branches clothed with short fulvous pubescence. Leaves ovate, ovate-oblong or suborbicular, 2.5–8 × 3.2–6.5 cm, glabrous above, fulvous tomentose beneath, apex subacute, margin entire, base rounded or acute; secondary nerves 3–4 pairs; petiole 2.5–3.2 cm. Flowers white, 1.2 cm long, rather dense in short fulvous-tomentose, cymose panicles; stamens exerted. Drupe yellow, ovoid-conical, apiculate; fruiting calyx cup shaped.

Rare in dry forests. Fls.: & Frs.: Jan.–May.

Yerramalais (KNL), *JSG* 33779 (MH) & 17680 (DD); Kondapalli (KSN), *CAB* 8134 (MH).

**Cordia gharaf** (Forssk.) Ehrenb. in Asch. Sitzungsber Ges. Naturf. Freunde Berlin 1879 (4): 46. 1879; Kazmi in J. Arnold Arbor. 51: 140. 1970. *Cornus gharaf* Forssk. Fl. Aegypt.-Arab 95. 1775. *Cordia rothi* Roem. & Schult. in Syst. Veg. 4: 798. 1819; FBI 4: 138; Gamble 2: 888 (624).

Tree, up to 10 m tall; bark grey or brownish-grey with deep longitudinal furrows; wood brown, streaked. Leaves oblanceolate to oblong, 4–7 × 2.8–3.8 cm, glabrous, apex obtuse or subacute, margin entire to slightly undulate, base slightly unequal, acute; secondary nerves ca. 5 pairs; petiole to 3 cm. Flowers cream-coloured in 6 cm long terminal cymes; branches recurved. Drupe yellow or reddish brown, ellipsoid, fruiting calyx recurved.

Rare, in dry forests. Fls.: June–Aug. Fruit pulp sweet and edible, bark is astringent. The wood is used as fuel.

Thumburatheertham TML (CTR), *SSR* 16170; Kuppam (CTR), *KGD* 10329 (MH); Gooty (ATP), *JSG* 16393 (MH); Adoni (KNL), *SL* 33801 (MH).

INDIA : Tamil Nadu.
WORLD : Pakistan, Sri Lanka, Arabia, N. Africa.

**Cordia macleodii** Hook. f. & Thoms. in Journ. Linn. Soc. London 2:128. 1858; FBI 4: 139; Gamble 2: 888 (624); Brandis 479.

Deciduous tree, up to 10 m tall; bark grey, corky; wood light brown when fresh cut, later reddish brown, mottled, straight grained; branchlets coarse and uneven textured, pubescent. Leaves large, rough, ovate, cordate, 10–24 × 10–20 cm, pubescent above when young, grey or tawny woolly beneath, apex obtuse or bluntly acuminate, margin distantly sharply dentate, 3–5-nerved at the base; secondary nerves 4–5 pairs, nerves prominent; petiole to 5 cm. Flowers white, polygamous in dense paniculate, terminal and axillary tomentose cymes. Drupe ovoid, to 2 × 1 cm, yellowish, glabrous, acute, fruiting calyx tomentose, ribbed; 1–2 seeded.

Rare, in dry deciduous forests, 600–1200 m Leaf fall: Dec.–Jan. New foliage: May. Fls.: Mar.–Apr. Frs.: May–Aug. Vern.: Pedda Batana, Iriki, Pedda Batuku.

Wood moderately heavy to heavy, hard. The timber is used for spokes and felloes for construction of heavy timber carts.

Manchippa (NZB), *SSR* 19335; Anantagiri (VSKP), *SSR* 10609; Mudheli RF (NZB), *TP & BR* 6357 & 6358; Gundlabrahmeswaram (KNL), *JLE* 23722 (MH).

INDIA : Deccan and Carnatic.

**Cordia monoica** Roxb. Pl. Cor. 1: 43. t. 48. 1795; FBI 4: 137; Gamble 2: 888 (624); Brandis 479.

Small tree, up to 8 m tall; young branches with brownish tomentum. Leaves ovate or elliptic-ovate, 3.8–8 × 2.1–4.1 cm, scabrous above, smooth and thinly pubescent beneath, apex sub-acute with mucro, margin entire to slightly undulate, base acute or obtuse; secondary nerves ca. 5 pairs, prominent beneath; petiole 0.78–2.5 cm. Flowers light yellow, 5 mm long, dense in branched, lax cymes. Drupe orange-red, ellipsoid, 1 cm long, fruiting-calyx erect; seed one.

Occasional, in deciduous forests. Fls.: Oct.–Dec. Frs.: May. Vern.: Botuku Kuro, Pacha Botuku. Tender fruits used for dysentry. Fruits edible.

Chintala forest (PKSM), *SSR* 16125; Ethipothala falls (GTR), *VRK* 5796; Muchukota RF (ATP), *NY* 976.

INDIA : Peninsula.
WORLD : Sri Lanka.

**Cordia sebestena** L. Sp. Pl. 100. 1753.

Evergreen tree, up to 6 m tall. Leaves thick-coriaceous, broadly elliptic or lanceolate, 5.5–13 × 2.6–9 cm, rough-hairy above, smooth and puberulous beneath, apex acute, margin entire or undulate, base rounded or obtuse; secondary nerves 5–7 pairs, tertiary nerves and reticulations prominent beneath; petiole to 3.5 cm. Flowers scarlet-orange, 4.5 cm long, in few-flowered, 10 cm long cymes. Fruit lanceolate, to 4 cm long.

Occasionally cultivated in gardens, compounds in the plains. Fls.: March–May. Frs.: May–June.

Tirupati (CTR), *SSR* 19398; Andhra University campus (VSKP), *SSR* 19665.

**Cordia wallichii** Don, Gen. Hist. 4: 379. 1837; Gamble 2: 887 (624); Brandis 479; Beddome Fl. Sylv. S. India t. 245. 1872. *C. obliqua* Willd. var. *wallichii* (Don) C.B. Clarke in Hook. f. Fl. Brit. India 4: 137. 1883. *Cordia tomentosa* Wall. in Roxb. Fl. Ind. 2: 239. 1824.

Deciduous tree, up to 12 m tall; bark greyish; branchlets yellowish tomentose. Leaves thick, orbicular-ovate, 7.5–11 × 4.2–11 cm, glabrescent above, densely and softly fulvous tomentose beneath with stellate hairs, apex obtuse to subacute, margin slightly undulate, base truncate; usually 3-nerved from base; secondary nerves 5–6 pairs; petiole to 5.5 cm. Flowers white in terminal, paniculate cymes. Drupe ovoid, with long acumin, yellow.

Occasional, in dry deciduous forests. Fls.: Mar.–May. Frs.: June. Vern.: Peddari Khera. Wood is useful. Fruit is considered to be expectorant, astringent and demulcent.

Gundlabrahmeswaram (KNL), *SSR* 15997; Anantagiri (VSKP), *SSR* 19609; Sirnapally RF (NZB), *BR & GO* 9093.

INDIA : Peninsula.
WORLD : Pakistan.

## EHRETIA P. Brown

1. Leaf margin serrate. Drupe with 2-celled pyrene ........ *E. acuminata*
1. Leaf margin entire. Drupe with 4 pyrenes:
    2. Leaves scabrous beneath, apex acuminate, base attenuate, dark brown to nearly black when dry ........ *E. canarensis*
    2. Leaves not scabrous beneath, apex and base not as above:
        3. Leaves unsymmetric, base cuneate ........ *E. laevis*
        3. Leaves symmetrical, base rounded or obtuse, acute:
            4. Secondary nerves 4–5 pairs, distant and irregular ........ *E. pubescens*
            4. Secondary nerves 6–7 pairs, close and regular ........ *E. aspera*

**Ehretia acuminata** R. Br. Prodr. 497. 1810; FBI 4: 141; Gamble 2: 890 (626); Brandis 481.

Tree to 10 m tall; bark greyish-black, longitudinally fissured; wood light brown, mottled with white specks, rough; branchlets glabrous. Leaves ovate or elliptic, 8.5–16 × 3.5–6.5 cm, glabrous or minutely hispidulous above, apex acute or shortly acuminate, margin sharply serrate, base acute-obtuse; secondary nerves 6–8 pairs, 1.7 cm apart; petiole to 3.2 cm. Flowers white, sessile, fascicled on 9 cm long, terminal panicles. Drupe yellow, ellipsoid, 5 mm across with 2 pyrenes; seeds 4.

Rare, in moist deciduous forests of Visakhapatnam district. Fls.: Jan.–Mar. Frs.: Apr.–May.

The leaves are used as fodder. The ripe fruit is insipid and edible; the unripe fruit is pickled. Wood moderately hard, used for building, agricultural implements scabbards and gun stocks.

Vantlamamidi (VSKP), *SSR* 15944; G. Madugula (VSKP), *SSR* 15962; Pedavalasa (VSKP), *SSR* 19661.

INDIA : Sub tropical Himalaya, Khasi hills; N. Circars.
WORLD : Myanmar.

**Ehretia aspera** Willd. Phytogr. 4. t. 2. 1794; Gamble 2: 891 (626). *E. laevis* Roxb. var. *aspera* (Willd.) C.B. Clarke in Hook. f. FBI 4: 142. 1883.

Small tree, up to 5 m tall; branches brown; bark smooth, light grey to whitish yellow and soft inside; wood tough. Leaves elliptic or ovate, 7–8 × 5–6 cm, hispid-pubescent beneath, apex obtuse or subacute, margin entire, base obtuse; secondary nerves 6–7 pairs, close and regular. Flowers white in terminal rarely axillary, hispid-pubescent corymbose cymes. Drupe dark brown, glabrous, apiculate, pyrenes 4 with a cavity.

Occasional, in dry forests of Kurnool and West Godavari districts. Fls. & Frs.: Jan.–May. Vern.: Tella Juvi, Pogadi. The tree stands moderate shade. It coppices well and produces root-suckers. The wood is used for agricultural implements, but little used except for fuel.

Diguvametta/Nallamalais (KNL), *JLE* 32556 (MH).

INDIA : Throughout the greater part of India.
WORLD : Myanmar.

**Ehretia canarensis** (C.B. Clarke) Gamble Fl. Pres. Madras 2: 891 (626). 1923. *E. laevis* Roxb. var. *canarensis* C.B. Clarke in Hook f. FBI 4: 142. 1883.

Small tree, up to 7 m tall; wood greyish or brownish-white; branchlets brownish-grey, hairy. Leaves broadly oblong or elliptic-oblanceolate, 8–17 × 5–8.2 cm, glabrescent, scabrous beneath, dark brown to blackish when dry, apex acuminate, margin entire, base attenuate; secondary nerves impressive. Flowers white in axillary, unilateral cymes. Drupe orange, fleshy, indehiscent.

Rare in the dry forests at 500–800 m. Fls. & Frs.: Sept.–Apr.

Brahmagundum (CTR), *GVS* 46942 (CAL & MH); Gundlabrahmeswaram. RF (KNL), *JLE* 23835 (MH); Pakhal (WGL), *ANH* 15932 (MH).

**Ehretia laevis** Roxb. Pl. Cor. t. 56. 1796 & Fl. Ind. 2: 34. 1824; FBI 4: 141. Gamble 2: 891 (626); Brandis 481; Beddome Fl. Sylv. S. India t. 246. 1872.

Deciduous tree, up to 10 m tall; bark grey, smooth; wood greyish-white, turning pale yellowish grey, even-grained, fine and even textured. Leaves coriaceous, unsymmetrical, variable in shape, usually elliptic, 4–15.6 × 3–8.2 cm, glabrous above, small tufts of hairs in the axils of the nerves beneath, apex obtusely acuminate, margin entire, base oblique; main lateral nerves 5–6 pairs; petiole to 3.2 cm. Flowers light yellow in lateral, dichotomous scorpioid cymes. Drupe sub-globose, 5–6 mm in dia. with 1–4 pyrenes, black.

Common, in mixed deciduous forests at 800 m. Leaf fall: Dec. New foliage: Apr.–May. Fls.: Feb.–Mar. Frs.: Apr.–May. Vern.: Pagidi chettu, Chilla poku, Tella pisini, Pala Dantam, Pedda pikki, Bodidhi, Poka, Giduguri.

Leaves are used for fodder and the bark is chewed like pan colours the teeth red. Fruits are edible. Wood tough and durable used to limited extent for building and more commonly as fuel; suitable for brush backs, shoe lasts, match boxes and splints.

Pathangal (NZB), *BR & KH* 9648; Mahanandi (KNL), *JLE* 32722 (MH); Guvvalacheruvu (CDP), *R.V. Reddy* 8072; Lambasingi (VSKP), *GVS* 47381 (CAL & MH); Mamandur (CTR), *CECF* 4284 (CAL).

INDIA : Throughout.
WORLD : Tropical and subtropical regions of the old world.

**Ehretia pubescens** Benth. in Royle. Ill. Bot. Himal. 306. 1836; Gamble 2: 891 (626). *E. laevis* Roxb. var. *pubescens* (Benth.) C.B. Clarke in Hook. f. Flora Brit. India 4: 142. 1883.

Small tree; bark greyish, slightly rough; branchlets pubescent. Leaves chartaceous, elliptic-oblong, 5–8 × 1.8–4.5 cm, thinly hirsute above, sericeous beneath, apex obtuse to bluntly or narrowly acute, margin entire to slightly undulate, base acute; secondary nerves ca. 5 pairs; petiole 1–1.5 cm. Flowers white in few-flowered, small corymbs. Drupe red, globose, 5 mm across.

Occasional, in dry deciduous forests to 700 m. Fls.: Aug.–Oct. Frs. Dec.–Feb.

Tirumala (CTR), *SSR* 13761; Batrepalli RF (ATP), *TP & NY* 850; Lankamalai RF (CDP), *SRSR* 14628.

INDIA : Throughout.

## SOLANACEAE

### **SOLANUM** Linnaeus

**Solanum erianthum** D. Don Prodr. Fl. Nepal 96. 1825. *S. verbascifolium auct. non* L., FBI 4: 230; Gamble 2: 936 (657); Brandis 489.

Shrub-small tree, up to 6 m tall; trunk to 5 m tall; bark greyish-brown, lenticellate, smooth; wood light yellow; branchlets fulvous-tomentose. Leaves thick-chartaceous, elliptic-oblong, 11–20 × 4.8–8 cm, slightly pubescent above, densely, fulvous tomentose beneath, apex bluntly acuminate, margin entire, base acute; secondary nerves 8 pairs, reticulations prominent beneath; petiole to 3.5 cm. Flowers white, 1.6 cm across in long peduncled, densely tomentose cymes. Berry yellow, globose, 8 mm dia., tomentose when young.

Occasional in moist forests of East Godavari and Visakhapatnam districts. Fls.: Jan.–June. Frs.: Throughout the year. Vern.: Budama, Rasagadi, Pittu. Wood soft.

Bodhuluru (EG), *SSR* 15237; Galikonda (VSKP), *SSR* 19633.

INDIA : Throughout.
WORLD : Malaya, Indo-China, N. Australia, tropical America.

## BIGNONIACEAE

### KEY 1

1. Leaves simple ........ ***Tecomella***
1. Leaves compound:
   2. Leaves 2–4 pinnate:
      3. Perfect stamens 5; flowers in stout terminal racemes ........ ***Oroxylum***
      3. Perfect stamens 4; flowers in panicles:
         4. Flowers bright blue. Leaflets <2 cm long ........ ***Jacaranda***
         4. Flowers white. Leaflets > 3 cm long:
            5. Flowers very fragrant, panicles 25 cm across; anther cells one perfect, one reduced; corolla tubular ........ ***Millingtonia***
            5. Flowers slightly fragrant, panicles 15 cm across; anther cells two perfect; corolla funnel-shaped. ........ ***Radermachera***
   2. Leaves 1-pinnate:
      6. Leaves digitate ........ ***Tabebuia***
      6. Leaves not digitate:
         7. Calyx spathaceous:
            8. Flowers large, scarlet, spathaceous calyx ribbed, to 5 cm ........ ***Spathodea***
            8. Flowers not large, white; spathaceous calyx not ribbed, to 2 cm ........ ***Dolichandrone***
         7. Calyx tubular or campanulate:
            9. Ovary 1-locular; racemes on long pendulous stalks ........ ***Kigelia***
            9. Ovary 2-locular, sometimes 4-locular due to false septum; racemes or panicles on erect stalk:
               10. Calyx cupular, 5-toothed ........ ***Tecoma***
               10. Calyx campanulate, bilabiate:

11. Capsule flattened, septum 4-angled ........................... ***Heterophragma***
11. Capsule cylindrical or 4-angled, septum terete ............. ***Stereospermum***

## KEY 2 (Fruiting)

1. Fruit indehiscent, unilocular with a sausage fleshy berry ........................................ ***Kigelia***
1. Fruit dehiscent, 2-locular, often 4-locular due to false septum with loculicidal or septicidal capsule:
   2. Fruit usually cylindrical, often subterete, 4-angled:
      3. Septum flat and glossy, valves smooth ................................................ ***Tabebuia***
      3. Septum terete, corky or spongy, usually brittle, valves woody:
         4. Capsule not ribbed, to 90 cm long, valves tuberculate. Seeds multiseriate in each locule .................................................................................. ***Radermachera***
         4. Capsule ribbed, to 70 cm long, valves with white speckles. Seeds biseriate in each locule ...................................................................... ***Stereospermum***
   2. Fruit compressed rarely subcylindrical:
      5. Dehiscence septicidal, dehiscing through both the sutures:
         6. Capsule very broad and large, to 80 cm long, valves smooth, woody; seeds thinly discoid ................................................................................ ***Oroxylum***
         6. Capsule linear, to 35 cm long, valves thick, rough, not so woody; seeds not discoid ........................................................................................ ***Millingtonia***
      5. Dehiscence loculicidal or apparently septicidal rarely through the sutures only:
         7. Capsule 4-locular due to the presence of pseudo- septum:
            8. Valves glabrous ............................................................. ***Dolichandrone***
            8. Valves pubescent ............................................................ ***Heterophragma***
         7. Capsule 2-locular, pseudo-septum absent:
            9. Fruit long, linear, valves not woody:
               10. Fruit subterete ........................................................... ***Tecoma***
               10. Fruit compressed ........................................................ ***Tecomella***
            9. Fruit broad, valves woody
               11. Valves semi-woody, keeled, lanceolate ...................................... ***Spathodea***
               11. Valves very hard, not keeled, oblong to orbicular .................. ***Jacaranda***

## **DOLICHANDRONE** (Fenzl) B. Seemann

### KEY 1 (Vegetative and Fruiting)

1. Leaflets <3 cm long, petiolule 1–5 mm. Capsule compressed, without speckles ... ***D. falcata***
1. Leaflets >5 cm long, petiolule 0.6–2 cm. Capsule sub terete with white speckles:
   2. Seed with corky wings .......................................................... ***D. spathacea***
   2. Seed with membranous wings:
      3. Leaves and rachis glabrous, apex narrowly acuminate, base oblique. Capsule straight ............................................................................ ***D. atrovirens***
      3. Leaves and rachis floccose pubescent, apex rounded to subacute, base rounded or acute. Capsule arcuate ................................................................ ***D. arcuata***

### KEY 2 (Flowering)

1. Corolla long, tube 9–12 cm long ............................................... ***D. spathacea***
1. Corolla short, tube 1.5–5 cm long:
   2. Calyx spathe glabrous ............................................................ ***D. atrovirens***

2. Calyx spathe pubescent:
   3. Corolla tube 1.5 cm long; calyx spathe obtuse with short, stout mucro .... ***D. falcata***
   3. Corolla tube 5 cm long; calyx spathe acute with recurved point ............. ***D. arcuata***

**Dolichandrone arcuata** (Wt..) Clarke in Hook. f. Fl. Brit. India 4: 380. 1884; Gamble 2: 996 (700); Brandis 495; *Spathodea arcuata* Wt.. Icon. Pl. Ind. Orient. t. 1340. 1848; Beddome Fl. Sylv. S. India 169. 1872.

Densely foliaceous, deciduous tree, up to 15 m tall; bark brown, peeling off in flakes; wood white; young branches tomentose. Leaves imparipinnate, to 15 cm long, rachises grooved, tomentose, petiole to 7 cm; leaflets 5, thick, chartaceous elliptic, or ovate, 6–9 × 4.8–6 cm, pubescent on both surfaces, apex obtuse or with obtusely acute, margin entire, base rounded; secondary nerves 5 pairs, flattened above, raised below; petiolule floccose, to 1.5 cm. Flowers white or pale yellow, scented, 7 cm long in terminal racemes. Capsule sub-terete, 4–7.5 × 1.2–2.5 cm, arcuate with speckles; seeds numerous, 0.8 × 1 cm, greyish; wings 1.5 × 0.7 cm.

Common, in dry deciduous forests of Nallamalais, especially in the ravines at 500–700 m. Leaf fall: Jan.–Feb. New foliage: Mar. Red in colour. Fls.: Nov.–Feb. Frs.: Throughout the year.

Appapur (MBNR), *SSR* 15538; Rollapenta to Peccheruvu (KNL), *JLE* 16779, 32567 and 42268 (MH).

INDIA : Peninsula.

**Dolichandrone atrovirens** (Roth) Sprague in Bull. Misc. Inform. 304. 1919; Gamble 2: 996 (700). *Bignonia atrovirens* Heyne ex Roth, Nov. Pl. Sp. 284. 1821. *Dolichandrone crispa* Seem in Journ. Bot. 8. 781. 1870; FBI 4: 379; Brandis 494.

Deciduous tree, up to 15 m tall; bark brown, rough; wood yellowish-brown; young branches velvety pubescent. Leaves imparipinnate to 12 cm long, rachis tomentose, petiole to 7 cm; leaflets 3–7, thin-coriaceous, elliptic, 4.5–10.5 × 2.5–6 cm, glabrous, apex acuminate, margin entire, base obliquely truncate; secondary nerves 6 pairs, prominent on both sides; petiolule to 2 cm. Flowers white in 1–8-flowered corymbose racemes. Capsule subterete, 25–40 cm long, brown, 10-ribbed or streaked, slightly compressed, speckled, smooth, valves woody and coriaceous; seeds subrectangular, flattened, 0.8 × 0.6 cm, wings cream, 1 × 0.8 cm, membranous, situated at the ends.

Common, in the dry deciduous forests at 500–900 m. Fls.: July–Sept. and May–June. Frs.: Throughout the year. Vern.: Oddi, Niru oddi, Neeruddi, Niru uddi, Vanka Niroddi. Wood useful.

Tirumala hills (CTR), *SSR* 13741; Chitvel RF (CDP), *SSR* 15147; Narsapur RF (MDK), *BR* 11985; Kalasamudramu RF (ATP), *NY* 1194; North Dhone RF (KNL), *TP & RVR* 1319; Rangammagudem (RR), *MSM* 10316; Guvvalacheruvu (CDP), *RVReddy* 8432.

INDIA : Peninsula.

**Dolichandrone falcata** (Wall. ex DC.) Seem. Fl. Vill.-Nov. 151. 1880. var. **falcata;** FBI 4: 380; Gamble 2: 976 (760); Brandis 493. *Spathodea falcata* Wall. ex DC. Prodr. 9: 206. 1845; Beddome Fl. Sylv. S. India t. 71. 1870.

Deciduous tree, up to 10 m tall; bark bluish grey, exfoliating in irregular scales; wood white and close-grained; younger parts hoary pubescent. Leaves 1-pinnate to 13 cm, petiole to 3 cm, rachis villous; leaflets 5–7, chartaceous, suborbicular to obovate, glabrescent above, thinly pubescent on the nerves beneath, apex rounded with short acute tip, margin entire, base acute; secondary nerves 4–5 pairs; petiolule to 5 mm. Flowers white in 3–5-flowered, corymbose racemes. Capsule falcately curved, flat, 30–50 × 2–2.5 cm; brown, glabrous; seeds numerous, rectangular, greyish, 1 × 0.8 cm, wings 1.4 × 1 cm, cream.

Common, in dry localities, often on rocky slopes 300–600 m. Leaf fall: Nov.–Dec. New foliage: Apr. Fls.: Apr.–June. Frs.: Feb. Vern.: Chittu-Niruddi, Chittoddi, Oddi.

Wood is hard and used for building and agricultural purposes.

Diguvametta (PKSM), *SSR* 16104; Narsapur RF (MDK), *SSR & CP* 16190; Eturnagaram (WGL), *SSR & KSM* 18545; Rampur (ADB), *SSR* 19307; Nallamalais (KNL), *RVR* 1365 & 3044; Balapalli (CDP), *JSG* 10797 & 11200 (MH); Annavaram (EG), *JSG* 15204 (CAL); Lakshmipuram (VSKP), *CAB* 1760, 1858 (MH); Macherla RF (GTR), *VRK* 3864; Akkannapet (MDK), *BR & CP* 11565; Mothe RF (NZB), *BR & GO* 9066.

INDIA : Deccan Peninsula.

**Dolichandrone falcata** Seem. var. **lawii** (Seem) Haines Bot. Bihar & Orissa 2: 658. 1922; Gamble 2: 996 (700). *Dolichandrone lawii* Seem. in Journ. Bot. 8: 381. 1870; FBI 4: 380; Brandis 493.

Deciduous tree, up to 8 m tall; bark greyish, smooth, exfoliating in irregular scales. Leaves imparipinnate, rachis puberulous, to 4 cm, petiole 2.7–5 cm; leaflets 5–7, ovate or orbicular, 1.1–2 × 0.9–1.7 cm, glabrous above, grey-puberulous beneath, apex very shortly subacute, margin entire, base obtuse or rounded; secondary nerves ca. 4-pairs, curved towards apex; petiole 0 to 1 mm. Flowers white. Pod compressed, less curved, glabrous.

Rare, in dry deciduous forests at 500–700 m of Cuddapah and Chittoor. Vern.: Chittu-Niruddi, Chittuddi, Oddi.

SVU (CTR), *SSR* 19387; Macherla RF (GTR), *VRK* 3864.

**Dolichandrone spathacea** (L.f.) K. Schum. Fl. Kais. Wilh. Land. 123. 1889; Sprague in Kew Bull. 304. 1919; Gamble 2: 996 (700). *Bignonia* *spathacea* L.f. Suppl. Fl. 293. 178. *Dolichandrone rheedii* Seem. J. Bot. 8: 380. 1870; FBI 4: 379; Brandis 494.

Deciduous tree, 10–15 m tall; wood whitish or grey. Leaves imparipinnate, 5–9-foliolate, petiole to 7 cm; leaflets elliptic or rhomboid, 10 cm long, glabrous, black when dry, apex abruptly acuminate, margin entire, base oblique; secondary nerves 6 pairs; petiolule 6 mm. Flowers white, spathaceous, large, few in corymbs; corolla tube 9–12 cm long, slender below, widening above, lobes 5 × 4 cm. Capsule nearly straight, 30–40 × 1.2–2.5 cm; seeds numerous, rectangular with corky wings.

Very rare, along the margins of streams of Sriharikota Island of Nellore district. Fls.: Apr.–June. Frs.: Aug.–Dec.

The seeds are antiseptic. The bark yields a blackish fibre; decoction of the bark is used in preservative treatment of fishing nets. The wood is soft and light and is used for making floats and wooden shoes.

Sriharikota (NLR), *ASR* 378 & 470 (VV).

INDIA : Assam, Bengal, South India, Andaman Island, Malabar, Travancore, Sundarbans, mostly in Coastal swamps.

WORLD : From S.E. Asia and Indonesia to New Caledonia and the Soloman Islands.

## HETEROPHRAGMA A. de Candolle

**Heterophragma quadriloculare** (Roxb.) K. Schum. in Engl. & Prantl, Pfam. 4(3B): 243. 1895. *Bignonia quadriloculare* Roxb. Pl. Cor. 2: 24. 145. 1798. *Heterophragma roxburghii* DC. Prodr. 9: 210. 1845; FBI 4: 381; Gamble 2: 997 (701); Brandis 494.

Deciduous tree, up to 20 m tall; bark blackish or dark-brown, slightly rough; wood yellowish-white; branchlets glabrous. Leaves imparipinnate, whorled (3), rachis to 25 cm long, petiole to 8 cm; leaflets 7–11, thick-coriaceous, elliptic-oblong, 8–13.5 × 4.8–8.2 cm, glabrous, apex obtuse to short acuminate, margin serrulate, base unequal; secondary nerves ca. 5 pairs; petiolule 1–3 mm. Flowers creamish, (pinkish in bud) fragrant, to 4 cm long in densely fulvous tomentose, 15 cm long, cymose panicles. Capsule straight, thick, to 30 cm long.

Occasional, in deciduous forests. Fls.: Jan.–Mar. Vern.: Kapu-Gargu, Kala-Goru, Baray-Kalikod, Bondugu.

Rampur (ADB), *SSR* 19322; Racherla (ADB), *GO & DAM* 5158; Kadam dam (ADB), *GO & DAM* 5191; Kunavaram (EG), *JSG* 15806 (MH).

INDIA : Maharashtra, Western Ghats (Peninsula), Central India.

## JACARANDA A.L. Jussieu

**Jacaranda mimosifolia** D. Don, Bot. Reg. 8: t. 631. 1822; Blatter *et al.* Beautif. Ind. Trees (ed. 2) 93. 1954. *J. ovalifolia* R.Br. in Curtis Bot. Mag. t. 2327. 1822.

Tree, up to 12 m tall; bark brownish, slightly rough. Leaves odd-pinnate to 20 m long, petiole to 3 cm, pinnae 5–10, leaflets 10–15 pairs, coriaceous, 0.7 × 0.3 cm, terminal leaflet to 1.5 × 0.4 cm, apex narrowly acuminate, margin entire, base acute; petiolule to 1 mm. Flowers blue, many, showy in 25 cm long, terminal panicles. Capsule square shaped, 6 cm dia., deep brown, woody, compressed at right angles to the septum; seeds numerous, winged.

Hills above 600 m, and planted in gardens and road sides. Fls.: Mar.–May. Frs.: Throughout the year. Eng. Mimosa leaved Jacaranda.

Tirumala (CTR), *SSR* 15974.

INDIA : Planted throughout for ornamental purpose.
WORLD : Native in N.W. Argentina, Tropical America, Bolivia.

## KIGELIA A. de Candolle

**Kigelia africana** (Lam.) Benth. in Hook. Niger Fl. 463. 1849. *Bignonia africana* Lam. Encycl. 1: 424. 1785. *Kigelia pinnata* DC. Prodr. 9: 247. 1845; Gamble 2: 1000 (705); Brandis 492.

Deciduous, spreading tree, up to 15 m tall; trunk short; bark greyish-brown, rough. Leaves 1-pinnate, to 50 cm, petiole to 7 cm; leaflets 5–7, coriaceous, elliptic-oblong, rarely obovate, glabrous, rough, dark green, apex retuse, rarely emarginate, margin entire to minutely toothed, base obliquely acute, rounded or subcordate; secondary nerves ca. 7 pairs; subsessile. Flowers large, foetid, dark maroon with yellow stripes in 30–70 cm long, pendulous racemes. Fruit pale green, oblong (gourd-like), to 50 × 12 cm, 1–4 at the ends of the pendulous peduncles, indehiscent; seeds many, embedded in pulp.

Planted in gardens and along road sides in plains. It thrives in deep open soils. Fls.: Mar.–May. Frs.: Throughout the year. Leaf fall: Feb. New foliage: Mar.–Apr. Vern.: Yenugula chettu, Vatavrukshamu, Kijili, Nagamalle. Eng.: African Calabash tree, Sausage tree.

Wood is hard and of good quality. The bark is used in rheumatism, dysentry and veneral diseases.

Sunkesula (KNL), *GO* 3357; Vijayapuri South (GTR), *VRK* 6821; Jannaram (ADB), *GO & DAM* 5413; Kotgiri (NZB), *BR & KH* 9643.

INDIA : Cultivated throughout.
WORLD : Native of Tropical Africa.

## **MILLINGTONIA** Linnaeus f.

**Millingtonia hortensis** L.f. Suppl. Pl. 291. 1781; FBI 4: 377; Gamble 2: 994 (669); Brandis 496; Beddome Fl. Sylv. S. India t. 249, 1872; Blatter *et al.,* Beautiful Ind. Trees (ed. 2) 106. 1954. *Bignonia suberosa* Roxb. Pl. Cor. t. 214. 1811 & Fl. Ind. 3: 111. 1852.

Evergreen tree, up to 20 m tall; bark dark yellowish grey, corky; wood soft, yellowish-white; young branches slightly pubescent. Leaves bipinnate, to 30 cm long, pinnae 2–4 pairs; petiole to 8 cm; leaflets chartaceous, ovate, 2–8 × 1–4.5 cm, glabrous and dark green, apex acuminate, margin entire, base often oblique, rarely rounded; secondary nerves 4–5 pairs, petiolule to 8 mm. Flowers numerous, white, fragrant in 15–25 cm long corymbose panicles on pendent branches. Capsule linear, to 35 × 2 cm, compressed, woody; seeds numerous, surrounded by 3 transparent wings on 3 sides.

Cultivated in plains. Common near railway stations, villages and towns. The tree is easily propagated by root suckers. Fls.: Aug.–Feb. Frs.: Mar. Vern.: Akasa Malli, Punnaga Malli, Boddu Malli, Manu Sampangi, Karaku, Kavuk.

Neelavai (ADB), *GO & DAM* 5148; Madakasira town (ATP), *TP & NY* 797; Patha Rajampet Railway gate (NZB), *BR & CPR* 7152.

INDIA : Largely planted in almost all areas.
WORLD : S.E. Asia and Malaysia; probably also native in the Lesser Sunda Isles.

## **OROXYLUM** Ventenat

**Oroxylum indicum** (L.) Benth. ex Kurz. For. Fl. Br. Burma 2. 337. 1877; FBI 4: 378; Gamble 2: 994 (698 ); Brandis 496. *Bignonia indica* L. Sp. Pl. 625. 1753. *Calosanthes indica* (L.) Bl., Bijdr 761. 1826.

Deciduous tree, up to 8 m tall; branched at top; bark light brown, smooth with green juice and often with numerous corky lenticels; wood yellowish-white; branchlets glabrous. Leaves very large, 1.5–2 m long, 2–3 pinnate with opposite pinnae; rachis stout; leaflets 2–4 pairs, coriaceous, ovate or elliptic, 3–10 × 3–7 cm, glabrous, apex acuminate, margin entire, base rounded; secondary nerves 5–7 pairs; petiolule 0.6–1 cm. Flowers large, fleshy, dark pink or purplish, foetid in terminal, stout racemes. Capsule straight, flat, 30–80 × 3–5 cm, brown, thick, valves semi-woody, tapering at both ends, glabrous; seeds numerous, wings white, papery.

Chiefly along streams and ravines and on the cool sides of the hills in the forests at 500–700 m. Leaf fall: Dec.-Feb. New foliage: July. Fls.: July–Aug. Frs.: Dec.–Mar. Vern.: Mokka vepa, Dundilam, Nemali chettu, Pampini.

Wood soft and of no use. Seeds purgative. Bark used for tanning and dyeing.

Upper Ahobilam (KNL), *SSR* 15288; Polavaram (WG), *DCSR* 523 (CAL); Sunkarimetta (VSKP), *NPBK* 10956 (MH); Rollapenta RF (KNL), *JLE* 42244 (MH).

INDIA : Throughout.
WORLD : Tropical Himalaya, Indo-China, Malaysia, West and South China.

## RADERMACHERA Zollinger et Moritzi

**Radermachera xylocarpa** (Roxb.) Schum. in Engl. Nat. Pflanzenfam. 4 (36). 243. 1895; Gamble 2: 999 (702). *Bignonia xylocarpa* Roxb. Fl. Ind 3: 108. 1832; Wt. Icon. t. 1835; Beddome Fl. Sylv. S. India t. 70. 1870. *Stereospermum xylocarpum* (Roxb.) Benth. & Hook. f. Gen. Pl. 2: 1047. 1876; FBI 4; 383.

Large deciduous tree, up to 20 m tall; bark thick, light grey, rather smooth and flaky; wood orange-brown, straight or irregularly inter-locked-grained, medium-coarse and fairly even-textured; branchlets lenticellate. Leaves bipinnate, 40–70 cm long, large, rachis angular; leaflets 7–9, thin-coriaceous, elliptic-oblong, 6–14 × 3–7 cm, glabrous, apex acute or acuminate, entire, base acute or rounded; secondary nerves 5–7 pairs; petiolule 5 mm. Flowers white, tinged with yellow, fragrant in erect, silky tomentose, 15 cm across, corymbose panicles. Capsule cylindric or terete, elongate, 30–70 × 1.5–2 cm, tuberculate, slightly curved, valves woody; seeds subquadrangular, compressed, 0.6 cm, winged on either side.

Occasional, in hilly slopes of dry deciduous forests above 650 m. It is light demander and produces root-suckers. Leaf fall: Feb. New foliage: May–June. Fls.: Apr.–May. Frs.: Apr.–May (following year). Vern.: Naguru, Warawaili, Naga-Dundilam. Eng.: Padri-tree.

The tender fruits are eaten as a vegetable. The plant is credited with antiseptic properties, and a resin is extracted from the wood, which is used in the treatment of skin troubles. Wood very hard, tough, moderately heavy and used for furniture, cart-poles, house-building, carriages and agricultural implements, it is also suitable for panelling, packing cases and sleepers. It is a good fuel.

Gundlabrahmeswaram. (KNL), *SSR* 15996; Tirumala (CTR), *SSR* 16185; Upper Sileru (VSKP), *GVS* 47417 (MH & CAL); Horsely hills (CTR), *JSG* 15753 (MH).

INDIA : Deccan Peninsula, Central India.

## SPATHODEA Palisot de Beauvois

**Spathodea campanulata** P. Beauv. Fl. Oware 1: 47. t. 27. 1805; Gamble 2: 1000 (703); Blatter *et al.*, Beautiful Ind. Trees (ed. 2) 136. 1954.

Tree, up to 15 m tall; bark pale; branchlets rufous tomentose, lenticellate. Leaves imparipinnate; petiole to 8 cm; leaflets 9–19, elliptic-ovate, 4–10 × 2.5–6 cm, dark green, shining, apex acute to narrowly, shortly acuminate, margin entire, base acute with 2–3 fleshy glands. Flowers scarlet, showy, large in long, loose, terminal racemes; calyx leathery, spathaceous. Capsule brown, flat, 20–30 cm long, glabrous, erect.

Cultivated in gardens, road sides, and also grown as avenue tree. Fls.: July. Frs.: Mar. Eng.: Scarlet-bell tree, Fountain tree or squirit tree.

Tirumala gardens (CTR), *DAM* 3691.

INDIA : Cultivated throughout.
WORLD : Native of tropical Africa.

## STEREOSPERMUM Chamisso

### KEY 1

1. Filaments of the stamens with a tuft of woolly hair at the base; panicles glabrous .................. *S. personatum*
1. Filaments of the stamens without a tuft of woolly hair at the base, sometimes villous. Panicles more or less pubescent:
   2. Flowers purple or with yellow streaks in lax panicles .................. *S. colais*
   2. Flowers dull crimson in dense panicles .................. *S. suaveolens*

### KEY 2 (Fruiting)

1. Capsule stout, nearly terete, ribbed, to 20 mm in dia., straight .................. *S. suaveolens*
1. Capsule slender, angled, not ribbed to 13 mm in dia., curved or twisted:
   2. Capsule obscurely angular, linear, often curved .................. *S. colais*
   2. Capsule clearly 4-angular, spirally twisted .................. *S. personatum*

**Stereospermum colais** (Dillw.) Mabb. in Taxon 27: 553. 1978. var. *angustifolium* (Haines) Bennet & Raizada in Ind. Journ. For. 4: 67. 1981. *Bignonia colais* Buch-Ham. ex Dillw. Rev. Ref. Hort. Mal. 28. 1839. *Stereospermum chelonoides* DC. var. *angustifolium* Haines. Deser. List. Trees etc. South circle of C.P. 169. 1916. *S. angustifolium* Haines in Kew. Bull. 121. 1922; Gamble 2: 998 (702).

Deciduous tree, up to 12 m tall; bark grey to brown, fairly smooth; wood hard, grey. Leaves 1-pinnate, to 45 cm long, petiole to 8 cm; leaflets 5–7, oblong, elliptic to lanceolate, 6–16 × 4–6.5 cm, glabrous above, pubescent beneath, apex acuminate, margin entire and villous, base cuneate; petiolule to

1.3 cm. Flowers purple with yellow streaks in 20–40 cm long, terminal panicles. Capsule linear, to 50 × 1.3 cm, woody, obscurely quadrangular.

Occasional, in deciduous forests in north coastal districts and Telangana at 400–800 m. Fls.: Mar. Frs.: May. Wood is used for building furniture, tea-boxes, canoes etc.

Eturnagaram (WGL), *SSR* 18537 & 18539; Minumuluru (VSKP), *SSR* 15955; Manchippa RF (NZB), *BR* 9026; Narsapur forest (MDK), *KMS* 8016 (MH & CAL); Lakshmipuram (VSKP), *CAB* 1742 (MH).

INDIA : Assam, Eastern Himalaya, Chittagang, Orissa, Maharashtra, Konkan, North Kanara.
WORLD : Myanmar.

**Stereospermum personatum** (Hassk.) Chatterjee in Bull. Bot. Soc. Bengal 2: 70. 1948. *Dipterosperma personatum* Hassk. Flora 25. Pt. 2, Beibl. 28. 1842. *S. tetragonum* DC. Prodr. 9: 210. 1845; Gamble 2: 998 (701). *S. chelonoides auct. non* (L.f.) DC; FBI 4: 381; Brandis 495; Beddome Fl. Sylv. S. India t. 72. 1870.

Densely foliaceous, deciduous tree, up to 20 m tall; bark brown to grey, fairly smooth; wood greyish brown with darker patches. Leaves 1-pinnate, 15–20 cm long, petiole to 1 cm; leaflets 9–11, coriaceous, elliptic, 7.5–10 × 2–3 cm, glabrous, apex caudate-acuminate, margin entire to minutely toothed, base acute, main lateral nerves 8 pairs, flattened below, raised above; petiolule to 1 cm, 2 cm for terminal leaflet. Flowers yellow, veined with red in large glabrous panicles. Capsule slender, terete or 4-gonous, 30–40 cm long, curved, valves coriaceous, epicarp thick with speckles; seeds trigonous, or wedge shaped, compressed, inserted in deep notches of the dissepiment.

Common, in dry deciduous forests above 750 m. Leaf fall: Feb.–Mar. New foliage: Apr. Fls.: Apr.–June. Frs.: May (following year). Vern.: Pulkevalamu, Pisul, Isikirasi, Tagada, Kala-kod, Maga vepa.

The root bark and the flowers along with sugar are used in the preparation of cool drink. The juice of the leaves boiled with oil used in diseases of ear and teeth and in rheumatism. The leaves are lopped for fodder. Wood light to heavy, hard, used to a limited extent for dugouts, building and tea-boxes.

Tirumala hills (CTR), *SSR* 13704; Mothugudem (KMM), *SSR* 15950; Avagondapalem (ATP), *SSR & KSM* 19563; Pothurajupenta (KNL), *SSR* 15985; New Srivarimettu (CTR), *SSR* 18593; Krishnapuram (VSKP), *CAB* 1968 (MH); Kadam (ADB), *GO & DAM* 5191.

INDIA : Tropical Himalaya.
WORLD : Myanmar, Thailand, Indo-China, Malaysia, Sri Lanka.

**Stereospermum suaveolens** (Roxb.) DC. in Bibl. Univ. Geneve 17. 124. 1838 et. Prodr. 9: 211. 1845; FBI 4: 382; Gamble 2: 998 (701); Brandis 493. *Bignonia suaveolens* Roxb. Fl. Ind. 3: 104. 1832.

Large deciduous tree, branches many, spreading; bark grey, exfoliating in flat scales; wood yellowish-brown, straight grained, coarse and uneven-textured; branchlets glabrous, lenticellate. Leaves imparipinate, rachis to 20 cm, petiole to 10 cm; leaflets 7–9, broadly elliptic, 7–17 × 3–12 cm, glabrous and rough above, puberulous beneath, apex narrowly to bluntly acute, margin entire, base unequal; secondary nerves 6–8 pairs, reticulations very prominent beneath; petiolule 2–5 mm, stout. Flowers dull purple, fragrant in large, trichotomous, viscidly hairy panicles. Capsule dull brown, straight, cylindric, stout, speckled, to 50 × 2 cm; seeds pale yellowish brown, winged, 2 × 0.8 cm, furrowed across the middle.

Occasional, in semi-evergreen and dry deciduous forests, often found on clayey lands. Fls.: Mar.–Apr. Frs.: Following Jan. Vern.: Kokkesa, Kalagoru, Kuberakshi, Tedla-pala, Paalal. The tree is useful for reclothing bare hill-sides.

The seeds yield dark green, non drying fatty oil. Bark yield dark coloured gum. Leaves are much lopped for fodder. Wood hard but little used; useful for building, constructional work, furniture and carts; it is an excellent fire wood and makes good quality charcoal.

Mothugudem (KMM), *SSR* 15247; Rampur (ADB), *SSR* 19302; Ramanapenta (KNL), *JLE* 23829 (CAL); Chintapalli (VSKP), *GVS* 28142 (MH).

INDIA : Rajaputana, Chota Nagapur, Central India, Indian Peninsula.
WORLD : Myanmar.

## TABEBUIA Goes ex A. de Candolle

1. Flowers golden yellow. Leaves silvery scale on both surfaces ............................ ***T. argentea***
1. Flowers pink, rose or white. Leaves lepidote ........................................................ ***T. rosea***

**Tabebuia argentea** (Bur. & K. Schum) Britt. Sc. Surv. Porto Rico & Virgin Islands. 6: 197. 1925. *Tecoma argentea* Bur. & K. Schum. in Mart. Fl. Bras. 8(2): 322. 1840–1906.

Tree, up to 8 m. Leaves digitate, petiole long; leaflets 5–7, oblong, 5–15 cm long, densely silvery scaly on both surfaces, apex acute to subobtuse, margin entire, base acute; petiolule long. Flowers yellow, 6–6.5 cm long. Capsule grey, lined with black.

Planted in gardens. (Thammanna *et al.* 1996) Fls.: Mar.–Apr.

INDIA : Planted throughout.
WORLD : Native of Paraguay.

**Tabebuia rosea** (Bertol.) DC. Prodr. 9: 215. 1845. *Tecoma rosea* Bertol. Fl. Guatemal. 25. 1840. *Tabebuia pentaphylla* Hemsl. Biol. Cent. Amer. 2: 495. 1882, non L. 1763.

Deciduous tree, bark scaly. Leaves digitate, petiole long; leaflets 5, thinly coriaceous, oblong-obovate, 8–10 × 3–5 cm, apex obtuse, margin enitre, base rounded to acute; petiolule to 4 cm. Flowers rose, pink or white. Capsule thick.

Cultivated. Anantapur town (ATP), *SSR* 19690.

INDIA : Cultivated throughout.
WORLD : S. Mexico to Venezuela, cultivated in the tropics.

## **TECOMA** A.L. Jussieu

**Tecoma stans** (L.) Kunth in H.B.K. Nov. Gen. Pl. 3:144. 1819. *Bignonia stans* L. Sp. Pl. (ed. 2). 871. 1763. *Stenolobium stans* (L.) Seem. Ann. Mag. Nat. Hist. (Ser. 3) 10: 30. 1862; Gamble 2: 1000 (703).

Small tree, 6–10 m tall; bark greyish, rough; wood is yellowish-brown, mottled, close grained; branchlets glabrous. Leaves imparipinnate, rachis to 6 cm tall; petiole to 7 cm; leaflets 7–9, coriaceous, lanceolate, 5.5–11 × 1.1–3 cm, glabrous, apex acuminate, margin dentate, base obtuse; secondary nerves 15–16 pairs, parallel; petiolule 2–3 mm. Flowers golden yellow, large in racemes. Fruit curved, 6–8 cm long.

Cultivated in gardens and temples. It coppices fairly well, and is easily grown from seed or cutting. Fls. & Frs.: Throughout the year. Vern. Suvarna ganneru, Swarna ganneru, Pacha ganneru.

The root is reported to be a powerful diuretic, vermifuge and tonic. The wood is highly prized for furniture, carving and agricultural implements.

S.K. University (ATP), *NY* 538 & *VRK & EC* 3220.

INDIA : Cultivated and selfsown throughout.
WORLD : From Florida through Central and South America to North Argentina, widely cultivated in the tropics.

## **TECOMELLA** Seem

**Tecomella undulata** (Sm.) Seem. in Ann. Mag. Nat. Hist. ser. 3, 10: 30, 1862; Blatt. et Hallb. in J. Bombay Nat. Hist. Soc. 26: 551, 1919. *Bignonia undulata* Sm. Exot. Bot. 1: 35, 1805.*Tecoma undulata* G. Don. Gen. Syst. 4: 223. 1837; FBI 4: 378.

Small tree; trunk mutilated. Leaves thick, coriaceous, oblong, apex obtuse to retuse; midrib prominent, fine reticulately veined. Flowers orange with scarlet specks near to throat, turns dark brown when dried.

Cultivated in gardens (Thammanna *et al.* 1996).

Leaves are used as fodder, the wood is used for toys and other pieces of carving, it takes good lacquer polish.

INDIA : Cultivated.

## VERBENACEAE

1. Flowers in capitate heads. Fruit capsule. Mangrove plants ............ ***Avicennia***
1. Flowers not in heads. Fruit drupaceous. Terrestrial plants:
   2. Flowers in long pendent spikes ............ ***Citharexylum***
   2. Flowers in elongate cymose or racemose panicles — Terrestrial plants:
      3. Flowers large, yellow. Calyx deciduous in fruit. ............ ***Gmelina***
      3. Flowers small, not yellow. Calyx persistent in fruit:
         4. Flowers actinomorphic, all stamens equal in length:
            5. Flowers 4-merous, purple; cymes to 8 cm long ............ ***Callicarpa***
            5. Flowers 5–6 merous, white; panicle to 30 cm long ............ ***Tectona***
         4. Flowers zygomorphic; stamens didynamous:
            6. Leaves simple. Ovary 2-ovuled ............ ***Premna***
            6. Leaves digitate. Ovary 4-ovuled ............ ***Vitex***

### **AVICENNIA** Linnaeus

**Avicennia officinalis** L. Sp. Pl. 110. 1753; FBI 4. 606; Gamble 2: 1105 (774); Brandis 514.

Evergreen, small to medium-sized tree; crown hemispherical; bark greyish-brown, smooth; wood brown or grey, coarse-grained; branchlets white-tomentose. Leaves thick-chartaceous, obovate or elliptic-oblong, 7–10.5 × 4–5 cm, glabrous and shining above, minutely and closely pubescent beneath, apex obtuse or rounded, margin entire, base attenuate; secondary nerves 10–12, parallel, forming on intramarginal vein, intersecondary nerves numerous, conspicuous beneath. Flowers yellow in axillary and terminal, peduncled, 3-chotomous panicles. Fruit greenish, ovoid, 2.5–3.5 cm long, wrinkled; hypocotyl densely fulvous hairy.

In the mangrove swamps of coast. Fls.: June–July. Frs.: Aug. Vern.: Madha. Eng.: White Mangroove.

Wood hard, used only for fuel and also used for cheap beams and door frames, rice pounders and in oil mills. The bark is astringent. A green, bitter and somewhat aromatic resin which oozes from the bark is used in medicine.

Nizampatnam (GTR), *VRK* 6905.

The flesh of unripe fruits is used for healing skin lesions of small pox. The leaves are rich in protein and are used as cattle fodder.

DISTRIBUTION : Sri Lanka, India to shores of Indian Pacific oceans.

### **CALLICARPA** Linnaeus

1. Branchlets and undersurface of leaves softly fulvous, yellow tomentose. Leaves ovate, apex shortly acuminate, base truncate; secondary nerves to 1.5 cm apart. Calyx shortly 4-lobed ............ ***C. tomentosa***

1. Branchlets and undersurface of leaves greyish tomentose. Leaves elliptic-lanceolate, apex narrowly acuminate, base acute; secondary nerves to 3.5 cm apart. Calyx sub-truncate. ...... .................................................................................................................. ***C. arborea***

**Callicarpa arborea** Roxb. Fl. Ind. 1. 390. 1832; FBI 4.567; Gamble 2: 1092 (764); Brandis 511.

Evergreen tree, up to 14 m tall; crown spreading; bark light coloured, rough; wood light brownish-white, even-grained; branchlets densely tomentose. Leaves chartaceous, ovate or elliptic-lanceolate, 15–22 × 8–10.5 cm, glabrous above, close, hard, finely stellate pale tomentose beneath, apex obtusely acuminate, margin slightly denticulate; secondary nerves 10–12 pairs; petiole to 5.5 cm. Flowers small, lilac-purple in dichotomous corymbose cymes. Drupes purple, 4 mm dia., rough, with persistent calyx.

Common on the slopes of moist deciduous forest of Visakhapatnam district at 800–1200 m. Fls.: Oct. and Apr.–June. Frs.: Aug.–Nov. and Dec.–Jan. Vern.: Gandagummadi, Sanka Gochi. Wood only used for fuel, charcoal and at best it might produce rough boarding.

Anantagiri-Galikonda (VSKP), *SSR* 7367; Borra caves (VSKP), *DAM* 5977; Cherukonda (VSKP), *GVS* 2159 (MH).

INIDA : Eastern sub-Himalayan tract.
WORLD : Nepal.

**Callicarpa tomentosa** (L.) Murray. Syst. Veg. ed 13: 130. 1774. *Tomex tomentosa* L.Sp. Pl. 118. 1753. *Callicarpa lanata* L. Mant. Pl. 331. 1771; FBI 4: 567; Gamble 2. 1092 (764); Brandis 512.

Evergreen tree, up to 12 m tall; bark grey, rough and corky; wood pale brown, smooth; branchlets with a thick grey or fulvous stellate hairy, young branches stout. Leaves coriaceous, crowded towards the end of branches, broadly elliptic-ovate, 17–30 × 10–20 cm, glabrescent and shining above, dense stellate tomentose below, apex acuminate, margin entire or slightly toothed, base truncate; secondary nerves 8–10 pairs, impressed above, very prominent beneath; petiole to 4 cm. Flowers purple in axillary, divaricately branched cymes with stellate hairs. Drupe globose, 4 mm across, black, smooth, shining.

Rare, in dry deciduous forests of Srikakulam district, above 800 m. Fls.: Mar.–Apr. Frs.: Throughout the year. Wood moderately hard, used for carving and legs of cots. Leaves boiled in milk, used as a wash for apthae of the mouth.

Kusimi RF (SKLM), *SSR* 15598; Donubai (SKLM), *GVS* 62428 (MH): Anantagiri (VSKP), *GVS* 19558; 44267 (MH).

INDIA : West side of Peninsula from Bombay southwards.
WORLD : Sri Lanka.

## **CITHAREXYLUM** Linnaeus

**Citharexylum subserratum** Sw. Nov. Gen. Pl. 091. 1788.*C. fruticosum* L. var. *subserratum* (Sw.) Mold. in Phytologia 1. 17. 1935 & 6: 383. 1958.

Small to medium sized tree; branchlets glabrous. Leaves coriaceous, narrowly elliptic, 9–18 × 2.6–6.6 cm, glabrous, apex obtuse to shortly acuminate, margin entire, sometime serrulate, base attenuate; secondary nerves 6–8 pairs, to 2.8 cm apart; petiole to 1 cm. Flowers white in slender to 35 cm long spikes.

Cultivated in gardens. Fls.: July–Sept. Vern.: Ananta Padma Lakshmi chettu, Maryada jati.

Tirumala hills (CTR), *DAM* 10116; Yedapally (NZB), *BR & KH* 6468; Near Hussenapuram (KNL), *RVR* 3131.

WORLD : Argentina northwards to Southern USA.

## **GMELINA** Linnaeus

1. Armed, small trees. Leaves small, base acute, glabrous beneath. Flowers in racemes; calyx-teeth 5. Drupe obovoid ........ ***G. asiatica***
1. Unarmed, medium-sized trees. Leaves large, base cordate-decurrent, fulvous tomentose beneath. Flowers in panicles, calyx teeth 4. Drupe ovoid. ........ ***G. arborea***

**Gmelina arborea** Roxb. Pl. Cor. t. 246. 1815 & Fl. Ind. 3: 84. 1832; FBI 4: 581; Gamble 2: 1097 (768); Brandis 509; Beddome Fl. Sylv. S. India t. 253. 1872.

Deciduous tree, up to 15 m tall; bark yellowish-grey, rather corky; wood greyish-white, glossy, smooth and even-grained, medium coarse-textured; young branches fulvous tomentose. Leaves sub-coriaceous, broadly ovate-cordate, 3–20 × 3.6–15.5, glabrous above when mature, stellately fulvous tomentose beneath, apex acuminate, margin entire, base cordate; secondary nerves 5–6 pairs; petiole to 10.5 cm. Flowers brownish-yellow in terminal and axillary panicles. Drupe ovoid, 2.5 × 1 cm, orange-yellow when ripe, fleshy, smooth.

Common, in moist deciduous forests at 400–600 m. It coppices well. Leaf fall: Nov.–Dec. New foliage: Feb.–Mar. Fls.: Feb.–Apr. Frs.: May–June. Vern.: Gamar tek, Gummadi, Gummadia teku, Pasini, Pedda Gumudu. Eng.: Gumhur.

Whole plant is used as antidote for snake bite and scorpion sting. The fruit is said to be sweetish-bitter and cooling; root is bitter, tonic and stomachic and useful in fever and indigestion. Wood is very light, and it is one of the best and most reliable timbers, used for planking, furniture, panels of doors, carriages, decks of boards, dug outs, grain-measures, carving, musical instruments, clogs, foot board-bottoms, roofs and bent parts, panelling of upper class, railway carriages etc.

Gundlabrahmeswaram. (KNL), *SSR* 15527; Simhachalam (VSKP), *SSR* 15562; Penchalakona (NLR), *SSR* 193493.

WORLD : Nepal to Bhutan, Sri Lanka, Phillippines.

**Gmelia asiatica** L. Sp. Pl. 626. 1753; FBI 4: 582; Gamble 2: 1098 (768); Brandis 509.

Armed shrub to tree; bark brownish-white, thin; wood grey. Leaves chartaceous, elliptic, 1.6–3.5 × 1.4–2.5 cm, glabrous, apex obtuse to subacute, margin entire, base acute; secondary nerves 1–3 pairs, impressed below; petiole to 1.4 cm. Flowers yellow, 2.8 cm long in axillary and terminal racemes. Drupe yellow, obovate, 1.8 cm long; seeds 2.

Common, in plains and scrub jungles. Fls.: May–Aug. Frs.: Nov.–Feb. Vern.: Kavva Gummudu, Chiru Gummudu, Salla Gummadu, Chiru Nelli, Nela Gumudu.

Fruit paste applied on head to remove dandruff and also on wounds. Wood is used for musical instruments.

Tirumala (CTR), *SSR* 13772; Simhachalam (VSKP), *SSR* 15567; Taduku RF (CDP), *DAM* 5938, Addatigala (EG), *HS* 20922 (BLAT); Kammavaripalli (KNL), *GO & PVP* 3384; Diguvametta (PKSM), *JLE* 42163 (MH).

INDIA : Peninsula.
WORLD : Sri Lanka, Bangladesh.

## **PREMNA** Linnaeus

1. Calyx shortly 5-lobed. Leaves with tawny, yellow stellate tomentum ............ ***P. tomentosa***
1. Calyx 2-lipped. Leaves with softly, greyish pubescence:
   2. Calyx conspicuously 2-lipped ............ ***P. latifolia***
   2. Calyx obscurely 2-lipped:
      3. Leaves glandular, apex acute to shortly acuminate, densely tomentose beneath, petiole 3–10 cm ............ ***P. hamiltonii***
      3. Leaves eglandular, apex obtuse, glabrous or slightly villous beneath, petiole to 1.2 cm ............ ***P. serratifolia***

**Premna hamiltonii** J.L. Ellis in Bull. Bot. Surv. India 22: 193. 1982. *P. flavescens* Buch.-Ham. ex Hook. f. Flora of British India 4: 578. 1885 (*auct. non* Juss.); Gamble 2: 1096 (767); Brandis 511.

Evergreen tree, up to 10 m tall; bark pale, smooth; wood pale with yellow or brown streaks; branchlets striate, pubescent. Leaves ovate or elliptic or ovate-oblong, 8–25 × 4.5–13.5 cm, yellow gland dotted, sparsely puberulous above, densely tomentose beneath, apex acute to shortly acuminate, margin coarsely toothed, base obtuse or subcordate; secondary nerves 7-pairs, conspicuous beneath, tertiaries subparallel, prominent; petiole 3–10 cm. Flowers pale green, fragrant in tomentose or pubescent, 15 cm long, corymbose panicles; bracts filiform. Drupe black, globose, 5 mm long, endocarp of 1 pyrene.

Rare, in moist deciduous forests at 600–1000 m. Fls.: May. Frs.: Aug.–Jan. Vern.: Naurua, Nelli, Pedda Naruga.

Anantagiri (VSKP), *SSR* 19612; Gundlabrahmeswaram. (KNL), *JLE* 16952 (MH); Rollapenta (KNL), *JLE* 42246 (MH); Lambasingi (VSKP), *HS* 20785 & *SKW* 2842 (BLAT).

INDIA : Sikkim terai, Chota Nagapur.
WORLD : Bangladesh.

**Premna latifolia** Roxb. Fl. Ind. 3: 76. 1832; FBI 4: 577; Gamble 2: 1096 (767); Brandis 511.

Tree, up to 10 m tall; bark thin, light-grey or brown, smooth; Leaves chartaceous, ovate to sub-orbicular, 3–8 × 2–5 cm, pubescent on both surfaces, black when drying, apex obtuse to acute, sometimes shortly acuminate, margin entire, base rounded; secondary nerves 3–4 pairs; petiole to 1.7 cm. Flowers small, greenish-white in terminal corymbs. Drupe depressed globose, 7 mm, with subentire persistent calyx.

Common, in foot hills and lower slopes of deciduous forests. Fls.: Apr.–May. Frs.: May–June. Vern.: Takli, Nelli, Pedda Nelli kura, Konda Manga.

Gulamallebad (KNL), *RVR* 1433; Palakonda hills (CDP), *CS* 9354; P.B.V. Palem Coast (EG), *DAM* 6567; Nagarjuna Konda (GTR), *VRK* 6717.

INDIA : Peninsula.
WORLD : Bangladesh.

**Premna serratifolia** L. Mant. Pl. 253. 1771; Gamble 2: 1096 (767). *Premna integrifolia* L. Mant. Pl. 252. 1771; FBI 4: Brandis 510.

Tree, to 6 m tall; bark thin, pale; wood light brown, scented. Leaves thin-coriaceous, elliptic-ovate, 2–4.5 × 1.5–2.3 cm, apex obtuse, margin entire to crenate, base rounded or subcordate. Flowers greenish-white in 8 cm long, terminal corymbs. Drupe black, ca. 4 mm across.

Rare, in plains as well as forests. Fls. with two peaks during Jan.–Apr. and July–Oct. Frs.: Throughout the year.

Kothapatnam (PKSM), *CAB* 7935 (MH & CAL).

INDIA : Karnataka, Tamil Nadu.
WORLD : Sri Lanka, Malaysia.

**Premna tomentosa** Willd. Sp. Pl. 3: 314. 1800; FBI 4: 576; Gamble 2: 1095 (767); Brandis 510; Beddome Fl. Sylv. S. India t. 251. 1872.

Deciduous tree, up to 10 m tall; bark grey or light brown, smooth; wood light brown, smooth and close and even-grained, branchlets with densely, soft, tawny tomentum of stellate hairs. Leaves thick, chartaceous, 2.5–15.5

× 2–9 cm, ovate, tomentose among nerves above, densely tawny-yellow, stellate tomentose, apex acuminate, margin entire, base cordate; secondary nerves 8–10 pairs; petiole to 3.2 cm. Flowers greenish yellow, scented in terminal, corymbose cymes. Drupe sub-globose, 4 mm across, yellowish, puberulous.

Common, in deciduous forests of most districts at 600–1350 m. Leaf fall: Dec.–Jan. New foliage: Mar. Fls.: Mar.–Apr. Frs.: May. Vern.: Naguru, Kokkiti, Narava, Kampu Gummuda.

Wood of good quality but little used. Fruits edible. An extract of the inner bark is stated to be used to arrest diarrhoea, and a decoction of the roots is given in stomachache. The roots yield an aromatic oil which is also used in stomach disorders. Wood is durable, used for house building, furniture, combs, weaving shuttles, turnery, carving and fancy work. It is also suitable for rafters and temporary structures.

Talakona (CTR), *SSR* 13727; Kalasamudramu (ATP), *SSR* 13758; Kusimi (SKLM), *SSR* 15905; Pakhal (WGL), *SSR & KSM* 18520.

INDIA : Peninsula.
WORLD : Bangladesh.

## **TECTONA** Linnaeus f. *nom. cons.*

**Tectona grandis** L.f. Suppl. 151. 1781; FBI 4. 570; Gamble 2: 1092 (765); Brandis 509; Beddome Fl. Sylv. S. India t. 250. 1872.

Large deciduous tree, up to 20 m tall; bark grey or light greyish-brown, fibrous with shallow longitudinal fissures, exfoliating in long, thin, narrow somewhat corky flakes; wood golden yellow, turning brown, scented, and oily to the touch, coarse and uneven textured; branchlets quadrangular, stellatelly tomentose. Leaves broadly elliptic, 15–30 × 9–25 cm, rough above, dense grey tomentose beneath, apex acute or obtuse, margin entire, base rounded or cuneate; secondary nerves 8–9 pairs; petiole to 3 cm. Flowers white in terminal, large, dichotomously branched panicles. Drupe brown, globose, 1.6 cm across, enclosed in the enlarged calyx; exocarp thick, spongy; endocarp bony, thick; seeds numerous, erect, oblong.

Very common, in hilly forests of Nallamalais, Telangana and Northern areas at 500–1200 m. It grows on a great variety of soils, but requires perfect drainage. It is light demander. It is extensively used as a leading timber in view of its unique wood properties. Leaf fall: Nov.-Dec. or early Jan. New foliage: May. Fls.: Apr.–June. Frs.: Nov.–Jan. Vern.: Teku, Pedda teek. Eng.: Teak.

Wood is light to moderately heavy and hard, especially used for doors, windows, flooring, stair cases, wains, cotting, trusses and tie-beams; it also employed for bridges, harbour work as piles and balks in wells and other purposes, where strength and durability are essential. Seeds are reported to be efficaceous in the treatment of stones in bladder.

Rampur (ADB), *SSR* 19315; Jyothi RF (CDP), *DAM* 8307; Bukkapatnam (ATP), *NY* 628; Gullemallaba RF (KNL), *RVR* 1671; Ghanapur RF (NZB), *BR* 7114; Mutaypally (MDK), *CP* 11783; Anantasagar (RR), *MSM* 10407.

INDIA : Peninsula, Archipelago, Central India, Bengal, Assam, Bihar, Orissa, The Andamans.

WORLD : Myanmar, Java, Siam and Malaysia.

## **VITEX** Linnaeus

### KEY 1(Vegetative)

1. Leaves glabrous on both surfaces (except in *V. leucoxylon*, pubescent only along the midrib beneath) leaflets petioluled:
   2. Leaves 3-foliolate, yellow gland-dotted beneath ........ ***V. peduncularis***
   2. Leaves 5(3)-foliolate, eglandular:
      3. Terminal leaflet to 25 cm long , apex long acuminate ........ ***V. quinata***
      3. Terminal leaflet to 15 cm long, apex obtuse or shortly acuminate ....... ***V.leucoxylon***
1. Leaves glabrous above, thinly or densely pubescent or tomentose beneath; leaflets sessile or subsessile.
   4. Small trees. Leaf apex subobtuse, rarely acute ........ ***V. trifolia***
   4. Medium to large trees. Leaf apex acuminate:
      5. Leaflets lanceolate or oblanceolate, thinly pubescent beneath; petiole winged in young state ........ ***V. altissima***
      5. Leaflets elliptic or ovate, fulvous pubescent beneath; petiole not winged ... ***V. pinnata***

### KEY 2 (Flowering)

1. Inflorescence axillary:
   2. Flowers in long, lax panicles; corolla glabrous with yellow palate. Drupe ovoid, or globose, 5 mm long ........ ***V. peduncularis***
   2. Flowers in short, divaricate, corymbose cymes; corolla with long purple hairs on the palate. Drupe ellipsoid, to 2.5 cm long. ........ ***V. leucoxylon***
1. Inflorescence terminal:
   3. Flowers white or cream coloured, tinged with blue or violet:
      4. Flowers sessile; rachis grey, pubescent; calyx fulvous-villous. ........ ***V. altissima***
      4. Flowers clearly pedunculate; rachis brown, puberulous; calyx slightly puberulous. ........ ***V. quinata***
   3. Flowers blue or purple:
      5. Bract minute. Panicle with cymose branches. ........ ***V. trifolia***
      5. Bracts foliaceous, conspicuous. Panicle with short racemose branches. ... ***V. pinnata***

**Vitex altissima** L.f. Suppl. Pl. 294. 1781; FBI 4: 584; Gamble 2: 1102 (772); Brandis 504; Beddome Fl. Sylv. S. India t. 252. 1872. *V. alata* Willd. Ges. Naturf. Freunde Berlin Neue Schriften 4: 203. 1803.

Densely foliaceous, large tree, up to 20 m tall; crown dense; bark yellowish-grey, scaly, fibrous; wood grey to yellowish brown, often streaked with dark brown, heavy, straight or interlocked grained, medium-fine textured, close grained; young shoots villous with soft hairs. Leaves 3–5-foliolate, petiole to 5 cm, slightly winged, leaflets thick, chartaceous, central leaflet obovate, to 15 × 4 cm, lateral ones oblanceolate, 4–9.5 × 1.5–3.4 cm, glabrescent above, thinly pubescent especially on the nerves beneath, apex acuminate, margin entire, base acute; secondary nerves 9–15 pairs; petiolule subsessile. Flowers numerous, in 10–25 cm long, racemose panicles. Drupe globose, 0.5 cm across.

Common, in moist deciduous forests along streams, less on the slopes. Fls.: June–Sept. Frs.: Ripening by Jan. Vern.: Nemali Adugu, Busichettu, Mayurapau, Gandarupu.

The winged petioles are most conspicuous in young plants and coppice shoots. Wood is very hard, rigid, entirely tough, strong and heavy, very durable. Wood valuable for building construction, furniture, carts, fellows of wheels, and occasionally for sleepers. It is also a good fuel wood.

Talakona (CTR), *SSR* 13728; Tirumala (CTR), *SSR* 13781; Palakonda hills (CDP), *DAM & CS* 7709; Maripakala (EG), *SKW* 1306 (BLAT).

INDIA : Southern & Western India, Deccan, Karnataka.
WORLD : Sri Lanka, Malayasia.

**Vitex leucoxylon** L.f. in Suppl. Pl. 293. 1781; FBI 4: 587; Gamble 2: 1103 (772); Brandis 504. *Wallrothia leucoxylon* (L.f.) Roth, Nov. Pl. Spec. 319. 1821.

Large, deciduous tree, up to 20 m tall; crown spreading, bark grey, smooth; wood dark purple, straight grained, fine-medium textured; young branches slightly puberulous. Leaves 5-foliolate, petiole to 10 cm, leaflets thin, coriaceous, obovate-oblanceolate, 7–15 × 3–4 cm; glabrescent, apex subobtuse, margin entire and toothed, base acute; secondary nerves 8–10 pairs, with close reticulate veins; petiolule 2 cm. Flowers cream coloured in axilllary, corymbiform, dichotomous cymes. Drupe ellipsoid, oblong, 2.5 × 1.3 cm, dark purple, supported on the enlarged flattened calyx, 4-seeded.

It is chiefly distributed along the banks of rivers, streams and tanks of Eastern Ghats up to 700 m. Fls.: Feb.–Apr. Frs.: Throughout the year. Vern.: Luki, Kokks, Konda Vavili, Gaja vavila, Nevaledi, Jinna Koyi.

Wood moderately hard and useful for furniture and carts. The root and bark are astringent; the root is reported to be used as febrifuge. The leaves are smoked for relieving headache and catarrh.

Tirumala (CTR), *SSR* 13722; Appapur RF (MBNR), *SSR* 15536; Chintala forest (PKSM), *SSR* 16123; Kambakkam (CTR), *SSR* 16164; Sidduliah kona (NLR), *SSR* 19361; Sidhout (CDP), *SKW* 5900; Toduku Vagu bank (KNL), *DAM* 5941; Mamandur (VSKP), *GVS* 31950 (MH); Cement Nagar Ghat (KNL), *DAM* 5691.

INDIA : Peninsula.
WORLD : Sri Lanka.

**Vitex peduncularis** Wall. ex. Schauer in DC. Prodr. 11: 687. 1847; FBI 4. 587; Gamble 2: 1103 (772); Brandis 505. *Vitex peduncularis* var. *roxburghiana* C.B. Clarke in Hook. f. Fl. Brit. India 4: 587. 1885.

Tree, up to 20 m tall; bark thick, whitish, exfoliating in irregular flakes; wood reddish-grey, straight or sometimes wavy grained, fine and even textured; young branches cinerous. Leaves 3-foliolate, petiole to 6.2 cm; leaflets submembranous, lanceolate-oblanceolate, 8–14 × 2–4.3 cm, glabrous, apex acuminate, margin entire-coarsely serrate, base acute; petiolule to 1.8 cm. Flowers cream in axillary, lax, dichasial cymes. Drupe black, ovoid or globose, 6 mm across.

Occasional, in the moist deciduous forests, often on the banks of streams 600–1300 m. Fls.: Apr. Frs.: May-July.

Wood hard, heavy to very heavy used for posts, beams, building, wells, sugar-cane crushers, rice-pounders, oars, and carvings. An infusion of the leaves and of the root bark or young stem bark is useful in malaria and black water fever. The bark is also said to be used as an external application to relieve pain in the chest.

Minumuluru (VSKP), *SSR* 15947; Gundlabrahmeswaram. (KNL), *SSR* 16121; Mamandur (CTR), *SKW* 5807 (BLAT); Sidhout (CDP), *SKW* 5901 (BLAT); Maredumilli (EG), *GVS* 24264 (MH).

WORLD : Himalaya (Nepal), Myanmar; Indo-China, Malaysia.

**Vitex pinnata** L. Sp. Pl. 630. 1753; Mold. in Phytologia 5: 61. 1957. *V. pubescens* Vahl, Symb. Bot. 3: 85. 1794; FBI 4: 585; Gamble 2: 1103 (772).

Tree, trunk-fluted, bark greyish, somewhat fissured, flaking in long thin pieces; wood reddish brown or olive brown, close-grained, smooth; branchlets 4-angled with tawny pubescence. Leaves coriaceous, ovate or elliptic, to 16 × 7 cm, pubescent, apex obtuse or acuminate, margin entire, base rounded to cuneate; secondary nerves 6–7 pairs; petiolule 0. Flowers pale blue in terminal, pyramidal panicles; bracts foliaceous. Fruit globose, glabrous.

Occasional, in dry forests 400–800 m. Fls.: May–July. Frs.: Oct.–Dec. Vern.: Nemaliadugu, Rachangi, Busi chettu, Raatangi.

Wood very hard, heavy and durable; employed for various purposes; and is useful for constructional work, furniture, agricultural implements, tool handles and for turnery.

Guvvalacheruvu RF (CDP), *R.V. Reddy* 8417; Addatigala (EG), *HS* 20906 & *SKW* 3900, 3901 (BLAT); Kotaganti Konda (VSKP), *GVS* 44248 (MH); Baruva konda (VSKP), *KCJ* 17135 (MH); Karaka (VSKP), *CAB* 1690 (MH).

**Vitex quinata** (Lour.) F.N. Williams in Bull. Herb. Boiss. II 5: 431. 1905. *Cornutia quinata* Lour. Fl. Cochinch. 887. 1790. *Vitex heterophylla* Roxb. Fl. Ind. 3: 75. 1832 & in Wall. Pl. As. Rar. 3: 15. t. 226. 1832; FBI 4: 585; Brandis 504.

Tree, up to 8 m tall; branchlets 4-angled. Leaves 5(3) foliolate; petiole 6.5–11 cm; leaflets membranous, elliptic-oblong or oblanceolate; 6–24 × 2.3–8.6 cm, (laterals are smaller than terminal one) glabrous, glandular, apex long-acuminate, margin entire or very slightly undulate, base acute to shortly narrowed; secondary nerves 6–11 pairs, reticulations prominent beneath; petiolule 3 mm (basal ones) to 3 cm (terminal one). Flowers cream coloured with bluish throat in 20 cm long, terminal, puberulous panicles. Drupe black, globose or ellipsoid, 5 mm dia., glabrous; seed 1.

Rare, in semi-evergreen forests near streams. Fls.: May-June. Frs.: Sept.-Oct.

Wood hard, heavy and durable; suitable for building, boxes etc.

Rompulu forest (VSKP), *SSR* 19663; Minumuluru (VSKP), *GVS* 28261 & 30040 (MH).

INDIA : Assam, E. Bengal.
WORLD : Java.

**Vitex trifolia** L. Sp. Pl. 638. 1753 (*trifolius*); FBI 4: 583; Gamble 2: 1102 (771); Brandis 504.

Small tree, up to 4 m tall; bark pale-grey, smooth; young branches softly pubescent. Leaves 3-foliolate; leaflets obovate-rhomboid, 4–8 × 2–4 cm, glabrous above, very densely white tomentose, apex subobtuse or acute, margin entire, base cuneate, (acute in lateral ones); petiolule 0. Flowers bluish, white spotted without, in terminal panicles. Drupe purplish-black, enclosed by the base of enlarged calyx, globose, 6 mm dia.

Occasional, in plains and dry deciduous forests. It is very hardy and is met with even in poor soils and dry localities. Fls.: & Frs.: Throughout the year. Vern.: Chiru vavili, Nochali.

The leaves are very much used as medicine, chiefly in poultice for rheumatic pains, inflammation, sprains etc. The powdered leaves and their infusion are given in fever. The flowers are prescribed with honey in fevers accompanied with vomiting and severe thirst.

Andhra University campus (VSKP), *DAM* 6338; Vijayawada (KSN), *SKW* 7180, 7181 (BLAT).

INDIA : Throughout.
WORLD : Sri Lanka, China, Japan, Malaysia, Australia.

## LAURACEAE

1. Inflorescence dense, clustered or umbelliform; all anthers introrse:
   2. Fertile stamens 6. Leaves 3-plinerved ............ ***Neolitsea***
   2. Fertile stamens 9 or 12:
      3. Bracts persistent until anthesis. Fruits subtended by unchanged calyx tube. Leaves penninerved, non-subverticillate ............ ***Litsea***
      3. Bracts caducous before anthesis. Fruits subtended by cyathoform calyx tube appandaged by residue of lobes. Leaves 3-plinerved, usually verticillate ............ ***Actinodaphne***
1. Inflorescence laxly cymose or panicle; inner most whorl of perfect stamens with extrorse anthers:
   4. Anthers 2-celled; perianth deciduous ............ ***Beilschmiedia***
   4. Anthers 4-celled; perianth persistent:
      5. Leaves 3-plinerved. Fruiting perianth enlarged into a cupule embracing the fruit ............ ***Cinnamomum***
      5. Leaves penninerved. Fruiting perianth hardly enlarged:
         6. Perianth lobes spreading or reflexed, not hardened ............ ***Persea***
         6. Perianth lobes erect, hardened ............ ***Phoebe***

### ACTINODAPHNE Nees

**Actinodaphne madraspatana** Bedd ex Hook. f. Fl. Brit. India 5: 149. 1886. *A. hookeri* Beddome Fl. Sylv. S. India t. 296.1872; Gamble 2: 1231 (845).

Evergreen tree, up to 12 m tall; branchlets pubescent. Leaves chartaceous, elliptic or oblong-lanceolate, 13–21 × 4.7–6.7 cm, glabrous, shining above, galucous below, apex acute, margin entire, base acute, lowest pair of nerves just above the base; secondary nerves 7–10 pairs; petiole to 2 cm. Flowers yellow, dioecious; male flowers ca. 8 in clusters; female flowers in umbellate cymes. Fruit ellipsoid.

Common, in Seshachalam hills of Cuddapah and Chittoor districts at 700–900 m. Fls.: Aug.–Sept. Frs.: Feb. Vern.: Kovangutti.

Tirumala hills (CTR), *SSR* 13783; *KJ & SSR* 12561 & 12562; Dharmagiri, Tirumala (CTR), *DR* 1236 (MH); Talakona (CTR), *DR* 471 (MH); Mangalipenta (CDP), *JSG* 1924 (MH); Tirumala hills (CTR), *SKW* 5635 (BLAT).

It is endemic to Cuddapah and Chittoor districts.

### BEILCHMIEDIA Nees

1. Leaf apex and base obtuse. Fruit ellipsoid ............ ***B. sikkimensis***
1. Leaf apex acuminate, base cuneate. Fruit oblong-obovoid ............ ***B. roxburghiana***

**Beilchmiedia roxburghiana** Nees in Wall. Pl. As. Rar. 2: 69. 1831; FBI 5: 121; Gamble 2: 1221 (855); Brandis 528; Beddome Fl. Sylv. S. India t. 263. 1872.

Evergreen, moderate-sized tree; wood white, even-grained; branchlets blackish-brown. Leaves chartaceous, lanceolate or elliptic-oblong, 10–12 × 4–8.7 cm, apex obtuse or bluntly acuminate, margin entire, base rounded; triplinerved, tertiaries parallel; petiole to 2 cm. Flowers whitish in long-pedunculate racemes. Berry oblong, 3.3 cm long.

Rare, and cultivated. Fls.: and Frs.: Apr.–May.

Tirumala hills (CTR), *KJ & SSR* 12557.

INDIA : Assam, Khasi hilla, Manipur, Orissa.
WORLD : Myanmar.

**Beilschmiedia sikkimensis** King ex Hook. f. Fl. Brit. Ind. 5: 122. 1886; Brandis 529.

Tree, branchlets whitish. Leaves coriaceous, lanceolate or elliptic-oblong, 7.5 × 2.5–6 cm, glabrous, apex acuminate, margin entire, base obtuse; secondary nerves 10–12 pairs, irregular; petiole to 8 mm. Flowers in axillary, tomentose racemes. Berry ellipsoid, 2.7 × 1.4 cm.

Occasional, in forests of East Godavari district. Fls. & Frs.: Apr.–May.

From Ramavaram to Nulakamaddi (EG), *GVS* 27449 (MH).

WORLD : Sri Lanka, Myanmar.

## CINNAMOMUM Schaefter

1. Leaf apex abruptly acuminate, petiole long .................... ***C. camphora***
1. Leaf apex caudate-acuminate, petiole short .................... ***C. caudatum***

**Cinnamomum camphora** (L.) J.S. Presl, Prir. Rostlin. 2: 47. t. 8. 1825; FBI 5: 134; Gamble 2: 1225 (858); Brandis 534.

Evergreen tree, up to 10 m tall; much branched with a strong persistent camphoraceous odours, wood yellowish brown. Leaves ovate or elliptic-lanceolate, 3–10 × 1.5–4 cm, glabrous, dark green above, glaucous-waxy beneath, apex abruptly acuminate, margin entire, base acute; 3–5-nerved, bullate at nerve-axils; petiole to 2.5 cm, reddish. Flowers whitish, very sweet scented in pseudo-terminal panicles, to 7 cm. Berry black, subglobose.

Cultivated, in Agency areas of Visakhapatnam. Fls.: Nov. Vern.: Gamburamu, Karpuramu, Kappuramu, Himavaluka.

The leaves have a pleasant smell when crushed. The plant is considered sedative, antiseptic, diaphoretic, anthelmintic and is useful in insecticidal properties. The oil obtained by the distillation of wood gives a higher yield of camphor and safrole, a highly valued constituent used in the preparation of the expensive perfumes. Wood is moderately hard, it is too expensive.

Araku valley (VSKP), *GVS* 22665 (MH).

INDIA : Widely cultivated.
WORLD : Native of Japan, China and Taiwan.

**Cinnamomum caudatum** Nees in Wall. As. Rar. 2. 76. 1831; FBI 5: 134; Gamble 2: 1225 (358); Brandis 534.

Evergreen tree, 3–6 m tall. Leaves ovate or obricular, 6–10 × 3–5 cm, glabrous, apex caudate-acuminate, margin entire, base acute or rounded; 3–5 ribbed; petiole short. Flowers small in short, axillary, few to 2-flowered, pubescent cymes. Fruit globose or ellipsoid, fleshy; persistent perianth lobes reflexed.

Rare, in dense moist forests of Visakhapatnam district at above 1000 m.

Paderu (VSKP), *GVS* 42594 (MH); Anantagiri (VSKP), *NPBK* 10998 (MH); Gudem (VSKP), *SKW* 2055 (BLAT).

## **LITSEA** Lamark *nom. cons.*

1. Perianth incomplete or aborted:
   2. Leaves thin, glabrous beneath or nearly so. Umbels corymbose or racemose, few-flowered, puberulous ........ ***L. glutinosa***
   2. Leaves thick, fulvous tomentose beneath. Umbels solitary, many-flowered, dense tomentose ........ ***L. deccanensis***
1. Perianth lobes complete, segments 6:
   3. Leaves subcoriaceous, glabrous and glaucous beneath, base narrowed ........ ***L. laeta***
   3. Leaves chartaceous, tomentose, not glaucous beneath, base rounded ...... ***L. monopetala***

**Litsea deccanensis** Gamble Fl. Pres. Madras 2: 1235 (864). 1925. *Tetranthera tomentosa* Nees in Wall. Pl. As. Rar. 2: 65. 1830. *p.p. Litsea tomentosa* Heyne ex Wall. Pl. As. Rar. 2: 65. 1861; non Bl., 1826; FBI 5: 157; Brandis · 536.

Evergreen tree, up to 12 m tall; bark thin, smooth; wood yellowish, wavy-grained, branchlets stringy, fulvous tomentose. Leaves coriaceous, crowded towards the end of branches, elliptic-oblong, 10–22 × 5–7 cm, glabrous above, softly and densely tomentose, apex acute, margin entire, base obtuse-acute; secondary nerves to 14 pairs, conspicuous beneath; petiole to 2 cm, tomentose. Flowers in globose, pedunculate, many-flowered umbels. Berry globose, 1 cm across, black.

Rare, in the hills of Cuddapah and Chittoor districts at 700-800 m. Fls.: Oct.–Dec. Frs.: Mar. Vern.: Nara Mamidi. Wood is heavy and difficult to work; it is suitable for rafters, wall plates and tea boxes.

Horsely hills (CTR), *SSR* 13796; Tirumala hills (CTR), *KJ & SSR* 12591; Horsely hills (CTR), *JSG* 20984 (DD).

WORLD : Myanmar and Sri Lanka.

**Litsea glutinosa** (Lour.) C.B. Robinson in Philipp. Journ. Sci. 6: 321. 1911. *Sebifera glutinosa* Lour. Fl. Cochin. 2: 783. 1790. *Litsea chinensis* Lamk.

Encycl. Meth. 3. 574. 1789; Gamble 2: 1235 (864). *Litsea sebifera* Pers. Syn. 2:4 1807; FBI 5: 157; Brandis 536.

Evergreen tree, up to 12 m tall; bark brown, somewhat corky; wood greyish-brown, fairly straight grained, medium and even textured; branchlets pubescent. Leaves coriaceous, very variable in shape elliptic, obovate, or elliptic-oblong, glabrous, apex obtuse or acute, margin entire, base acute; secondary nerves 7–12 pairs; petiole 1–3.5 cm. Flowers white or yellowish, small; umblels compound, few-flowered. Fruit globose, black, 1.2 × 0.8 cm.

Occasional, in the hill of moist deciduous forests at 650–1100 cm. It stands a fair amount of shade and grows fast. Fls.: June. Frs.: Sept. Vern.: Meda Nara-Naliki, Narva alagi, Kanuju, Burra Manu, Kanuja Nalike.

The leaves are mucilagenous and considered emotient and antispasmodic and also used as cattle fodder. The fruit is edible. The seeds yield 35% of a fat with a strong aromatic odour and disagreeable taste. The root is sweetish bitter, astringent and tonic. Wood moderately hard and heavy, used for house building, furniture, oars and agricultural implements. Bark is demulcent, astringent and it is having brown colouring matter.

Tirumala hills (CTR), *SSR* 13715, 13773 ; *KJ & SSR* 12597; Beemaram RF (ADB), *TP & GO* 5462; Talakona (CTR), *GVS* 91464 (MH); Mogalipenta (CDP), *JSG* 21276 (DD); Tadepalli (EG), *NRR & TRS* 1988 (MH); Krishnapuram (VSKP), *CAB* 44673 (MH).

INDIA : Central and Southern regions.
WORLD : South China, Malaysia and tropical Australia, Sri Lanka, Myanmar.

**Litsea laeta** (Nees) Hook. f. Fl. Brit. Ind. 5: 169. 1886; Gamble 2: 1236 (865); Brandis 538. *Tetranthera laeta* Nees in Wall. Pl. As. Rar. 2: 67. 1831.

Small tree; branchlets slender, pubescent. Leaves sub-coriaceous, elliptic-oblong or lanceolate, 15–25 cm long, green above, glaucous beneath, apex acute, margin entire, base narrowed; secondary nerves 12 pairs, reticulations faint. Flowers small, sessile in short peduncled, puberulous umbels. Berry globose, seated on the perianth tube.

Rampa hills of East Godavari at about 800 m (Gamble, 1925).

INDIA : Eastern Himalaya, Sikkim, Assam.
WORLD : Bhutan.

**Litsea monopetala** (Roxb.) Pers. Syn. Pl. 2: 4. 1806. *Tetranthera monopetala* Roxb. Pl. Cor. 2: 26. t. 148. 1798. *Litsea polyantha* Juss. in Ann. Mus. Hist. Paris B: 211. 1805; FBI 5: 162; Gamble 2: 1237 (866); Brandis 536.

Evergreen tree, up to 15 m tall; bark dark grey, exfoliating in corky scales; wood olive-grey, smooth, fairly straight grained, medium-fine and quite even-textured; branchlets pubescent. Leaves chortaceous, elliptic-oblong, rarely obovate, 5–20 × 5–10 cm, glabrous above, brownish tomentose beneath, reddish brown when dry, apex obtuse or slightly acute, rarely emarginate, margin entire, base acute or rounded; secondary nerves 7–12 pairs, transverse nervules parallel, prominent beneath; petiole to 1.2 cm. Flowers small, sessile, 5–6 in each umbellate head. Berry black or deep purple, ovoid, 0.7 cm long, seated on 1 mm long, saucer shaped perianth.

Occasional, in valleys, chiefly near streams and shady places of moist forests of Visakhapatnam to Godavari and Srikakulam at 1000–1200 m. Fls.: Feb.–Mar. Frs.: Apr.–May. The leaves renewed in May. Vern.: Chiru Mamidi, Bhunjada.

The tree is reported to be a fast grower. Powdered bark and roots are used as external applications for pains and bruises; they are also used to heal fractures in animals. The leaves have a cinnamomum like odour when bruised. The seed yield an oil which is used medicinally. Wood is light, not durable, occasionally used in agricultural implements, furniture, oars, house building. Wood is also used as good fuel.

Anatagiri (VSKP), *SSR* 18605; way to Borra hills (VSKP), *SSR* 15940; Kusimi forest (SKLM), *SSR* 15597; Maredumilli (EG), *GVS* 47328 (MH); Minumuluru (VSKP), *GVS* 92231; Lambasingi (VSKP), *HS* 20748 (BLAT); Sunkarimetta (VSKP), *HSP* 20564 (BLAT).

INDIA : Orissa, Chota Nagapur.
WORLD : Myanmar, Thailand and Malaysia.

## NEOLITSEA Merrill

1. Leaves elliptic or oblong lanceolate, apex acutely acuminate. Fruiting perianth very small. ........ ***N. zeylanica***
1. Leaves elliptic or oblong-lanceolate, apex obtusely acute. Fruiting perianth rather large. ........ ***N. foliosa*** var. ***caesia***

**Neolitsea foliosa** (Nees) A.M. & J.M. Gowan. Trees of North Bengal, 111. 1929. var. *caesia* Gamble 2: 1240 (868). 1925. *Litsea foliosa* Nees, Syst. Lour. 622. 1836.

Evergreen, medium-sized tree. Leaves subcoriaceous or chartaceous, elliptic-lanceolate, 5–11 × 3–5 cm, glabrous, glaucous beneath, apex acute, margin entire, base acute; triplinerved; petiole 1–2 cm. Flowers yellow, small in axillary and lateral, sessile clusters of 4–5-flowered umbels. Berry globose, 8 mm in dia.

Occasional, in Northern Circars and Seshachalam hills at 600–800 m. Fls.: Sept.–Nov. Frs.: Jan–Mar. Vern.: Bodeda, Aku Patrikamu, Buroyida.

Mamandur to Thumburatheertham (CTR), *SSR* 16177; Tirumala (CTR), *KJ* 12554; Seshachalam hills (CDP), *CECF* 4881 (CAL); Rampa hills (EG), *MSR* 1536 (CAL).

**Neolitsea zeylanica** (Nees) Merr. in Philipp. J. Sci. (Bot.) 1 (Suppl.) 57. 1906; Gamble 2: 1239 (867). *Litsea zeylanica* Nees Cinnam. Disp. 1: 58. 1823 (*ceylanica*).

Evergreen tree, up to 9 m tall; bark thick, grey with speckles; wood light brown-yellow. Leaves thickly coriaceous, elliptic or oblong-lanceolate, 11–16.5 × 2.5–5.1 cm, glabrous above, glaucous beneath, apex acutely acuminate, margin entire, base acute or cuneate; secondary nerves to 5 pairs, basal lateral nerves 1 cm above the base; petiole to 2.2 cm. Flowers subsessile, dioecious, 2–5 in umbells. Fruit globose, 8 mm in dia., dark purple on persistent perianth.

In moist deciduous and evergreen forests of Visakhapatnam district at 1000–1350 m. Fls.: Jan.–Feb. Frs.: Apr.

Galikonda (VSKP), *SSR* 15926; Anantagiri (VSKP), *GVS* 22689 (MH).

INDIA : Khasia mountains, Western Peninsula.
WORLD : Sri Lanka, Sumatra, Java, Bangladesh.

## **PERSEA** P. Miller *nom. cons.*

**Persea macrantha** (Nees.) Kosterm. in Reinwardtia 6(2): 193. 1962. *Machilus macrantha* Nees in Wall. Pl. As. Rar. 2: 70. 1831; FBI 5: 140. 1886; Gamble 2: 1227 (859); Brandis 530; Beddome Fl. Sylv. S. India t. 264. 1872.

Evergreen tree, up to 20 m tall; bark pale brown with darker spots, rough; wood reddish-white, darkening on exposure, straight grained, medium and even textured; branchlets glabrous. Leaves coriaceous, elliptic-oblong, 8–20 cm, glabrous and shining above, glaucous beneath, apex acute or rounded, margin entire, base slightly oblique and rounded; secondary nerves 4–8 pairs; petiole to 3.8 cm. Flowers greenish-white, 1 cm across, pedicellate in terminal or axillary, 17 cm long panicles. Berry dark purple or black, globose, 1.3 cm across, glabrous with a basal persistent lobed calyx; seed 1, globose, smooth, 1.5–2 × 0.83 cm, black.

Rare, in semi-evergreen and moist deciduous forest. It is a shade bearer. Fls.: Nov.–Jan. Frs.: Apr.–June. Vern.: Nara.

Wood is moderately hard, light and durable, used for house building as planks, rafters, scantlings, flooring and ceiling boards, dug-outs; suitable for commercial and tea chest plywood, dadoes, cabinet panels, slate frames and matches; it may be used for railway sleepers after treatment.

Sunkarimetta (VSKP), *SSR* 15913; Talakona (CTR), *SSR* 15554 & 19602; Vanadurga temple, Minumuluru (VSKP), *GVS* 47354 (MH).

INDIA : Southern region.
WORLD : Sri Lanka.

### **PHOEBE** C.G.D. Nees

**Phoebe wightii** Meissner in DC. Prodr. 15(1): 38. 1864; Gamble 2: 1228 (860); Beddome Fl. Sylv. S. India t. 292. 1872. *P. paniculata auct. non* Nees. FBI 5: 142. 1886 in part.

Densely foliaceous, evergreen tree, up to 10 m tall; bark thin, brown; wood brown, smooth with brown medullary patches. Leaves coriaceous, lanceolate-elliptic, 9–17 × 4–5.8 cm, glabrous above, glaucous and minutely puberulous beneath, apex acuminate, margin entire, base cuneate to acute; secondary nerves 6 pairs, the nerves and transverse nervules distant; petiole to 4 mm. Flowers yellowish, 1 cm across in terminal or axillary cymose panicles; peduncle slender, 7–10 cm long, usually lateral from the branches below the upper most axils. Berry red, globose, 1 cm across; persistent perianth lobes erect, cupular; seed 0.7 cm across.

Occasional, in semi-evergreen forests. Fls.: Apr.–May. Frs.: Throughout the year.

Galikonda (VSKP), *GVS* 19631 (MH).

INDIA : Peninsula.

## HERNANDIACEAE

### **GYROCARPUS** N.J. Jacquin

**Gyrocarpus americanus** Jacq. Select. Stip. Amer. 282, t. 178. 1763; Gamble 1242 (869); Brandis 542; *G. jacquini* Gaertn. Fruct. 2: 92, t. 97. 1791; FBI 2: 461; Beddome Fl. Sylv. S. India t. 196. 1872.

Deciduous tree, up to 6 m tall; bark smooth, shining, greenish-white; wood grey; branchlets brittle, glabrous, tomentose in very young. Leaves ovate, entire or lobed, 8–15 × 8–13 cm, glabrous above, soft and thinly tomentose beneath, apex acute or acuminate, margin entire, base broad, truncate, cordate; 3-nerved from the base, veins closely reticulate; petiole 3 to 13 cm. Flowers small, white or pale yellow, unisexual in large branched cymes; males numerous; females few. Drupe obovoid-globose, 1–2 cm in dia, dark brown, slightly pubescent, wrinkled in the lower part, furrowed in the upper, crowned by 2-spathulate wings; seed solitary.

Common, in dry deciduous forests and along the roadsides. to 500 m. Leaf fall: Dec. New foliage: Jan.–Feb. Fls: Jan.–Mar. Frs.: May–Sept. Vern.:

Tella poniki, Kummara Poliki, Thanaku, Thanaki, Poliki. Wood soft and used for making boxes and toys.

Kovur (WG), *SSR* 15257; Gundam RF (KNL), *TP & RVR* 3013 & 3017; Chelama RF (KNL), *JLE* 23732 (CAL).

WORLD : Tropical West and South Africa, tropical Asia, Polynesia and South America.

## PROTEACEAE

### GREVILLEA R. Brown ex R.A Salisbury

**Grevillea robusta** A. Cunn. ex R. Br. Suppl. Prodr. Fl. Nov. Holl. 24. 1830; Gamble 2: 1243 (870).

Evergreen tree, very tall; sparingly branched at the top; young branches hoary tomentose. Leaves 2-pinnatifid, fern like, the secondary lobes entire or again lobed, lanceolate, margins recurved, silvery beneath. Flowers orange with slender pedicelled in unilateral racemes. Fruit broad, very oblique, 2 to 2.5 cm long, dehiscing on one side; seeds 1–2, often winged, ca. 1 cm in dia.

Cultivated, originally introduced in the Agencies as a shade tree for coffee plantations. It is now well established in Northern Circars and Tirumala hills. It is also a fast growing tree and reproduces naturally. Fls.: May–June. Frs.: July and Aug. Vern.: Parana, Perane. Eng.: Silver Oak.

Toys are made from the wood. The milky juice is used in rheumatism, toothache, asthma etc. and also used as fish poison.

Tirumala (CTR), *KJ & SSR* 12564.

INDIA : Commonly cultivated throughout.

WORLD : Native of E. Australia, cultivated in other tropical and subtropical countries.

## SANTALACEAE

### SANTALUM Linnaeus

**Santalum album** L. Sp. Pl. 349. 1753; FBI 5: 231; Gamble 2: 1261 (883); Brandis 553; Beddome Fl. Sylv. S. India t. 256. 1872.

Evergreen, semiparasitic tree, up to 12 m tall; bark grey, cracked; wood yellowish-brown, turning to dark brown when exposure; close-grained; young branches purplish, striate, drooping. Leaves chartaceous, elliptic-ovate, to 9 × 3 cm, glabrous, thin, apex subacute, margin entire, base acute; secondary nerves ca 5 pairs; petiole to 1.5 cm. Flowers maroon or brownish-purple within, green without in 6 cm long axillary pedunculate cymes. Drupe dark purplish, globose, 8 mm in dia, endocarp hard, ribbed; seed 1.

Occasional, in open dry deciduous forests and scrubs, grows mostly on red, ferruginous loam, overlying metamorphic rocks. It reproduces plentiful from seeds dispersed by birds also to certain extent from root-suckers and also coppices well. Fls.: June–Dec. Frs.: Jan.–Mar. Vern.: Tella chandanamu, Malayajamu, Chandanamu, Srigandham, Sree chandanamu, Hari chandanamu, Krishna chandanamu.

Sapwood is unscented; heart wood strongly, characteristically scented, last for many years and great demand for carvings. Wood contains essential oil and is used for mainly distillation, the remaining wood being used for carving incense making, the darker the wood the higher is the oil content. Wood powder used as cosmetic, and paste applied externally for relieving headache. Wood is moderately hard to heavy, used for burning as incense in temples. The wood is used for a variety of small carved articles, such as small boxes, cabinet panels, walking sticks, combs, picture frames, fan handles, pen holders, card cases, trays and similar articles. Both oil and wood credited with cooling, diaphoretic, diuretic and expectorant properties. Sandal wood finds several applications in house remedies. Fresh leaves yield a pale yellow wax.

INDIA : Karnataka and Tamil Nadu; introduced into Rajasthan, Uttar Pradesh, Madhya Pradesh and Orissa.

## EUPHORBIACEAE

1. Male flowers reduced to a single pedicellate stamen, enclosed in an involucre with a solitary central 3-carpelled pedicelled female flower without perianth or perianth rarely as 1–3 minute scales .......... ***Euphorbia***
1. Flowers not reduced as above:
    2. Cells of ovary 2-ovuled:
        3. Leaves small, 0.6–3 cm .......... ***Phyllanthus***
        3. Leaves medium to large, 3–20 cm:
            4. Stamens numerous .......... ***Dimorphocalyx***
            4. Stamens 3–16:
                5. Fruit drupe or fleshy:
                    6. Inner whorl of perianth present .......... ***Bridelia***
                    6. Inner whorl of perianth absent .......... ***Drypetes***
                5. Fruit capsule:
                    7. Disc present:
                        8. Stamens connate by their connectives in a column; style connate in a column, lobed at tip .......... ***Glochidion***
                        8. Stamens not connate; style connate below, spreading, bifid and then divided into plumose arms .......... ***Epiprinus***
                    7. Disc absent:
                        9. Flowers dioecious; stamens 4. .......... ***Margaritaria***
                        9. Flowers monoecious; stamens 5–6 .......... ***Cleistanthus***
    2. Cells of ovary 1-ovuled:
        10. Leaves peltate:

11. Leaves entire or shallowly lobed. Floral bracts present. Capsule 6–7 mm across ... ........ ***Macaranga***
11. Leaves lobed more than half of their length. Floral bracts absent. Capsule 3 cm across -cultivated. ........ ***Manihot***

10. Leaves not peltate:
    12. Inner whorl of perianth present at least in the male flowers:
        13. Leaf base not ribbed; eglandular:
            14. Leaves shallowly lobed ........ ***Jatropha***
            14. Leaves entire:
                15. Plants monoecious. Stamens bent in bud condition ........ ***Croton***
                15. Plants dioecious. Stamens straight ........ ***Homonoia***
        13. Leaf base 3–5-ribbed, glandular:
            16. Flowers in spikes or racemes. 4 glands only in between basal nerves of leaf base ........ ***Alchornea***
            16. Flowers in cymose panicles. Petiole glandular:
                17. Stamens not borne on a receptacle. Latex red ........ ***Givotia***
                17. Stamens borne on a receptacle. Latex watery ........ ***Aleurites***
            14. Inner whorl of perianth absent in both male and female flowers:
                18. Stamens 2–3:
                    19. Capsule deeply lobed, crustaceous ........ ***Exoecaria***
                    19. Capsule not deeply lobed, fleshy ........ ***Sapium***
                18. Stamens 10-numerous:
                    20. Flowers in axillary cymes or clusters; calyx imbricate ........ ***Suregada***
                    20. Flowers in spikes; calyx valvate:
                        21. Male inflorescence pendulous or catkin like; style linear. Fruit drupaceous ........ ***Trewia***
                        21. Inflorescence erect; style fimbriate. Fruit capsular ........ ***Mallotus***

## ALCHORNEA Swartz

**Alchornea mollis** Muell.-Arg. in Linnaea 34: 168. 1865; FBI 5: 420; Gamble 2: 1324 (926); Brandis 587.

Small tree, up to 5 m tall; branches pubescent. Leaves coriaceous, broadly ovate, 6–22 × 3.5–16.5 cm, pubescent, apex caudate-acuminate, margin dentate, base cuneate or rounded, 4 glandular, in between the basal nerves; 3-ribbed; secondary nerves 4-pairs above basal nerves; petiole to 15 cm. Flowers small in terminal, long spikes or racemes; males solitary or in clusters, females solitary. Capsule globose, 9 mm, grey-pubescent; seeds subglobose.

Occasional, in the hills especially in moist vallyes above 700 m. Fls. & Frs.: June–Aug. Vern.: Badiki.

Pedavalasa (VSKP), *SSR* 19667;Korapur (VSKP), *GVS* 44401; Minumuluru (VSKP), *GVS* 30042; Vankachinta (VSKP), *GVS* 47345 (MH).

INDIA : Assam, Sikkim.
WORLD : Tropics.

## ALEURITES J.R. Forster & J.G.A. Forster

**Aleurites moluccana** (L.) Willd. Sp. Pl. 4: 590. 1805; FBI 5: 384; Gamble 2: 1318 (922); Brandis 576; Beddome Fl. Sylv. S. India t. 276. 1872. *Jatropha moluccana* L. Sp. Pl. 1006. 1753.

Evergreen tree; young shoots densely clothed with brownish or grey stellate tomentum. Leaves subcoriaceous, broadly elliptic or ovate to deltoid, tomentose when young, apex shortly acuminate, base truncate; petiole long, 2-glandular at apex. Flowers white in large, cymose, terminal panicles. Drupe globose, fleshy, 4 cm across with hard stone; seeds large.

Cultivated in the gardens (Pullaiah & Ali Moulali, 1997). Fls.: Dec.–Feb. Frs.: Feb.–May.

Lumbang oil is extracted from seeds, which is used in painting and varnishing industry. Medically the oil is used as purgative, and externally it is employed in rheumatism and ulclers. The oil is used for culinary purpose as an illuminant and as a hair tonic. The oil cake is rich in protein, but it is toxic, used as manure. The wood is soft and is used for making packing-cases; it is reported to be suitable for match-sticks and as a raw material for pulp. The bark yields tannin.

Native to the Indo-Malayan region, but now widely distributed throughout the tropics. It is naturalized in India and found wild in South India and Assam.

WORLD : China to Polynesia, New Zealand.

## BRIDELIA Willdenow

1. Secondary nerves numerous, parallel, straight:
    2. Flowers in clusters in spikes ........ ***B. airy-shawii***
    2. Flowers in clusters in the leaf axils:
        3. Secondary nerves forked or reticulate within the margin ........ ***B. pubescens***
        3. Secondary nerves not forked, joining in a marginal thickened one:
            4. Leaf apex obtuse or emarginate, brown beneath ........ ***B. crenulata***
            4. Leaf apex acute, cinerous-glaucous beneath ........ ***B. cinerascens***
1. Secondary nerves 6–10 pairs, not markedly straight or parallel:
    5. Leaves oblong-lanceolate, secondary nerves unbranched up to the marginal nerve. Flowers white, numerous ........ ***B. monoica***
    5. Leaves rhombic-obovate, secondary nerves branched near margin. Flowers pale green, few ........ ***B. montana***

**Bridelia airy-shawii** P.J. Li in Acta Phyt. Sin 20(1): 117. 1982. *B. retusa* (L.) Spreng. Syst. Veg. 3: 48. 1826; Muell.-Arg. in DC. Prodr. 15: 493. 1865; FBI 5: 268 p.p; Gamble 2: 1280 (896); Brandis 560; Beddome Fl. Sylv. S. India t. 260. 1870.

Deciduous tree, up to 12 m tall; spiny when young; bark grey, longitudinally cracked; wood grey to crimson, shallowly or deeply interlocked grained in narrow bands, medium textured; branchlets tomentose. Leaves chartaceous, elliptic-oblong or obovate, 6–19.5 × 3–9 cm, glabrous above, puberulous and glaucous beneath, apex acute, or obtuse, margin entire, slightly crenulate, base obtuse; secondary nerves 15–25 pairs, conspicuous with finely reticulate veins. Flowers greenish-yellow or purple in 5–10 cm long, terminal spikes or in axillary clusters. Drupe purple or black, seated on enlarged calyx, pyrene furrows curled and deep, 1–2 seeds with fairly thick bony shells, with brownish papery testa.

Common, in mixed and deciduous forests at 600–1200 m. It is drought hardy, produces root-suckers and good coppicer. Leaf fall: Feb.–Mar. New foliage: May–June. Fls.: May–July. Frs.: Nov.–Jan. Vern.: Kora Moddi, Mulu Maddi, Putta karaka, Anemu, Koramanu, Koderu, Durva Maddi, Pia Maddi, Bonta yepi.

Leaves used as cattle fodder. Fruits edible. Bark with gingelly oil used as liniment in rheumatism and it is said to be used in tanning. Wood hard, moderately heavy, durable; it is in fair demand for house building or rafters, posts, and floor boards; it is also used for carts, shafts, yokes and agricultural implements.

Upper Ahobilam (KNL), *SSR* 15209; Pulikonda (ADB), *PSPB* 9159; Tummala kunta (CTR), *GVS* 4960; Gandi (CDP), *JSG* 10979 (DD); Addateegala (EG), *SKW* 4007 (BLAT), *GVS* 68570 (CAL); Narsapur (MDK), *KMS* 6732 (CAL); Mudheli RF (NZB), *BR* 9537; Anantagiri (VSKP), *TP & EC* 7379; Pakhal RF (WGL), *KMS* 13200 (MH).

INDIA : Throughout the greater parts.
WORLD : Sri Lanka.

**Bridelia cinerascens** Gehrm.in Bot. Jahrb. Syst. 41 Beible. 95: 30. 1908; Gamble 2: 1280 (896). *B. retusa* var. *glauca* Hook. f. FBI 5: 268. 1887.

Tree, up to 8 m tall; branchlets puberulous. Leaves coriaceous, elliptic-oblong, 4.5–11 × 2.5–6 cm, glabrous above, cinereous-glaucous beneath, apex obtuse-acute, margin slightly crenulate, base obtuse or acute; secondary nerves 13–22 pairs; petiole to 1 cm. Flowers greenish-yellow, dioecious, axillary in clusters in the leaf axils. Drupe globose, 8 mm in dia, purple-black, pyrene furrows rather shallow, the ridges curved.

Common, in mixed deciduous forests. Fls.: Apr. Frs.: July–Sept. Vern.: Kosengi, Korra Maddi, Kora Manu, Yerra sittuga.

Japali, Tirumala hills (CTR), *SSR* 15274; Pothurajupenta (KNL), *SSR* 15989; Chintala RF (PKSM), *SSR* 16119; Guvvalacheruvu (CDP), *PSPB* 8425.

INDIA : Tamil Nadu.

**Bridelia crenulata** Roxb. Fl. Ind. 3: 934. 1832. *B. roxburghiana* (Muell.-Arg.) Gehrm. Bot. Jahrb. Syst. 41: 30. 1908; Gamble 2: 1280 (896); *B. retusa* Spreng. var. *roxburghiana* Muell.- Arg. in DC. Prodr. 15. 493. 1866; FBI 5: 268.

Deciduous, dioecious tree, up to 10 m tall; bark red when cut; branchlets fulvous-tomentose. Leaves coriaceous, elliptic-oblong or elliptic-obovate, 7.5–12.5 × 3–7 cm, glabrous above, brown tomentose beneath, apex emarginate or obtuse, margin crenulate, base truncate to rounded; secondary nerves 15–20 pairs; petiole to 1 cm. Flowers greenish-purple, pedicellate in clusters in the leaf axils. Drupe depressed-globose, 4 mm in dia., pyrene furrows deep, the ridges branching from the middle one and curved.

Rare, in moist evergreen forests of Kurnool and Srikakulam districts. Fls. & Frs.: Apr.–July. Vern.: Aanemchettu.

Wood used for agricultural implements and house construction.

Chelama (KNL), *PSPB* 11819.

INDIA : Peninsula.
WORLD : Sri Lanka, Nepal, Myanmar, Indo-China, Malay, Sumatra.

**Bridelia monoica** (Lour.) Merr. in Philipp. Journ. Sci. Bot. 13. 142. 1919. *Cluytia monoica* Lour. Fl. Cochinch. 2: 638. 1790. *Bridelia tomentosa* (Bail.) Bl. Bijdr. 597. 1825; FBI 5: 271; Gamble 2: 1281 (896); Brandis 560.

Small tree, up to 8 m tall; bark white, smooth; wood red; branches brown tomentose, drooping, slender. Leaves coriaceous, oblong-lanceolate to linear lanceolate, 4.5–13 × 2.8–5.5 cm, apex acuminate, margin slightly wavy, base acute, thinly puberulous above when young, brown tomentose beneath; secondary nerves 8 pairs; petiole to 1 cm. Flowers white, very small in axillary clusters. Berry globose to 7 mm dia., blue-black, puberulous, pericarp slits into 6 valves; seed black.

Occasional, in the hills of East Godavari and Visakhapatnam districts above 800 m.. Fls.: Sept.–Oct. Frs.: Nov.–Jan. Vern.: Talla Balli.

Galikonda (VSKP), *SSR* 15928; Annavaram (EG), *KCJ* 12582; Rajavommangi (EG), *GVS* 68641; Barnakonda (VSKP), *KCJ* 17134 (MH); Paderu (VSKP), *SKW* 4173 (BLAT).

WORLD : Australia, Myanmar, China, Malay Peninsula and Archipelago.

**Bridelia montana** Roxb. Willd. Sp. Pl. 4: 978. 1805; FBI 5: 269, p.p; Gamble 2: 1281 (896). *Cluytia montana* Roxb. Pl. Cor. 2: 36. t. 171. 1802. *Bridelia hamiltoniana* Muell.-Arg. in Linnaea 34. 77. 1865; FBI 5: 271. 1887.

Small tree, up to 8 m tall; bark thin, rather rough; branchlets glabrous. Leaves coriaceous, obovate or oblanceolate, 4–10.5 × 2–5.8 cm, glabrous, green above, paler beneath, apex acute or obtuse, margin entire, base cuneate; secondary nerves 7 pairs; oblique; petiole to 5 mm. Flowers green, few in leaf axils. Drupe globose, 0.5 cm in dia, pyrene furrows shallow, the ridges rounded, curving from a central one.

Common, in dry deciduous and degraded forests. Fls.: Sept.–Oct. Frs.: Jan.–Mar. Vern.: Panchavoni, Tella Yegise, Pantangi, Balli, Kanchep; Sanna Koctari, Gundu bigalu. Fruit edible.

Ramannapenta (KNL), *TP & SSR* 15211; Vanipenta (KNL), *SSR* 15299; Wankidi (ADB), *PSPB* 9154; Alurukona (ATP), *TP* 932; Kalasamudram (ATP), *NY* 1200; Guvvalacheruvu (CDP), *KMS* 6412 (CAL); Palakonda (CDP), *CS* 7628, *KMS* 2386 (DD); Addatigala (EG), *SKW* 3930; Gollarayikonda (EG), *GVS* 62535 (CAL); Kondapalli (KSN), *SKW* 3237 (BLAT); Velutla (NZB), *BR* 9526; Sikharam (PKSM), *RKM* 422 (CAL).

**Bridelia pubescens** Kurz. in J. As. Soc. Beng. 4: 241. 1873 et. For. Fl. Brit. Burma 2: 1877; FBI 5: 270; Brandis 560.

Evergreen tree, up to 15 m tall; branchlets tubercled, pubescent. Leaves membranous, elliptic or obovate, 7–15 × 0.5–5 cm, thinly puberulous above, tawny pubescent beneath, apex shortly acuminate, margin entire, base acute-rounded; secondary nerves 9–12 pairs, arching; petiole 6 mm. Flowers pubescent in axillary compact clusters. Drupe black, ovoid or oblong with persistent outer tepals.

Occasional, along streams in semi-evergreen forests 1000-1500 m. Fls. & Frs.: Aug.–Nov.

Minumulur (VSKP), *GVS* 28262; Madalamma gudi (VSKP), *GVS* 42583; Sunkarimetta (VSKP), *GVS* 22647 (MH).

INDIA : Assam, Sikkim, Bengal, and Orissa
WORLD : Myanmar, Yunnan and Taiwan, Nepal.

## CLEISTANTHUS J.D. Hooker ex Planchon

1. Leaves elliptic-oblong, apex retuse or obtuse. Petals linear; ovary glabrous. Capsule 2–2.5 cm dia., shining; seeds globose ........ *C. collinus*
1. Leaves ovate-lanceolate, apex acuminate. Petals obovate; ovary villous. Capsule 0.8–1 cm dia., not shining; seeds cordate ........ *C. patulus*

**Cleistanthus collinus** (Roxb.) Hook. f. Fl. Brit. Ind. 5: 274. 1887; Gamble 2: 1282 (897); Brandis 561. *Cluytia collina* Roxb. Pl. Cor. 2: 37, t. 169. 1802. *Lebidieropsis orbiculalaris* Muell.-Arg. in DC. Prodr. 15: 509. 1866.

Deciduous, monoecious tree, up to 8 m tall; bark dark brown, rough; wood dark reddish brown; branchlets terete, glabrous. Leaves coriaceous, elliptic, orbicular or oblong-obovate, 4–12 × 3–7.5, apex obtuse or retuse, margin entire, base obtuse, glabrous, bright green; secondary nerves 5–6 pairs; petiole to 0.7 cm. Flowers yellowish-green in small, axillary, silky clusters. Capsule globose, 2 cm across, 3-valved, dark brown, shining; seeds 3, globose, testa smooth, shiny.

Very common in dry forests, up to 800 m. It coppices well. Leaf fall: Jan. New foliage: May. Fls.: Feb. Frs.: May onwards dehiscing by next February. Vern.: Lodisa, Vodisa, Nalla-Kordtha. Eng.: Star gooseberry.

The leaf, the root and the fruits are irritants, and they are used as fish poision. Tender leaves are deadly poisonous. Wood is durable, used for house posts, fences etc.

Bairluti (KNL), *SSR* 5213; Seethampet RF (SKLM), *SSR* 15576; Narsapur RF (MDK), *SSR* 18502; Vempalli (ADB), *GO & PVP* 4794; Kalasamudram (ATP), *EC* 2885; Addatigala (EG), *SKW* 2991; Papi hills (WG), *DCSR* 73 (CAL); Vinukonda (GTR), *VRK* 6902, Urlakonda (NLR), *MSRS* 1304 (CAL); Malkapur (NZB), *TP & BR* 6106; Moulatta (PKSM), *RKM* 406; Madugala (VJN), *RKM* 406, *GVS* 47260 (CAL).

INDIA : Almost all States.
WORLD : Sri Lanka.

**Cleistanthus patulus** (Roxb.) Muell.-Arg. in DC. Prodr. 15 505. 1866; FBI 5: 279; Gamble 2: 1282 (897), Brandis 561. *Cluytia patula* Roxb. Pl. Cor. 2: 37. t. 170. 1802. *Lebidiera patula* (Roxb.) Muell.-Arg. in Linnaea 32: 21. 1864.

Evergreen tree, up to 8 m, tall; monoecious; branches horizontal; bark smooth, grey, exfoliating in thin plates; wood pink or crimson, then yellowish, close-grained; branchlets terete, glabrous. Leaves coriaceous, 3.5–9.2 × 1.9–4 cm, glabrous, red when young, ovate-lanceolate, apex acuminate, margin entire, base acute or rounded; secondary nerves 5 pairs; petiole to 5 mm, thick. Flowers yellowish, small in clusters in axillary spikes. Capsule 0.7 cm across, pilose-glabrous; seeds reniform, albumen thin, cotyledons large, fleshy, truncate.

Occasional, in Rayalaseema and Northern circars. Chiefly in ravines or dry evergreen forests 600–1400 m. Fls.: Jan.–Mar. Frs.: Apr.–July. Vern.: Jegura, Jiguru, Tella vadisae.

Gundlabrhmeswaram (KNL), *SSR* 15529; Simhachalam (VSKP), *SSR* 15566; Tadepally (WG), *NRR* and *TRS* 86119 (MH); Ramavaram (EG), *GVS* 27452 (MH).

INDIA : Karnataka, Orissa.
WORLD : Sri Lanka.

## CROTON Linnaeus

1. Staminal filaments glabrous. Leaves ovate, base cordate. Capsule to 0.8 cm long ................ ***C. scabiosus***
1. Staminal filaments villous at base. Leaves oblong-lanceolate, base acute. Capsule to 1.2 cm long ................ ***C. roxburghii***

**Corton roxburghii** Balak. in Bull. Bot. Surv. Ind. 3: 39. 1962. *C. oblongifolius* Roxb. Fl. Ind. 3: 685. 1832; FBI 5: 386; Gamble 2: 1315 (919); Brandis 577.

Small to medium-sized tree; bark grey or brownish; wood yellowish white; branchlets tomentose. Leaves crowded at the ends of branchlets, oblong-lanceolate, 12–25 × 6–10.5 cm, glabrous when mature, apex subacute, margin crenate, serrate, base acute; secondary nerves numerous, slender; petiole to 4 cm. Flowers pale yellowish-green, in elongated terminal racemes; male above, females below on the axis. Capsule subglobose, slightly 3-lobed, clothed with small orbicular scales; seeds ellipsoid, rounded and quite smooth on the back.

Rare, in dry forests of Visakhapatnam and Srikakulam districts. Fls. & Frs.: Mar.–Dec. Bhutala-bhairi, Bhutan kusamu.

Donubai (SKLM), *NRR* 23614; Suttiguda (VSKP), *NRR* 83795 (MH).

INDIA : Karnataka, Orissa.
WORLD : Sri Lanka.

**Croton scabiosus** Beddome. Fl. Sylv. S. India 2. t. 283. 1872; FBI 5: 386; Gamble 1314 (919); Brandis 577.

Deciduous tree, up to 6 m tall; bark dark brown, tough; wood yellowish white, close grained; branchlets yellowish hairy. Leaves coriaceous, ovate, 4–10.5 × 7.8–8 cm, clothed with minute orbicular scales on both surfaces, pale and silvery beneath, green and very much rough above, apex obtuse or slightly acute, margin crenate, base cordate, with 2 sessile glands; 3–5-ribbed; petiole 3–5 cm. Flowers yellowish-white in terminal, 4–10 cm long racemes. Capsule globose, 3-valved, 0.8 × 0.3 cm, yellowish.

Common on dry rocky hills of Kurnool and Cuddapah districts at 350–1350 m. Fls.: Mar.–June. Frs.: July. Vern.: Chilla, Verri chilla. Wood hard.

Pulivendala RF (CDP), *SSR* 10148; Guvvalacheruvu (CDP), *PSPB* 9197; Sidhout (CDP), *SKW* 5909, 5910 (BLAT).

## DIMORPHOCALYX Thwaites

1. Branchlets glabrous. Leaves elliptic or oblong-oblanceolate, base obtuse-acute. Stamens 11–16, sepals, ovary and capsules ochraceous puberulous. ................ ***D. glabellus***
1. Branchlets pubescent. Leaves obovate, base attenuate. Stamens 16–22, sepals, ovary and capsules adpressed pilose ................ ***D. kurnoolensis***

**Dimorphocalyx glabellus** Thw. Enum. Pl. Zeyl. 278. 1861; FBI 5: 403; Gamble 2: 1337 (934).

Evergreen tree, 5–10 m tall; bark reddish-brown, deeply furrowed, spongy; wood red; branchlets striate, glabrous, whitish. Leaves elliptic or elliptic-oblong, 3–15 × 2–6 cm, glabrous, apex obtuse or acute, margin entire, base rounded; secondary nerves 5–7 pairs; petiole to 2 cm..Flowers white, 6 mm long, solitary or few; male flowers axillary, 1–few in branched cymes; female flowers axillary or terminal, 1–few in umbellate racemes. Capsule 1–1.5 cm long, subglobose, slightly hairy, scarcely lobed; fruiting calyx obovate, 0.5–1.2 cm long; seeds ovoid, 3.

Occasional, in semievergreen forests and scrubs near the sea. Fls.: Feb.–Aug. Frs.: Aug.–Dec.

Siddulaiahkona (NLR), *SSR* 19676; Narasimhulu Konda (NLR), *SSR* 19679; Dolphin's nose (VSKP), *JSG* 16105 (CAL); Tadepalli (EG), *NRR & TRR* 86117 (MH).

INDIA : Southern region, Orissa.
WORLD : Sri Lanka.

**Dimorphocalyx kurnoolensis** Venkataraju & Pullaiah in Bot. Bull. Aca. Sin. 35: 201. 1994.

Dioecious tree, up to 5 m tall; bark greyish; branchlets terete, pubescent, striate with lenticels. Leaves coriaceous, elliptic-lanceolate or obovate, 6–14.5 × 3–5.5 cm, glabrous on both surfaces, green above, paler beneath, apex obtuse or subacute, margin slightly undulate, base acute or attenuate, often 3-ribbed; secondary nerves 7–9 pairs; petiole to 2 cm, channelled above. Flowers dioecious, white, male flowers subsessile in terminal cymes; Female flowers few, peduncled in racemes. Capsule subglobose, 1–1.5 × 1.3–1.7 cm, dark brown, 3-lobed, adpressed-pilose, with accrescent calyx; seed yellowish, glabrous, 4 × 7 mm.

Rare, along the sides of the streams at the foot hills of Erramalai hills. Fls.: Jan. Frs.: Mar. Vern.: Erra tandra.

Erramalai hills near Owk (KNL), *RVR* 6661; *PSPB* 11859, 11860, 11847.

## DRYPETES M. Vahl

Charkrabarty *et al.*, (1997) have reviewed the genus *Drypetes* (Euphorbiaceae) in the Indian subcontinent of which 26 indigenous species are recognised, among these 3 species are found in Andhra Pradesh.

1. Branchlets pendulous. Disc wanting, stamens 2–3. Drupe yellowish, pendulous. ........................ ***D. roxburghii***
1. Branchlets spreading. Disc present, stamens 6–12. Drupe red or scarlet, not pendulous:
    2. Leaves pubescent along nerves beneath, 6.7–18 × 4.5–7 cm. Ovary 2-celled. Drupe 2.5 × 1.2 cm. Female flowers solitary, sessile. ........................ ***D. assamica***

2. Leaves glabrous, 4–7 × 2.2–3 cm. Ovary 1-celled. Drupe 8 mm across. Female flowers in fascicles, pedicellate. ........ ***D. sepiaria***

**Drypetes assamica** (Hook. f.) Pax & Hoffm. in Pfrich. 5. 147 (15): 241. 1922. *Cyclostemon assamica* Hook. f. Fl. Brit. India 5: 342. 1887; Gamble 2: 1301 (910); Brandis 567.

Small, evergreen tree, up to 8 m tall, bark light; wood white, slightly yellow streaked; branchlets pubescent. Leaves subcoriaceous, elliptic, lanceolate or oblong, 6.7–18 × 4.5–7 cm, shining, pubescent along nerves beneath, apex obtuse or shortly acuminate, margin entire, base acute; secondary nerves 11–15 pairs, inter-costal reticulations between the nerves; petiole 8 mm. Flowers yellowish-green; male flowers clustered in axils; female flowers solitary, sessile. Drupe scarlet, ovoid-oblong, bilocular, 2.5 × 1.2 cm.

Rare, in moist places of Visakhapatnam district. Fls.: June–Aug. Frs.: Sept.–Nov.

Sunkarimetta (VSKP), *GVS* 22667 (MH); *NPBK* 10936 (MH).

INDIA : Tropical region.
WORLD : Africa, Malay.

**Drypetes roxburghii** (Wall.) Hurursawa in J. Fac. Sci. Univ. Tokyo Sect. 3. Bot. 6: 337. 1954. *Putranjiva roxburghii* Wall. Tent. Fl. Nap. 61. 1826; FBI 5. 1336; Gamble 2: 1311 (917); Brandis 566. Beddome Fl. Sylv. S. India t. 275. 1872.

Moderate-sized, dioecious evergreen tree; bark dark grey with horizontal lenticels; wood greyish-white, close-grained; branchlets puberulous, lenticellate. Leaves coriaceous, elliptic- oblong, 5–12.5 × 1–8 cm, glabrous, apex shortly acuminate, margin distantly serrulate, base obtuse, somewhat oblique; secondary nerves ca. 11 pairs, not prominent; petiole to 1 cm. Flowers greenish-yellow, dense, sessile in axillary clusters around branchlets. Drupe yellowish, globose or ellipsoiod, 2 × 1.5 cm, pubescent; seed 1, crustaceous.

Occasional, usually grows on alluvial soil along the rivers or in swamps or evergreen forests at 600–1000 m; sometimes cultivated in gardens for hedges and on roadsides. Fls.: Mar.–Apr. Frs.: Oct.–Dec. Vern.: Kudrajuvi, Kaduru, Putrajivika, Veku, Mahaputra-jivi.

The seed yields a fatty oil, used for burning. Leaves with fruit decoction used for cold and fever. Leaves are lopped for fodder. Wood hard, strong, heavy and durable, used for house building, agricultural implements, tool handles and turnery. The nuts are strung in necklaces and rosaries.

Mamandur-Thumbura theertham (CTR), *SSR* 16172; Jodevagu (ADB), *GO & DAM* 5178; Salivendrakona (CDP), *JSG* 16785 (DD).

INDIA : Peninsula.
WORLD : Sri Lanka to W. Himalaya, Java, Myanmar, Moluccas, E. New Guinea.

**Drypetes sepiaria** (Wt. & Arn.) Pax & Hoffm. in Engl. Pflanzenr 81: 27. 1922. *Hemicyclia sepiaria* Wt. & Arn. in Edinburgh New Philos. J. 14: 297. 1833. FBI 5: 387, Gamble 2: 1299 (909); Brandis 566.

Evergreen, small tree, up to 7 m tall; branchlets whitish, puberulous, stiff. Leaves thick-chartaceous, elliptic-oblong, 4–7 × 2.2–3 cm, glabrous, apex obtuse-retuse, margin entire, sometimes repand-toothed, base rounded; secondary nerves 5–6 pairs, impressed beneath, irregular; petiole to 7 mm. Flowers yellowish, pedicellate, 6–8 in axillary clusters. Drupe red, globose, 8 mm across; seed 1 or 2.

Occasional, along streams of moist forests to 600 m. Fls.: Feb.–Mar. Frs.: April.–July. Vern.: Bira, Manchi-Bira.

Chintala forest (PKSM), *SSR* 16150; Sidduliah Kona (NLR), *SSR* 19367 & 19376; Bakharapeta (CTR), *SKW* 5716 (BLAT); Dornal (KNL), *JSG* 18685 (CAL); Narsipatnam (VSKP), *GVS* 29650 (MH).

INDIA : Peninsula.
WORLD : Sri Lanka.

## EPIPRINUS W. Griffith

**Epiprinus mallotiformis** (Muell.-Arg.) Croizat in J. Arnol Arbor. 23: 52. 1942. *Symphyllia mallotiformis* Muell.-Arg. Linnaea 34: 156. 1865; Gamble 2: 1323 (925). *Cephalocroton indicum* Beddome Fl. Sylv. S. India t. 261. 1872.

Evergreen, monoecious tree, 6–10 m tall. Leaves coriaceous, elliptic or oblong, ovate-elliptic, 5.5–13 × 2–6.5 cm, glabrous, apex acute or shortly acuminate, margin entire, base cuneate or acute, oblique; tri-nerved, secondary nerves 6–7 pairs; petiole to 3 cm. Flowers greenish, small, in interrupted, subterminal to 11 cm long, spiciform racemes; male flowers many, clustered; female flowers few. Capsule grey, depressed globose, rugose, 1.8 cm in dia., 3-lobed; seeds globose, crustaceous.

Rare, in valleys of evergreen forests at 700–900 m. Fls.: Nov.–Dec. Frs.: Jan.

Near Papanasanam in Tirumala (CTR), *DR* 1257 (SVU).

INDIA : Western Peninsula.

## EUPHORBIA Linnaeus

1. Branchlets unarmed ........ ***E. tirucalli***
1. Branchlets armed with stipular spines:
    2. Branchlets cylindrical ........ ***E. nivulia***
    2. Branchlets angled:
        3. Style 2-fid ........ ***E. antiquorum***

3. Style simple:
   4. Branchlets exclusively 3-angled. Leaves deciduous, small. .............. ***E. barnhartii***
   4. Branchlets 5-angled. Leaves not deciduous, large. .......................... ***E. ligularia***

**Euphorbia antiquorum** L. Sp. Pl. 450. 1753; FBI 5: 255; Gamble 2: 1277 (894); Brandis 558.

Small tree, up to 8 m tall; trunk stout, cylindric or fluted; bark thick, rough, brown; branches numerous, curving upwards, stout, fleshy, green, jointed, 3–5 sinuate winged. Spines stipular, divaricate, sharp, short. Leaves small, obovate-oblong, glabrous, fleshy, margin entire, base cuneate, deciduous early; petiole 0. Cyathia terminal, subsessile. Capsule compressed, 1.2 cm dia.; seeds globose, small.

Common, in dry hilly regions especially in rocky areas. Fls.: July–Dec. Frs.: Feb.–Mar. Vern.: Bommajemudu, Peddajemudu.

The plant is considered to be purgative, digestive, pungent and bitter. The juice of the plant is used as an irritant in rheumatism. It is popular application for warts and other cutaneous affections. The latex is used as fish poison.

Muchukota (ATP), *TP & NY* 208; Bugganipalli (KNL), *TP & RVR* 1815; Macherla (GTR), *VRK* 3816; Mypadu (NLR), *PSPB* 9101; Rampa (EG), *VNS* 6756 (CAL).

INDIA : Throughout.
WORLD : Sri Lanka.

**Euphorbia barnhartii** Croizat Euphorb. 54. 1934. *E. trigona* Roxb. Fl. Ind. 2: 256. 1832; FBI 5: 256. *p.p.* Gamble 2: 1277 (893); Brandis 558.

Small tree, up to 7 m tall; trunk thick; branches ascending; branchlets 3-winged, spines straight, paired, sharp. Leaves obovate, spathulate, deciduous, glabrous, pale beneath, apex mucronate, margin entire, base obcuneate. Cyathia in axillary, lax cymes; involucre cupular. Capsule trigonus.

Occasional, in rocky areas and foot hills in scrub-jungles. Fls. & Frs.: Feb.–Apr. Vern.: Kattimandu.

The latex is acrid and vericant. The juice from heated leaves is used for earache. Powdered leaves alone or with turmeric are used as poultice for boils.

Gavisiddeswara hills (ATP), *PSPB* 11845; Srisailam (KNL), *TRS & PVS* 9535 (MH).

INDIA : Orissa, Karnataka.
WORLD : Sri Lanka.

**Euphorbia ligularia** Roxb. Fl. Ind. 2: 465. 1832; Sm. in Kew Bull. 36: 296. 1981. *E. nerifolia auct. non* L. 1753; FBI 5: 255; Gamble 2: 1277 (893); Brandis 558.

Small tree, up to 6 m tall; branchlets 5-angled, stipular spines paired, sharp, perisistent on spirally arranged tubercles. Leaves towards the ends of branches, fleshy, obovate or obovate-oblong, 14–20 × 4–5.5 cm, glabrous, smooth, apex rounded with apicule, margin entire, base cuneate to tapering; midrib prominent beneath, secondary and other nerves obscure, petiole 0. Cyathia in axillary, subterminal, short pedunculate cymes; involucre hemispheric; central flower male, sessile; lateral flowers bisexual, pedicellate. Capsule sharply 3-lobed, 3 mm.

Occasional in village hedges. Fls. & Frs.: Feb.–Apr. Vern.: Aku-Jemudu.

The latex is acrid, rubefacient, purgative and expectorant. It cures dermatitis. It is used to remove warts and cutaneous eruptions. The juice is employed in earache.

Kondapudi (EG), *CAB* 537 (MH); Alisagar (NZB), *BR* 7281; Sivanoor (MDK), *BR & CP* 11549; Surzapur (ADB), *TP & PVP* 4109; Pecheruvu (KNL), *PSPB* 9135.

INDIA : Western Peninsula.
WORLD : Sri Lanka, Baluchistan, Malay Islands.

**Euphorbia nivulia** Buch.-Ham. in Trans. Linn. Soc. London 14: 286. 1825; FBI 5: 255; Gamble 2: 1297 (893); Brandis 558.

Armed, deciduous, succulent tree, up to 10 m tall; trunk straight; branches whorled, branchlets cylindrical with stipular, well developed, divaricate, persistent, sharp spines on small corky bases arranged in regular rows; wood white, soft, even-grained. Leaves obovate to oblanceolate, 8–20 × 3–4 cm, glabrous, apex obtuse, entire, base cuneate, nerves not conspicuous. Cyathia in 2–4 cm long, terminal, lax cymes. Capsules trigonous, 6 x8 mm, glabrous; seeds 4 mm, 4 angular, smooth.

Common along the rocky areas in open dry forests. Fls.: Jan.–Mar. Frs.: Apr.–May. Vern.: Bontha jemudu, Aku-jemedu, Erra-kalli. Modu-remudu, Jemudu.

The juice of the leaves is given as a purgative and diuretic. It is said to be used for earache, mixed with neem oil. It is applied externally in rheumatism. The bark is suitable for making composite cork boards with Hevea rubber latex as binder.

Upper Ahobilam (KNL), *SSR* 15286; Tiryani RF (ADB), *PSPB* 9159; Annavaram (EG), *SKW* 1480 (BLAT); Dummagudem (EG), *JSG* 500 (CAL); Macherla RF (GTR), *VRK* 3819; Laxmapur RF (NZB), *TP & BR* 6263.

INDIA : Throughout.
WORLD : Myanmar.

**Euphorbia tirucalli** L. Sp. Pl. 452. 1753; FBI 5: 254. Gamble 2: 1276 (893); Brandis 558.

Small, succulent tree, up to 5 m tall; bark greenish-brown, smooth; wood white or grey, close grained; branches spreading, unarmed, cylindric; branchlets terete, polished, whorled or fascicled, articulate. Leaves small, linear-oblong, deciduous, 3–12 × 1–2 mm, glabrous, apex obtuse to subacute, margin entire, base cuneate, petiole 0 or very short. Cyathia few, terminal. Capsule subglobose, glabrous, smooth, 6 × 8 mm; seeds ovoid, 3–4 mm across, smooth, caranculate.

Introduced, in field hedges and roadsides. Fls.: Jan.–Mar. Frs.: Apr.–May. Vern.: Kalli, Tirukalli, Manchi Jamudu, Kanchi Jemudu, Sanna Jamudu.

The latex is vesicant, rubefacient and is used as an application for warts, rheumatism and toothache. It is also used in ear-ache and toxic to rats and fishes. Wood is fairly hard used for rafters, toys, venering purposes. It gives a charcoal, suitable for use in gun powder.

Bukkapatnam (ATP), *TP & NY* 2102; Bheemaram (ADB), *GO & DAM* 5184; Palakonda (CDP), *CS* 8280; Mamidipalli (NZB), *BR & KH* 9013.

INDIA : Cultivated and then naturalized throughout.
WORLD : Native of Africa.

## **EXCOECARIA** Linnaeus

**Excoecaria agallocha** L. Syst. Nat. ed. 10: 1288. 1759; FBI 5: 472; Gamble 2: 1344 (941); Brandis 585.

Evergreen tree, up to 5 m tall; bark grey, smooth, shining, lenticellate; wood white, soft, fine to medium and even-textured; branchlets terete, interrupted. Leaves alternate, thickly-coriaceous, elliptic or ovate-elliptic, orbicular, 3–6 × 1.5–4 cm, glabrous, apex shortly acuminate, margin entire or shallowly sinuate-crenate, base acute; secondary nerves ca. 8 pairs, rather obscure; petiole 1–2.5 cm. Flowers greenish, minute, fragrant; bract rounded, fleshy; male flowers sessile, numerous in catkin-like, to 10 cm long spikes; female flowers pedicellate in 2–3.5 cm long racemes. Capsule deeply lobed, depressed, 1.6–2.5 cm in dia.; seeds sub-globose, smooth.

Common in tidal forests and swamps on coast. Fls. & Frs.: May–Aug. Vern.: Thilla, Chilla, Tella.

Wood is used for rough building and carpentry and it is also used as fair packing case; used also for making charcoal and as fuel. Sap milky and poisonous, which is injurious to skin and eyes. It is also called the blinding tree, producing blisters on the skin. Leaves are poisonous. The bark has strong emetic and purgative properties, used as fish poison.

Indalwai (EG), *PGS & NRM* 9312; Kothapalem (GTR), *VRK* 5763 & 5775; Ipur (NLR); *SKW* 6649 (BLAT); Harbour area (VSKP), *TAR* 7188 (CAL); Coringa (EG), *SKW* 1481 (BLAT).

INDIA : Peninsula.
WORLD : Sri Lanka, Malaysia to North Australia and Pacific.

## GIVOTIA Griffith

**Givotia moluccana** (L.) Sreem. in Taxon 24: 696. 1975. *Croton moluccanum* L. Sp. Pl. 1005. 1753 excl. ref. Burmann. *Givotia rottleriformis* Griff. in Calcutta J. Nat. Hist. 4: 388. 1844; FBI 5: 395; Gamble 2: 1342 (938); Brandis 578; Beddome Fl. Sylv. S. India t. 285. 1875.

Deciduous tree, up to 20 m tall; bark smooth, peeling off in circular thick bosses; wood white; branchlets fulvous tomentose. Leaves chartaceous, broadly ovate or orbicular, 5–15 × 5–20 cm, glabrous above, yellowish tomentose beneath, apex acuminate, margin irregulary dentate, glandular, base cordate, 5-nerved; petiole to 10 cm with 2 glands. Flowers yellow in 10–15 cm long, axillary or subterminal racemose or paniculate cymes. Drupe oblong or subglobose, to 3 × 2.5 cm, brown, smooth; seed one, globose, hairy.

Occasional, in dry deciduous forests at 400–1000 m. Leaf fall: Feb.–Mar. New foliage: Apr.–May. Fls.: May–July. Frs.: Jan.–Mar. Vern.: Konda Ponaku, Tella Poliki, Yella Dabba.

Wood soft, very light, used for carved images, toys, lacqured articles and catamerans. The seeds give an oil.

Kalasamudram RF (ATP), *SSR* 13752; Mannanur (MBNR), *SSR* 15532; Diguvametta (PKSM), *SSR* 16103; Madharam (ADB), *GO* 5467; Palakonda (CDP), *DAM* 5959; Chelama (KNL), *JSG* 78056 (DD); Both RF (NZB), *BR* 9065.

INDIA : Western Peninsula.
WORLD : Sri Lanka.

## GLOCHIDION J.R. Forster & J.G.A. Forster *nom. cons.*

### KEY 1 (Flowering)

1. Anthers 4 or more:
   2. Flowers in axillary fascicles ........ ***G. tomentosum***
   2. Flowers in umbellate clusters on short usually supra- axillary peduncles ... ***G. zeylanicum***
1. Anthers 3:
   3. Female flowers pedicellate ........ ***G. velutinum***
   3. Female flowers sessile ........ ***G. ellipticum***

### KEY 2 (Vegetative)

1. Branchlets hairy:
   2. Branchlets zig-zag. Secondary nerves strong, pubescent beneath ........ ***G. velutinum***

2. Branchlets straight. Secondary nerves slender, densely tomentose beneath ........................ *G. tomentosum*

1. Branchlets glabrous:
   3. Leaves oblong or ovate-lanceolate, up to 17 × 8 cm. .................. *G. zeylanicum*
   3. Leaves elliptic-oblong, up to 11 × 4 cm .................. *G. ellipticum*

KEY 3 (Fruiting)

1. Capsule glabrous, not clearly lobed .................. *G. zeylanicum*
1. Capsule puberulous to pubescent, clearly lobed:
   2. Capsule 4-locular .................. *G. velutinum*
   2. Capsule 5-locular:
      3. Capsule with long stalk .................. *G. tomentosum*
      3. Capsule with sessile or very short stalk .................. *G. ellipticum*

**Glochidion ellipticum** Wt. Ic. Pl. Ind. Orient. t. 1906. 1852; FBI 5: 321; Gamble 2: 1308 (915); Brandis 574.

Evergreen, monoecious tree, up to 8 m tall; branchlets slender, 4-angular, glabrous. Leaves coriaceous, elliptic-oblong, 3–11 × 1.5–3.5 cm, glabrous on both surfaces, apex acuminate, margin entire, base acute, often oblique; secondary nerves 4–6 pairs; petiole to 4 mm. Flowers in axillary fascicles, male and female flowers mixed, greenish yellow; male flowers pedicellate, female flowers sessile. Capsule depressed globose, 1 cm across, 4–5-lobed, pubescent; seeds red.

Occasional on the slopes of moist dry forests of Visakhapatnam district. Fls. & Frs.: Throughout the year. The paste of the seed is antiallergtic.

Anantagiri (VSKP), *PSPB* 11805. Birmisala (VSKP), *GVS* 47334 (MH).

INDIA : Peninsula.

**Glochidion tomentosum** Dalz. in Hooker's J. Bot. Kew. Gard. Misc. 2: 38. 1850; FBI 5: 309; Gamble 2: 1307 (914); Brandis 573.

Small tree, up to 8 m tall; monoecious; bark black, rough; branchlets tomentose. Leaves elliptic-oblong, 8–14 × 3.5–6 cm, apex acute-acuminate, margin entire, base obliquely rounded, pubescent on nerves above, finely grey tomentose beneath; secondary nerves ca. 7 pairs; petiole to 3 mm. Flowers pedicellate in axillary fascicles. Capsule depressed globose, 7 mm across, conical, 10-12 lobed, pubescent, beaked.

Occasional, in the hills of Visakhapatnam and East Godavari at 800–1200 m. Fls.: Jan.–Feb. Frs.: Apr.–July.

Galikonda (VSKP), *SSR* 15929; Maredumilli (EG), *GVS* 24273 & 24695 (MH).

INDIA : Western Peninsula.

**Glochidion velutinum** Wt. Ic. Pl. Ind. Orient. t. 1907. t. 2. 1852; FBI 5: 322; Gamble 2: 1307 (914); Brandis 574.

Small tree, up to 8 m tall; monoecious, bark brown and rough; wood crimson; branchlets zig-zag, velutinous. Leaves thinly coriaceous, elliptic-oblong or elliptic-lanceolate, 3.5–10 × 2–4 cm, apex acute, margin entire, undulate, base unequal, acute, thinly pubescent above, densely pubescent beneath; secondary nerves 6–8 pairs; petiole to 3 mm. Male flowers yellow; female green. Capsule orbicular, depressed, 8-lobed, 0.4–0.7 cm, pubescent; seeds bright red in colour, flattened.

Occasional, along the streams in moist deciduous forests, at 600–1000 m . Fls.: Apr.–May. Frs.: June–Aug. Bark used in tanning industry.

Tirumala (CTR), *SSR* 13776; Gundlabrahmeswaram (KNL), *JLE* 16941 (MH); Chintapalli (VSKP), *SKW* 2844 (BLAT); Galikonda (VSKP), *GVS* 11596 (MH); Sunkarimetta (VSKP), *HS* 20531 (BLAT).

WORLD : Myanmar, Bhutan.

**Glochidion zeylanicum** (Gaertn.) Juss. Euphorb. Gen. Tent. 107. 1824; FBI 5: 310; Gamble 2: 1306 (914); Brandis 573. *Bridelia zeylanica* Gaertn. Fruct. 2: 128. t. 109. 1790.

Tree, up to 8 m tall; bark grey thin, corky; branchlets zig-zag, coppery, glabrous. Leaves coriaceous, ovate-lanceolate or oblong, 13–17.5 × 5–7.5 cm, glabrous, glaucous above, paler beneath, apex sub-acuminate, entire, base unequal, truncate-rounded; secondary nerves 6–10 pairs; petiole 5 mm, thick. Flowers greenish, in axillary subumbellate clusters. Capsule globose, 0.6 × 0.8 cm, obscurely lobed, glabrous, often beaked; seeds shining, wedge-shaped.

Common, along the streams and in moist places of semi-evergreen forests. Fls. & Frs.: Throughout the year. Vern.: Pageri, Itepulla, Neeru mamidi.

Fruits used as cooling, restorative; Leaves used in itches. Bark used as stomachic.

Tirumala (CTR), *SSR* 15269 & 15284; Gundlabrahmeswaram (KNL), *SSR* 15995; Saddicudumadugu (ATP), *PSPB* 9173; Dummakonda (EG), *GVS* 68631; Sapparla gadda (VSKP), *GVS* 42781 (MH).

INDIA : Tamil Nadu, Orissa, Karnataka.
WORLD : Sri Lanka, Malaysia, Bangaladesh.

## HOMONOIA Loureiro

**Homonoia comberi** (Haine) Merr. in Lingnan Sci. J. 19: 188. 1980; Raiz. in Ind. For. 24: 488. *Lasiococca comberi* Haine in Kew Bull. 1920: 70. 1920; Gamble 2: 1334 (933).

Evergreen tree, up to 8 m tall; dioecious, often buttressed; bark pale or white; wood brown and white; branchlets puberulous, greyish. Leaves coriaceous, slightly panduri-form or obovate, 7–17 × 2.3–5 cm, glabrous on both surfaces, glandular especially on the veins of upper surface, dark green above, paler beneath, apex abruptly-acuminate, margin slightly undulate; base cordiform, primary lateral nerves 6–12 pairs rather strong; petiole 4 to 8 mm, tomentose. Male flowers short, pedicellate in axillary racemes; female flowers solitary, long peduncled in axils of terminal leaves. Capsule 3-lobed, 1.2 cm dia., depressed, covered with bristles or conical tubercles; seed subglobose, brown, smooth.

Occasional along rocky ravines, Visakhapatnam and Chittoor districts above 700 m. Fls.: June–Aug. Frs.: Sept.

Kambakkam (CTR), *PSPB* 11856; Minumuluru (VSKP), *GVS* 30007 (MH); Paderu (VSKP), *SKW* 2324, 2325, 2326 (BLAT).

INDIA : Orissa.

## **JATROPHA** Linnaeus

**Jatropha curcas** L. Sp. Pl. 1006. 1753; FBI 5: 383; Gamble 2: 1340 (937); Brandis 576.

Small tree, up to 5 m tall; monoecious; bark greenish-white, smooth, peeling off in thick flakes; wood white, very soft; branchlets glabrous. Leaves orbicular-cordate, 5–12 × 3.5–13 cm, glabrous, shallowly 3 to 5-lobed, lobes triangular, acuminate, entire, base cordate; petiole to 15 cm. Flowers greenish-white in axillary cymes. Capsule globose, 3 cm across, rugose.

Common, semi-wild in forests and villages; and also grown as hedge plant along roadsides and out skirts of villages. Fls.: Apr.–June. Frs.: Dec.–Jan. Vern.: Adavi amudam, Nepalamu, Adavi-Nepalamu, Pedda-Nepalamu, Kondamudamu.

The seeds give a medicinal oil. The seeds are anthelmintic and are used as rat poision; the plant juice relieves tooth-ache and is also a purgative; the oil extracted from the seeds is considered to be a good substitute for petrol. The leaf juice is known to releive piles, inflamation, while the sap is used for wound dressing and other skin infections.

Devipatnam (EG), *SSR* 15254; G. Madugula (VSKP), *SSR* 15958; Nizampatnam (GTR), *VRK* 6800; Narsingi (MDK), *BR & CP* 11528; Kamareddy (NZB), *PSPB* 9160; Kadam (ADB), *GO & BR* 9411; Bukkapatnam (ATP), *NY* 1036.

Indigenous in tropical America, cultivated in most tropical countries, naturalized in India.

## **MACARANGA** Du Petit-Thouars

1. Male inflorescence axis zig zag; bracts ovate, with a glandular appendage. Stamens 6–8, basally connate. Leaf margin clearly toothed .......... ***M. indica***

1. Male inflorescence axis not zig-zag; bracts spathulate, without a glandular appandage. Stamens 3, free. Leaf margin entire or minutely toothed ........................................ ***M. peltata***

**Macaranga indica** Wt., Icon. Pl. Ind. Orient. t. 1883, 1852 & t. 1949. f. 2. 1853; Beddome Fl. Sylv. S. India t. 237 A. 1872; FBI 5: 446; Gamble 2: 1326 (927); Brandis 592.

Evergreen tree, up to 12 m; bark grey, smooth; wood greyish-red; branchlets stout, glaucous; young shoots tawny pubescent. Leaves broadly ovate or sub-orbicular, peltate, 12–19 × 11–18 cm, glabrous above, tawny-pubescent and glandular beneath, apex acuminate, margin toothed; base 3-ribbed with 6 basal nerves; secondary nerves 6–10 pairs, prominent, running nearly to margin and giving short branching ending in a tooth; tertiary nerves scalariform, conspicuous; petiole 6–12 cm. Stipules ovate, ciliate, caducous. Female panicles pyramidally branched, 8 cm long, hairy, flowers 1–3 at the node; 2.5 mm across. Capsule globose, 7 mm across, densely pubescent when young.

Rare, in ravines of semievergreen forests at 1000–1400 m. Fls.: Sept.–Oct. Frs.: Oct.–Nov.

Wood soft and suitable for fencing, huts and tea boxes. The bark exudes a gum 'kino' used to cure sores by external application.

Sunkarimetta (VSKP), *GVS* 44354 (CAL).

INDIA : Peninsula, E. Himalaya, Assam, Andaman Islands, Bengal.
WORLD : Myanmar.

**Macaranga peltata** (Roxb.)Muell.-Arg. in DC. Prodr. 15: 1010; Gamble 2: 1326 (928). *Osyris peltata* Roxb. Fl. Ind. 3: 755. 1832. *Macaranga roxburghii* Wt. Ic. t. 1949. f. 4. 1858; FBI 5: 448; Brandis 592.

Evergreen tree, up to 13 m tall; dioecious, bark pale with lenticels; wood pale brown, soft; branchlets striate, densely grey-tomentose, glaucous. Leaves coriaceous, ovate, peltate, 12.5–30 cm dia, puberulous or villous with red glandular below, glabrous above, apex acuminate, margin finely toothed, acute, with usually 9 radiating ribs at the base, the midrib with 7–12 pairs of nerves, the two next nerved on the outerside, transverse nervules parallel; petiole to 20 cm. Flowers greenish yellow in panicles. Capsule globose, 6 mm, glandular, echinate when young; seeds black.

Occasional, along moist valleys and hills above 700 m.. The tree is resinous and fast growing. Fls.: Oct.–Nov. Frs.: Apr.–June. Vern.: Boddi, Putta-tamara, Konda japhera, Konda tamara.

A reddish gum "kino" exudes from the cut branches bases of petioles, young shoots and fruits, used for sizing paper and for taking impressions of leaves, coins medallions etc., it is used also substitute for gum arabic. Fruits edible. Wood soft and of no value. Leaves used as meal plates.

Muthyalu forest (SKLM), *SSR* 15599; way to Thumburatheerthum (CTR), *SSR* 16167; Anantagiri (VSKP), *PSPB* 11804.

INDIA : Peninsula, Andamans.
WORLD : Sri Lanka, Tenasserin.

## MALLOTUS Loureiro

1. Leaves penninerved .......... ***M. resinosus***
1. Leaves 3-ribbed at the base:
   2. Leaves alternate, red glandular beneath. Capsule red glandular .......... ***M. philippensis***
   2. Leaves opposite and alternate, yellow glandular beneath. Capsule stellate pubescent. .......... ***M. rhamnifolius***

**Mallotus philippensis** (Lam.) Muell.-Arg in Linnaea 34: 1196, 1865; FBI 5: 442; Gamble 2: 1322 (924); Brandis 590; Beddome Fl. Sylv. S. India t. 289. 1872 ('philippinensis') *Croton philippensis* Lam. Encycl. 2: 206. 1786.

Evergreen tree, up to 12 m tall; bark grey, nearly smooth, exfoliating in small woody plates; wood grey to light red, smooth, fairly straight-grained, medium fine and even-textured; branchlets striate, densely grey-tomentose. Leaves elliptic-ovate or ovate-lancecolate, 4–19 × 2–10 cm, glabrous above, pubescent and numerous red orbicular glands beneath, apex acuminate, margin entire, base cuneate; strongly 3-nerved, secondary nerves 4–7 pairs above basal ones; petiole to 5.5 cm, with 2 small sessile glands one on each side of the summit. Flowers pale yellow in axillary or terminal spikes, male spikes long; female spikes short. Capsule 3-lobed, red glandular, loculiciadally 3-valved, 1 cm across, smooth; seeds globose, black, 4 mm across with a moderately thick brittle testa.

Common, on hills of deciduous forests and open scrub jungles above 700 m. It is a good soil improver; sensitive to drought; coppices well and produces root suckers. Fls.: Oct.-Dec. Frs.: Feb.–Mar. Vern.: Kumkuma chettu, Sindhuramu, Vasantha Gandhamu, Chandiramu, Hanumanthuni-Bottu, Naga-rakta.

The red dye known as "Kamala dye" obtained from the red glands on the fruit extensively used for dyeing wool and silk. Kamala dye has been used for colouring food stuffs and beverages. Dye used also as Sindhur or Kumkum by women. The fruit is bitter, anthelmintic and styptic, used in external applications for parasitic affections of the skin. The wood of no value as timber but forms a good fuel; it is suitable for rafters, tool handles, match boxes and small turnery articles like bobbins, cotton reels, pen holders and rulers. Leaves are used as fodder.

INDIA : Tamil Nadu, Karnataka, Orissa.

WORLD : Western Himalaya to Taiwan, throughout Malaysia to Australia and Melanesia.

**Mallotus resinosus** (Blanco) Merr. Sp. Blanco 222. 1918 & Enum. Philip. Fl. Pl. 2: 436. 1923; Airy Shaw in Kew Bull. 26: 294. 1972 & 35: 656. 1980. *Adelia resinosa* Blanco. Fl. Filip. (2): 562. 1845. *Mallotus muricatus* (Wt.) Beddome Fl. Sylv. S. India t. 208. 1873, in part, non Muell. Arg. 1865; FBI 5: 436; Gamble 2: 1322 (925); Brandis 589. *Mallotus intermedius* (Baillon) Balakr. in Bull. Bot. Surv. India 10: 245. 1968.

Tree. Leaves opposite, rhombic-lanceolate or oblanceolate; up to 12.5 × 3.2 cm, apex acuminate, margin crenate, entire with age, acute; secondary nerves 8–10 pairs; petiole to 1 cm. Flowers in spikes. Capsule yellow-glandular, with spines.

Rare, Horsely konda in Chittoor (cf Fischer in Gamble), Tirumala (Thammanna *et al.* 1976).

INDIA : Peninsula.

**Mallotus rhamnifolius** Muell.-Arg. in Linnaea 34: 196. 1865; FBI 5: 440; Gamble 2: 1322 (924); Brandis 589.

Small tree; branchlets fulvous tomentose. Leaves alternate and opposite, oblong or lanceolate, 7.5–11.8 × 4.3–6.8 cm, glabrous above, pubescent beneath; apex acuminate with mucro, margin entire, base rounded or acute; 3-ribbed, secondary nerves 4–7 pairs above basal-ones, tertiary nerves parallel to the secondary veins, prominent beneath; petiole to 4.5 cm. Spikes erect, axillary with brown tomentum; male flowers in clusters; female flowers solitary. Capsule globose, 3-valved, finely stellate-pubescent.

Occasional, in hills of moist forests, especially near the streams above 700 m. Fls.: Jan.-Feb. Frs.: Mar. Vern.: Konda-kunkuma.

Galikonda (VSKP), *SSR* 19642, 19614; Sirivaka (EG), *CAB* 5014; Anantagiri (VSKP), *GVS* 21729 (MH); Kambakkam hills (CTR), *PSPB* 9180; Saddicudumadugu (CTR), *PSPB* 9176.

INDIA : Southern districts of Western Peninsula.
WORLD : Sri Lanka.

## MANIHOT P.Miller

**Manihot glaziovii** Muell. Arg. in C. Martius Fl. Bras. 11(2): 446. 1874; Gamble 2: 1346 (942); Airy Shaw, Euphorb. Borneo 174. 1975.

Tree, up to 8 m tall. Leaf-lobes 3 to 6, elliptic to obovate, 8–18 × 4.5–8 cm, glabrous, dark green above, glaucous beneath, apex obtuse with narrow apicule, margin entire; secondary nerves regular, 12–15 pairs, midvein raised and prominent above; petiole peltate, 5–20 cm. Flowers greenish-yellow in racemose, lax panicles. Capsule green, globose, 3 cm across, pendulous.

Introduced and cultivated, planted in gardens. Fls.: & Frs.: Throughout the year. Vern.: Siyara-Rabbaru, Sima-Pittati, Badama.

Rubber is obtained from the latex of the plant and yield is poor than that of *Hevea*. Leaves, flowers, fruits and root bark are cyanogenetic. Kernels yield a greenish yellow drying oil with slight bitter taste.

Anantapur town (ATP), *SSR* 18583.

INDIA : Nilgiri hills, Malabar, Assam and Orissa.
WORLD : Native of North Brazil.

## **MARGARITARIA** Linnaeus f.

**Margaritaria indica** (Dalz.) Airy Shaw in Kew. Bull. 20: 387. 1966. *Prosorus indicus* Dalz. in Hooker's J. Bot. Kew. Gurd. Misc. 4: 346. 1852; Gamble 2: 1294 (905). *Phyllanthus indicus* (Dalz.) Muell.-Arg. in Linnaea 32: 52. 1863; FBI 5: 305; Brandis 571.

Evergreen tree, up to 20 m tall; bark smooth, whitish; branchlets glabrous, lenticellate. Leaves membranous, elliptic, 7.5–12.5 × 3.5–5 cm, glabrous, glaucous beneath, apex acute-subacuminate, margin entire, base rounded-acute; secondary nerves 8–9 pairs; petiole to 8 mm. Male flowers very small, green, in fascicles; female flowers larger pedicellate in cluster. Capsule green, smooth, depressed globose, with 3 vertical furrows, 0.8–1.2 cm; seeds blue, subtrigonous.

Rare, in the semi-evergreen forests along streams 600–900 m. Fls.: Apr.–June. Frs.: Aug.–Oct.

Siddulaiah kona (NLR), *SSR* 19675; Mamandur (CTR), *GVS* 31953 (MH).

INDIA : Tamil Nadu, Orissa.

WORLD: Sri Lanka, S.E. Asia, scattered through Malaysia to New Guinea and Queensland.

## **PHYLLANTHUS** Linnaeus

1. Fruit a dry capsule. Anthers connate without apiculus; dehiscence vertical ... ***P. polyphyllus***
1. Fruit a fleshy drupe. Anthers cohering by the connectives, which end in an apiculus, dehiscence longitudinal:
   2. Leaves > 4 × 1.8 cm. Drupe angular ........ ***P. acidus***
   2. Leaves < 2.5 × 1.5 cm. Drupe not angular:
      3. Leaves on the branchlets very many, 100 or more, apex apiculate. Style-arms spreading. Fruit depressions shallow. ........ ***P. emblica***
      3. Leaves on the branchlets many up to 50, apex rounded or retuse. Style arms recurved. Fruit depressions deep. ........ ***P. indofischeri***

**Phyllanthus acidus** (L.) Skeels, U.S.D.A. Bur. Pl. Industr. Bull. 148: 17. 1909. *Averrhoa acida* L. Sp. P. 428. 1753. *Cicca acida* (L.) Merr. interpr. Herb. Amb. 314. 1917. *Cicca distycha* L. Mant. 1: 124. 1767. *Phyllanthus distichus* Muell.-Arg. in DC. Prodr. 15(2): 413. 1866; FBI 5: 304.

Small to moderate-sized tree; branches thick, tuberculate and closely marked with the scars of the deciduous, leafy, slender branchlets. Leaves ovate or elliptic to obovate, 5–7.5 × 2–3 cm, glabrous, pale-glaucous beneath, base rounded; petiole to 3 mm; stipules minute, toothed. Flowers in slender racemes on leafless branchlets; male flowers red, minute; females green, larger. Fruit yellowish, depressed-globose, 6–8-lobed, 1.6 cm dia., pericarp fleshy, acid.

Cultivated, in gardens, house compounds etc. Fls.: May. Frs.: June–July. Vern.: Racha usiri, Chinna usiri. Fruit is sour, edible; cooked or pickled or is made into jam or jelly.

SK University (ATP), *NY* 997.

**Phyllanthus emblica** L. Sp. Pl. 982. 1753; FBI 5: 289; Brandis 570; Beddome Fl. Sylv. S. India t. 258. 1872. *Emblica officinalis* Gaertn. Fruct. 2: 122. t. 108. f. 2. 1790; Gamble 2: 1295 (906).

Deciduous, monoecious tree, up to 15 m tall; branches spreading; bark greyish-white, smooth, peeling off in flakes; wood red or reddish-brown, more or less shallowly and irregularly interlocked grained, coarse and uneven textured; branchlets glabrous and finely pubescent. Leaves oblong, 1.3–2.3 × 0.6–1 cm, glabrous, light green, apex obtuse, apiculate, margin entire, base truncate; petiolule 1 mm. Flowers greenish-yellow in axillary fascicles on the leaf bearing branchlets, often on the naked portion below the leaves. Drupe pale yellow, depressed globose, 1–2 cm in dia., 3-celled, 6-seeded; seeds reniform, 3-gonous, shining, reddish-brown, small, 6 × 7 mm.

Very common, in dry deciduous forests above 600 m. It is recommended for afforestation on dry rocky areas and refractory sites. It is a light demander and sensitive to drought; coppices well and produces root suckers. Leaf fall: Mar.–Apr. Fls.: Mar.–May. Frs.: Oct.–Feb. Vern.: Pedda usiri chettu, Triphalamu, Pulla usirika, usiri. Eng.: Indian gooseberry.

Fruits acidic; cooling, refrigerent, diuretic, and laxative; contains much gallic acid, and are made into pickles. Wood hard, heavy, durable under water, useful for poles, agricultural implements and furniture; it yields excellent charcoal and firewood.

Tirumala (CTR), *SSR* 13782; Chintala forest (PKSM), *SSR* 16124; Eturnagaram (WGL), *SSR & KSM* 18528; SK University Campus (ATP), *NY* 996; Cuddapah *JSG* 11092 (DD); Balugram (KNL), *TP & RVR* 2751; Anupu (GTR), *VRK* 3964; Sirpur RF (ADB), *GO & PVP* 4911; Bhimnagar (NZB), *TP & BR* 6207.

INDIA : Common throughout.
WORLD : China, North Myanmar, Malay Islands, Sri Lanka.

**Phyllanthus indofischeri** Bennet in Ind. For. 109: 221. 1983. *Emblica fischeri* Gamble in Kew. Bull. 1925: 330. 1925 and Fl. Pres. Madras 2: 1295 (906).

Tree, up to 7 m tall; monoecious; branchlets puberulous. Leaves many on the branchlets, up to ca.60, oblong, 0.4–1.1 × 0.3–0.5 cm, glabrous, green above and glaucous below, apex retuse, margin entire, base truncate; secondary nerves obscure; subsessile. Flowers greenish, subsessile in axillary fascicles; male flowers many in upper axils, female flowers few in lower axils. Drupe globose, fleshy, with 3 pyrenes; seeds trigonous.

Occasional in Chittoor district. Fls.: Nov. Frs.: Jan.

Balapalli (CDP), *JLE* 15018; Mahanandi (KNL), *JLE* 25475 (MH); Talakona (CTR), *PSPB* 11861.

INDIA : Peninsula.

**Phyllanths polyphyllus** Willd. Sp. Pl. 4: 586. 1805; FBI 5: 290; Gamble 2: 1288 (901).

Deciduous small tree, up to 8 m tall; branchlets terete. Leaves oblong, 0.5–2 × 0.2–0.5 cm, glabrous, apex apiculate, margin entire, base rounded; secondary nerves 6–8 pairs; petiole to 8 mm. Flowers small, light yellow, monoecious, along the leaf rachis on ventral side. Capsule greenish, globose, 3-valved, 4 mm across, 3-lobed; seeds triquetrous.

Occasional, in moist deciduous forests. Fls.: Dec.–Feb. Frs.: Feb.–Mar.

Talakona (CTR), *SSR* 13791; Satrasala (GTR), *VRK* 6777; Tammapalli (KNL), *JSG* 10903 (DD).

INDIA : Peninsula.
WORLD : Sri Lanka.

## SAPIUM R. Brown

1. Leaves crowded at the ends of the branchlets, elliptic or obovate, margin crenate-serrate; very young leaves green; petiole eglandular. Flower sessile, reddish on robust spikes .................. ***S. insigne***
1. Leaves not crowded, elliptic-lanceolate, margin entire; very young leaves red; petiole with 2 glands at the top or base of lamina. Flowers pedicellate, greenish on slender spikes .................. ***S. eugeniifolium***

**Sapium eugeniifolium** Buch.-Ham. (in Wall. Cat.no. 7970. 1847) (as *eugeniafolium*) ex Hook. f. FBI 5: 470. 1880; Brandis 585.

Tree, monoecius, up to 15 m tall; crown spreading; bark greyish, smooth. Leaves coriaceous, elliptic-lanceolate, 3–11.5 × 2.3–4.6 cm, red when very

young branches fulvous-pubescent. Leaves broadly ovate, 10–17 × 6–11 cm, glabrous, apex acuminate, margin entire-irregularly crenate, base truncate or cordate, 3-ribbed; secondary nerves 5-pairs above basal ones; petiole to 12 cm. Male flowers in pendulous, many-flowered racemes; female flowers on long pedunculate racemes. Drupe oblong, woody; seeds smooth, dark brown.

Rare, along streams and moist hilly regions of Visakhapatnam and Srikakulam, also cultivated for ornament. It coppices vigorously and produces root suckers. Fls.: Feb. Frs.: March–May. Vern.: Botku, Eru Ponaku. Eng.: False white teak.

Wood is extremely light, not durable, used for packing cases, tea boxes and yokes, drums, barrels, carved images. The fruit pulp is reported to be sweet and edible. A decoction of the shoots is said to relieve flatulence and is also used for the treatment of swelling.

Regadi (SKLM), *GVS* 62480; Narsipatnam (VSKP), *GVS* 30078 (MH).

INDIA : Throughout.
WORLD : Sri Lanka, Indo-China, S. China, W. Malaysia.

## STILAGINACEAE

As stated by Airy Shaw (Kew. Bull. 23: 277. 1969) "Taxonomically *Antidesma* is a trouble some genus". Airy Shaw (Willis; A Dictionary of Flowering Plants and Ferns ed. 7: 1076. 1966) reinstated the family Stilaginaceae for the isolated genus *Antidesma* Linn. (*Stilago* Linn.), long included in Euphorbiaceae. He states that this group forms connecting link between the Euphorbiaceae and the Icacinaceae. Airy Shaw (l.c) add "The Small Drupaceous fruits with their characteristic foveolate-reticulate often flattended endocarp differ widely from anything found in the Euphorbiaceae, but show much similarity to those of several genera of Icacinaceae. The genus *Antidesma* may be regarded as to some extent intermediate between Icacinaceae and Euphorbiaceae, and fully merits the family rank given it by Agardh, Lindley and others.

### **ANTIDESMA** Linnaeus

#### KEY 1

1. Spikes villous in panicles; ovary pubescent, perianth lobes 5–7 ................ ***A. ghaesembilla***
1. Spikes glabrous or pubescent, simple (often compound male spikes in *A. bunius*); ovary glabrous; perianth lobes 4:
   2. Male flowers pedicellate:
      3. Stamens 2 ................ ***A. acidum***
      3. Stamens 3–4 ................ ***A. menasu***
   2. Male flowers sessile:
      4. Male spikes slender, pistillode acute ................ ***A. zeylanica***
      4. Male spikes thick, pistillode truncate ................ ***A. bunius***

## KEY 2

1. Leaves tomentose beneath, apex obtuse ........ ***A. ghaesembilla***
1. Leaves glabrous beneath, apex acute or acuminate:
   2. Apex with mucro:
      3. Leaves elliptic-oblong, 9–20 × 2–5 cm ........ ***A. menasu***
      3. Leaves obovate-rhomboid, 4–9 × 3–4.5 cm ........ ***A. zeylanicum***
   2. Apex without mucro:
      4. Leaves dull, secondary nerves 8 pairs, conspicuous ........ ***A. bunius***
      4. Leaves shining, secondary nerves ca. 6 pairs, faint ........ ***A. acidum***

**Antidesma acidum** Retz. Obs. 5: 30. 1789; Sharma *et al.*, in Biol. 2: 127. 1977. *A. diandrum* (Roxb.) Roth Nov. Pl. Sp. 369. 1821; FBI 5: 361. Gamble 2: 1298 (908); Brandis 565.

Deciduous tree, up to 10 m tall; crown flat; bark smooth, grey; wood pinkish-grey, close grained; branchlets terete, glabrous. Leaves elliptic-lanceolate to obovate-lanceolate, 4–10 × 2–3.5 cm, glabrous, shining, pale beneath, apex acute to obtusely acuminate, margin entire, base acute; secondary nerves 6 pairs, faint; petiole to 4 mm. Flowers pedicellate in 8 cm long racemes. Drupe globose, 3 mm across, purplish-red, epicarp wrinkled when dry .

Common, in moist deciduous forests, usually near streams at 700–1100 m. Leaves turn red from January to March. Leaf fall: Mar.–Apr. New foliage: June–July. Fls.: May–June. Frs.: Nov.–Jan. Vern.: Pulleru, Gumadipella, Pullaru, Gumudu, Gurrapujalli.

Fruits edible, wood hard. Leaves used as vegetable and made into a preserve. Seeds yield a fatty oil.

Tirumala (CTR), *SSR* 13711; Guvvalacheruvu (CDP), *PSPB* 9196; Pangidi (KNL), *RVR* 2393; Kota (EG), *MSRS* 501 (CAL); Addatigala (EG), *SKW* 4019 (BLAT); Diguvametta (PKSM), *JLE* 22189 (MH); Minumuluru (VSKP), *STP* 20700 (BLAT); Ramjiguda (ADB), *TRS & TL* 83168; Jyothimonidi (VJN), *GVS* 68599 (MH); Paderu (VSKP), *SKW* 4234 (BLAT).

INDIA : Tamil Nadu, Karnataka, Orissa, and Maharashtra.
WORLD : Himalaya, Myanmar, S. China, Indo-China, Java.

**Antidesma bunius** Spreng. Syst. Veg. 1: 826. 1825; FBI 5: 358. Gamble 2: 1298 (980); Brandis 564.

Small evergreen tree, up to 6 m tall; bark smooth, grey; wood pinkish or pink; branchlets pubescent. Leaves elliptic-lanceolate or oblanceolate, 17.5 × 0.5 cm, shining, apex acuminate, margin entire, base acute; secondary nerves ca. 8 pairs, curved and looped. Flowers greenish, in pubescent, axillary, 22 cm long spikes; male flowers often compound; female flowers simple. Drupe reddish, globose-ovoid.

Rare, in semievergreen forests at 900–1500 m. Fls. & Frs.: June–Aug. Vern.: Anepu.

Fruits used for jams and jellies and in the preparation of wine. Leaves are acidic and edible, diaphoretic. Wood hard and yields pulp for card boards. Bark poisonous.

Chintapalli (VSKP), *SKW* 2821; Paderu (VSKP), *HS* 20770 (BLAT); Lambasingi (VSKP), *SKW* 2821 (MH).

INDIA : Orissa, Karnataka.
WORLD : China, Sri Lanka.

**Antidesma ghaesembilla** Gaertn. Fruct. 1: 189, t. 39. 1788; Muell-Arg. in DC. Prodr. 15: 251. 1866; FBI 5: 357; Gamble 2: 1298 (908); Brandis 564. *A. paniculata* Roxb. Fl. Ind. 3: 770. 1832. *A. pubescens* Roxb. Pl. Cor. t. 167. 1802 & Fl. Ind. 3: 770. 1832.

Deciduous tree, up to 10 m tall; bark grey, smooth or slightly cracked; wood red; branchlets greyish tomentose. Leaves chartaceous, elliptic, orbicular-oblong, 4–10 × 3–5.8 cm, glabrous above, tomentose beneath, apex obtuse, rounded or sub-acute, margin entire, base obtuse, subcordate; main lateral nerves 5 pairs; petiole to 1.8 cm. Flowers greenish in slender, tomentose, paniculate spikes. Male spikes to 8 cm; female spikes to 5 cm. Drupe subglobose, 3 mm across, reddish-purple.

Common, in moist deciduous forests at above 600 m. Fls.: Apr.–June. Frs.: Sept.–Oct. Vern.: Giridi chettu, Pulleru, Polari, Pulsur, Kathwpulitsi, Tella-pulseru, Nallaballi.

Leaves are eaten, also used as a flavouring. Fruit edible used for seasoning meat and fish preparation. Wood employed for light rafters in house construction.

Tirumala (CTR), *SSR* 13724; Seethampet (SKLM), *SSR* 15574; Gundla-brahmeswaram (KNL), *SSR* 15994 & 15995; Bijjur RF (ADB), *GO & PVP* 4881; Maredumilli (EG), *GVS* 24404; Chintapalli (VSKP), *GVS* 30083 (MH).

INDIA : Throughout.
WORLD : Himalaya, North Australia, Malaysia, S. China, Sri Lanka.

**Antidesma menasu** Miq. (in Hohen. Pl. Ind. exsicc. 1847. h. 104. nomen) ex Tul. Ann. Sci. Nat. Bot. 15: 215. 1851; FBI 5: 564; Gamble 2: 1297 (908); Brandis 565.

Tree, up to 13 m tall; bark rough, black; branchlets sparsely tomentose. Leaves chartaceous, elliptic-oblong, 9–20 × 2–5 cm, shining, apex obtuse or shortly acuminate with mucronate, margin entire, base oblique, acute; secondary nerves 6–8 pairs, prominent, looped with transverse nervules; petiole to 7 mm. Flowers greenish in terminal racemes. Drupe red, globose.

Occasional, in hills of semievergreen forests. Fls. & Frs.: Feb.–June. Fruits edible.

Minumuluru (VSKP), *SSR* 15964; Sapparla (VSKP), *SSR* 19647; Nulakamaddi (EG), *GVS* 24337 (MH).

INDIA : Peninsula.

**Antidesma zeylanicum** Lamk. Encycl. 1: 207. 1782; Muell.-Arg. in DC. Prodr. 15: 256. 1866; Gamble 2: 1297 (908); *A. alexitera* L. Sp. Pl. 1027. 1753 in part; FBI 5: 359; Brandis 565.

Small tree, up to 7 m tall; bark black, smooth; branchlets terete, glabrous. Leaves subcoriaceous, obovate-rhombiod or elliptic, 4–9 × 3–4.5 cm, glabrous, shining, apex obtusely or acutely apiculate, margin entire, base acute or cuneate; secondary nerves 6 pairs; petiole to 6 mm. Flowers cream coloured in slender spikes, male spikes to 9.5 cm, axillary or terminal, female spikes subterminal to 5 cm. Drupe globose, 5 × 4 mm, bluish red.

Occasional, on either sides of the streams in the deciduous forests at above 500–1400 m. Fls.: With two peaks during Mar.–Apr. and Sept.–Nov. Frs.: Throughout the year.

Galikonda (VSKP), *SSR* 15919, *PSPB* 11813; Sathnella river bed (ADB), *GO & PVP* 4218; Mudheli RF (NZB), *TP & BR* 6373; Dhaniyani cheruvu (ATP), *TP & NY* 839; Gulemallabad RF (KNL), *RVR* 1418.

INDIA : Peninsula.
WORLD : Sri Lanka.

## BISCHOFIACEAE

Genus *Bischofia* Bl. is recently excluded from Euphorbiaceae and placed in a separate family Bischofiaceae by Airy Shaw (Kew. Bull. 18: 252. 254. 1965 & 21(3): 327–329, 1967). According to him, the essentially pinnate nature of the leaves of *Bischofia* provides concrete evidence for the exclusion of the genus from Euphorbiaceae wherein the leaves are digitately compound.

### **BISCHOFIA** Blume

**Bischofia javanica** Bl. Bijidr. 1168. 1827; FBI 5: 345; Gamble 2: 1312 (918); Brandis 558; Beddome Fl. Sylv. S. India t. 259. 1872.

Densely foliaceous, deciduous tree, up to 20 m tall; bark grey with a brown tinge, exfoliating in angular scales; wood red, straight or shallowly and irregularly interlocked grained; branchlets glabrous. Leaves trifoliolate, petiole to 11.5 cm; leaflets coriaceous, elliptic or ovate-oblong, 10–16 × 5–8.2 cm, glabrous, green above, paler beneath, apex caudate-cuspidate, margin crenulate-serrate, base acute or rounded; secondary nerves 7–9 pairs, 1.3 cm apart; petiolule to 1 cm, for upper leaflet to 4.5 cm. Flowers green; male panicles 9–14 cm long, much branched; female panicles looser and less branched. Drupe reddish-brown, globose, fleshy, 0.7–1.6 cm across; endocarp 2-valved; style persistent; seeds 3 or 4, smooth, shining, brown.

Occasional, a characteristic tree of shade ravines of swamps and river banks and of valleys in hills of Visakhapatnam up to 1400 m. Fls.: Feb.-Mar. Frs.: May-June. Vern.: Nalupu-Mushthi, Panta.

Juice of leaves considered cure for sores. Bark of twigs contains tannin; seed contain a drying oil. Wood moderately hard and heavy, moderately durable, excellent for planking and other household purposes.

Sunkarimetta (VSKP), *SSR* 15914; Chintapalli (VSKP), *SSR & KSM* 16141.

INDIA : Southern and North Western region.
WORLD : Malaysia, Southern China, the Rhukyus, Polynesia and Australia, Taiwan.

## ULMACEAE

1. Leaves penninerved. Stamens twice the number of perianth lobes. Fruit samara. Plants deciduous ........ ***Holoptelia***
1. Leaves 3-nerved from base. Stamens as many as perianth lobes. Fruit a drupe. Plants evergreen:
   2. Plants dioecious. Leaf scabrous above, glaucous beneath ........ ***Trema***
   2. Plants with polygamous flowers. Leaves glabrous on both surfaces, green beneath .. ***Celtis***

## **CELTIS** Linnaeus

1. Flowers 4-merous. Leaves thin, base oblique, margin serrate at the upper half ........ ***C. tetrandra***
1. Flowers 5-merous. Leaves thick, base rounded or acute, margin entire throughout:
   2. Flowers few in racemes; style arms entire, pistillode absent. Twigs reddish-brown. Leaves dark brown when dry ........ ***C. timorensis***
   2. Flowers numerous in panicles; style arms 2-fid, pistillode present. Twigs grey-brown. Leaves green as fresh leaf when dry ........ ***C. philippensis***

**Celtis phillippensis** Blanco, Fl. Filip. 197. 1857, var. *wightii* (Planchon) Soep. in Steenis. Fl. Males. 1.8: 62. 1977. *C. wightii* Planchon Ann. Sci. Nat. Bot. (Ser. 3) 10: 307. 1848; FBI 5: 483; Fischer 3: 1349 (944); Brandis 596.

Evergreen tree, up to 12 m tall; bark brownish, smooth; branchlets appressed pubescent when young. Leaves thick coriaceous, ovate or elliptic-oblong, glabrous, 7.5–13 × 3.2–5.2 cm, apex acuminate, margin entire, base acute or cuneate, 3-ribbed, lateral veins conspicuous, running throughout the leaf, reticulately veined with transverse parallel nervules; petiole to 2 cm. Flowers yellow, pedicellate in axillary puberulous, 1 cm long, cymose; panicles bisexual flowers above, males below. Drupe elliposid, 1 cm across, red, obtuse.

Common, in dry forests of almost all districts to 1000 m. Fls.: Aug.–Sept. Frs.: Oct.-Nov. Vern.: Koti-Bira, Chipura, Kaki-Mushti.

Tirumala foot-path (CTR), *SSR* 13762; Tirumala hills (CTR), *KJ & SSR* 12577; Siddulaiah Kona (NLR), *SSR* 19570; Nallamalais (KNL), *JLE* 23753 (MH); Dornal (KNL), *JSG* 18695 (DD).

INDIA : Tamil Nadu.
WORLD : Tropical Africa, Madagascar, Myanmar, Indo-China, Hainan, Malaysia, North-Eastern Australia, Taiwan.

**Celtis tetrandra** Roxb. Fl. Ind. 2: 63. 1832; FBI 5: 482; Fischer 3: 1348 (944); Brandis 596. *C. serotina* Planchon. Ann. Sci. Nat. Bot. (Ser. 3) 10: 301. 1848.

Evergreen tree, up to 15 m tall; bark greyish-black, smooth; wood greyish-white with brown dots and streaks; young branches pubescent. Leaves coriaceous, obliquely ovate-elliptic, 3.5–10 × 1.4–5 cm, glabrous and green above, greyish puberulous beneath, apex acuminate, margin serrate above the middle, base obliquely acute, 3-ribbed, lateral nerves 2-pairs; petiole to 6 mm. Male flowers 2–5 borne on leafless new shoots, female flowers 1–2 in leaf axils. Drupe ellipsoid, ridged, 0.8 cm across, 1–2 on slender stalks, ripening yellowish orange.

Occasional in moist deciduous forests of Visakhapatnam at 1000–1500. Fls.: Feb. Frs.: Apr.-May. Vern.: Jabjaabal. Wood not hard.

Minumulur (VSKP), *SSR* 15943; Minumulur towards Paderu (VSKP), *GVS* 30001 (MH).

INDIA : Almost all the States.
WORLD : Malaysia, Myanmar, China.

**Celtis timorensis** Spanoghe in Linnaea 15: 343. 1841. *C. cinnamomea* Lindley ex Planchon, Ann. Sci. Nat. Bot. (Ser. 3) 10: 303. 1848; FBI 5: 482; Fischer 3: 1349 (944); Brandis 596.

Evergreen tree, up to 10 m tall; bark pale grey, smooth; wood dull white; young branches pubescent. Leaves membranous, ovate-elliptic, 5.5–11 × 2.2–5.2 cm, glabrous, blackish when dry, apex acute-acuminate, margin entire, base obtuse or rounded, 3-ribbed; basal lateral nerves nearly reaching the apex; petiole to 1.4 cm. Flowers small, whitish, pedicellate on 2 cm long, axillary, dichotomous racemes. Drupe ovoid, globose, 0.6 cm across, orange when ripe.

In dry and semi-evergreen forests of Nallamalais and hills of Visakhapatnam at 600–1500 m. Fls.: Mar.–Apr. Frs.: May. Vern.: Koti-Bira, Chipuru, Kaki-Mushti. Wood not useful.

Chintala (PKSM), *SSR* 16111 & 16113; Galikonda (VSKP), *RHB* 48761 (MH); Nagarjuna Konda (GTR), *KT* 9871 (CAL).

INDIA : Sikkim, Orissa, Tamil Nadu, Karnataka.
WORLD : Sri Lanka, Myanmar, Malay islands, Australia and Africa.

## **HOLOPTELIA** J.E. Planchon

**Holoptelia integrifolia** (Roxb.) Planch. in Ann. Sci. Nat. Bot. (Ser. 3) 10: 266. 1848; FBI 5: 481; Fischer 3: 1348 (943); Brandis 595. *Ulmus integrifolia* Roxb. Pl. Cor. t. 78. 1798 & Fl. Ind. 2: 68. 1832; Beddome Fl. Sylv. S. India t. 370. 1873.

Deciduous tree, up to 16 m tall; bark whitish-grey, pustular smooth, exfoliating in quadrilateral scales; wood light yellowish-grey, interlocked grained, medium and quite even-textured; branchlets terete, glabrous, lenticellate. Leaves chartaceous, elliptic-ovate, 6–14 × 3–8 cm, puberulous along nerves above when young, glabrous later, apex abruptly acuminate, margin entire, base rounded; secondary nerves 5–6 pairs; petiole to 1.3 cm. Flowers yellow in 2 cm long, axillary, racemes or in fascicles at the scars of fallen leaves. Samara orbicular, ca. 3 cm in dia., flattened, yellow, wings reticulately-veined.

Common, in deciduous forests and in villages along roadsides to 1200 m. It thrives in deep porous soil with good drainage, but becomes stunted and crooked on poor shallow soil. It is moderate light-demander, coppices well and is not frost hardy. Leaf fall: Jan.–Feb. New foliage: Apr. Fls.: Jan.–Feb. Frs.: Apr.–May. Vern.: Nauli, Pulari, Pedda Nevili, Tapasi, Tappachi, Nevili, Namlinara.

The bark when cut, and the leaves and twigs when crushed emit an unpleasant odour. Wood light to moderately hard, used for cheap grades of furniture, match boarding, panelling, door frames, wall plates, carts and carriages, combs, shoe heels, mathematical instruments; suitable for plywood, packing cases, slate frames, match boxes, splints, paper pulp and also for fuel. Twigs used as tooth brushes.

Rollapenta (KNL), *SSR* 15511; Srisailam (KNL), *SSR* 15520; Thummala bailu (PKSM), *SSR* 16118; Balarajupalli (CDP), *SSR* 16156; Mondisadak (NZB), *BR & TP* 6366; way to Kunthala (ADB), *GO & DAM* 5124.

INDIA: Throughout.
WORLD : Sri Lanka, Myanmar and South Vietnam.

## **TREMA** Loureiro

**Trema orientalis** (L.) Bl. in Mus. Bot. Lugd. Bat. 2: 62. 1856; FBI 5: 484; Fischer 1350 (945); Brandis 597. *Celtis orientalis* L. Sp. Pl. 1044. 1753. *Sponia wightii* Planchon, Ann. Sci. Nat. Bot. (Ser. 3) 10: 322. 1848; Beddome Fl. Sylv. S. India t. 311. 1873.

Evergreen, dioecious tree, up to 12 m tall; trunk straight with spreading branches; bark greyish brown, smooth, with numerous lenticels; wood light reddish-grey; branchlets terete, sparsely pubescent. Leaves chartaceous, ovate or ovate-lanceolate, 7–18 × 2–7 cm, somewhat rough above, whitish beneath, clothed with soft tomentum, apex acuminate, margin serrulate, base unequal, rounded or sub-cordate; secondary nerves 3–5 pairs, basal pairs reaching to the half the length of the leaf; petiole to 1.5 cm, pubescent. Flowers greenish in 2 cm long, axillary, pubescent cymes; the male cymes rather more compact than the female and usually on different trees. Drupe ovoid, 0.3 cm long, black when ripe, glabrous.

Common, in dry and evergreen forests, along perennial streams in moist deciduous forests. It is rapid growing tree, quickly appearing in forest clearings. The tree is useful for the rapid reclothing of cut-over areas and as a shade plant. Fls. & Frs.: Throughout the year. Vern.: Morali, Boggu chettu, Priyalu, Aava, Konda Jonna, Ada Nelli. Eng.: Charcoal tree.

Wood is soft and is employed for making charcoal for the manufacture of gun powder. Used for tea-chests, match-boxes, and splints, wooden shoes and fishing floats.

Way to Upper Ahobilam (KNL), *TP & SSR* 15205;Pamuleru (EG), *TP & KJ* 12523;Talakona (CTR), *KJ & SSR* 12552; Peccheruvu (KNL), *RVR* 2183; Tirumala (CTR), *KSM* 6930 (MH).

INDIA : Almost all States.
WORLD : Sri Lanka, Malaysia, S.E. and E. Asia, Polynesia, New Guinea, Taiwan.

# MORACEAE

1. Flowers in a hollow receptacle; flowers hidden .......... ***Ficus***
1. Flowers not in a hollow receptacle; flowers patent:
    2. Leaves coiled in bud; stipules surrounding the stem, stamens erect.......... ***Artocarpus***
    2. Leaves not coiled in bud; stipules not surrounding the stem; stamens inflexed:
        3. Flowers solitary or pedunculate heads. Sap milky .......... ***Streblus***
        3. Flowers in spikes. Sap watery.......... ***Morus***

## **ARTOCARPUS** J.R. Forster & J.G.A. Forster *nom. cons.*.

1. Fruit tubercled, very large 30–80 cm long. Leaves glabrous, apex obtuse to shortly acuminate; stipules 5–8 cm long, spathaceous. Trees evergreen ..........***A. heterophyllus***

1. Fruit smooth, 3–8 cm long. Leaves pubescent beneath, apex cuspidate; stipules small, not spathaceous. Trees deciduous ........................................................................ ***A. lakucha***

**Artocarpus heterophyllus** Lam. Encycl. 3: 209. 1789. *A. integrifolius auct. non* L.f. 1781; Gamble 2: 1369 (957).

Large, evergreen tree; crown dense, round; bark black, mottled green, smooth; wood bright yellow, darkening on exposure; branchlets glabrous. Leaves coriaceous, 10–17 × 5.6–10.1 cm, dark green and shining, apex obtuse with shortly acumen, margin entire, base rounded; secondary nerves ca. 7 pairs, prominent, 2 cm apart, curved upwards; petiole to 3 cm. Flowers greenish, crowded on globose, oblong or cylindrical, solitary, axillary and cauliflorus receptacles. Fruit oblong-globose, greenish-yellow, large, clothed with fleshy and enlarged perianth, carpels spiny; seeds reniform, greyish, and oily.

Cultivated, in hills of high elevations, especially in northern Circars; in some areas of Visakhapatnam. It is also grown as a wild state. It is grown by tribals and farmers for its fruits. It grows on variety of soils. Fls.: Nov.-Dec. Frs.: July-Aug. Vern.: Panasa; Eng.: Jack fruit tree.

Rampachodavaram (EG), *TP & KJ* 12540; Maredumilli (EG),*TP & KJ* 12517; Anantagiri (VSKP), *GVS* 19528 (MH).

INDIA : Widely cultivated throughout.
WORLD : S.E. Aisa, Sri Lanka.

**Artrocarpus lakucha** Buch.-Ham. in Mem. Wern. Soc. 5. 333. 1826; Mabb. in Taxon 26: 529. 1977. *A. lakoocha* Roxb. Fl. Ind. 3: 524. 1852. 1852; FBI 5:543; Fischer 3: 1369 (959); Brandis 612.

Deciduous tree, up to 15 m tall; crown dense; bark grey, flaky in small, round wood scales; wood red; interlocked grained, very coarse and quite even textured; branchlets tomentose with milky juice. Stipules small, lanceolate, grey pubescent. Leaves coriaceous, oblong or elliptic, 10–30 × 5–15 cm, glabrous and shining above, pubescent beneath, apex cuspidate, margin entire-serrate, base rounded or truncate; main lateral nerves 13 pairs reaching the margin, conspicuous beneath; petiole to 2 cm. Flowers monoecious, globose, axillary; male receptacles orange-yellow, spongy, ovoid to 2.5 cm, sessile, deciduous, sepals 2–3, puberulous, stamen 1; female receptacles irregularly subglobose, pedunculate, smooth. Fruit globose, 6 cm in dia., velvety, yellow; seeds oblong, few.

Occasional, in hills of evergreen forests (1300 m) along the banks of streams and of moist ravines. Commonly culitivated for its edible fruits. Leaf fall: Mar. Fls.: Dec. Frs.: May and Oct.–Nov. Vern.: Nakkarenu, Kamma regu, Lakuchamu. Eng.: Monkey-Jack.

The fruit is much used for culinary purposes. Ripe fruit is used for chutneys. It is sliced, dried and used in place of tamarind particularly when tamarind is forbidden during the course of medicinal treatment. Fibre is obtained from the bark and is used for cordage. But in powder form is applied to sores to draw out purulent matter. Wood is fairly strong, durable popular for house construction work, medium weight furniture and boat building. It is also used for keels of sea-going ships.

Minumuluru (VSKP), *SSR* 15943; Foot of Galikonda hills (VSKP), *GVS* 19620 (MH).

INDIA : Orissa, Kerala, Tamil Nadu, Eastern and North India, Andaman and Nicobar Islands.

WORLD : Myanmar.

## **FICUS** Linnaeus

1. Receptacles in dense clusters or fascicles on tubercles or on leafless racemes or panicled branches from the trunk or larger branches:
  2. Receptacles in fascicles on long peduncles:
    3. Leaves broadly elliptic ovate, pubescent beneath, base cordate. Receptacles 5–6.2 cm dia. .......... ***F. auriculata***
    3. Leaves ovate-elliptic, glabrous glabrescent base sub-cuneate or rounded. Receptacles 2–3 cm dia. .......... ***F. oligodon***
  2. Receptacles dense on leafless branches on short peduncles:
    4. Leaves glabrous on both surfaces, margin entire .......... ***F. racemosa***
    4. Leaves scabrid-hispid above, pubescent or tomentose beneath, margin toothed or serrate:
      5. Leaves alternate, base semicordate. Receptacled branches vary long. .......... ***F. semicordata***
      5. Leaves opposite, base rounded or sub-cordate. Receptacled branches short .......... ***F. hispida***
1. Receptacles solitary, paired or 2–6 together in the axils of leaves:
  6. Leaves with very many fine lateral nerves—cultivated .......... ***F. elastica***
  6. Leaves not as above:
    7. Receptacles distinctly peduncled:
      8. Leaves lobed—cultivated .......... ***F. carica***
      8. Leaves entire or not lobed:
        9. Leaves scabrous above:
          10. Receptacles solitary. Leaf base equal .......... ***F. exasperata***
          10. Receptacles paired. Leaf base inequilateral .......... ***F. tinctoria***
        9. Leaves smooth above:
          11. Receptacles obovoid, pubescent .......... ***F. dalhousiae***
          11. Receptacles subglobose, glabrous .......... ***F. nervosa***
    7. Receptacles sessile:
      12. Apices of leaves caudate:
        13. Caudate-acumen short, leaf base narrowed .......... ***F.talboti***
        13. Caudate-acumen long, leaf base broad:
          14. Leaf base cordate .......... ***F. arnottiana***
          14. Leaf base truncate or subcordate:
            15. Bark greyish-black. Apical caudate-acumen one-sixth as long as blade. Receptacles black when ripe .......... ***F. rumphi***
            15. Bark greenish. Apical caudate acumen one-third as long as the blade. Receptacles dark purple when ripe .......... ***F. religiosa***

12. Apices of leaves obtuse, blunt or acute-shortly bluntly acuminate:
    16. Leaves tomentose or pubescent beneath even at maturity:
        17. Midrib usually with an elongated gland at base. Receptacle scarlet red, puberulous ........ ***F. benghalensis***
        17. Midrib eglandular. Receptacles green tomentose ........ ***F. mollis***
    16. Leaves glabrous:
        18. Lateral nerves above 20 pairs, inter secondaries as prominent as secondaries:
            19. Branches drooping. Aerial roots numerous. Leaves shining, base indistinctly 3-veined at base ........ ***F. benjamina***
            19. Branches not drooping. Aerial roots few. Leaves not shining, base distinctly 3-veined at base ........ ***F. microcarpa***
        18. Lateral nerves less than 20 pairs, inter secondaries not prominent as secondaries:
            20. Receptacles very shortly peduncled. Aerial roots few ........ ***F. virens***
            20. Raceptacles sessile. Aerial roots absent:
                21. Raceptacles 2–6 together on tubercles from leafless axils ........ ***F. tsjahela***
                21. Raceptacles in pairs in leafy axils. ........ ***F. amplissima***

**Ficus amplissima** Smith in Rees, Cyclop. 14. n. 68. 1810. *F. tsiela* Roxb. ex Buch.-Ham. Trans. Linn. Soc. London. 15: 149. 1827; FBI 5: 515; Fischer 3: 1362 (953); Brandis 603; Beddome Fl. Sylv. S. India 223. t. 314. 1873.

Spreading tree, up to 15 m tall; no aerial roots, trunk cylindric; bark greenish-grey, smooth. Leaves coriaceous, ovate-elliptic, 7–12 × 5–6 cm, apex short bluntly acuminate, margin entire, base rounded; lateral nerves 10–13 pairs; petiole to 7 cm. Receptacles monoecious, 1 cm dia., sessile in axillary pairs on leafless branchlets, depressed-globose or pyriform, purple when ripe; basal bracts 3; orifice circular, closed by 3 flat apical scales.

Common, in deciduous and evergreen forests up to 1000 m. Syconia Jan.–Mar. Vern. Pedda Juvi, Pitta juvvi. Leaves and twigs are lopped for fodder. The wood is used for furniture and fuel.

Bairluti (KNL), *SSR* 15517; Gangapur RF (MDK), *TP & CP* 12075; Way to Appapur (MBNR), *SSR & KJ* 15541; Kalasamudram (ATP), *EC* 2947; Macherla (GTR), *VRK* 3943; SK University Campus (ATP), *KJ* 12571; Kadamvagu (ADB), *TRS* 85150 (MH); Balapalli (CDP), *JLE* 15750 (MH); Duggeru (SKLM),*GVS* 62519 (MH); Malakondapet (PKSM), *RKM* 6804 (CAL).

INDIA : Sikkim, Assam, Tamil Nadu, Orissa, Karnataka.
WORLD : Sri Lanka, Maldive.

**Ficus arnottiana** (Miq.) Miq. Ann. Mus. Bot. Ludg.-Bat. 3: 287. 1867; FBI 5: 513; Fischer 3:1363 (953); Brandis 602. *Urostigma arnottiana* Miq. in Hook. Lond. J. Bot. 6: 564. 1847.

Tree, up to 7 m tall; glabrous, often epiphytic, also creeping over rocks, without aerial roots; bark pale, smooth; wood white. Leaves subcoriaceous, broadly ovate, 5–11 × 4–7 cm, apex shortly caudate-acuminate, margin entire or undulate, base rounded or cordate, 5–7- ribbed; lateral primary nerves 5–7 pairs, flattened above, raised below; petiole 5–12 cm. Receptacles 6 mm across, sessile in pairs or clusters on tubercles in axils of fallen leaves, depressed-globose, reddish when ripe, smooth; orifice plane; bracts 3, ovate, membranous.

Common, in all districts in rocky areas in the hills above 600 m. Syconia: Aug. onwards. Leaf fall: Mar.-Apr. New foliage: May. Old leaves often turn copper-coloured. Vern.: Kalla-Ravi, Banda Ravi. The leaves and twigs are lopped for fodder.

Kalasamudram (ATP), *SSR* 13755; Talakona (CTR), *KJ & SSR* 12549; Bhupatipalem (EG), *TP & KJ* 12508; Near Sikharam, Srisailam (KNL), *KJ & SSR* 12579; Battigudam (KNL), *NRR & TRS* 86587 (MH); Kondapalli forest (KSN), *PVK* 5744 (MH); Karaka (VSKP), *CAB* 1681 (MH).

INDIA : Throughout.
WORLD : Sri Lanka.

**Ficus auriculata** Lour. Fl. Cochinch. 666. 1790. *F. roxburghii* Wall. ex Miq. in Ann. Mus. Lugd. Bat. 3: 296. 1867; FBI 5: 534. *F. macrophylla* Roxb. Fl. Ind. 3: 556. 1832 (non Des.).

Evergreen, low spreading tree; bark greyish brown, smooth; branchlets hollow; young shoots pubescent. Leaves broadly elliptic-ovate, large, 20–40 × 15–28 cm, glabrescent above, shortly-softly pubescent beneath, apex shortly acuminate, margin repand-toothed or coarsely serrate, base cordate, 5–7-ribbed; secondary nerves 5–7 pairs, tertiaries sclariform, prominent; petiole 6–10 cm. Receptacles purplish, pyriform, 5–6.2 cm dia., clustered, pendulous, cauline from tubercles on short, lateral leafless branches and trunk; peduncle 4–7 cm long, pubescent, bracteate at their apex.

Rare, along streams in semi-evergreen forests of Visakhapatnam and East Godavari at 800–1500 m. Fls. & Frs.: Throughout the year. Vern.: Raach bodda. Raj dummiri.

Fruits are edible, and made into jam and curries. The bark yields a coarse fibre. Leaves are lopped for fodder.

Galikonda (VSKP), *SSR* 15916; Sapparla (VSKP), *GVS* 30092; Sunkarimetaa (VSKP), *GVS* 22644 (MH).

INDIA : Chota Nagapur, Orissa, Himalaya from Punjab to Bhutan.
WORLD : Bangladesh, Myanmar, Thailand and China.

**Ficus benghalensis** L. Sp. Pl. 1059. 1753: FBI 5: 499; Fischer 3: 1361 (952). *F. indica* L. Sp. Pl. 1060. 1753. *Urostigma benghalensis* (L.) Gasp. Nov. Gen. Fic. 7. 1844.

Evergreen tree, up to 25 m tall; sending down many aerial roots from the spreading branches and form support, thus extending the growth of the tree indefinitely; bark greyish white, smooth; wood grey or greyish white, more or less irregularly grained, rather coarse and uneven textured; young branches softly pubescent. Leaves thick coriaceous, ovate or ovate-oblong, 8–16 × 6–11 cm, glabrescent above, puberulous beneath, apex obtuse, margin entire, base subcordate, rounded, or rarely acute, 5–7-ribbed; 4–7 pairs of lateral nerves above the basal nerves; petiole to 6 cm. Receptacle monoceious, globose, 1.7 cm dia., red dotted, sessile in axillary pairs, puberulous, basal bract 3, obtuse; orifice slightly raised, circular.

Very common, on roadsides near temples, wild in village surroundings and old walls up to 1000 m. It is hardy and drought resistant and withstands mild frost. It is sacred tree of Hindus and most popular shade providing tree. Syconia: Sept.–Mar. Vern.; Marri, Pedda Marri, Rohi, Vatairilsjam. Eng.: The Banyan. Indian Fig.

Fruits are eaten by birds, monkeys and also by man in time of scarcity. Leaves are lopped as fodder for domestic animals and elephants. Sewn leaves are used as wrappers for light articles. Wood light and durable underwater and it is used for furniture. The tree is host for Indian lac insects. Milky juice, leaves, bark, infusion of young buds, seeds and tips of young roots are used medicinally. The aerial branches are used for palanquin poles, cart yoaks, tent poles, and carrying shafts. A coarse rope is prepared from the bark and aerial roots of the tree.

Tirumala (CTR), *KJ & SSR* 12558; V.P. South (GTR), *VRK* 3941; Mahanandi (KNL), *JLE* 42344 (MH); Nasrullabad (NZB), *TP & BR* 6179; Dunna (KSN), *PVK* 5903 (MH); Mahanandi (KNL), *JLE* 42344 (MH); Narsapur (MDK), *KMS* 6726 (MH); Simhachalam hill (VSKP), *GVS* 19365 (MH).

INDIA : Throughout the sub-Himalayan forests, Deccan hills and plains of India; Sacred tree and widely planted.

WORLD : Bangladesh and Pakistan.

**F. benghalensis** var. **krishnae** (DC.) Corner is found in Tirumala gardens and considered as Horticultural variety of *F. benghalensis*.

**Ficus benjamina** L. Mant. Pl. 129. 1767; King in Ann. Roy. Bot. Gard. (Calcutta) 1. t. 52A. 1857. FBI 5: 508; Fischer 3: 1367; Brandis 604.

An evergreen tree, up to 20 m tall; crown spreading, aerial roots numerous, trunk cylindric; bark whitish-grey, smooth; branches drooping, branchlets glabrous. Leaves subcoriaceous, elliptic-ovate rarely obovate, 5.5–8.5 × 2.8–4.8 cm, glabrous and shining, drying grey-green to olivaceous brownish, apex abruptly acuminate, margin entire, base rounded or acute, 3-nerved; lateral nerves 6–14 pairs, secondary nerves almost as prominent, raised on both sides, connected by an intramarginal nerves from base; petiole to 2.2 cm. Receptacles monoecious, obovoid, ellipsoid or subglobose, sessile or stalked, pyriform, axillary, paired, ripening yellow to orange, basal bracts 2–3, crescentic; orifice slightly raised, circular, closed by 3 small flat apical bracts.

Often planted for shade in avenues and as an ornamental tree, rarely near water canals or moist localities up to 1200 m. Syconia: Mar.–May. Vern.: Konda-Juvvi., Putra-Juvvi, Konda golugu.

The wood is reported to be suitable for match boxes.

Anantagiri (VSKP), *SSR* 15933; Talakona (CTR), *KJ & SSR* 12551; Srisailam (KNL), *KJ* 12573; Maredumilli (EG), *GVS* 27297 (MH).

WORLD : South China to the Solomon Islands.

**Ficus carica** L. Sp. Pl. 1753; Bailey Man. Cult. P1. 339. 1949.

Much branched, small deciduous tree, branching from the base. Leaves thick, broad-ovate nearly orbicular, 3–5-lobed, 10–20 cm long, rough above, pubescent beneath, apex acuminate, entire, base cordate, 3–5 nerved; secondary nerves 6–8 pairs above basal pairs; petiole to 6 cm. Receptacles solitary or paired, axillary and subterminal, pear-shaped, variable in size, yellowish-purple when ripe.

Introduced and cultived; commerically grown in Anantapur. It is usually propagated by cuttings. Syconia: Jan.–June. Vern.: Anjur, Anjuram.

Fresh figs are delicious, mildly laxative and of high nutrient value; it is diuretic, demulcent, emotient and are used in the prevention of nutritional anaemias; latex is used as an anthelmintic. Leaves are used as fodder. The wood is used occasionally in cabinet work.

INDIA : Commercially grown in some States.
WORLD : It is considered to be native of Carica in Asia minor and is grown in nearly all tropical and subtropical countries.

**Ficus dalhousiae** Miq. in London J. Bot. 4: 571. 1845; FBI 5: 499; Fischer 3: 1364 (954); Brandis 601.

Tree to 15 m tall, spreading; young branches softly pubescent. Leaves coriaceous, ovate-elliptic, broadly ovate or ovate-oblong, 14.5–26 × 7.5–15 cm, glabrous above, white pubescent beneath, apex obtuse to shortly acuminate, margin entire, base subcordate; lateral nerves 12–14 pairs, inter-secondary nerves absent, petiole 3–6 cm long. Receptacles yellowish, 0.8 cm across, pubescent, obovoid, pedunculate in pairs.

Rare, in rocky ravines of deciduous forests of Chittoor and Cuddapah districts at 700–1000 m. Fls.: & Frs.: Mar.–Sept.

Tirumala-Seshatheertham (CTR), *DR* 1197 (SVU).

**Note** : Endemic to Chittoor and Cuddapah districts.

**Ficus elastica** Roxb. Hort. Beng. 65, 1814 nom. nud. & Fl. Ind. 3:541. 1832; FBI 5: 508; Brandis 603.

Small to large, evergreen tree with numerous aerial roots; trunk much buttressed and fluted. Leaves thick, coriaceous, oblong or elliptic, 15–25 cm long, glossy green, apex with sharp point, margin entire, base rounded or somewhat narrowed; lateral nerves numerous, fine parallel, petiole to 4 cm; stipules pink, about half as long as leaf lamina. Receptacles yellow, paired, sessile on old leaf axils.

Grown as an ornamental tree in gardens for its glossy ornamental leaves or for shade (Thammanna *et al.*, 1996).

Vern.: Rubber tree. Eng.: Indian Rubber tree.

In ancient times latex of Indian rubber is obtained by tapping the stem and main branches, but presently it is replaced by *Hevea brasiliensis*; it is now valued mostly for shade and ornament. Wood is used as fuel. This tree is considered as a good source of fodder.

INDIA : Indigenous. Native of the outer Himalaya, Assam, Khasi hills.
WORLD : Myanmar, Malaysia to Indonesia, often cultived.

**Ficus exasperata** Vahl Enum. Pl. 2: 197 & 402. 1806. *F. asperrima* Roxb. Fl. Ind. 3: 554. 1832; FBI 5: 522; Fischer 3: 1366 (955). 1928; Brandis 605.

Deciduous tree, 8–10 m tall; without aerial roots, bark whitish-grey, smooth; wood whitish; branchlets scabrous. Leaves subcoriaceous, oblong-elliptic to obovate, 10.2–14.5 × 5.2–6.8 cm, scabrid with stout stiff white hairs on both surfaces, raised dots conspicuous on upper surface, more or less densely tomentose beneath, drying grey-green, apex blunt to shortly acuminate, margin serrate, toothed, base often inequilateral, acute, 3-ribbed, elongate; lateral nerves 3–5 pairs above basal ones, prominent on the lower surface; petiole to 1.5 cm. Receptacles dioecious, globose, 1.3–1.5 × 0.9–1.3 cm, solitary , axillary, pedunculate, hispid without, yellow when ripe; basal bracts 0, orifice circular, raised.

In mixed dry deciduous forests at 800–1200 m. Syconi [illegible] Mar.–May. Vern.: Karaka boddu, Siribodda, Karasaana. Wood soft and [illegible]ss. Leaves are used as sand paper to polish wood.

Kusimi RF (SKLM), *SSR* 15904; Pamuleru (EG), *TP & KJ* 12521; Borra caves (VSKP), *KJ* 12587; Seethampet (SKLM), *GVS* 62444 (MH); Atmakur (NLR), *KCJ* 383 (MH); Rampa hills (EG), *JSG* 16011 (MH).

INDIA : South India and Orissa.
WORLD : E. Africa, Sri Lanka, Arabia.

**Ficus hispida** L.f. Suppl. Pl. 442. 1781; FBI 5:522; Fischer 3:1367 (956); Brandis 606.

Tree, up to 8 m tall; without aerial roots, trunk straight with root suckers, laxly branched, bark grey, rough; wood dark grey; branchlets stout, hollow. Leaves membranous, obovate, or ovate-oblong 15–25 × 6–10.5 cm, hispid-scabrid above, hispid tomentose, and glaucous beneath, apex shortly apiculate, margin toothed, base rounded or subcordate, 3-ribbed, lateral nerves 6–10 pairs, petiole to 6 cm. Receptacles obovoid, to 2.5 cm across, yellowish, hispid on leafless branches from trunk; bracts triangular. Achene ovoid.

Common, in hedges, thickets, vacant plots up to 900 m. Syconia: Dec.–Mar. Vern.: Kakimedi, Boddu Marri, Kukka bodda, Brahmamedi, Bomma Medi.

Immature fruits are used in curries and ripe fruits are edible. Leaves and twigs are lopped as fodder. The wood is soft and light, not useful. Bark yields a rough fibre. The leaves are used for poulticing boils.

Tirumala (CTR), *KJ & SSR* 12565; Maredumilli (EG), *TP & KJ* 12519; Gowraram (NZB), *TP & BR* 6390; Kalvabugga (KNL), *RVR* 3106; Narsapur RF (MDK), *TP & CP* 12015; Jannaram (ADB), *TP & PVP* 4190; Palakonda hills (CDP), *CS* 8274; Javuku (GTR), *VRK* 3865.

INDIA : Central and South India.
WORLD : Sri Lanka, Myanmar, Bhutan, Malaysia, China to Australia.

**Ficus microcarpa** L.f. Suppl. 442. 1781; *F. retua auct. non* L.; King in Ann. Roy. Bot. Gardn. Cal. 1:50 1887; FBI 8:511; Fischer 3: 1362; Brandis 603; *F. retusa* (var. *nitida* Thunb.) Hook. f. FBI loc. cit.; Fischer loc. cit.

Deciduous tree, up to 15 m tall; aeial roots few; bark brown, nearly smooth; wood light reddish-grey; branchlets glabrous, aerial roots few. Leaves coriaceous, elliptic or obovate, 3–9.5 × 1.5–5 cm, glabrous, shining, apex rounded to acute or retuse, margin entire, base narrowed or cuneate, 3-nerved; lateral nerves 10–13, closely and reticulately veined; petiole to 1.6 cm. Receptacles monoecious, 8 mm dia., sessile in axillary pairs, rarely solitary, globose, yellowish red when ripe.

Common, often planted in avenues above 500 m. Syconia ripening by October. Vern.: Juvvi, Konda Juvvi, Yerra-Juvvi, Konda pillara, Nandireka, Billa juvi, Hema vudaga.

Leaves and twigs are lopped for fodder. The wood is used for furniture or fuel.

Tirumala (CTR), *SSR & KJ* 13740; Horsely hills (CTR), *SSR* 13792; Upper Ahobilam (KNL), *SSR* 13792, *SSR & KJ* 15206; way to Tumbura Theertham (CTR), *SSR* 1678; Araku forest (VSKP), *KJ* 12583; Kalasamudram (ATP), *NY* 1193; Palakonda hills (CDP), *CS* 9333; Manchiryal (ADB), *GO &*

*DAM* 5189; Bison hill (EG), *CAB* 5160 (MH); Bhadrachalam (KMM), *JSG* 15809 (MH).

INDIA : Throughout.
WORLD : Sri Lanka, South China, Ryukya Isles to New Britain.

**Ficus mollis** Vahl, Symb. Bot. 1:82. 1790. *F. tomentosa* Roxb. ex Willd. Sp. Pl. 4: 1136. 1806; FBI 5:501; Fishcer 3: 1361 (952).

Evergreen tree, up to 15 m tall; bark greenish-white, smooth, scaly; wood white; young branches densely, yellowish tomentose; aerial roots several. Leaves crowded towards the ends of branches, coriaceous, ovate-elliptic or obovate-elliptic, 4–12 × 3–7 cm, glabrous or nearly so and dotted above, densely rusty-grey tomentose beneath, apex obtuse to sub-acute, margin entire, base cordate with a gland at the base of the midrib beneath; main lateral nerves 5–6 pairs; petiole to 3 cm. Figs subglobose, sessile, axillary in pairs with rusty tomentum, green, 1 cm in dia.

Common, in forests and villages in dry localities. Fls. and Frs.: Almost throughout the year. Vern.: Jivi, Juvvi, Kal Juvi. Wood is not useful.

Tirumala (CTR), *SSR* 13742; Kalyani dam (CTR), *KJ & SSR* 12553; B. Velamala Kota (EG), *TP & KJ* 12536; Potchera water falls (ADB), *GO & DAM* 5043; Gundumala (ATP), *TP & NY* 1234; Madhavaram RF (KNL), *TP & PVR* 1384; Velutla RF (NZB), *BR* 9035; Nagarjuna Konda (GTR), *VRK* 6707; Pakhal (WGL), *KMS* 13154 (MH); Kakarapadu (VSKP), *GVS* 65645 (MH); Sangam (NLR), *MSR* 1148 (CAL).

INDIA : Almost all States.
WORLD : Sri Lanka.

**Ficus nervosa** Heyne ex Roth in Roem. & Schult. in L. Syst. Veg. 1: 513. 1817; FBI 5: 512; Fischer 3: 1364 (954); Brandis 600. *F. angustifolia* Roxb. Fl. Ind. 3: 554. 1832.

Evergreen, medium to large tree; aerial roots absent;bark brown, mottled white; wood white, young branches minutely, appressedly pubescent; latex white turning yellow. Leaves thinly coriaceous, elliptic, oblong-lanceolate or oblanceolate, 10–16 × 5–8 cm, glabrous, shining, apex abruptly and narrowly acuminate, margin entire to slightly undulate, base narrowed, subequilateral, 3-nerved, not stronger, secondary nerves 9–11 pairs; petiole 0.8–2.5 cm. Receptacles subglobose, smooth or slightly verrucate, peduncle to 1.5 cm, yellow to red when ripe; achene ovoid, acuminate, smooth.

Occasional, moist deciduous forests above 1000 m. Common, in almost all the districts. Fls. & Frs.: Mar.–Dec. Vern.: Vonjar.

Vaiyar (VSKP), *PSPB* 1181.

INDIA : Assam, Chota Nagpur, Karnataka, Tamil Nadu, Orissa.
WORLD : Sri Lanka, Myanmar, S. China, Taiwan, Tonkin, Annam.

**Ficus oligodon** Miq. in Ann. Mus. Bot. Lugd. Bat. 3: 234. 297. 1867; Corner in Gard Bull. Singap. 18: 43. 1960. *F. pomifera* Wall. ex King in Ann. Roy. Bot. Gard. Calcutta 1: 215. 1887; FBI 5: 538. 1888; Fischer 3: 1365 (954); Brandis 609.

Tree, up to 12 m tall; bark grey; wood spongy. Leaves lanceolate or ovate to broadly elliptic, 10–25 × 7–12 cm, glabrous above, minutely papillose or glabrescent beneath, apex acute to acuminate, margin entire to coarsely toothed, base subcuneate or rounded, 3(5)-ribbed; secondary nerves 5–8 pairs, subregular, tertiaries transverse, prominent beneath; petiole to 12 cm. Receptacle sub-globose, 2–3 cm dia. on tubercles from branchlets or branches; peduncle 3–6 cm; reddish when ripe.

Rare, in semievergreen forests of Visakhapatnam and East Godavari district at 650–1000 m. Fls. & Frs.: Mar.–Aug. Fruits edible. Wood soft.

Gudem (VSKP), *SSR* 15966; Sapparla (VSKP), *SSR* 19640.

WORLD : Malay Peninsula and Archipelago.

**Ficus racemosa** L. Sp. Pl. 1060. 1753. *F. glomerata* Roxb. Pl. Cor. 2: 13. t. 123. 1799; FBI 5: 535; Fischer 3: 1364 (954); Brandis 609.

Deciduous tree, up to 15 m tall with few or no aerial roots; bark smooth, grey; wood greyish brown, soft grained or shallowly or irregularly interlocked grained, very coarse and uneven textured; branchlets minutely pubescent, reddish-brown. Leaves coriaceous, elliptic-oblong, 4–13.5 × 2.5–6 cm, glabrous, apex acute or obtuse, margin entire, base oblique, 3-nerved; secondary nerves 6–8 pairs; petiole to 4 cm. Receptacles monoecious, globose, 2.5–4 cm in dia., reddish, white pilose on tubercled leafless branchlets from trunk; orifice slightly sunken; achene tuberculate.

Common, along river banks, roadsides, village surrounds up to 1500 m. Leaf fall:Nov.–Dec. New foliage: Jan.–Apr. Syconia: Throughout the year. Vern.: Medi, Atti, Moida, Bodda, Paidu.

Twigs and leaves are used as fodder for cattle and elephants. Bark, leaves, fruits and leaf galls used medicinally. The receptacles are used in vegetable, after removal of the minute flowers. Wood very light, moderately soft and fibrous, durable underwater used for cheap furniture, carts, frames, ploughs, oars, yokes, bellows etc.

Siddulaiah kona (NLR), *SSR* 19383; Nerudugunda (ADB), *GO & DAM* 5105; Itukalapalli (ATP), *VRK* 3944; Ghani RF (KNL), *TP & RVR* 1794; Yella Reddy (NZB), *TP & RVR* 1794; Maredumilli (EG), *TP & KJ* 12503.

INDIA : Throughout.
WORLD : Bangladesh, China to Australia, Sri Lanka.

**Ficus religiosa** L. Sp. Pl. 1059. 1753; FBI 5: 513; Fischer 3; 1363 (953); Brandis 601; Beddome Fl.Sylv. S. India t. 314. 1873.

Deciduous, fast growing tree, up to 20 m tall; often epiphytic when young; aerial roots absent; bark greenish, smooth, wood greyish-white or yellowish white, fibrous, broadly and irregularly interlocked grained, coarse and uneven textured; branchlets yellowish grey, glabrous. Leaves coriaceous, ovate, 10–18 × 5–13 cm, glabrous, long caudate-acuminate, margin wavy, base truncate or sometimes cordate; secondary nerves ca. 8 pairs, inter-secondaries present, petiole to 12 cm. Receptacles sessile, paired, smooth, globose, 1.2 cm dia., dark purple when ripe.

Common, along roadsides, near temples, wild in village surrounding, river banks up to 1000 m; it is a sacred tree of Hindus, and Buddhists. It is also planted for shade. Partly deciduous in summer. Fls. & Frs.: May–Aug. Vern.: Ravi, Ragi, Raigu, Ashwarthamu. Eng.: The peepal tree.

The fruits and tender leaf buds are occasionally eaten in times of scarcity. The leaves and twigs are lopped for cattle and elephant fodder. Wood moderately durable, used for yokes, boxes, cheap boarding etc., it is also used for wooden spoons and bowls.

Talakona (CTR), *KJ & SSR* 12550; SK University (ATP), *KJ* 12567; Nirmal (ADB), *GO & BR* 9403; Owk (KNL), *RVR* 3052; Aklaspur (KNR), *GVS* 20232 (MH).

INDIA : Throughout.
WORLD : Pakistan, Bangladesh, China to Thailand.

**Ficus rumphi** Bl., Bijdr. 437. 1825; FBI 5: 512; Brandis 601. *F. cordifolia* Roxb. Fl. Ind. 3: 548. 1832 (non Bl.).

Evergreen tree, up to 20 m tall; without aerial roots, at first usually epiphytic; bark greyish-black, smooth; branchlets glabrous. Leaves subcoriaceous, broadly ovate, 10–15 × 6–8.5 cm, glabrous, shining, apex acuminate, margin undulate, base subcordate, 3–5 ribbed; lateral nerves 6–8 pairs above basal ones, conspicuous; petiole to 9.2 cm, lenticel like structure are in the base of the petiole; stipules ovate-lanceolate. Receptacles globular, 1.3 cm dia., smooth, cream with reddish spots, sessile in axillary pairs on old leaf scars, black when ripe, basal bracts 3, rotund; orifice circular, slightly raised, covered by 3 apical bracts. Achene minute.

Occasional, along river banks and moist deciduous forests. at 800–1000 m. Syconia: Jan.–Apr. Ripen in May–June.

The ripe fruits are occasionally eaten. Juice and fruit are emetic, the juice is given internally as a vermifuge and for the relief of asthma. The twigs and leaves are lopped for cattle and elephant fodder. The wood is soft and spongy and is used as fuel or for charcoal.

Anantagiri (VSKP), *SSR* 15934, *KJ & SSR* 12586; Satwa near Araku (VSKP), *NPBK* 10858 (MH)

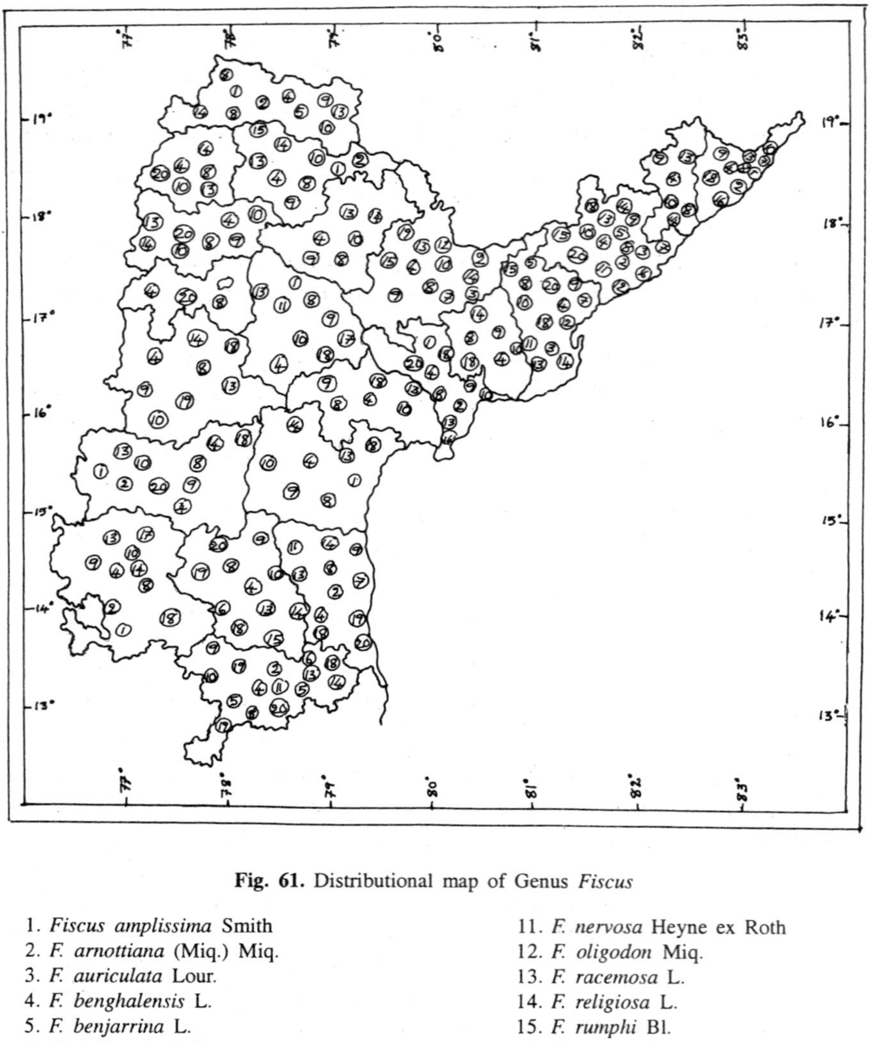


**Fig. 61.** Distributional map of Genus *Fiscus*

1. *Fiscus amplissima* Smith
2. *F. arnottiana* (Miq.) Miq.
3. *F. auriculata* Lour.
4. *F. benghalensis* L.
5. *F. benjarrina* L.
6. *F. dalhousiae* Miq.
7. *F. exasperata* Vahl
8. *F. hispida* L.f.
9. *F. microcarpa* L.f.
10. *F. mollis* vahl
11. *F. nervosa* Heyne ex Roth
12. *F. oligodon* Miq.
13. *F. racemosa* L.
14. *F. religiosa* L.
15. *F. rumphi* Bl.
16. *F. semicordata* Buch.-Ham.
17. *F. talboti* King
18. *F. tinotoria* Forst. F.
19. *F. tsjahela* Rheede ex Burm f.
20. *F. virens* Ait

INDIA : Outerhills of sub-Himalayan tract. Punjab eastwards to Assam, Bengal, Central and South India, Andaman Islands.

WORLD : Myanmar eastwards to South-east Asia.

**Ficus semicordata** Buch.-Ham. ex Smith in Rees Cyclop. 14: n. 71. 1810. *F. cunia* Buch.-Ham. ex Roxb. Fl. Ind. 3: 561. 1832. FBI 5;523; Fischer 3: 1366 (956); Brandis 606.

Evergreen tree, up to 12 m tall; without aerial roots, bark rough, reddish brown; wood greyish-brown; branchlets pubescent. Leaves coriaceous, semisagittate, one side with a large 3–4-ribbed lobe, the other cuneate, elliptic, oblong-lanceolate, 4–22 × 4–7 cm, scabrid above, more or less pubescent beneath, apex shortly acuminate, margin entire or slightly serrate, base semicordate; lateral nerves 9–14 pairs, raised above, flattened below; petiole to 2.5 cm. Receptacles 2.5 cm in dia., paired, pear-shaped, peduncled on long leafless scaly branchlets from the trunk near base, reddish-brown when ripe; basal bracts 3, ovate, pubescent; orifice circular covered by 3 apical bracts.

Common near streams and on cool, damp slopes of hills of Visakhapatnam and East Godavari districts above 900 m. Syconia: Most of the year and usually ripe by May-June, and Geocarpic, sometimes leafless in May. Vern.: Erubodda, Bomma Marri.

The twigs are valued as fodder and the tree is host of Indian Lac Insects. Fruits are edible. The juice of the roots is given in bladder ailments. Wood is moderately hard. Bark yields a strong fibre suitable for ropes.

Galikonda (VSKP), *SSR* 15915; Pamuleru (EG), *TP & KJ* 12524; Anantagiri (VSKP), *GVS* 19557 (MH); Gamperkot (SKLM), *KCJ* 1732 (MH).

INDIA : Assam, Bengal, Orissa.
WORLD : Myanmar, Bangladesh, China, Malaysia, Thailand to South-east Asia.

**Ficus talboti** King Ann. Roy. Bot. Gard. (Calcutta), 1: t. 63. 1887; FBI 5:512; Fischer 3: 1363 (953); Brandis 604.

Evergreen tree, 10–20 m tall; aerial roots absent; bark green, smooth; wood very white, turning yellow. Leaves subcoriaceous, narrow elliptic to broad ovate, 7–11 × 4–6 cm, glabrous, apex shortly, acutely caudate, margin entire, base rounded, 3-nerved; secondary nerves 8–10 pairs, raised on both surfaces; petiole to 3 cm, glandular at apex below. Receptacles obovoid, sessile, ripening yellowish, spotted with red colour dots, in axillary pairs; achenes smooth.

Occasional in forests. Fls. & Frs.: Dec.–Feb. Wood is not useful.

Bairluty (KNL). *SSR & KJ* 15517; Gundumala RF (ATP), *NY* 1236.

INDIA : Peninsula.
WORLD : Sri Lanka, Myanmar, Laos, Cambodia, N. Thailand.

**Ficus tinctoria** Forster f. Fl. Ins. Austr. 76. 1786. *F. gibbosa* Bl. var. *parasitica* (Willd.) King in Ann. Roy. Bot. Gard. (Calcutta) 1: t. 2a-b. 1887; FBI 5: 497; Fischer 3: 365 (955); Brandis 599.

Tree, up to 10 m tall; often epiphytic, enclosing the trunks of trees in a perfect network of branches and creeping along the wells and on the side of wells, without aerial roots, bark greyish, smooth; wood brownish grey. Leaves thinly coriaceous, rhomboid or sub-rhomboid or elliptic ovate, 4–12 × 3–6 cm, hispid and dark green above, scabrid and minutely hispid and paler beneath; apex acute, margin entire, base oblique, cuneate; main nerves 4–7 pairs, petiole to 0.7cm. Receptacles dioecious, 5–7 mm across, paired, globose, peduncled, hairy; male receptacles yellow; female receptacles red when ripe.

Common in open places 1200 m. Syconia: Feb.–Apr. Vern.; Konda Juvvi, Pakki.

Leaves are used to polish ivory and the root bark is employed in native medicine.

Appapur (MBNR), *SSR* 15541; Ramavaram (EG), *TP & KJ* 12526; Araku (VSKP), *KJ* 12584; Tirumalai hills (CTR), DR 1173 (MH); Duggeru forest (SKLM), *GVS* 62153 (MH); Cuddapah, *JSG* 10823 (DD).

INDIA : Almost all the States.
WORLD : Sri Lanka, Myanmar, S. China, Indo-China, Andaman & Nicobar Islands.

**Ficus tsjahela** Rheede ex Burm. f. Fl. Ind. 227. 1768. King in Ann. Roy. Bot. Gard. Calcutta, 1:57, t. 70 & 84. 1887. 88 (*tsjakela*); FBI 5; 514; Fischer 3: 1862 (953); Brandis 603; Corner in Dassan. & Fosb. Rev. Handb. Fl. Ceylon 3: 237. 1981.

Deciduous tree, up to 18 m tall; aerial roots absent; bark dark coloured, rough; wood brown; branchlets glabrous, dark brown. Leaves coriaceous, oblong-elliptic, or elliptic-ovate, 6–19 × 3–10.5 cm, glabrous, dark green above, pale beneath, ashy grey when dry, apex shortly acuminate, margin entire, base truncate or rounded, 3-ribbed; 7–12 pairs above basal ones, conspicuous, to 2.5 cm apart from basal ones to just above lateral nerves; petiole to 8 cm. Receptacle monoecious, depressed-globose, 6 mm across, greyish-purple, sessile, 2–6 together on tubercles in fallen leaf branches, basal bracts 3, bifid, orifice plane, circular; apical bracts 3.

In moist regions of moist deciduous forests above 600 m. Syconia: Ripen by May. The wood is suitable for match boxes.

Tirumala (CTR), *SSR* 16184; Farahabad (MBNR), *DC* 13664.

INDIA : Peninsula
WORLD : Sri Lanka.

**Note** : The correct spelling of this species is '*tsjahela*' as given by Corner (l.c.); Fischer gives it as *tjakela* (l.c).

**Ficus virens** Ait. Hort. Kew.3: 451. 1789. *F. infectoria* Roxb. Fl. Ind. 3: 551. 1832. non Willd; 1806; FBI 5; 515; Fischer 3: 1362 (953).

Large, spreading, deciduous tree, epiphytic in early stages, with a few aerial roots; bark greenish-grey, smooth; wood grey. Leaves subcoriaceous, 7–18 × 2.6–5.7 cm, glabrous, apex abruptly acuminate, margin entire to slightly undulate, base acute, truncate or subcordate; secondary nerves 7–12 pairs, basal pair shorter, intercostals zig zag; petiole 2.5–7 cm, articulate, glandular at the apex below. Receptacles axillary, paired, shortly peduncled, globose, whitish with red dots.

Common, in plains to hills up to 1200 m. Fls. & Frs.: Oct.–May. Vern.: Banda Juvvi, Jati Juvvi. It forms a good shade tree for coffee plantations.

Anantagiri (VSKP), *SSR* 15935; Tirumala (CTR), *SSR* 16186; Narsapur RF (MDK), *SSR* 16194; Anantagiri (RR), *SSR* 18517; Gandhari RF (NZB), *TP & BR* 6333; Balapalle (CDP), *JLE* 14327 (MH); Vempali (KNL), *GVS* 20188 (MH); Rampachodavaram (EG), *GVS* 2724 (MH); Pakhal (WG), *ANH* 15943 (MH).

INDIA : Throughout.
WORLD : Sri Lanka, Myanmar, Malaysia, South east Asia to Philippines and Pacific Islands.

## MORUS Linnaeus

**Morus alba** L. Sp. Pl. 986. 1753; FBI 5: 492; Fischer 3: 1370 (958); Brandis 612.

Deciduous, medium-sized tree; crown round; branches grey or greyish yellow; young shoots pubescent. Leaves thin, sub-coriaceous, broadly ovate, 7–12 × 4–7 cm, glabrous and shining above, slightly pubescent along nerves beneath, apex acute, margin dentate or c[illegible] toothed, base subcordate, basal nerves 3–5; secondary nerves 3–5 p[illegible]; petiole 1–2.5 cm. Flowers monoecious, greenish, aggregated into spike[illegible] Fruit red, then purple, sweet.

Usually cultivated, for rearing the silk worms. Fls. & Frs.: Throughout the year. Vern.: Resham chettu. Eng.: Mulberry.

Nizamabad outskirts (NZB), *BR & KH* 6490; SK University Campus (ATP), *TP & NY* 1250.

INDIA : Cultivated in almost all States.
WORLD : Cultivated in Baluchistan, Afghanistan, Europe, Western and Central Asia and in China.

## STREBLUS Loureiro

1. Branchlets thorny. Leaves smooth, margin distantly serrate towards the very upper half. Male flowers subsessile. Fruit obliquely ovoid .......... ***S. taxoides***
1. Branchlets not thorny. Leaves rough on both surfaces, margin irregularly dentate. Male flowers pedicellate. Fruit globose .......... ***S. asper***

**Streblus asper** Lour. Fl. Cochinch. 615. 1790; FBI 5: 489. Fischer 3: 1353 (947). 1928; Brandis 615. *Epicarpurus orientalis* Bl.. Bijdr. 488. 1825.

Evergreen scraggy or bushy tree to 7 m tall; latex milky, bark smooth, grey; wood yellowish or brownish-white; branchlets hairy. Leaves thick-coriaceous, elliptic-obovate or rhomboid, 1.7–6.8 × 1–3 cm, rough both surfaces, apex acute, margin irregularly dentate, base acute or cuneate, lateral nerves 4–6 pairs, raised on under surface; petiole to 2 mm. Stipules linear and caducous. Flowers dioecious, male flowers in globose heads, cream coloured, axillary, peduncle to 5 mm; female flowers solitary on axillary, usually fascicled peduncles. Fruit globose, 6 mm across, yellow, enclosed in the enlarged fleshy sepals, style persistent; seed 1, globose, smooth, 3 mm across.

Common, in shrubberies, hedges in dry deciduous forests up to 1000 m. New foliage: Mar. Fls.: Jan.–Mar. Frs.: Apr. Vern.: Stianike, Barranki, Barniki, Pakki, Bamigi.

Wood hard and tough used for making small articles and cart wheels. The rough leaves are used to polish wood.

Way to Upper Ahobilam (KNL), *SSR* 15292; Kusimil (SKLM), *SSR* 15600; Kamareddy (NZB), *BR* 9272; Pocharam sanctuary (MDK), *BR & CP* 11613; Neelwai (ADB), *GO & DAM* 5159; Balapalli (CDP), *JLE* 15745 (MH); Putula (KSN), *PVK* 5714 (MH); Punyagiri (VSKP), *GVS* 23327 (MH); Gokavaram (EG), *JSG* 1216 (DD).

INDIA : Throughout.
WORLD : Sri Lanka, Myanmar, S. China, Hainan, Indio-China, Thailand, Andaman Islands, Malaysia.

**Streblus taxoides** (Roth) Kurz. Forest Fl. Burma 2; 465. 1877. *Trophis taxoides* Heyne ex Roth Nov. Pl. Spec. 368. 1821. *Phyllochlamys spinosa* (Roxb.) Bureau in DC. Prodr. 17: 218. 1873; FBI 5: 488; Fischer 3: 1352 (946); Brandis 614.

Evergreen, small tree; bark peeling off in thin flakes; branchlets glabrous, minutely puberulous, ending with a spine. Leaves obovate or oblanceolate, 3–8.5 × 1–3 cm, glabrous, apex acuminate, margin distantly serrate towards the very upper half, base cuneate, or slightly cordate; secondary nerves 8–10 pairs, intersecondaries numerous, united near the margin and form an intra-marginal vein; petiole 3–4 mm. Male flowers brownish, 5 mm across, sessile to subsessile, aggregated in clusters; female flowers solitary, pedicellate, sometimes two together with bracts at the base. Drupe obliquely ovoid, 5 mm, enclosed by enlarged foliaceous perianth.

Occasional, in semievergreen forests, especially banks of streams at 700–1200 m. Fls.: Feb.–Mar. Frs.: Apr.–June. Vern.: Barivenka, Sukali.

Talakona (CTR), *SSR* 19680; Maredumilli (EG), *GVS* 24274; Chintapalli (VSKP), *JLE* 51194; Krishnapuram (VSKP), *CAB* 1987 (CAL).

INDIA : Orissa, Tamil Nadu, Maharashtra.
WORLD : Sri Lanka, Myanmar, Indo-China, Hainan, Thailand, Malaysia.

## URTICACEAE

1. Leaves narrow-lanceolate, to elliptic, 12.5–32 × 5–13 cm, glabrous. Flowers in divaricate cymes.—Fruit inflated ........ ***Laportea***
1. Leaves ovate-lanceolate, 6–12 × 3–5 cm, rugose above, white tomentose beneath. Flowers in capitate heads in dichotomous cymes or branched spikes. ........ ***Debregeasia***

### DEBREGEASIA Gaudhichaud-Beaupre

**Debregeasia longifolia** (Burm. f.) Wedd. in DC. Prodr. 16: 235. 1869. *Urtica longifolia* Burm. f. Fl. Ind. 197. 1768. *Debregeasia velutina* Gaud. Voy. Bonite Bot. t. 90. 1844–46; FBI 5: 590; Fischer 3: 1389 (971); Brandis 618.

Dioecious small tree; young branchlets hispid-pubescent. Leaves thick-chartaceous, ovate-lanceolate, 6–12 × 3–5 cm, rugose above, white-tomentose beneath, apex acuminate, margin serrate, base narrowed or truncate; penninerved, lowest pair arising from base; petiole to 5 cm. Flowers small, subsessile in axillary, capitate heads in dichotomous cymes or branched spikes. Fruit bright orange-yellow.

Rare in moist deciduous to evergreen forests. Fls.: Nov.–Jan. Frs.: Dec.–Mar. Vern.: Kerangi.

Araku forest (VSKP), *PSPB & KJ* 12560; Sunkarimetta (VSKP), *GVS* 22651 (MH).

INDIA : Western Ghats from the Konkan southwards and Eastern Ghats.
WORLD : Sri Lanka, Himalaya, Myanmar, China, Malaysia.

### LAPORTEA Gaudichaud *nom. cons.*

**Laportea crenulata** Gaud. in Freye. Voy. Bot. 498. 1866; FBI 5: 550; Fischer 3: 1373 (595); Brandis 616.

Small tree; bark white, smooth, very thin; wood pale-brown; branchlets covered with minute stinging hairs. Leaves narrow, lanceolate to oblong-elliptic, 12.4–32 × 5–13 cm, glabrous, apex acuminate, margin entire or crenulate in the upper half, base narrowed, rarely rounded; secondary nerves 10–14 pairs, prominent beneath; petiole to 3.8 cm. Flowers greenish in axillary divaricate cymes. Utricle white, inflated.

Occasional, in moist deciduous forests. Fls.: Sept.-Oct. Frs.: Nov.-Dec.

Galikonda (VSKP), *SSR* 19641; Padalammagudi (VSKP), *GVS* 42585 (MH); Rampa hills (EG), *JSG* 15940 (MH).

INDIA : Orissa.
WORLD : Sri Lanka.

## CASUARINACEAE
### CASUARINA Adanson

**Casuarina equisetifolia** Forst. & Forst. f. Char. Gen. Pl. 104. F. 53. 1776; FBI 5: 598; Fishcher 3: 1389 (972).

Tree, up to 15 m tall; bark brown, rough, fibrous, peeling in vertical strips; wood yellowish-pink to reddish-brown, straight or somewhat wavy grained, medium-fine and even textured; branchlets 10–12 × 0.2 cm arising from the axils of small, recurved scales. Leaves reduced to subulate scales, 0.5 to 8 mm long. Male spikes 3.7 cm long; female spikes up to 1 cm long. Fruit woody globose nutlets, compressed, winged, tipped by persistent style; seeds minute, exalbuminous.

Common, in Coastal areas, especially planted in sandy soils near the sea and also planted in gardens for its ornamental purposes. It is light demander, sensitive to water logging and fire. Fls.: Feb.–Apr. and Sept.–Oct. Frs.: June and Dec. Vern.: Sarugudu, Chavuka, Sarvi. Eng.: Horse tail Oak.

The wood is very hard, heavy to very heavy, and it makes an excellent fuel. It is periodically felled for poles, fuel etc., used in poor class houses as posts and rafters.

S.K. University campus (ATP), *NY* 514; Ethipothala water falls; Macherla (GTR), *VRK* 3566; Neelvai ADB), *GO & DAM* 5160.

INDIA : Planted throughout.
WORLD : Malay Islands, Australia, Pacific.

## SALICACEAE
### SALIX Linnaeus

**Salix tetrasperma** Roxb. Pl. Cor. t. 97. 1795; Gamble 3: 1389 (973); Beddome Fl. Sylv. S. India t. 302. 1873.

Deciduous, small or medium-sized tree; bark greyish-brown, rough with deep, vertical fissures; wood red, straight or somewhat interlocked grained, medium-fine and mostly even textured; branchlets silky-pubescent. Leaves membranous, lanceolate or oblanceolate to ovate-lanceolate, 5–16 × 1.5–6 cm, glabrous above, sparsely hairy or glabrescent and pale glaucous beneath, apex subacuminate, margin glandular-serrate, base narrowed or rounded; petiole to 2.5 cm. Flowers cream coloured in lax or dense-flowered catkins. Capsule 3–4 mm long, 2-valved, pubescent; valves recurved; seeds ovate-lanceolate, flattened, 2 mm across with brownish hairs.

Rare, along the banks of rivers, streams and in wet swampy places; in the Agencies of North Circars above 1000 m. Leaf fall: Dec.–Jan. New foliage: Feb.–Mar. Red. Fls.: Oct.–Feb. Frs.: Dec.–May. Vern.: Iroda, Etipala, Konda Ganneru, Vodime. Eng.: Indian willow, Willow.

The branches are used for basket making. The shoots are lopped for fodder. The bark is used as a febrifuge. Wood is very light to heavy and soft to moderately hard; it is fairly durable, used for posts, planks, house building, bent-wood furniture, ploughs and agricultural implements; suitable for cabinet and fancy work and match boxes. It may also be employed for pencils and cricket stumps.

Sunkarimetta (VSKP), *SKW* 20587 & *HS* 20524 (BLAT); Chintapalli (VSKP), *JLE* 37110 (CAL).

INDIA : Tamil Nadu, Orissa, Karnataka.
WORLD : Afghanistan, Indo-China, S. China, Malay Peninsula, Archipelago.

## PALMAE (ARECACEAE)

1. Trees dioecious:
    2. Leaves simple, flabelliform, stalks shortly spinous. Carpels 3, fused. Drupe medium-sized ........ ***Borassus***
    2. Leaves 1-pinnate, leaflets spinous. Carpels 3, free. Drupe small ........ ***Phoenix***
1. Trees monoecious:
    3. Leaves 2-pinnate. Stamens many ........ ***Caryota***
    3. Leaves 1-pinnate. Stamens 3–9:
        4. Spadices infra-foliar. Ovary 1-celled. Drupe 3.5 cm long. Leaf sheath tubular ... ***Areca***
        4. Spadices interfoliar. Ovary 3-celled. Drupe 10–30 cm long. Leaf sheath 0. ....... ***Cocos***

### **ARECA** Linnaeus

**Areca catechu** L. Sp. Pl. 1189. 1753; FBI 6: 405; Blatter Palms Brit. Ind. 471. 1926; Fisher in Gamble 3; 1555 (1085); Brandis 646.

Tree; 12–18 m tall; trunk slender, straight, uniform; wood brownish-grey. Leaves to 1.5 m in a terminal crown, prominently arching; leaflets numerous, linear-laceolate, 15–30 cm long, close; glabrous, upper confluent, lower ones plicate. Spathe glabrous, compressed; spadix much branched, rachis stout, compressed, branches with filiform tip bearing more or less distichous, minute male flowers with 6 stamens; female flowers solitary at the base and axils of the branches. Fruit orange or scarlet, ovoid, 3.5 × 2 cm, smooth with fibrous mesocarp.

Cultivated, in the Coastal regions. Vern.: Gautupoka, Kolapoka, Oppa valkal, Vaktia. Eng.: Betal nut palm, Areca nut.

Areca nuts are extensively used as a masticatory either alone or with betal leaves. Powdered nuts are prescribed in diarrhoea and urinary disorders. Dried nuts are said to sweeten the breath, strengthen the gums and produce a stimulant and exhilarent effect on the system. The leaves are used for thatching and making rough bags and mats; midribs for making brooms. The wood is strong and elastic; ideally suited for making stationary articles, scales, rulers, paper cutters, waste paper baskets etc. The stems are useful for making chip boards, fibre boards and also plastic sheets. The bark and wood are used for tanning.

Cultivated from India to the Solomon Islands and less commonly in Africa and tropical America.

## **BORASSUS** Linnaeus

**Borassus flabellifer** L. Sp. Pl. 1187. 1753; FBI 6: 482; Fischer 3: 1562 (1090). Brandis 657. *B. flabelliformis* Murr. in L. Syst. Veg. (ed. 13). 827. 1774.

Large tree, bark bluish with horizontal, annular depressions; wood dark brown with black striations. Leaves simple, palmate, 30–60 cm dia., with 60–80 linear-lanceolate segments folded along the midrib, petiole 30–60 cm, semiterete, stout, edges spinousely serrate, base broad, split. Spadices large, interfoliar, dioecious, branched; peduncle sheathed with numerous open spathes; male spikes 1–3 at the ends of branches, cylindric, densely clothed with thick closely imbricating bracts; unilateral spikelets numerous, minute, concealed by the bracts; female spikes large, globose, terminating the branches of spadix. Fruit subglobose, 10–12 cm in dia., smooth, reddish-black when ripe; pyrenes 1–3, compressed in soft yellowish fibrous pulp.

Common, wild and cultivated in plains from the Coast; an indicator species for the more arid plains. It can grow in a variety of soils. Fls.: Mar.-Apr. Frs.: May–Aug. Vern.: Thati. Eng.: Palmyra palm.

It is one of the most useful trees of India. Its leaves are used for thatching cottages, fans and umbrellas; its stems are used as house posts of roof beams. Portions of the fruits and seed are edible; these are small and jelly-like, but is very refreshing in the hot weather. The cut spadices yield sugar and toddy, which is almost as famous for its use.

Anantapur town (ATP), *KNR* 19626 (MH); Gangineni (KSN), *PVK* 5572 (MH); Gulemallebad RF (KNL), *RVR* 1421.

INDIA : Throughout.
WORLD : Africa, Madagascar, Sri Lanka, Indo-China and Siam, E. Java, Lesser Sunda Islands, Near New Guinea, Queensland.

## CARYOTA Linnaeus

**Caryota urens** L. Sp. Pl. 1189. 1753; FBI 6: 422; Fischer 3: 1560 (1089); Brandis 654.

Tree, up to 20 m tall; stem smooth, grey, annulate, shining, bark dark brown, stout. Leaves bipinnate, 2 m long; pinnae ca. 10 pairs, sub-opposite, curved; leaflets induplicate, 30 × 10 cm, margin irregularly serrate, upper serrations caudate, strongly nerved with parallel nerves; petiole stout. Inflorescence 3–4 m long, spadices interfoliar; branches slender, pendulous; spikes 3–5, to 40 cm long. Berry brownish-black, globose, 1.8 cm across, stony, smooth; seeds 1 or 2.

Common, hills above 800 m in shady valleys. It is having good regenerating capacity. Fls. & Frs.: Throughout the year. Vern.: Bee bee jadalu, Konda jeelugu. Eng.: Fish-tail or Sago palm.

Leaf and flower stalks are commerically employed to make very durable ropes or for the preparation of rough brushes. The pith yields a good sago. The flowering peduncles are tapped for liquor. The timber is strong and durable. The young inflorescence is used to obtain cordage.

SK University (ATP), *TP & NY* 1290; Nallamalais (KNL), *JLE* 32645 (MH); Neelakantapuram (VJN), *MV* 19827; Srunagavarapukota (VSKP), *BR & KNR* 9883.

INDIA : Indigenous; throughout.
WORLD : Sri Lanka, Myannar, Malay.

## COCOS Linnaeus

**Cocos nucifera** L. Sp. Pl. 1188. 1753; FBI 6: 482. 1893; Fischer 3: 1557 (1086); Brandis 648.

Tree, unbranched, straight or slightly curved: marked with ring-like leaf scars, bark grey with broad annualr rings. Leaves pinnatisect, feathery, 2–6 m long, rachis stout, leaflets numerous, reduplicate, leathery, greenish, apex tapering, or acute, margin entire, base narrow; petiole stout, elongate, 1–2 m. Spadix stout, monoecious, divided into numerous, drooping spikes bearing at their base female with few male flowers, the upper portion being densely covered with male flowers. Drupe obtusely trigonous, 20–30 cm long, greenish yellow; epicarp shining, very hard; mesocarp thick, fibrous, endocarp hard, stony with 3 basal pores; endosperm fleshy, and living inside the endocarp; the cavity inside the flesh is filled with a watery fluid.

Common, planted near the houses, fields etc. It grows well in sandy soils in Coastal districts. Vern.: Kobbari, Tenkai, Narikelam. Eng.: Coconut palm.

It is very useful tree for mankind. Wood is useful for house posts; leaf ribs are useful to prepare brooms; coir is obtained from the fibrous pericarp. Fruits yield a most refreshing cool drink is also called coconut water or coconut milk. Coconut oil is extracted from the endosperm of the fruit. Coconut meat is used as condiment.

INDIA : Throughout.

## **PHOENIX** Linnaeus

**Phoenix sylvestris** (L.) Roxb. Fl. Ind. 3: 787. 1832; Gamble 3: 1559 (1088); Brandis 645. *Elate sylvestris* L. Sp. Pl. 1189. 1753.

Tree, up to 12 m tall; stem cylindrical; trunk densely covered with the bases of fallen petioles, bark dark grey; wood yellow. Leaves greyish-green, 2–4 m long; leaflet numerous, 10–40 × 1–2.3 cm, rigid, glabrous, apex acuminate, spinous; lowest leaflets spiny. Spadices to 1 m; spathes coriaceous, woody; spikes numerous, in clusters, 10–30 cm long; female flowers dense; male flowers distant. Drupe orange to reddish brown, oblong-ellipsoid, to 1.8 cm; flesh sweet, scanty.

Common on bunds of paddy fields, banks of streams and tanks, along road sides, waste lands, scrub jungles etc. Fls.: Jan.-Apr. Frs.: May–June and Sept.–Oct. Vern.: Pedda-ita. Eng.: Wild Date palm; Date Sugar palm.

The leaves are used for thatching and for making mats, baskets, fans, bags, brooms, fishing nets, etc. They also yield a soft fibre which bleaches well. The tree yields a light brown, durable, wood consisting of a hard and rough outer cylinder and soft inner portion, it is used for construction bridges, and piers and for tent pegs; trunk freed from pith make excellent water conducts. Toddy is also tapped from the tree. Fruits edible.

Racherla RF (KNL), *RVR* 1361; Indampalli (ADB), *GO* 5410; Nallapadu (GTR), *VRK* 6775; SK University (ATP), *NY* 516; Karakamoksa (VSKP), *KCJ* 449 (MH); Tanuku (WG), *KCJ* 449 & 457 (MH).

INDIA : Throughout India.
WORLD : Sri Lanka, Myanmar.

## POACEAE (GRAMINAE)

1. Branches unarmed at nodes. Spikelets in globose heads. Rachilla internodes very short, not disarticulating ........ ***Dendrocalamus***
1. Branches armed at nodes. Spikelets not in globose heads. Rachilla internodes elonagate up to half the lenghth of lemma, promptly or tardily disarticulating ........ ***Bambusa***

## BAMBUSA Schreber

**Bambusa arundinacea** (Retz.) Roxb. Pl. Cor. t. 79. 1796; Fischer 3: 1859. (1286); FBI 7: 395; Brandis 671. *Bambosa arundinacea* Retz. Obs. Bot. 5: 24. 1788.

Gregarious, densely tufted, armed; culms 20–40 m tall; prominent with 2–3 recurved spines. Culm sheath coriaceous, variable, orange-yellow, black or purplish hairs within; blade to 10 cm long, triangular. Leavaes linear-lanceolate, 10–14 × 1–2.5 cm, glabrous or puberulous beneath, apex stiff, margin entire, base rounded; secondary nerves 4–6 with 7–9 intersecondary nerves; petiole short. Panicle compound; spikelets oblong, 3–6 flowered, clustered in heads. Caryopsis oblong, 2–8 mm.

Common, in deciduous and semievergreen forests; very frequent in Nallamalais. Vern.: Veduru, Mulla veduru, Bongullu.

Bhupatipalem (EG), *TP & MSG* 12249; Talakona (CTR), *DR* 1303 (SVU), Kanuparth (NLR), *JSG* 21813 (DD); Siglabyle (CDP), *JSG* 20860 (CAL & DD).

INDIA : Throughout.

## DENDROCALAMUS Nees

**Dendrocalamus strictus** (Roxb.) Nees in Linnaea 9: 476. 1834; FBI 7. 404; Fischer 1858 (1256); Brandis 675. *Bambusa stricta* Roxb. Pl. Cor. t. 80. 1798.

Middle sized, deciduous tree; culms greyish-green, often blotched, solid or with small cavity. Leaf blade linear-lanceolate, stiff, erect, 8–12 × 0.5-0.8 cm, shortly petiolate, sheaths glabrescent. Inflorescence a compound panicle; spikelets clustered in globose heads in long spikes. Caryopsis ovoid-subglobose.

Common, in hilly forests at 400–1000 m. Fls. & Frs.: Nov.–Apr. Vern.: Sasanum veduru.

Peecheruvu RF G(KNL), *PVP & RVR* 2177; Pamuleru (EG), *TP & MSG* 12169; Kalasamudram RF (ATP), *NY* 398 & *PVP* 2601; Sapparla peak (VSKP), *GVS* 29698 (CAL).

INDIA : Throughout.
WORLD : Pakistan, Myanmar, ɩ ɔore and Java.

# REFERENCES


Ahamed, M. and M.P. Nayar. 1984. Red Sanders tree. Bot. Surv. India **2**: 142–143.

Ahmedulla, M. and M.P. Nayar. 1987. *Endemic Plants of the Indian region*. Vol. I, *Peninusular India*, Bot. Surv. India, Calcutta.

Airy-Shaw, H.K. 1972. *Bridelia*. The Euphorbiaceae of Siam. Kew Bull. **26**: 227–231.

Airy-Shaw, H.K. 1972. *Glochidia*. The Euphorbiaceae of Siam. Kew Bull. **26**: 271–281.

Airy-Shaw, H.K. 1972. *Macaranga*. The Euphorbiacece of Siam. Kew Bull. **26**: 286–292.

Aiton, W. 1789. *Hortus Kewensis* (ed. 1), (2 Vols.), London.

Aiton, W.T. 1810–1913. *Hortus Kewensis* (ed. 2), London.

Alan, Butterworth. 1911. *Some Madras Trees*. Madras.

Anonymous. 1948–76. *The Wealth of India—Raw Materials*, CSIR, New Delhi.

Babu, C.R. 1974. The genus *Zanthoxylum* L. (Rutaceae) in India. Bull. Bot. Surv. India **16**: 48–71.

Bailey, L.H. 1949. *Manual of Cultivated Plants*. New York.

Baker, C.A. and R.C. Bakhuizen Van der Brink. 1963, 1965 & 1968. *Flora of Java* (3 Vols.), The Netherlands.

Balakrishnan, N.P. 1954. New Plant records from South India. Bull. Bot. Surv. India **6**: 81–92.

Balakrishnan, N.P. 1966. Nomenclatural notes on some flowering plants. J. Bombay Nat. Hist. Soc. **63**: 327–331.

Balakrishnan, N.P. 1980. Nomenclature of some flowering plants. Bull. Bot. Surv. India **22**: 173–179.

Balakrishnan, N.P. and A.N. Henry. 1961. *Boswellia ovalifoliolata*: a new species of *Boswellia* from South India. J. Bombay Nat. Hist. Soc. **58**: 546–548.

Bannerjee, L.K., A.R.K. Sastry and M.P. Nayar. 1989. *Mangroves in India* BSI, Calcutta.

Beddome, R.H. 1874. *Icones Plantarum Indiae Orientalis*. Madras.

Beddome, R.H. 1869–74. *The Flora Sylvatica for Southern India* (3 Vols.), Madras.

Bennet, S.S.R. 1971. The taxonomic status of the genus *Pongamia* Vent. (Papilionaceae). J. Bombay Nat. Hist. Soc. **68**: 302–303.

Bennet, S.S.R. 1987. *Name changes in Flowering plants of India and adjacent regions*. Dehradun.

Benthall, A.P. 1946. *Trees of Calcutta and its neighbourhood*, Calcutta.

Bentham, G. and J.D. Hooker. 1862–83. *Genera Plantarum*. (3 vols.), London.

Bentvelzen, P.A.J. 1962. Reduction of the genus *Heynea* Roxb. ex. Sims to *Trichilia*. P. Br. (Meliaceae). Acta. Bot. Neerl. **11**: 11–20.

Bhandari, M.M. 1978. *Flora of the Indian Desert*. Jodhpur.

Bhirava Murthy, P.V. 1984. *Proceedings of the Seminar on Resources: Development and Environment of the Eastern Ghats*: Andhra University, Waltair.

Bhirava Murthy, P.V. and R. Krishna Mohan. 1991. A contribution to the flora of Prakasam district, Andhra Pradesh. J. Econ. Tax. Bot. **15**: 49–63.

Bisset, N.G., P.W. Leenhouts, A.J.M. Leewenberg, D. Philcox, C. Triel Roudel and J.R. Videl. 1973. The Asian species of *Strychnos*. Part III. Typification, miscellaneous notes, synpotic key and sectional classification. Lloydia. **36**: 179–201.

Biswas, K. 1943. Systematic and taxonomic studies on the flora of India and Burma Proc. 30th Indian Sci. Congr. Associ. Calcutta. 101–152.

Blatter, E. 1926. *Palms of Biritish India and Ceylon*, Oxford University Press.

Blatter, E. 1929. The Indian species of *Terminalia* Linn. Journ. Indian Bot. Soc. **8**: 245–262.

Blatter, E. ând C. McCann. 1934. Revision of the flora of Bombay Presidency. J. Bombay Nat. Hist. Soc. **37**: 15–35, 265–277, 764–779.

Blatter, S.J. 1937. *Some beautiful Tress of India*; John Bale Sons & Cornola Ltd., Great litchfield St. London, W.I.

Blatter, Elhelbert and Millard, W.S. 1954. *Some beautiful Indian Trees*, Ed. 2 Rev. by W.T. Stearn.Bombay.

Boulton, E.M.B. 1944. *Pocket Book of British Trees*. London.

Bourdillion, F.T. 1993. (Rep. ed.) *The Forest Trees of Travancore*.

Bor, N.L. 1953. *Manual of Indian Forest Botany*. Oxford University Press.

Borssum Walkes, J.V. 1966. Malesian Malvaceae revised, Blumea **14**: 1–213.

Bradley, H. 1849. Statistics of the circar of Dowlutabad, Hyderabad, Deccan, Madras J. Lit. & Sci. **15**: 481–551. (Plant products, pp 489–499).

Brandis, D. 1899. The Indian species of *Anogeissus*. Indian forester **25**: 286–287.

Brandis, D. 1906. *Indian Trees*. Archibald Constable, London.

Buchanan-Hamilton, F. 1807. *A. Journey from Madras through the countries of Mysore, Canara, Malbar*. (3 vols.), London (repr. ed. in 2 vols. 1870).

Bullock, A.A. 1960. Nomenclatural notes XII. The types of some generic names. Kew Bull. **14**: 40–45.

Burea. E. 1861. Ramarques surla classification des Bignoniaceae et observations surles generes *Radarmachera et Stereospermm* Adansonia **2**: 182–197.

Burkhill, I.T. 1965. *Chapters on the history of Botany in India*. Calcutta.

Burman, N.L. 1768. *Flora Indica*, Leiden, Amsterdam.

Calder, C.C., V. Narayanaswamy and M.S. Ramaswamy. 1926. List of species and genera of Indian Phanerogams not included in J.D. Hooker's flora of British India. Rec. Bot. Surv. India. **11**: 1–157.

Cameron, J. 1894. *Forest trees of Mysore & Coory*. Mysore Govt. Press, Bangalore.

Candolle, A.L.P.P. de and Candolle A.P. de. 1878–1896. *Monographiae phanerogamarum*. (9 vols.), Paris.

Candolle, A.P. de. 1824–1874. *Prodromus Systematis Naturalis Regni Vegetabilis*. 17 vols. & Index, Paris (vols. 8–17 by A.L.P.P. deCandolle).

Chakrabarthy, T. and M. Gangopadhyay. 1996. The Genus *Acacia* P. Miller (*Leguminosae: Mimosoideae*) in India. J. Econ. Tax. Bot. 20: 499–633.

Chakrabarthy, T. and M. Gangopadhyay. 1996. The Genus *Albizia* Durazz. (*Leguminosae: Mimosoideae*) in India. J. Econ. Tax. Bot. 20: 581–597.

Chakrabarty, T., M. Gangopadhyay and N.P. Balakrishnan. 1997. The genus *Drypetes* (*Euphorbiaceae*) in the Indian Sub-continent. J. Econ. Tax. Bot. 21: 251–780.

Champion, H.G. and S.K. Seth. 1968. *A revised survey of the forest types of India*. Govt. of India, New Delhi.

Chandra Bose, M. 1976. Notes on some rare and interesting plants from South India. Bull. Bot. Surv. India. **18**: 236–238.

Chatterjee, D. 1948. Notes on the affinities of Moringaceae. Sci. Cult. **14**: 253–254.

Chaudhuri, A.B. 1967. Studies on the Meliaceae of India and neighbouring countries. Bull Bot. Soc. Bengal **21**: 1–24.

Chopra, R.N. 1956. *Glossary of Indian Medicinal Plants*. New Delhi.

Colthurst, I. 1937. *Familiar Flowering Trees in India*, Thacker Spink & Co., Calcutta.

Cooke, T. 1901–1908. *The flora of the Presidency of Bombay*, London (3 vols.), repr. ed. 1958. Calcutta.

Corner, E.J.H. 1965. Check list of *Ficus* of Asia and Australia with keys to identification. Gard. Bull. Singapore **21**: 1–186.

Cowen, D.V. 1950. *Flowering Trees and Shrubs in India*. Thacker & Co., Bombay.

Crete, P. 1963. Embryo. In: P. Maheshwari (ed.). *Recent Advances in the Embryology of Angiosperms*. Delhi, pp. 171–220.

Croizat, L. 1947. Nomenclatural transfer and corrections in the *Euphorbiaceae*. Trop. Woods **88**: 30–32.

Cronquist, A.J. 1968. *The Evolution and Classification of Flowering Plants*. Boston.

Cronquist, A.J. 1981. *An Integrated System of Classification of Flowering Plants*, London.

Cunhiff, K.M. 1973. Trees for Tommorrow. Trees S. Afr. **24**: 107–111.

Dahlgren, R.M.T. 1980. A revised system of classification of the Angiosperms. Bot. J. Linn. Soc. **80**: 91–124.

Dalzell, N.A. and A. Gibson. 1861. *The Bombay Flora* (2 vols.) Bombay.

Das, V.S.R., K.N. Rao and J.V.S. Rao. 1966. *Chemotasonomical investigaion of Nyctanthes*. Die Natur Wissensechaften. 3. 439.

Datta, R.M. and J.N. Mitra. 1947. The systematic position of the family Moringaceae based on a study of *Moringa pterigosperma* Gaertn. (*M. oleifera* Lam.). J. Bombay Nat. Hist. Soc. **47**: 355–357.

Davis, T.A. and A. Kundu. 1965. Floral structure and stamens in *Ceiba pentandra* (L.) Gaertn. J. Bombay Nat. Hist. Soc. **62**: 399–411.

Daniel, P. 1980. The fast vanishing umbrella thorn tree (*Acacia planifrons* Wt. & Arn.), Ind. J. For. **3**: 349–352.

Dayananda, P. 1981. Conserving the flora of Peninsular hills. Bull. Bot. Surv. India, **23**: 250–253.

Deb, D.B. 1965. Plants new to India. Ind. For. **91**: 193–194.

Deb, D.B. 1989. Taxonomic revision of the Genus *Hymenodictyon* (Rubiaceae) in India. J. Econ. Tax. Bot. **13(3)**: 675–684.

Delile, a.R. 1813. *Description de l'egypte. Histororie naturella* (2 vols.), Paris.

Desfontaines, R.L. 1799. *Flora Atlantica* etc. (2 vols.) Paris.

Ding Hou, L. 1962. Celastraceae-I, In: *Flora Malesiana* Ser. **1. 6**: 226–291.

Don, D. 1825. *Prodromus Florae Nepalensis*, London.

Don, G. 1831–1838. *A general history of Dichlamydeous Plants*. London.

Duthie, J.F. 1903–1929. *Flora of the upper gangetic plains and of the adjacent Siwaliks and Sub-Himalayan tracts* (3 vols.) Calcutta.

Edlin, H.L. 1935. A critical revision of certain taxonomic groups of the Malvales, New Phyton. **34**: 1–20; 121–143.

Elliot, W. 1859. *Flora Andhrica*. Madras.

Ellis, J.L. 1966. A contribution to the Vascular Flora of Balapalle range of Seshachalam Reserve Forest, Cuddapah district. Bull. Bot. Surv. India **23**: 126–128.

Ellis, J.L. 1968. The Flora of Nallamalais on the Eastern ghats of India. I. A preliminary list. Bull. Bot. Surv. India **10**: 149–160.

Ellis, J.L. 1981. Observation on the vegetation of Nallamalais. Bull. Bot. Surv. India **23:** 126–128.

Ellis, J.L. 1982 a: Wild Plant resources of Nallamalais on the Eastern Ghats. In: *Proceedings of Seminar Resources, Development of Environment in the Eastern Ghats*, P.V. Bhirava Murthy (ed.) pp. 65–67.

Ellis, J.L. 1982 b. Nomenclatural notes on three plants from South India. Bull. Bot. Surv. India. **22**: 193.

Ellis, J.l. 1987–1990. *Flora of Nallamalais*. Calcutta.

Endlicher, S.L. 1836–1840. *Genera planatarum*. Vienna.

Engler, A. 1983. *Burseraceae* In: DC Monogr Phan **4**: 1. 169, p. 1–3.

Engler, H.G.A. 1900–1937. Das Pflanzenreich Regni vegetabilis conspectus. Leipzig.

Engler, H.G.A. and K.A.E. Prantl. 1897–1915. *Die Naturlichen Pflanzenfamilien*. (23 vols.) ed. 1. Leipzig.

Engler, H.G.A. and Gilg. 1924. *Syllabus der Pflanzenfamilien*. Berlin.

Excell, A.W. and C.A. Stace. 1972. Patterns of distribution in the *Combretaceae*. In: D.H. Valentine, ed., Taxonomy, Phytogeography.

Forster, J.R. and J.G.A. Forster. 1775. *Characters Genera Plantarum* etc. London.

Fosberg, F.R. and M.H. Sachet. 1972. *Thespesia populnea* (L.) Solander ex Correa and *T. populnoides* (Roxb.) Kostelets. Smithsonian Controb. Bot. No.7: 1–173.

Fosskal, P. 1775. *Flora Aegyptiaco-Arabica*. Copenhagen.

Fortade, C.X. and M. Srisuku. 1969. A revision of *Lagerstroemia* L. (Lythraceae) Gard. Bull. Singapore **24**: 185–335.

Fyson, P.F. 1931. *The Flora of South Indian Hill Stations*, Madras.

Gaertner, J. 1788–1807. *De Fructibus et seminibus plantarum*, (3 vols.) Stuttgart.

Gamble, J.S. and C.E.C. Fischer. 1915–1935. *Flora of the Presidency of Madras*. London Rep. ed. 1957. Calcutta.

Gamble, J.S. 1881. *A Manual of Indian Timbers*. Calcutta.

Gangopadhyay, M. and T. Chakrabarty. 1997. The Family Combretaceae of Indian subcontinent. J. Econ. Tax. Bot. **21**: 281–364.

Gibson, A.S. 1863. *Forest trees of India*. Education Society Press, Byculla.

Gmelin, J.F. 1791–1792. *Linnaeus systema Nature*. ed. 13. Leipzig.

Gray, A. 1848. *A Manual of Botany of the northern United States*, Boston & Cambridge (ed. 2, New York).

Greuter, W. (ed.). 1988. *International code of Botanical Nomenclature*. Konigstein (Germany).

Haines, H.H. 1921–24. *The Botany of Bihar and Orissa*. London.

Hanumantha Rao, V. (ed.). 1985. *Andhra Pradesh Year Book*. Hyderabad.

Haworth, A.H. 1812. *Synopsis plantarum succulentarum*. London.

Heipko, P. 1984. Opiliaceae. In: *Flora Malesiana* Ser. 1. **10**: 31–52.

Henry, A.N., K. Vivekanathan and N.C. Nair. 1978. Rare and threatened flowering plants of South India. J. Bombay Nat. Hist. Soc. **75**: 684–697.

Henry, A.N. and Bela Roy. 1968. Nomenclatural notes on Indian flowering plants. Bull. Bot. Surv. India. **10**: 274.

Heywood, V.V. 1980. Glimpses of Plant Research. Modern Trends in Plant Taxonomy, Vol. V. Vikas Pub. House. New Delhi.

Hiern, W.P. 1873. A monography of the Ebenaceae. Trans. Cambridge Phios. Soc. **12**: 27–300, p. 1–11.

Hill, A.W. 1917. The genus *Strychnos* in India and the East. Kew Bull. 1917: 121–219.

Hills, G.A. 1960. *Comparison of Forest ecosystem in different climatic zones*. Bull. Internat. Soc. Trop. Ecol. **1**: 67–71.

Hooker, J.D. (ed.). 1872–1897. *Flora of British India*. (7 vols.). London.

Hooker, J.D. and T. Thomson. 1855. *Flora Indica*. London.

Hooker, W.J. and G.A.W. Arnott. 1830–1841. *The Botany of Captain Beechey's voyage*. London.

Howard, R.A. and T. Norlindh. 1962. The typification of *Diospyros ebenum* and *D. ebenaster*. Journ. Arn. Arb. 43: 94–102.

Huber, H. 1973. In: Abeywickrama, B.A. (ed.) *A Revised Handbook of Flora of Ceylon*. Vol. 1. University of Ceylon, Peradenia.

Humboldt, F.H.A., A.J.A. Bonpland and C.S. Kunth. 1816–1825. *Nova Genera et species plantarum* etc. Paris.

Hunter, J.R. 1981. Tendu (*Diospyros melanoxylon*) leaves bidi cigarettes and resource management. Econ. Bot. **35**: 450–459.

Hutchinson, J. 1959. *Evolution and Phylogency of Flowering Plants*. London.

Hutchinson, J. 1964. *The Genera of Flowering Plants*. Vol. I. Claredon Press, Oxford, England.

Hutchinson, J. 1969. *The Families of Flowering Plants*. Vol. 1. Dicotyledons. Oxford.

Iltis, H.H. 1960. Studies in the Capparidaceae. VII. Old World *Cleomes* adventive in the new world. Brittonia **12**: 279–294.

Ireland, S.M. 1934. *Trees and Shrubs of Punjab*. Uttar Chand Kapur & Sons, Lahore.

Jacobs, M. 1960. Capparidaceae. In: *Flora Malesianna* Ser. 1. **6**: 226.

Jacobs, M. 1964. The genus *Crataeva* (Capparaceae). Blumea **12**: 177–208.

Jackson, D.B. 1953. (Red. Ed.) *A Glossary of Botanic terms*. Gerald Buckwork & Co. Limited, London.

Jain, S.K. 1978. Welcome Address. *125th Anniversary Souvenir of Madras Herbarium, Botanical Survey of India, Southern Circle*, Coimbatore, pp. 1–2.

Jain, S.K. 1982. *Flora of India Tasks before 2000 AD*. Panchanan Maheshwari Memorial Medal lecture, Indian Botanical Society.

Jain, S.K. and R.R. Rao. 1977. *Field and Herbarium methods*. New Delhi.

Jain. S.K. and A.R.K. Sastry. 1980. *Threatened plants of India. A State of the Art Report*, New Delhi.

Janaki Ammal, E.K. 1954. The scope and function of the reorganised Botanical Survey of India, Sci. & Culture **22**: 278.

John, S.R.J. 1979. A new approach to the construction of field keys for the identification of tropical trees. Austr. J. Ecol. **3**: 403–416.

Jussieu, A.L. de 1789. *Genera plantarum secundum Ordines Naturales Disposita* (ed. 2. 1791). Paris.

Kapil, R.N. and R.S. Vani. 1966. *Nyctanthes arbor-tristis* L. Embryology and relationships. Phytomorphology **16**: 533–563.

Kapoor, S.L. and L.D. Kapoor. 1973. Further contribution to the flora of the Karimnagar district of A.P. Bull. Bot. Surv. India. **15**: 76–84.

Karthikeyan, S. and B.D. Sharma. 1983. A catalogue of species added to Gamble's *Flora of the Presidency of Madras*. J. Bombay Nat. Hist. Soc. **80**: 63–70.

Karthikeyan, S., M.P. Nayar and R.S. Raghavan. 1981. *An annotated Bibliography of Taxononmic Botany of Peninsular India*. 1959–1978. Botanical Survey of India. Howrah.

Kashyap, G. 1961. *Shorea talura* Roxb. A synonym of *S. roxburghii* G. Don. J. Bombay Nat. Hist. Soc. **58**: 543–544.

Khan, A.A. 1954. The Andhra Forest. Ind. For. **80**: 753–758.

Khan, M.S. 1953. *Forest Flora of Hyderabad State*. Hyderabad.

Kostermans, A.J.G.H. 1977. Notes on Ceylonese Ebony trees (Ebenaceae). Cey. J. Sci. (Bio-Sci.). **12**: 89–108.

Krishna Mohan, P. 1986. *Flora of Prakasam district*. Ph.D. Thesis, Andhra University, Visakhapatnam.

Kumar, S.V. and Manoranjan Bhanja. 1992. *Forestry Seed Manual of Andhra Pradesh*. Andhra Pradesh Forest Department, Hyderabad.

Kumari, G.R. and G.V. Subba Rao. 1982. The genus *Picrasma* Bl. (Simaroubaceae) in South India. J. Econ. Tax. Bot. **3**: 249–250.

Kunth, C.S. 1833–1850. *Enumeratio plantarum ombium hucusque cognitarium*. Stuttgart.

Kuntze, C.E.O. 1891–98. *Revisio Generum Plantarum*. 3 vols. Leipzig.

Lakshminarayana, K., P. Venkanna and T. Pullaiah. 1997. *Flora of Krishna district, Andhra Pradesh, India*. M.D. Publications, New Delhi.

Lamarck, J.B. and A.P.M. de 1791–1823. *Tableau Encyclopedique et Methodique; Botanique*. 3 vols. & supp. 1. Paris (Vol. 2 part 5, 2 & Vol. 3 by J.L.M. Poiret.).

Larsen, K. and Larsen, S.S. 1979. Nomenclatural notes on some old world *Bauhinia*. Taxon **28**: 591–592.

Lawrence, G.H.M. 1951. *Taxonomy of Vascular Plants*. New York.

Legris, P. & V.M. Mehr-Homji. 1982. The Eastern Ghats; Vegetation and bioclimatic aspects. Proc. Seminar on Resources Development and Environment in Eastern Ghats. 1–7, Andhra University, Waltair.

Linnaeus, C. 1753. *Species Plantarum* (2 vols.) Stockholm.

Linnaeus, C. 1755–56. *Centuria Plantarum*, (2 vols.), Stockholm.

Linnaeus, C. 1778–79. *Systema Naturae*. ed. 10 Vol. 2. Botany, Stockholm.

Linnaeus, C. 1767. *Mantissa Plantarum* pages 1–142. Stockholm.

Linnaeus, C. 1771. *Mantissa Plantarum Altera*, pages 143–587. Stockholam.

Linnaeus, Fil. 1781. *Supplementum Plantarum Systamatis vegetabilium* etc. Brunsvigae.

Livera, E.J. 1925. The genus *Chukrasia* Ann. Roy. Bot. Gard. Peradeniya. **9**: 307–308.

Lushington, A.W. 1915. *Vernacular list of trees, shrubs and woody climbers in the Madras Presidency*, Madras.

Mabberly, D.J. 1978. The latin name of the Padri tree (Bignoniaceae). Taxon **27**: 553.

McCann, C. 1947. *Trees of India.* Probasthain.

Maheshwari, J.K. 1960. Taxonomic studies on Indian Guttiferae - the genus *Calophylluym* L. Bull. Bot. Surv. India **2**: 139–148.

Maheshwari, J.K. 1962. Studies on the naturalised flora of India. Proc. Summer School, Botany, Darjeeling, 156–170.

Maheshwari, J.K. 1976. Plant resources in India. Sci. & Cult. **42**: 18–23.

Majumdar, N.C. 1979. Dilleniaceae. In: *Fascicles of Flora India* **2**: 1–16.

Mathew, K.M. 1981–88. *The Flora of Tamil Nadu—Carnatic.* Vol. 6. Tiruchirapalli.

Mehra, K.L. 1965–66. Portuguese introduction of fruit plants in India I–III, Indian Hort. **10**: 8–12, 23–25.

Meher-Homji, V.M. (1974). 1975. Buttress like structures on the upper part of the trunk of *Ceiba pentandra* (L.). Gaertn. Journ. Bombay Nat. Hist. Soc. **71**: 336–338.

Merril, E.D. 1917. *An interpretation of Rumphius Herbarium Amboinese.* Bur. Sci. Publ. Manila **9:** 1–595.

Merril, E.D. 1951. Notes on *Elaeocarpus* Linnaeus. Journ. Arn. Abr. **32:** 157–200.

Michaux, A. 1803. *Flora Borealiamericana.* (2 vols.). Paris.

Miller, P. 1768. *The Gardener's Dictionary.* (ed. 8). London.

Mitra, R.L. and S.C. Srivastava. 1979. The genus *Bruguiera* Lam. (Rhizophoraceae) in India. Bull. Bot. Surv. India **21**: 142–150.

Mooney, H. 1938. *Supplement to the Botany of Bihar and Orissa.* Ranchi.

Mukherjee, S.K. 1968. The genus *Alangium.* Lam. in India. Bull. Bot. Surv. India **10**: 330–331.

Mukherjee, S.K. 1972 a. Indian *Allophylus* (Sapindaceae), Ind. For. **98**: 489–496.

Mukherjee, S.K. and Dali Chandra. 1983. An outline of the revision of Indian Anacardiaceae. Bull. Bot. Surv. India **25**: 52–61.

Naidu, K.C. 1983. Angiosperm flora of Sambarukonda of Gudem, Visakhapatnam district. Phytologia **54**: 299–301.

Naidu, K.V. and R.S. Prakasa Rao. 1966. *Flora of Cuddapah district.* Dicotyledons. Diocesan Press, Madras.

Naidu, K.V. and R.S. Prakasa Rao. 1967. A contribution to the Flora of Tirupati hills. Part–I. Ind. For. **93**: 123–125.

Naidu, K.V. and R.S. Prakasa Rao. 1969. A contribution to the Flora of Tirupati hills, Part II. Ind. For. **95**: 618–628.

Naidu, K.V., P.M. Swamy and K. Naryana Rao. 1971. A contribution to the Flora of Tirupati hills. Part III. Ind. For. **97**: 89–100.

Nair, N.C. and A.N. Henry. 1983. *Flora of Tamil Nadu, India.* Ser. 1. Analysis Vol. 1. Coimbatore.

Narasimha Rao, G. and N. Ramayya. 1987. Biological spectrum of Narsapur, Medak district (A.P.), India. J. Swamy Bot. Cl. **4**: 183–187.

Narayana Rao, K., Thammanna and K.S.R. Das. 1981. *Plant Wealth of Tirumala.* T.T.D. Press.

Nayar, M.P. 1968. Nomenclatural notes on the genus *Sonerila.* Roxb. (Melastomataceae). J. Bombay Nat. Hist. Soc. **65**: 805–807.

Nayar, M.P. 1980 a. Endemism and patterns of distribution of endemic genera (Angiosperms) in India. J. Econ. Tax. Bot. **1**: 99–110.

Nayar, M.P. 1980 b. Endemic Flora of Peninsular India and its significance. Bull. Bot. Surv. India **22**: 12–23.

Nayar, M.P., M. Ahmed and D.C.S. Raju. 1984. Endemic and rare plants of Eastern Ghats. Ind. J. For. **7**: 35–42.

Nayar, M.P., D.C.S. Raju and M. Ahmedulla. 1986. The dwindling flora of Andhra Pradeh. A call for conservation. Ind. J. For. **9**: 283–286.

Nayar, M.P., K. Thothathri and M. Sanjappa. (ed.). 1989. *Fascicles of Flora of India* No. 19. Howrah.

Ngan, P.T. 1965. A revision of the genus *Wrightia (Apocynaceae).* Ann. Missouri Bot. Gard. **52**: 114–175.

Nicolson, D.H. 1979. Nomenclature of *Bombax ceiba* (Bombacaceae) and *Cochlospermum* (Cochlospermaceae) and their type species. Taxon **28**: 367–373.

Pandey, Y.N. 1971. *Cassia* commonly occurring or cultivated in India. Journ. Bombay Nat. Hist. Soc. **68**: 311–318.

Panigrahi, G. 1977. Proposal to reject the name *Limonia acidissima* nom.*ambia rejc.* in favour of *L. elephantum* (Corr.) *comb. nov.* Taxon **26**: 576–577.

Parkinson, C.E. 1934. The Indian species of *Xylocarpus*. Indian Forester **66**: 136–143.

Parkinson, C.E. 1936. On some little known and confused Indian Terminalias. Indian Forester **62**: 406–409.

Patridge, E.A. 1911. *Forest Flora of H.H. The Nizam's Dominions*, Hyderabad, Deccan, Hyderabad.

Payens, J.P.D.W. 1967. A monograph of the genus *Barringtonia* (Lecythidaceae). Blumea **15**: 157–263.

Pennington, J.D. and B.T. 1975. A generic Monograph of the Meliaceae. Blumea **22**: 419–540.

Plowman, T. 1976. Orthography of *Erythroxylon* (Erythroxylaceae). Taxon **254**: 141–144.

Prain, D. 1890–1905. *Bengal Plants*. 2 vols. (repr. ed. 1963), Calcutta.

Prain, L.C. 1909. *Botany in 'Imperial Gazetter of India' Provincial Series, Hyderabad State*. Calcutta.

Pullaiah, T. 1997. *Flora of Andhra Pradesh*, Vol. III. Monocotyledons, Scientific Publishers, Jodhpur.

Pullaiah, T. & D. Ali Moulali. 1997. *Flora of Andhra Pradesh,* Vol. II. Rubiaceae-Ceratophyllaceae. Scientific Publishers, Jodhpur.

Pullaiah, T. & E. Chennaiah. 1997. *Flora of Andhra Pradesh*, Vol. I. Ranunculaceae-Alangiaceae. Scientific Publishers, Jodhpur.

Pullaiah, T., C. Prabhakar and B. Ravi Prasad Rao. 1997. *Flora of Medak District*, Daya Publishing House, New Delhi.

Pullaiah, T., P.V. Prasanna and G. Obulesu. 1992. *Flora of Adilabad Distrct.* C.B.S. Publishers. Delhi.

Pullaiah, T. and B. Ravi Prasad Rao. 1995. *Flora of Nizamabad*, Bishen Singh Mahendra Pal Singh, Dehra Dun.

Pullaiah, T. and N. Yesoda. 1989. *Flora of Anantapur District.* Dehradun.

Radford, A.E. *et al.*, 1974. *Vascular plant systematics*. New York.

Raghava Rao, S.R. 1987. *Flora of Mahabubnagar District*. Ph.D. Thesis, Osmania University, Hyderabad.

Raghava Rao, S.R. and T. Rajagopal. 1990. New records of flowering plants from Andhra Pradesh, India. Indian. J. Bot. **13**: 197–198.

Raghava Swamy, B.V. 1986. *Flora of Nalgonda District.* Ph.D. Thesis. Nagarjuna University, Guntur.

Raghava Swamy, B.V. and P.N. Rao. 1987. Further additions to the flora of Andhra Pradesh. J. Econ. Tax. Bot. **9**: 489–492.

Rajagopal, T. 1973. *Flora of Hyderabad including study of the foliar epidermal characters of species as an aid to taxonomy*. Ph.D. Thesis, Osmania University, Hyderabad.

Raju, D.C.S. 1966 a. Plants of Papi Hills. Ind. For. **92**: 482–492.

Raju, D.C.S. 1966 b. Vegetation and Flora of Polavaram Agency Tracts. Trop. Ecol. **7**: 114–124.

Raju, D.C.S. 1966 c. *Excursion flora of Simhachalam hill*. Penumantra 1–28. Howrah.

Raju, D.C.S. 1968. The Vegetation of West Godavary: Study of Tropical Delta. In: Proc. Sym. Recent Adv. Trop. Ecol. Int. Soc. Trop. Eco., Varanasi **1**: 348–358.

Raju, R.R.V. and T. Pullaiah. 1985. *Flora of Kurnool.* Bishen Singh Mahendra Pal Singh, Dehra Dun.

Raju, R.R.V. and T. Pullaiah. 1994. *Dimorphocalyx kunoolensis*—a new species of Euphorbiaceae from India. Bot. Bull. Acad. Sin. **35**: 201–204.

Raju, V.S. 1983. Some additions to the Flora of Andhra Pradesh. J. Econ. Tax. Bot. **4**: 295–297.

Raju, V.S. 1984. New Angiospermous taxa described from the State of Andhra Pradesh after its formation. J. Econ. Tax. Bot. **5**: 469–472.

Raju, V.S. 1986. Three interesting plant records for Andhra Pradesh. Ind. Bot. Reptr. **5**: 113–114.

Rakshit, S.C. and B.C. Kundu. 1970. Revision of the Indian species of *Hibiscus*. L. Bull. Bot. Surv. India. **12**: 151–175.

Ramachandrachari, S.T. 1980. *Floristic studies of Achampeta taluk (Mahaboobnagar dt.)*. Ph.D. Thesis, Osmania University, Hyderabad.

Ramachandra Chary, S.T. 1986. Additions to the forest, Flora of Telangana Region in Andhra Pradesh. Indian J. Bot. **9**: 156–158.

Ramachandran, A. 1975. *Inaugural address. 125th Anniversary Souvenir of Madras Herbarium, Botanical Survey of India, Southern Circle,* Coimbatore, pp 9–12.

Ramakrishnaiah, V. 1990. *Flora of Guntur District, Andhra Pradesh, India*. Ph.D. Thesis, S.K. University, Anantapur.

Rama Rao, N., D. Narasimhan and T. Ravishankar. 1987. On the occurrence of *Erythrina fusca* Lour. (Papilionaceae) in Southern India. J. Econ. Tax. Bot. **11**: 246–247.

Rama Rao, N. and A.N. Henry. 1996. *The Ethnobotany of Eastern Ghats in Andhra Pradesh, India*. BSI, Calcutta.

Ramayya, N. and T. Rajagopal. 1973. Taxonomy, forest ecology and anatomy of the flowering plants of the erstwhile State of Hyderabad – A review. Ind. Nat. Sci. Acad. **45**: 302–313.

Rangacharyulu, D. 1991. *Floristic studies of Chittoor district of Andhra Pradesh*, Ph.D. Thesis, S.V. University, Tirupati.

Rao, A.M. 1989. *Flora of Cuddapah district*, Ph.D. Thesis, S.V. University, Tirupati.

Rao, P.N., V.S. Raju and R.S. Rao. 1983. Floristic Research work in Andhra Pradesh. Bull. Bot. Surv. India **9**: 493–494.

Rao, Rolla S. 1958. Observation on the vegetation of the Rampa and Gudem Agency tracts of the Eastern Ghats—I. J. Bombay Nat. Hist. Soc. **55**: 429–499.

Rao, Rolla S. 1964. Observations on the vegetation of the Rampa and Gudem Agency tracts of the Eastern Ghats–II. J. Bombay Nat. Hist. Soc. **61**: 309–329.

Rao, Rolla S. and S. Hara Sreeramulu. 1986. *The Flora of Srikakulam district, Andhra Pradesh, India*. Meerut.

Rao, Rolla S. and K. Hemadri. 1979. *Medicinal Plants in Andhra Pradesh* (in Telugu). Telugu Academy, Hyderabad.

Rao, Rolla S. and R.S. Raghavan. 1965. Critical notes on three species of *Capparis* Linn. from Peninsular India. J. Bombay Nat. Hist. Soc. **62**: 412–424.

Rao, Rolla S. and P. Venkanna and T. Appi Reddy. 1986. *Flora of West Godavary district, Andhra Pradesh, India*. Meerut.

Rao, T.A. and A.R.K. Sastry. 1970. New distributional records for coastal plants of Andhra Pradesh. J. Bombay Nat. Hist. Soc. **67**: 614–615.

Rao, T.A. and P.G. Shanware. 1972. Analysis of the basic patterns of an esturine shore in the vicinity of the Coringa Bay. Andhra Pradesh, India. Proc. Ind. Acad. Sci. (Plant Sci.). **75**: 40–50.

Rathore, J.S. 1973. *D. melanoxylon*, a bread winner tree of India. Econ. Bot. **26**: 333–339.

Reddy, R.D., M.K. Prasad and K. Venkaiah. 1991. Forest *Flora of Andhra Pradesh*, Hyderabad.

Reddy, G.P. 1982. The tribal communities of Eastern Ghats, Culture and Ecology. *Proc. Seminar on Resources, Development and Environment in Eastern Ghats,* 173–180. Andhra University, Waltair.

Retzius, A.J. 1779–91. *Observations Botanicae*, Leipzig.

Ridley, H.N. 1930. *The dispersal of plants throughout the world*. Kent.

Roemer, J.J. and J.A. Schultes. 1817–30. *Carolia Linne Equitis systemas vegetabilicum Secundum classes etc.* (7 vols.). Stuttgart.

Roth, A.W. 1821. *Novae Plantarum Species Prasertium Indiae Orientalis Collectione Doct. B. Heynes cum descriptionibus et observationibus*, Halberstedt.

Roth, A.W. 1827. *Enumeratio Plantarum Phanerogamarum* etc. Leipzig.
Roxburgh, W. 1795–1810. *Plants of the coast of Coromandal* (3 vols.) London.
Roxburgh, W. 1814. *Hortus Benghalensis*, Serampore.
Roxburgh, W. 1832. *Flora Indica* (3 Vols.) Serampore.
Sahni, K.C. and S.S.R. Bennet. 1975. A new species of *Albizia* from Andhra Pradesh. Ind. For. **101**: 333–339.
Sanjappa, M. 1987. Revision of the genus *Butea* Roxb. ex Willd. and *Meizobiopies* Voight (Fabaceae). Bull. Bot. Surv. India **29**: 199–225.
Sanjappa M. 1991. *Legumes of India*. Bishen Singh Mahendra Pal Singh. Dehradun.
Saldanha, C.J. 1980. Angiosperm Floristics in India and its relation to Taxonomy—Glimpses of Plant Research. Vol. V: 143–147.
Saldanha, C.J. 1986–1996. *Flora of Karnataka*. 2 Vols. New Delhi.
Saldanha, C.J. 1989. Angiosperm Floristics in India and its relation to Taxonomy—Glimpses of Plant Research. Vol. **V**: 143–147.
Saldanha, C.J. and D.H. Nicolson. 1976. *Flora of Hassan district, Karnataka, India*. New Delhi.
Santapau, M. 1952. The need for a revision of Hooker's Flora of British India. J. Sci. Ind. Res. **71**: 529–532.
Santapau, H. 1953. The Flora of Khandala on the Western Ghats of India. Rec. Bot. Surv. Ind. **16**: 1–396. (3rd rev. ed. 1967).
Santapau, H. 1954. Annual excursion of the Indian Botanical Society, Hyderabad, Deccan. J. Ind. Bot. Soc. **33**: 78–92.
Santapau, H. 1955. *Botanical collector's Manual*. Calcutta.
Santapau, H. 1956. The Botanical exploration of India. J. Ind. Bot. Soc. **35**: 1–6.
Santapau, H. 1958. *Flora of Purandhar*. New Delhi.
Santapau, H. 1959. *Salmalia malabarica* and *S. insignis* in Bombay J. Bombay Nat. Hist. Soc. **56**: 364–365.
Santapau, H. 1962. The present state of Taxonomy and Floristics in India after Independence. Bull. Bot. Surv. India **4**: 209–216.
Santapau, H. 1966. *Common Trees*. New Delhi.
Santapau, H. and A.N. Henry. 1973. *A dictionary of flowering plants in India*. New Delhi.
Santapau, H. and S.K. Wagh. 1964. Nomenclatural changes in Indian plants. Bull. Bot. Surv. India 5: 105–109.
Sarkar, B.K. 1963 a. A note on Bidi leaf (*Diospyros melanoxylon*). Indian forester **89**: 39–45.
Sasidharan, N. 1987. *Forest trees of Kerala. A Check list an Index to important cities*. Peechi.
Sasidharan, N. 1996. Field identification of Forest trees. *Taxonomy and Plant conservation*. 179–187.
Sastry, A.R.K. and T.A. Rao. 1973. Studies on the Flora and vegetation of Coastal Andhra Pradesh. Bull. Bot. Surv. India. **15**: 92–107.
Sayeeduddin, M. 1935. Some of the common flowering plants of the Hyderabad state; their distribution, economic and medicinal importance. J. Asiat. Soc. Bengal Sci. **1**: 9–22.
Sayeeduddin, M. 1938. A further contribution to some of the common flowering plants of Hyderabad state; their distribution and economic importance. J. Bombay Nat. Hist. Soc. **40**: 190–212.
Sayeeduddin, M. 1941 a. Preliminary notes of a recent tour to Amrabad forest reserve. J. Bombay Nat. Hist. Soc. **41**: 907–910.
Sayeeduddin, M. 1941 b. Additions to our knowledge of the flowering plants of H.E.H. The Nizam's Dominions, Hyd. Deccan. J. Bombay Nat. Hist. Soc. **42**: 903–924.
Sayeeduddin, M. 1954. A brief sketch of the flora of Hyderabad, Deccan, Souvenir, 41st Session, Indian Science Congress, Hyderabad.
Scott, A.J. 1979. A revision of *Anogeissus* (Combretaceae). Kew. Bull. **33**: 556–566.
Sebastine, K.M. and A.N. Henry. 1966. A contribution to the flora of Pakhal and surrounding regions of Narasampet taluk, Warangal district, Andhra Pradesh. Bull. Bot. Surv. India **8**: 304–311.

Sebastine, K.M., T. Thothathri and N.P. Balakrishnan. 1960. Observations on the Flora of Narsapur, Medak district. Andhra Pradesh. Bull. Bot. Surv. India **2**: 275–285.

Sebastine, K.M., T. Thothathri and K. Rama Murthy. 1966. Some additions to the Flora of the Presidency of Madras II. Bull. Bot. Surv. India **8**: 80–81.

Sharma, B.D. and N.P. Balakrishnan. 1993. *Flora of India*, Vol. 2 (Papavaraceae—Caryo Phyllaceae). Botanical Survey of India, Calcutta.

Sharma, B.D. and M. Sanjappa. 1993. *Flora of India*. Vol. 3 (Portulacaceae —Ixonanthaceae). Botanical Survey of India, Calcutta.

Sharma, B.D., N.P. Balakrishnan, R.R. Rao and P.K. Hajra. 1993. *Flora of India* Vol. 1 (Rananculaceae—Barclayaceae), Botanical Survey of India, Calcutta.

Silar Mohammed, M. 1991. *Flora of Mohammadabad reserve forest, Rangareddy district*, M.Phil. dissertation, S.K. University, Anantapur.

Singh, Nagendra. 1978. Critical taxonomic notes on some species of *Cassia* L. found in India. J. Bombay Nat. Hist. Soc. **67**: 614–615.

Sprengal, K. 1824–27. *Carol Linnaeus Systema vegetabilitdam*, ed. 16, Gottingae.

Sreenivasa Rao, A. 1991. *Floristic studies on the Flora of Nellore district, Andhra Pradesh*. Ph.D. Thesis, S.V. University, Tirupati.

Stace, C.A. 1980. Plant Taxonomy and Biosystematics. Edward Aronold, London.

Steenis, C.G.G.J. Van. 1950–74. *Flora Malesiana*. Noordhoff & T.D. Netherlands.

Subba Rao, G.V. 1977. Flora of Visakhapatnam district. Andhra Pradesh. Bull. Bot. Surv. India. **19**: 122–1126.

Subba Rao, G.V. 1991. *A contribution to the Dicotyledonous Flora of Visakhapatnam district*. Ph.D. Thesis. Andhra University, Visakhapatnam.

Subba Rao, G.V. and G.R. Kumari. 1967. New Plant records from South India. Bull. Bot. Surv. India **9**: 186–189.

Subba Rao, G.V. and G.R. Kumari. 1968. New Records of Plants from South India—II. Ind. For. **94**: 682–685.

Subba Rao, G.V. and G.R. Kumari. 1970. New records of plants from South India—III. Bull. Bot. Surv. India. **12**: 208–209.

Subba Rao, G.V. and G.R. Kumari. 1981. Some interesting plants common to Western Ghats and Eastern Ghats. Bull. Bot. Surv. India. **23**: 30–37.

Subba Rao, G.V., N.C. Nair and G.K. Kumari. 1984. Plant resources of the Eastern Ghats. In *Proceedings of Seminar on Development and Environment in the Eastern Ghats*. P.V. Bhirava Murthy (ed.) pp. 27–33.

Subba Rao, T.V. 1960. *Forest types of Andhra Pradesh*, Thesis.

Subbarayudu, C. 1989. *Flora of Palakonda hills, Cuddapah district*. M.Phil. Dissertation, S.K. University, Anantapur.

Sudhakar, P.R. 1988. *Taxonomic studies of the family Mimosaceae in India*. Vol. I–II. St. Xavier's College, Mumbai.

Sudhakar, S. 1979. *Flora of Upper East Godavari district*, Ph.D. Thesis, Andhra University, Visakhapatnam.

Sudhakar, S. and Rolla S. Rao. 1985. Medicinal plants of upper East Godavari district (A.P.) and need for establishment of medicinal farm. J. Econ. Tax. Bot. **7**: 399–406.

Suryanarayana, B. 1979. A contribution to the vascular flora of Venkatagiri hills and environs. J. Bombay Nat. Hist. Soc. **71**: 240–258.

Suryanarayana, B. 1984. Some noteworthy plants from the Eastern Ghats in Andhra Pradesh. J. Econ. Tax. Bot. 5: 51–53.

Suryanarayana, B. and A. Madhusudana Rao. 1985. On the occurrence of *Capparis roxburghii* in Andhra Pradesh. J. Econ. Tax. Bot. **7**: 659–660.

Suryanarayana, B. and A. Sreenivasa Rao. 1985. Phytogeographical Notes on certain rare plants from Andhra Pradesh. J. Econ. Tax. Bot. **7**: 651.

Suryanarayana, B. and A. Sreenivasa Rao. 1987. New records of Fabaceae for Andhra Pradesh. J. Econ. Tax. Bot. **11**: 465–466.

Suryanarayana, B. and A. Sreeinivasa Rao. 1996. Further contribution to the Flora of Sriharikota island with a few new records for Andhra Pradesh. *Tamonomy Conservation* 217–222, CBS Publishers, New Delhi.

Styles, B.J. 1972. The flowering biology of the *Meliaceae* and its bearing on tree breeding. Silvae Genet. **21(5)**: 775.

Talbot, W.A. 1909–1911. *Florest Flora of the Bombay Presidency and Sind.*

Thammanna and K. Narayana Rao. 1990. *Medicinal Plants of Tirumala*, TTD Publications, Tirupati.

Thammanna and K., Narayana Rao, and K. Madhava Chetty. 1994. *Angiospermic Wealth of Tirumala.* TTD, Tirupati.

Thothathri, K. 1964. The Flora of Nagarjunakonda valley and the surrounding hills. Indian For. **90**: 539–546.

Thothathri, K. 1987. *Taxonomic revision of the Dalbergieae in the Indian subcontinent*, BSI, Calcutta.

Thunberg, 1794–1800. *Prodromus plantarum capensium.* Upsala.

Thwaites, G.H.W. 1858–64. *Enumeratio plantarum Zeylaniae*. London.

Torrey, J. 1843. *A Flora of the State of New York.* (2 Vols.), Albany.

Trimen, H. 1893–1931. *A. Handbook of the Flora of Ceylon.* London.

Trimen, H. and J.D. Hooker. 1893–1900. *A Hand Book of the Flora of Ceylon.* Vol. 6 (suppl.) London (in 1931 by A.H.G. Alston).

Tirel-Roudet, C. 1972. Loganiaceae. Fl. Camb Loas and Vietn. No.13: 3–89. 14 pl.

Troup, R.S. 1921. *Silviculture of Indian Trees,* V. **3**. Lauraceae to Coniferae. Oxford.

Uhl, L. 1930. *Flora of Guntur.*

Van Steenis, C.G.G.J. Van. 1968. Taxonomic work outside the large centres. A guide to the practice of herbarium taxonomy. Leon houts, P.W. 7–10. International Bureau for Plant Taxonomy and Nomenclature. Utrecht.

Venkaiah, M. 1980. *Studies on the vegetation and Flora of Vizayanagaram district*, Ph.D. Thesis, Andhra University, Visakhapatnam.

Venkateswara Reddy, R. 1990. *Flora of Guvvalacheruvu forest, Cuddapah District,* M.Phil. dissertation, S.K. University, Anantapur.

Venkateswarlu, J., P.V. Bhirava Murthy and P.N. Rao. 1972. *The Flora of Visakhapatnam,* A.P. Academy of Sciences. Hyderabad.

Venkateswarlu, V. 1947. The Esturial Flora of Godavari. J. Bombay Nat. Hist. Soc. **44**: 431–435.

Verdcourt, B. 1985. A synopsis of the Moringaceae. Kew Bull. **40**: 1–23.

Voigt, J.O. 1845. *Hortus suburbanus Calcuttensis*, Calcutta.

Wagh, S.K. 1960. *Studies on the Flora of Andhra State.* Ph.D. Thesis. Bombay University, Bombay, India.

Walkar, A.L. 1849. Statistical report on the circar of Warangal (Hyderabad, Deccan), Madras J. Lit. & Sci. **15**: 219–301.

Wallich, N. 1829–32. *Plantae Asiaticae Rariores.* (3 vols.) London.

Wallich, N. 1829–49. A numerical list of dried specimens of plants in the East India Company Museum, London.

Wight, R. 1834. *Contributions to the Botany India.* London.

Wight, R. 1838–1853. *Icones plantarum Indiae Orientalis.* (6 vols.) Madras.

Wight, R. 1840–1850. *Illustrations to Indian Botany.* London.

Wight, R. and G.A. Walker-Arnott. 1834. *Prodromus Florae Penninsuale Indiae Orientalis.* London.

Willdenow, C.L. 1797–1830. *Species Plantarum* (ed. 5.) (5 vols.). Berlin.

Willis, J.C. 1973. *A dictionary of flowering plants and ferns.* (rev. ed. 8th by H.K. Airy Shaw), Cambridge.

Whitmore T.C. ed. 1972. *Tree Flora of Malaya.* A Manual for Foresters. 2 vols. London.

# INDEX TO FAMILIES, GENERA AND SPECIES

## Italicized names indicate Synonymy and Basionyms

C

H



I

J



K



L



M

R

# INDEX TO VERNACULAR NAMES

A



B

N



O



P

V



W



Y